Lecture Notes in Computer Science 1210

Edited by G. Goos, J. Hartmanis and J. van

Springer
Berlin
Heidelberg
New York
Barcelona
Budapest
Hong Kong
London
Milan
Paris
Santa Clara
Singapore
Tokyo

Philippe de Groote J. Roger Hindley (Eds.)

Typed Lambda Calculi and Applications

Third International Conference
on Typed Lambda Calculi and Applications
TLCA '97
Nancy, France, April 2-4, 1997
Proceedings

Springer

Series Editors

Gerhard Goos, Karlsruhe University, Germany

Juris Hartmanis, Cornell University, NY, USA

Jan van Leeuwen, Utrecht University, The Netherlands

Volume Editors

Philippe de Groote
INRIA-Lorraine & CRIN-CNRS
615 rue du Jardin Botanique, F-54602 Villers-les-Nancy Cedex, France
E-mail: Philippe.deGroote@loria.fr

J. Roger Hindley
University of Wales Swansea, Mathematics Department
Swansea SA2 8PP, UK
E-mail: j.r.hindley@swansea.ac.uk

Cataloging-in-Publication data applied for

Die Deutsche Bibliothek - CIP-Einheitsaufnahme

International Conference on Typed Lambda Calculi and Applications <3, 1997, Nancy>:
Typed lambda calculi and applications : proceedings / Third International Conference on Typed Lambda Calculi and Applications, TLCA '97, Nancy, France, April 2 - 4, 1997. Philippe de Groote ; J. Roger Hindley (ed.). - Berlin ; Heidelberg ; New York ; Barcelona ; Budapest ; Hong Kong ; London ; Milan ; Paris ; Santa Clara ; Singapore ; Tokyo : Springer, 1997
(Lecture notes in computer science ; Vol. 1210)
ISBN 3-540-62688-3
NE: Groote, Philippe de [Hrsg.]; GT

CR Subject Classification (1991): F.4.1, F.3.0, D.1.1

ISSN 0302-9743
ISBN 3-540-62688-3 Springer-Verlag Berlin Heidelberg New York

Typesetting: Camera-ready by author
SPIN 10549470 06/3142 – 5 4 3 2 1 0 Printed on acid-free paper

Preface

This volume is the proceedings of the Third International Conference on Typed Lambda Calculi and Applications, TLCA'97, held in Nancy, France, from April 2 to 4, 1997.

It contains 24 papers. The number submitted was 54, their overall quality was high, and selection was difficult. The Programme Committee is very grateful to everyone who submitted a paper.

The Committee is also very grateful to *Per Martin-Löf*, University of Stockholm, for accepting our invitation to talk at the conference, on

The Problem of Predicativity.

(This talk is not included in this volume.)

The editors wish to thank the members of the Programme Committee and the Organizing Committee, listed overleaf, for their hard work and support.

We also express our gratitude to all the referees listed overleaf, and also those who wish not to be listed, for their essential assistance and time generously given.

Swansea, January 1997

Roger Hindley

Philippe de Groote

Programme Committee

H. Barendregt, Nijmegen
C. Böhm, Rome
M. Dezani-Ciancaglini, Turin
G. Dowek, INRIA-Rocquencourt
R. Hindley, Swansea (Chair)
F. Honsell, Udine
P. Lescanne, INRIA-Lorraine & CRIN-CNRS
A. Pitts, Cambridge
P. Scott, Ottawa
J. Smith, Chalmers
M. Takahashi, Tokyo
V. Tannen, Pennsylvania
J. Tiuryn, Warsaw

Organizing Committee

Z. Benaissa, **D. Briaud**, **A.-L. Charbonnier**, **Ph. de Groote** (Chair),
P. Lescanne, **A. Savary**

(INRIA-Lorraine and CRIN-CNRS)

Referees

Y. Akama	F. Alessi	T. Altenkirch	R. Amadio
A. Asperti	R. Banach	F. Barbanera	E. Barendsen
J. v. Benthem Jutting	D. Bechet	H. Becht	N. Benton
S. Berardi	A. Berarducci	R. Bloo	R. Blute
G. Bierman	L. Boerio	V. Bono	K. Bruce
F. Cardone	G. Castagna	M. Coppo	P-L. Curien
A. Compagnoni	F. Damiani	V. Danos	W. Dekkers
P. Di Gianantonio	R. Di Cosmo	P. Dybjer	R. Dyckhoff
A. Felty	M. Fiore	K. Fujita	H. Geuvers
G. Ghelli	P. Giannini	A. Glenstrup	A. Gordon
B. Gramlich	C. Hankin	T. Hardin	J. Harrison
H. Herbelin	S. Hirokawa	M. Hofmann	D. Howe
M. Hyland	B. Intrigila	T. Jensen	J. Jeuring
J. Joinet	A. Jung	R. Kashima	D. Kesner
C. Kirchner	F. Lamarche	M. Lenisa	X. Leroy
U. de´ Liguoro	P. Lincoln	J. Lipton	J. McKinna
R. Machlin	H. Mairson	L. Maranget	I. Margaria
S. Martini	P-A. Mellies	M. Miculan	E. Moggi
C. Munoz	T. Nipkow	B. Nordstrom	J. Palsberg
C. Paulin-Mohring	P. Panangaden	A. Piperno	M. Plasmeijer
J. van de Pol	R. Pollack	A. Pravato	F. Prost
F. van Raamsdonk	L. Regnier	C. Retoré	L. Roversi
S. Ronchi della Rocca	K. Rose	J. Seldin	P. Selinger
H. Schwichtenberg	D. Suciu	B. von Sydow	I. Takeuti
C. Talcott	M. Tofte	P. Urzyczyn	B. Werner
M. Venturini Zilli	M. Zacchi		

Table of Contents

A λ-to-CL Translation for Strong Normalization

Yohji AKAMA

Department of Information Science,
The University of Tokyo, akama@is.s.u-tokyo.ac.jp

Abstract. We introduce a simple translation from λ-calculus to combinatory logic (CL) such that: A is an SN λ-term iff the translation result of A is an SN term of CL (the reductions are β-reduction in λ-calculus and weak reduction in CL). None of the conventional translations from λ-calculus to CL satisfy the above property. Our translation provides a simpler SN proof of Gödel's λ-calculus by the ordinal number assignment method. By using our translation, we construct a homomorphism from a conditionally partial combinatory algebra which arises over SN λ-terms to a partial combinatory algebra which arises over SN CL-terms.

1 Introduction

We often find some translations from λ-calculus to combinatory logic (CL) provide a pleasing viewpoint in the study of λ-calculus.

The most typical example can be found in the study of the equational theories and the model theories of λ-calculus. The translations from λ-calculus to CL have been investigated comprehensively by Curry school [8], and we come to know that there is a translation $(-)^*$ such that

$$\textbf{(EqCon)}\quad \lambda\beta\eta \vdash A = B \iff CL + (\textit{extensionality}) \vdash A^* = B^*,$$

where (*extensionality*) is an inference rule of the form: $XZ = YZ$ for all Z implies $X = Y$. This nice correspondence of the equational theories enabled us to find a first-order axiomatization of λ-model, and to apply a model theoretical method (see [9]).

Another example is found in the study of compilation of functional programming languages with the call-by-need evaluation strategy [14]. There the programs are compiled to a term (*combinator code*) built up from basic CL-terms, so that the programs can be evaluated without environment. This correspondence between programs ($\sim$ λ-terms) and combinator codes comes about partly because programs do not admit evaluation inside λ-abstraction.

Although the strong normalization (SN) properties of type theories are studied by various means (reducibility method, semantical method [11, 2], divide-and-conquer method [1], etc.), no simple translation is known that relates the strong normalizability of λ-terms to that of CL-terms. One of our main results is a simple translation $(-)^{\circledast}$ that satisfies not only **(EqCon)** but also

(Equivalence) *A λ-term A is* SN *iff the translation result $A^{\circledast}$ is an* SN CL-*term* (the reductions are β-reduction in λ-calculus and weak reduction in CL).

Even in a simple type-regime, the translation makes sense; the types of A and $A^{\circledast}$ are the same. Our translation $(-)^{\circledast}$ depends on neat use of the combinator **K**; note that it is exactly the presence of **K** in the system that causes the two concepts of SN and *weakly normalizing (*WN*)* to be separate (recall the conservation theorem (Corollary 11.3.5 of [3])).

Our translation occurred to the author, when the author attempted to prove the SN of an annoying type theory by adapting *Howard's* SN *proof by assigning ordinal numbers to typed λ-terms* [10]. Our translation can be regarded as his assignment to λ-terms of not ordinal numbers but SN CL-terms. To explain the first use of our translation, we first review Howard's proof, which is not well-known but is an important work.

The importance of his proof is that it relates two consistency proofs of Peano Arithmetic; Gentzen's cut-elimination method by using ordinal numbers, and Gödel's interpretation method with primitive recursive functionals (PRF).

Howard considered a typed λ-calculus for PRF, and he assigned to the λ-terms ordinal numbers, for the purpose of reducing the SN to the well-orderedness of ordinal numbers. But the proof is not manageable, because the assignment for abstraction terms is very complicated.

More succinct work with the same motivation was followed by Schütte (Chapter VI, VII of [13]) . He considered a typed CL for PRF, and he assigned to the CL-terms ordinal numbers, for the same purpose.

Our translation provides a more succinct ordinal assignment to λ-terms than Howard's, namely, to assign to A the ordinal number Schütte assigned to $A^{\circledast}$.

Finally, we consider a partial combinatory algebra $\mathsf{SN}_{\circledast}/=_H$ which arises over the set $\mathsf{SN}_{\circledast}$ of closed SN λ-terms, and a partial algebra $\mathcal{SN}_{\circledast}/=_w$ which arises over the set $\mathcal{SN}_{\circledast}$ of closed SN CL-terms. $\mathsf{SN}_{\circledast}/=_H$ was employed in the semantical SN proof in [11] for type theories, and, interesting enough, a universal algebraic study of $\mathcal{SN}_{\circledast}/=_w$ opens some novel reduction theoretic viewpoints [6]. We construct a homomorphism from $\mathsf{SN}_{\circledast}/=_H$ to $\mathcal{SN}_{\circledast}/=_w$.

2 Equivalence

Combinatory logic (CL) is determined by the following rewriting rules:

$$\mathbf{S}XYZ \to_w XZ(YZ), \quad \mathbf{K}XY \to_w X.$$

$\to_w$ is called *weak reduction*. Terms of CL are called CL-terms, and are represented by $X, Y, Z, U, V, W, \ldots$.

λ-calculus is determined by the rewriting rule $(\lambda x.A)B \to_\beta A[x := B]$. Terms of λ-calculus are called λ-terms, and are represented by $A, B, C, D, E, \ldots$.

For a term X of λ-calculus, or CL, the length of X is denoted by $|X|$. The α-equivalence between λ-terms (CL-terms) is denoted by $\equiv$. For $\to_l$ $(l = w, \beta, \ldots)$, the transitive closure is denoted by $\to_l^+$, the reflexive, transitive closure by $\twoheadrightarrow_l$, and the reflexive, transitive, symmetric closure by $=_l$.

Because $\#\{Y | X \to_l Y\}$ is finite for $l = w, \beta$, if X is SN with respect to $\to_l$, then there exists the maximum of the lengths of reduction sequences from X. It is denoted by $\nu_l(X)$, or simply $\nu(X)$.

A postponement argument proves that A is SN with respect to $\beta\eta$-reduction iff A is SN with respect to β-reduction. Therefore we won't discuss the SN of $\beta\eta$-reduction in this paper.

Some translations from λ-calculus to CL were introduced by Rosser and others to obtain a neat correspondence between the extensional equational theories of both.

These *translations*, and ours as well, are recursively described in the following way: $(a)^\# \equiv a$ for each atom a, $(AB)^\# \equiv A^\# B^\#$ and

$$(\lambda x.\, D)^\# \equiv \lambda^\# x.\, D^\#,$$

for a suitable abstraction algorithm $\lambda^\#$ such that

$$(\lambda^\# x.\, U)V \twoheadrightarrow_w U\,[x := V].$$

Then $(-)^\#$ satisfies **(EqCon)**. The typical translation $(-)^*$ is determined by

$$\lambda^* x.\, X := \begin{cases} \mathbf{SKK}, & \text{if } X \equiv x; \\ \mathbf{K}X, & \text{if } x \text{ is not (free) in } X; \\ \mathbf{S}(\lambda^* x.\, U)(\lambda^* x.\, V), & \text{if } x \text{ is (free) in } X \equiv UV. \end{cases}$$

Unfortunately, conventional translations do not reflect SN of terms. E.g. $\lambda f.\, \Delta'\Delta'$ with $\Delta' \equiv \lambda x.\, xxf$ is even not normalizable but $(\lambda f.\, \Delta'\Delta')^*$ is a normal CL-term.

Definition 2.1 *Let c be a closed* CL*-term. Define for a λ-term A a* CL*-term A^c by an abstraction algorithm $\lambda^c x.\, W \equiv \mathbf{K}(\lambda^* x.\, W)(W\,[x := c])$, where $\lambda^* x.\, W$ is an abstraction algorithm mentioned above.*

Define $A^\circledast$ to be A^c with c being a distinct non-redex constant $\circledast$.

We will prove the following:

Theorem 2.2 (Equivalence) *A λ-term A is* SN *iff the translation result $A^\circledast$ is an* SN CL*-term.*

For convenience, we call the if-part *reflection* and the only-if-part *preservation*.

2.1 The Proof of Reflection.

Definition 2.3 *Let c be a closed* CL*-term. For each λ-term A, the set $[\![A]\!]^c$ of* CL*-terms is defined by the rules:*

$$\frac{}{a \in [\![a]\!]^c}\ (atom) \quad \textit{if a is an atom,} \qquad \frac{Y \in [\![B]\!]^c \quad Z \in [\![C]\!]^c}{YZ \in [\![BC]\!]^c}\ (app)$$

$$\frac{U \in [\![E]\!]^c \quad V \in [\![D]\!]^c \quad E \twoheadrightarrow_\beta D}{\mathbf{K}(\lambda^* x.\, U)(V\,[x := c]) \in [\![\lambda x.\, D]\!]^c}\ (abs).$$

Lemma 2.4 *Suppose $A \to_\beta A'$. For every $X \in [\![A]\!]^c$, if X is* SN, *then some $X' \in [\![A']\!]^c$ satisfies $X \to_w^+ X'$.*

Proof. By induction on the pair $\langle \nu(X), |X| \rangle$ lexicographically ordered with $\langle >, > \rangle$.

When $A \equiv (\lambda x. D)C \to_\beta A' \equiv D[x := C]$. Then, we can assume that $X \equiv \mathbf{K}(\lambda^* x. U)(V[x := c])Z$ where $E \twoheadrightarrow_\beta D$ and (‡) $U \in [\![E]\!]^c$, $Z \in [\![C]\!]^c$. Therefore, there exists a reduction sequence from $E[x := C]$ to $D[x := C] \equiv A'$:

$$E[x := C] \to_\beta A_1 \to_\beta A_2 \to_\beta \cdots \to_\beta A'. \tag{1}$$

Here, we can derive from (‡)

$$U[x := Z] \in [\![E[x := C]]\!]^c,$$

by induction on the derivation of $U \in [\![E]\!]^c$. Because $X \to_w^+ U[x := Z]$, we have $\nu(X) > \nu(U[x := Z])$. So, we can apply the induction hypothesis to the above $U[x := Z] \in [\![E[x := C]]\!]^c$. Thus, we can find X_1 in $[\![A_1]\!]^c$ such that $X \to_w^+ U[x := Z] \to_w^+ X_1$. Therefore we can again apply the induction hypothesis to $X_1 \in [\![A_1]\!]^c$. By iterating this argument along the reduction sequence (1), we can construct

$$X \to_w^+ U[x := Z] \to_w^+ X_1 \to_w^+ X_2 \to_w^+ \cdots \to_w^+ X',$$

such that $X_i \in [\![A_i]\!]^c, X \in [\![A]\!]^c$, and $X' \in [\![A']\!]^c$.

When $A \equiv \lambda x. D \to_\beta A' \equiv \lambda x. D'$. Then we can assume that

$$X \equiv \mathbf{K}(\lambda^* x. U)(V[x := c]), \quad U \in [\![E]\!]^c, \quad E \twoheadrightarrow_\beta D \to_\beta D', \quad V \in [\![D]\!]^c.$$

We can apply the induction hypothesis to $V \in [\![D]\!]^c$, because $\nu(V) \le \nu(V[x := c]) < \nu(X)$. Therefore, $\exists V'(V' \in [\![D']\!]^c \;\&\; V \to_w^+ V')$. Then, by (*abs*)-rule, we have $X' \equiv \mathbf{K}(\lambda^* x. U)(V'[x := c]) \in [\![\lambda x. D']\!]^c$. Moreover because $V \to_w^+ V'$, we get $X \to_w^+ X'$.

The other cases are clear. □

Theorem 2.5 *A λ-term A is* SN, *if some* CL-*term X in $[\![A]\!]^c$ is* SN.

Proof. A reduction sequence of length n from A yields a reduction sequence of length $\ge n$ from X, by applying the above Lemma n times. □

Because $A^c \in [\![A]\!]^c$, we have the reflection part of the **Equivalence** theorem.

Remark 2.6 Someone may wonder whether $A \to_\beta B$ implies $A^\circledast \twoheadrightarrow_w B^\circledast$. But this does not hold: take $A \equiv \lambda f. (\lambda x. f)x$, $B \equiv \lambda f. f$.

Remark 2.7 The mapping $[\![-]\!]^c$ is inspired by Howard's SN proof of typed λ-calculus for PRF(Section 4 of [10]). He introduced a mapping from λ-terms to sets of *vectors of ordinals* based on a complicated application operator, a variable binder and a well-founded partial order $\triangleright$ on ordinal vectors. His assignment system consists of axioms for redex-constants, the (*atom*)-axiom and (*app*)-rule above, and the following:

$$\frac{U \in [\![E]\!]^c \quad V \in [\![D]\!]^c \quad E \twoheadrightarrow_\beta D}{(\lambda^* x.\, U) \oplus (V\,[x := c]) \in [\![\lambda x.\, D]\!]^c}\ (absH)$$

where U and V stand for vectors of ordinals, c for a fixed constant vector of ordinals, and $\oplus$ for the componentwise *natural sum* of the vectors of ordinal. Our (*abs*)-rule is of the same form as the (*absH*)-rule except that the binary operator $\oplus$ is replaced with the **K** combinator. We reduced the SN of *type-free* λ-terms to the SN of the assigned CL-*terms*.

Remark 2.8 The theorems and lemmas in this subsection hold for the following situations also.

1. a *typed* λ-calculus and a *typed* CL for PRF.
   Here, the two calculus contain recursors as constants. If the recursors are added as term-operators, then the abstraction algorithm $\lambda^\circledast$ does not make sense.
2. the type-free λ-calculus and the type-free CL with an *applicative term rewriting system* (15p. of [12]).

2.2 The Proof of Preservation.

Lemma 2.9 *Let X be a* CL*-term $C\{R_1, \ldots, R_n\}$ with $C\{\{\}_1, \ldots \{\}_n\}$ being an n-holed linear context* [1] *and each R_i being a redex. And let Y be a* CL*-term $C\{\overline{R}_1, \ldots, \overline{R}_n\}$ with $\overline{R}_i$ being the contractum of R_i. Then if Y and each of the R_i's are* SN*, so is X.*

Proof. We use induction on $\langle C\{\overline{R}_1, \ldots, \overline{R}_n\}, \{R_1, \ldots, R_n\}\rangle$ ordered lexicographically with $\langle \rightarrow_w^+, (\rightarrow_w^+)^{mult}\rangle$. Here $\{R_1, \ldots, R_n\}$ is the multiset of $R_1, \ldots, R_n$, and $(\rightarrow_w^+)^{mult}$ is the multiset extension of $\rightarrow_w^+$. This induction is valid because the premise states both the SN of the first component of the induction value and the SN of each R_i.

Let Q be a redex in $C\{R_1, \ldots, R_n\}$. We will derive from $C\{R_1, \ldots, R_n\} \stackrel{Q}{\rightarrow}_w Z$ the SN of Z. We will proceed by cases depending on the position of the redex Q relative to $R_1, \ldots, R_n$.

1. Q is R_j for some j: then we can find an (n-1)-holed linear context

$$C'\{\{\}_1, \ldots, \{\}_{j-1}, \{\}_{j+1}, \ldots, \{\}_n\}$$

[1] each hole occurs once.

such that

$$\begin{aligned} Z &\equiv C'\{R_1,\ldots,R_{j-1},R_{j+1},\ldots,R_n\}, \\ C'\{\overline{R}_1,\ldots,\overline{R}_{j-1},\overline{R}_{j+1},\ldots,\overline{R}_n\} &\equiv C\{\overline{R}_1,\ldots,\overline{R}_n\}, \\ \{R_1,\ldots,R_{j-1},R_{j+1},\ldots,R_n\} &(\overset{+}{\underset{w}{\leftarrow}})^{mult} \{R_1,\ldots,R_n\}. \end{aligned}$$

By the induction hypothesis, Z is SN.

2. Q is strictly contained in R_j for some j: then $Z \equiv C\{R_1,\ldots,R'_j,\ldots,R_n\}$ and $R_j \overset{Q}{\to} R'_j$ for some R'_j. We can see that R'_j is still a redex. Let $\overline{R'_j}$ be its contractum. Because $\overline{R'_j} \twoheadleftarrow_w \overline{R}_j$, we have

$$C\left\{\overline{R}_1,\ldots,\overline{R'_j},\ldots,\overline{R}_n\right\} \twoheadleftarrow_w C\{\overline{R}_1,\ldots,\overline{R}_n\}$$
$$\{R_1,\ldots,R'_j,\ldots,R_n\}\ (\overset{+}{\underset{w}{\leftarrow}})^{mult}\ \{R_1,\ldots,R_n\}.$$

By the induction hypothesis, Z is SN.

3. Otherwise: then for some linear-holed contexts $D\{\ldots\}$, $E\{\ldots\}$,

$$Q \equiv D\{R_{i_1},\ldots,R_{i_k}\},$$
$$C\{R_1,\ldots,R_n\} \equiv E\{D\{R_{i_1},\ldots,R_{i_k}\},R_{i_{k+1}},\ldots,R_{i_n}\}.$$

From above,

$$Z \equiv E\{D'\{R_{i_1},\cdots,R_{i_1};\ldots;R_{i_k},\cdots,R_{i_k}\},R_{i_{k+1}},\ldots,R_{i_n}\},$$

for some linear-holed context $D'\{\ldots\}$. Moreover

$$E\{D'\{\overline{R}_{i_1},\cdots,\overline{R}_{i_1};\ldots;\overline{R}_{i_k},\cdots,\overline{R}_{i_k}\},\overline{R}_{i_{k+1}},\ldots,\overline{R}_{i_n}\} \overset{+}{\underset{w}{\leftarrow}} C\{\overline{R}_1,\ldots,\overline{R}_n\}.$$

By the induction hypothesis, Z is SN □

Lemma 2.10 *1. Suppose (a)* $X \overset{R}{\to}_w Y$ *and (b) if the redex* R *is* $\mathbf{K}UV$ *then* V *is* SN. *Then the* SN *of* Y *implies the* SN *of* X.
2. If X *and* $Y[x := X]X_1\ldots X_n$ *are* SN CL*-terms, then so is* $(\lambda^{\circledast}x.Y)XX_1\ldots X_n$.

Proof. (1) is obvious from the above lemma with $n = 1$. For (2), we first note the existence of a reduction sequence $(\lambda^{\circledast}x.Y)X \twoheadrightarrow_w Y[x := X]$ which erases only X or $\mathbf{K}X$. The iterated application of (1) along this sequence establishes (2). □

Lemma 2.11 $(A[x := B])^{\circledast} \equiv (A^{\circledast})[x := B^{\circledast}]$.

Proof. By induction on the length of A. □

The proof of the preservation part of the **Equivalence** *theorem.* We will derive the SN of $A^{\circledast}$ from the SN of A, by induction on $\langle \nu(A), |A| \rangle$ ordered lexicographically by $\langle >, > \rangle$. The proof will proceed by cases depending on the form of A.

When $A \equiv (\lambda x. D)CB_1 \ldots B_n$: then $A^{\circledast} \equiv (\lambda^{\circledast} x. D^{\circledast})C^{\circledast} B_1{}^{\circledast} \ldots B_n{}^{\circledast}$. By the premise, $\nu(C), \nu(D[x := C] B_1 \ldots B_n) < \nu(A)$. By the induction hypotheses, (†): $C^{\circledast}$ is SN, and $(D[x := C] B_1 \ldots B_n)^{\circledast}$ is SN. From the last, by Lemma 2.11, $D^{\circledast}[x := C^{\circledast}] B_1{}^{\circledast} \ldots B_n{}^{\circledast}$ is SN. By applying Lemma 2.10(2) with (†), we obtain the SN of

$$A^{\circledast} \equiv (\lambda^{\circledast} x. D^{\circledast})C^{\circledast} B_1{}^{\circledast} \ldots B_n{}^{\circledast}.$$

When $A \equiv \lambda x. D$: then $A^{\circledast} \equiv \lambda^{\circledast} x. D^{\circledast}$. By induction hypothesis, $D^{\circledast}$ is SN. Then by Lemma 2.10(2), $(\lambda^{\circledast} x. D^{\circledast})x$ is SN. Therefore $A^{\circledast} \equiv \lambda^{\circledast} x. D^{\circledast}$ is SN.

The other case are clear. □

This completes the proof of **Equivalence**.

Our translation $(-)^{\circledast}$ depends on neat use of the combinator **K**, and it is exactly the presence of **K** in the system that causes the two concepts of SN and WN to be separate. However, we can prove the following result for a "**K**-less" fragment of CL.

Theorem 2.12 *We have a recursive mapping* $(-)^{+}$ *such that*

$$A \text{ is an SN } \lambda\text{-term} \iff A^{+} \text{ is an SN BCWI-term.}$$

Here, a BCWI-term is defined to be a term built up by term-application from only atomic combinators **B**, **C**, **W**, **I** *which are subject to the following rewriting rules*

$$\mathbf{B}XYZ \to X(YZ), \quad \mathbf{C}XYZ \to XZY, \quad \mathbf{W}XY \to XYY, \quad \mathbf{I}X \to X.$$

However, $(-)^{+}$ *does not satisfy (***EqCon***).*

Proof. We note that some Turing machine e satisfies that for every λ-term A, A is SN iff

(¶) e with the input $\ulcorner A \urcorner$ halts,

where $\ulcorner A \urcorner$ is the Gödel number of A. (e performs breadth-first search of the reduction tree of A.) Since the BCWI-calculus can represent any computable functions just as the λI-calculus can, we can find a BCWI-term E independently from A, such that (¶) holds iff $E \ulcorner A \urcorner$ has a normal form. By the conservation theorem (Corollary 11.3.5 of [3]), we know that a BCWI-term is SN iff it has normal form. Therefore it is sufficient to let A^{+} be $E \ulcorner A \urcorner$.

It is easy to see that $(-)^{+}$ does not satisfy (**EqCon**). □

Remark 2.13 Lemma 2.9 holds for any term rewriting system (TRS) with the critical pairs being *trivial*. Here, we say a critical pair is *trivial*, if it is of the form $\langle X, X \rangle$.

One such TRS is *parallel or*: $por(tt, x) \to tt$, $por(x, tt) \to tt$, $por(ff, ff) \to ff$.

3 Partial Combinatory Algebras

A *partial combinatory algebra* (PCA) is a partial applicative structure that satisfies the conditions

$$\mathbf{s}fg\downarrow,\quad (\mathbf{S}):\ \mathbf{s}fga \simeq fa(ga),\quad (\mathbf{K}):\ \mathbf{k}fa = f.$$

Here, an assertion $f\downarrow$ denotes that *"f is defined"*, and $fg\downarrow$ implies both $f\downarrow$ and $g\downarrow$. The symbol $\simeq$ means that if one side is defined, so is the other side, and that both are equal, while the symbol $=$ means that both sides are defined with equal values. For the relevance of PCA's to constructive logic, see Chapter VI of Beeson [4].

Definition 3.1 *Let $\mathcal{SN}_\circledast$ be the set of type-free closed* SN CL*-terms possibly containing a constant $\circledast$ distinct from* **S** *and* **K**.

The following result from [5] can be proved by using Lemma 2.9.

Lemma 3.2 ([5]) *A quotient structure $\langle \mathcal{SN}_\circledast / =_w, \cdot / =_w \rangle$ with $\cdot$ being a term application is a* PCA.

Proof. An element of the structure is of the form $[U]_{=_w} := \{X \in \mathcal{SN}_\circledast | X =_w U\}$ for some $U \in \mathcal{SN}_\circledast$. The partial application $[U]_{=_w}[V]_{=_w}$ is defined to be $[UV]_{=_w}$. The well-definedness (i.e., the independence of the choice of representatives U, V) is guaranteed by Lemma 2.9 with $n = 1$. Define the constants $\mathbf{s}, \mathbf{k}$ of the structure to be $[\mathbf{S}]_{=_w}$, $[\mathbf{K}]_{=_w}$. □

The above PCA turns out to be equivalent to the PCA of closed normal CL-terms defined by Beeson (Chapter V. Section 6. [4]), in view of Exercise 8 of V.4 in [4].

Lemma 3.2 is contrasted with Hyland and Ong's observation [11] in that the set $\mathsf{SN}_\circledast$ of closed SN λ-terms does not give rise to a PCA.

Definition 3.3 (p.182, p.184 of [11]) *1. Let* $\mathsf{SN}_\circledast$ *be the set of type-free closed* SN *λ-terms possibly containing $\circledast$.*

2. Let $\rightarrow_H$ be the restricted form of one-step β-reduction defined by contracting only a closed *redex.*

3. A conditionally partial combinatory algebra *(*C-PCA*) is a partial applicative structure that satisfies the above-mentioned axioms* (**S**) *and* (**K**).

Lemma 3.4 (Theorem 2.6 of [11]) *A quotient structure $\langle \mathsf{SN}_\circledast / =_H, \cdot / =_H \rangle$ with $\cdot$ being a term application is a* C-PCA.

We show the following:

Theorem 3.5 *Define $\varphi : \mathsf{SN}_\circledast / =_H \rightarrow \mathcal{SN}_\circledast / =_w$ by $[M]_{=_H} \mapsto [M^\circledast]_{=_w}$. Then*

1. *φ is well-defined. Moreover,*
2. *$\varphi(ab) \simeq \varphi(a)\varphi(b)$, where if ab is not defined, then neither is $\varphi(ab)$.*

Proof. (1) It is sufficient to show that $M \to_H N$ implies $M^* \twoheadrightarrow_w N^*$. Because the redex of $M \to_H N$ is closed, we can write $M \equiv L[y := (\lambda x.P)Q]$ and $N \equiv L[y := P[x := Q]]$. From it, we can prove $M^* \equiv L^*[y := (\lambda^* x.P^*)Q^*]$ and $N^* \equiv L^*[y := P^*[x := Q^*]]$. Therefore $M^* \twoheadrightarrow_w N^*$.

(2) We can write $a = [M]_H$ and $b = [N]_H$. Then $ab \simeq [M]_H[N]_H \simeq [MN]_H$, and $\varphi(a)\varphi(b) \simeq [M^{\circledast}]_{=_w}[N^{\circledast}]_{=_w} \simeq [(MN)^{\circledast}]_{=_w}$. If ab is defined, then MN is an SN λ-term. By the **Equivalence theorem**, $(MN)^{\circledast}$ is an SN CL-term. Then $\varphi(a)\varphi(b)$ is defined and is equal to $\varphi(ab)$. If $\varphi(a)\varphi(b)$ is defined, then $(MN)^{\circledast}$ is an SN CL-term. By **Equivalence**, MN is an SN λ-term. Therefore ab is defined. Thus $\varphi(ab)$ is defined and is equal to $\varphi(a)\varphi(b)$. □

In [11], Hyland and Ong carried out their semantical SN proof for $\lambda 2$ with a C-PCA $\mathsf{SN}_{\circledast}/=_H$. Here, we point out that their proof can be carried out with a PCA $\mathcal{SN}_{\circledast}/=_w$, if we use our translation $(-)^{\circledast}$.

Acknowledgement

The author thanks J. Roger Hindley and Masako Takahashi for their advice.

References

1. Y. Akama. On Mints' reduction for ccc-calculus. In Bezem and Groote [7], pages 1–12.
2. T. Altenkirch. Proving strong normalization of CC by modifying realizability semantics. In H. Barendregt and T. Nipkow, editors, *Types for Proofs and Programs*, volume 806 of *Lecture Notes on Computer Science*, pages 3 – 18. Springer, 1994.
3. H. Barendregt. *The Lambda Calculus, Its Syntax and Semantics.* North-Holland, second edition, 1984.
4. M. Beeson. *Foundations of Constructive Mathematics.* Springer, 1985.
5. I. Bethke and J. W. Klop. Collapsing partial combinatory algebras. In G. Dowek, J.Heering, K.Meinke, and B.Möller, editors, *Higher-Order Algebra, Logic, and Term Rewriting*, volume 1074 of *Lecture Notes on Computer Science*, pages 57 – 73. Springer, 1996.
6. I. Bethke, J. W. Klop, and R. de Vrijer. Completing partial combinatory algebras with unique head-normal forms. In *Logic in Computer Science*, pages 448 – 454. IEEE Computer Society press, 1996.
7. M. Bezem and J.F. Groote, editors. *Typed Lambda Calculi and Applications*, volume 664 of *Lecture Notes in Computer Science.* Springer, 1993.
8. H. B. Curry, J. R. Hindley, and J. P. Seldin. *Combinatory Logic, Volume II.* Studies in Logic and the Foundations of Mathematics. North-Holland, 1972.
9. J.R. Hindley and J.P. Seldin. *Introduction to Combinators and Lambda-calculus.* Cambridge University Press, 1986.
10. W.A. Howard. Assignment of ordinals to terms for primitive recursive functionals of finite type. In A. Kino, J. Myhill, and R. Vesley, editors, *Intuitionism and Proof Theory*, pages 443–458. North-Holland, 1970.
11. J.M.E. Hyland and C.-H. L. Ong. Modified realizability toposes and strong normalization proofs (extended abstract). In Bezem and Groote [7], pages 179–194.

12. J.W. Klop. Term rewriting systems. In S. Abramsky, D. Gabbay, and T. Maibaum, editors, *Handbook of Logic in Computer Science*, volume 2, pages 2–117. Oxford University Press, 1992.
13. K. Schütte. *Proof Theory*. Springer, 1977. (translated by J. Crossley from the German.).
14. D. A. Turner. A new implementation technique for applicative languages. *Software Practice and Experience*, 9(1):31–49, January 1979.

Typed Intermediate Languages for Shape Analysis

Gianna Bellè and Eugenio Moggi

DISI - Univ. di Genova, via Dodecaneso 35, 16146 Genova, Italy
phone: +39 10 353-6629, fax: +39 10 353-6699, e-mail: {gbelle,moggi}@disi.unige.it

Abstract. We introduce $S2$, a typed intermediate language for vectors, based on a 2-level type-theory, which distinguishes between compile-time and run-time. The paper shows how $S2$ can be used to extract useful information from programs written in the Nested Sequence Calculus $\mathcal{NSC}$, an idealized high-level parallel calculus for nested sequences. We study two translations from $\mathcal{NSC}$ to $S2$. The most interesting shows that shape analysis (in the sense of Jay) can be handled at compile-time.

1 Introduction

Good intermediate languages are an important prerequisite for program analysis and optimization, the main purpose of such languages is to make as explicit as possible the information that are only implicit in source programs (as advocated by [18]). A common feature of such intermediate languages is an *aggressive* use of types to incorporate additional information, e.g.: binding times (see [18]), boxed/unboxed values (see [20]), effects (see [23]). In particular, among the ML community the use of types in intermediate languages has been advocated for the TIL compiler (see [9]) and for region inference (see [24, 1]).

In areas like parallel programming, where efficiency is a paramount issue, good intermediate languages are even more critical to bridge the gap between high-level languages (e.g. NESL) and efficient implementations on a variety of architectures (see [2, 3, 22]). However in this area of computing, intermediate languages (e.g. VCODE) have not made significant use of types, yet.

This paper proposes a typed intermediate language $S2$ for vector languages, and shows how it may be used to extract useful information from programs written in the Nested Sequence Calculus $\mathcal{NSC}$ (see [22]), an idealized vector language very closed to NESL. For an efficient compilation of $\mathcal{NSC}$ (and similar languages) on parallel machines it is very important to know in advance the size of vectors (more generally the shape of data structures). We study two translations from $\mathcal{NSC}$ to $S2$. The most interesting one separates what can be computed at compile-time from what must be computed at run-time, in particular array bound-checking can be done at compile-time (provided the while-loop of $\mathcal{NSC}$ is replaced by a for-loop).

Section 2 introduces the two-level calculus $S2$ and outlines its categorical semantics. Section 3 summarizes the high-level language $\mathcal{NSC}$, outlines two translations from $\mathcal{NSC}$ in $S2$ and the main results about them. Sections 4 and 5 give the syntactic details of the translations. Appendix A gives a formal description of $S2$ and defines auxiliary notation and notational conventions used in the paper.

Related work. The language $S2$ borrows the idea of built-in phase distinction from HML (see [17]), and few inductive types at compile-time from type theory (e.g. see [19, 5]). There are also analogies with work on partial evaluation, in particular 2-level lambda calculi for binding time analysis (see [18, 6]). None of these calculi make use of dependent types. Also [16] deals with shape checking of array programs, but without going through an intermediate language.

The translation of $\mathcal{NSC}$ into $S2$ has several analogies with that considered in [8] to give a type-theoretic account of higher-order modules, and has been strongly influenced by ideas from shape theory and shape analysis (see [14, 15]). There are also analogies with techniques for constant propagation, but these technique tend to cope with languages (e.g. Pascal) where constant expressions are much simpler than compile-time expressions in $S2$.

This paper uses categorical semantics as a high-level language for describing what is happening at the syntactic level. Nevertheless, after each categorical statement we will provide an informal explanation. A systematic link between type theories and categorical structures is given in [12].

2 The 2-Level Calculus $S2$ and its Semantics

We give a compact description of $S2$ as a Type System a la Jacobs (see [12], Chap.2), and summarize its type-constructors. For a more detailed description of $S2$ we refer to appendix A.

- Sorts: c and r, c classifies compile-time types and r run-time types.
- Setting: $c < c$ and $c < r$, i.e. c- and r-types may depend on c-values, but they may not depend on r-values.
- Closure properties of c-types: dependent products $\Pi x{:}\, A.B$, unit 1, sums $A + B$, dependent sums $\Sigma x{:}\, A.B$, NNO N, finite cardinals $n{:}\, N$.
- Closure properties of r-types: exponentials $A \to B$, universal types $\forall x{:}\, A.B$, unit 1, sums $A + B$, products $A \times B$, weak existential types $\exists x{:}\, A.B$.

The setting of $S2$ is very close to that of HML (see [12, 17, 11]).

Proposition 1. *$S2$ satisfies the following properties:*
- Context separation: $\Gamma \vdash J$ implies $\Gamma_c, \Gamma_r \vdash J$, where Γ_α is the sequence of declarations $x{:}\, A{:}\, \alpha$ of sort α in Γ
- Phase distinction: $\Gamma \vdash J_c$ implies $\Gamma_c \vdash J_c$, where J_c is an assertion of sort c (i.e. nothing, $A{:}\, c$ or $M{:}\, A{:}\, c$)
- No run-time dependent types: $\Gamma \vdash A{:}\, r$ implies $\Gamma_c \vdash A{:}\, r$.

Proof. They follow immediately from the restrictions imposed by the setting.

We briefly outline a semantics of $S2$. This is important not only as a complement to the formal description, but also to suggest possible improvements to $S2$ and discuss semantic properties of translations (which rely on features of models not captured by $S2$ without *extensionality*). In general, a categorical model for $S2$ is given by a fibration $\pi: \mathcal{C} \to \mathcal{B}$ with the following additional properties:

1. the base category $\mathcal{B}$ is locally cartesian closed and **extensive** (see [4]), i.e. it has sums and the functors $+: \mathcal{B}/I \times \mathcal{B}/J \to \mathcal{B}/(I+J)$ are equivalences, and it has a natural number object (NNO);
2. the fibration $\pi: \mathcal{C} \to \mathcal{B}$ is bicartesian closed and has $\forall$- and **$\exists$-quantification** along maps in the base (see [12, 21]), i.e. for any $f: J \to I$ in $\mathcal{B}$ the substitution functor $f^*: \mathcal{C}_I \to \mathcal{C}_J$ has left and right adjoints $\exists_f \vdash f^* \vdash \forall_f$ and they commute with substitution;
3. all functors $\langle in_0^*, in_1^* \rangle: \mathcal{C}_{I+J} \to \mathcal{C}_I \times \mathcal{C}_J$ (and $!: \mathcal{C}_0 \to 1$) are equivalences.

Remark. Extensivity of sums and the last property are essential to validate the elimination rules for $+$ over sorts (rule $+$-E-α). In a locally cartesian closed base category it is possible to interpret also identity types (of sort c) and validate the rules for extensional equality. Finite cardinals and vectors are definable from the natural numbers using the other properties of the base category. One could derive that the fibration π has sums from the other properties. In fact, $A_0 + A_1 = \exists_{[id,id]:I+I \to I} A$, where $A \in \mathcal{C}_{I+I}$ is such that $in_i^*(A) = A_i$.

There is a simple way to construct a model of $S2$ starting from any cartesian closed category $\mathcal{C}$ with small products and small sums (e.g. **Set** and **Cpo**) using the **Fam**-construction. The category $\mathbf{Fam}(\mathcal{C})$ has objects given by pairs $\langle I \in \mathbf{Set}, a \in \mathcal{C}^I \rangle$ and morphisms from $\langle I, a \rangle$ to $\langle J, b \rangle$ given by pairs $\langle f: I \to J, g \in \mathcal{C}^I(a, f^* b) \rangle$, where $f^* b$ is the I-indexed family of objects s.t. $(f^* b)_i = b_{fi}$ for any $i \in I$. Identity and composition are defined in the obvious way.

The model is given by the fibration $\pi: \mathbf{Fam}(\mathcal{C}) \to \mathbf{Set}$, where π is the functor $\langle I, a \rangle \mapsto I$ forgetting the second component (this is the standard way of turning a category $\mathcal{C}$ into a fibration over **Set**). In particular, the fiber $\mathbf{Fam}(\mathcal{C})_I$ over I is (up to isomorphism) $\mathcal{C}^I$, i.e. the product of I copies of $\mathcal{C}$.

Remark. The first property for a categorical model is clearly satisfied, since the base category is **Set** . The second follows from the assumptions about $\mathcal{C}$, since in $\mathcal{C}^I$ products, sums and exponentials are computed pointwise. The third is also immediate since $\mathcal{C}^{I+J}$ and $\mathcal{C}^I \times \mathcal{C}^J$ are isomorphic (and therefore equivalent).

The interpretation of judgements in $\pi: \mathbf{Fam}(\mathcal{C}) \to \mathbf{Set}$ is fairly simple to describe:

- $\Gamma \vdash$ is interpreted by an object $\langle I, a \rangle$ in $\mathbf{Fam}(\mathcal{C})$, namely $I \in \mathbf{Set}$ corresponds to Γ_c and $a = \langle a_i | i \in I \rangle \in \mathcal{C}^I$ corresponds to Γ_r;
- $\Gamma \vdash A: c$ is interpreted by a family $\langle X_i | i \in I \rangle$ of sets;
- $\Gamma \vdash M: A: c$ is interpreted by a family $\langle x_i \in X_i | i \in I \rangle$ of elements;
- $\Gamma \vdash B: r$ is interpreted by a family $b = \langle b_i | i \in I \rangle$ of objects of $\mathcal{C}$;
- $\Gamma \vdash N: B: r$ is interpreted by a family $\langle f_i: a_i \to b_i | i \in I \rangle$ of morphisms in $\mathcal{C}$.

We summarize some properties valid in these models, which are particularly relevant in relation to the translations defined subsequently.

Proposition 2 ($\forall\exists$-exchange).
Given $n{:}\,N{:}\,c \quad i{:}\,n{:}\,c \vdash A(i){:}\,c \quad i{:}\,n{:}\,c, x{:}\,A(i){:}\,c \vdash B(i,x){:}\,r$
$n{:}\,N{:}\,c \vdash \exists f{:}\,(\Pi i{:}\,n.A(i)).\forall i{:}\,n.B(i,fi) \cong \forall i{:}\,n.\exists x{:}\,A(i).B(i,x)$
namely the canonical map $\langle f,g\rangle \mapsto \Lambda i{:}\,n.\langle fi, gi\rangle$ *is an isomorphism.*

Remark. This property can be proved formally in *extensional* $S2$ by induction on the NNO N. The key lemma is $\Pi i{:}\,sn.A(i) \cong A(0) \times (\Pi i{:}\,n.A(si))$ and similarly for $\forall i{:}\,sn{:}\,A(i)$. The property is the *internal* version of the following property (which can be proved in system F with surjective pairing):
$\exists\langle\overline{x}\rangle{:}\,(A_1 \times \ldots \times A_n).(B_1 \times \ldots \times B_n) \cong (\exists x_1{:}\,A_1.B_1) \times \ldots \times (\exists x_n{:}\,A_n.B_n)$
where $\Gamma \vdash A_i{:}\,c$ and $\Gamma, x_i{:}\,A_i{:}\,c \vdash B_i(x_i){:}\,r$ for $i = 1,\ldots,n$.

Proposition 3 (Extensivity). *Given* $f{:}\,A \to 2{:}\,c$, *then* $A \cong A_0 + A_1$, *where* $2 = 1+1 \quad A_i = \Sigma a{:}\,A.eq_2(i, fa) \quad x,y{:}\,2{:}\,c \vdash eq_2(x,y){:}\,c$ *is equality on* 2, *i.e.*
$eq_2(0,0) = 1 \mid eq_2(0,1) = 0 \mid eq_2(1,0) = 0 \mid eq_2(1,1) = 1$

Remark. Also this property can be proved formally in *extensional* $S2$. The key lemma is $i{:}\,2{:}\,c \vdash eq_2(0,i) + eq_2(1,i) \cong 1$.

3 Translations of $\mathcal{NSC}$ into $S2$

The Nested Sequence Calculus $\mathcal{NSC}$ (see [22]) is an idealized vector language. Unlike NESL, it has a small set of primitive operations and no polymorphism, therefore is simpler to analyze. For the purposes of this paper we introduce the abstract syntax of $\mathcal{NSC}$ and refer the interested reader to [22] for the operational semantics. The syntax of $\mathcal{NSC}$ is parameterized w.r.t. a signature Σ of atomic types D and operations $op{:}\,\overline{\tau} \to \tau$

- Types $\tau ::= 1 \mid N \mid D \mid \tau_1 \times \tau_2 \mid \tau_1 + \tau_2 \mid [\tau]$.
  Arities for operations are of the form $\sigma ::= \overline{\tau} \to \tau$, and those for term-constructors are $\overline{\sigma}, \overline{\tau} \to \tau$.
- Raw terms $e ::= x \mid op(\overline{e}) \mid f(\overline{e}) \mid c(\overline{f}, \overline{e})$, where c ranges over term-constructors (see Figure 1) and f over abstractions $f ::= \lambda\overline{x}{:}\,\overline{\tau}.e$.

Remark. $\mathcal{NSC}$ has a term-constructor $while{:}\,(\tau \to \tau), (\tau \to 1{+}1), \tau \to \tau$ instead of for. We have decided to ignore the issue of non-termination, to avoid additional complications in $S2$ and translations. One must be rather careful when translating a source language which exhibits non-termination or other computational effects. In fact, in $S2$ such effects should be confined to the run-time part, since we want to keep type-checking and shape-analysis decidable.

The following sections describe in details two translations of $\mathcal{NSC}$ in $S2$. In this section we only outline the translations and give a concise account of them and their properties in terms of categorical models.

Term-constructors marked with † can raise an error

c	arity	informal meaning
$err\dagger$	τ	error
0	N	zero
s	$N \to N$	successor
eq	$N, N \to 1+1$	equality of natural numbers
$for\dagger$	$(N, \tau \to \tau), \tau, N \to N$	iteration
$*$	1	empty tuple
$pair$	$\tau_1, \tau_2 \to \tau_1 \times \tau_2$	pairing
π_i	$\tau_1 \times \tau_2 \to \tau_i$	projection
in_i	$\tau_i \to \tau_1 + \tau_2$	injection
$case\dagger$	$(\tau_1 \to \tau), (\tau_2 \to \tau), \tau_1 + \tau_2 \to \tau$	case analysis
$[]$	$[\tau]$	empty sequence
at	$[\tau], [\tau] \to [\tau]$	concatenation
$flat$	$[[\tau]] \to [\tau]$	flattening
$map\dagger$	$(\tau_1 \to \tau_2), [\tau_1] \to [\tau_2]$	mapping
$length$	$[\tau] \to N$	length of sequence
$get\dagger$	$[\tau] \to \tau$	get unique element of sequence
$zip\dagger$	$[\tau_1], [\tau_2] \to [\tau_1 \times \tau_2]$	zipping
$enum$	$[\tau] \to [N]$	enumerate elements of sequence
$split\dagger$	$[\tau], [N] \to [[\tau]]$	splitting of sequence

Fig. 1. Term-constructors of $\mathcal{NSC}$

3.1 The Simple Translation

The simple translation $\_^*: \mathcal{NSC} \to S2$ has the following pattern

- types $\vdash_{\mathcal{NSC}} \tau$ are translated to r-types $\vdash_{S2} \tau^*: r$
- terms $\overline{x}: \overline{\tau} \vdash_{\mathcal{NSC}} e: \tau$ are translated to terms $\overline{x}: \overline{\tau}^*: r \vdash_{S2} e^*: T\tau^*: r$ where T is the error monad on r-types.

Definition 4 (Error monad). Given $A: r$ the type $TA: r$ is given by $TA = A + 1$. The corresponding monad structure is defined by

val	$A \to TA$
	$val(x) = in_0(x)$

let	$(A \to TB), TA \to TB$
	$let(f, in_0(x)) = f(x)$
	$let(f, in_1(*)) = in_1(*)$

We write $[M]_T$ for $val(M)$ and $(\text{let } x \Leftarrow M \text{ in } N)$ for $let([x: A]N, M)$.

3.2 The Mixed Translation

The mixed translation consists of a pair of translations $(\_^c, \_^r): \mathcal{NSC} \to S2$ s.t.

- types $\vdash_{\mathcal{NSC}} \tau$ are translated to families of r-types $x: \tau^c: c \vdash_{S2} \tau^r(x): r$
- terms $\overline{x}: \overline{\tau} \vdash_{\mathcal{NSC}} e: \tau$ are translated to pairs of *compatible* terms $\overline{x}: \overline{\tau}^c: c \vdash_{S2} e^c: T\tau^c: c$ and $\overline{x}: \overline{\tau}^c: c, \overline{x}': \overline{\tau}^r: r \vdash_{S2} e^r: T'([x: \tau^c]\tau^r, e^c): r$

where (T, T') is the error monad on families of r-types.

Definition 5 (Error monad on families of types). Given $x{:}A{:}c \vdash A'{:}r$ the family $x{:}TA{:}c \vdash T'([x{:}A]A', x){:}r$ is given by $TA = A + 1$ and
$T'([x{:}A]A', in_0(x)) = A'(x)$
$T'([x{:}A]A', in_0(*)) = 1$
We write $T'(A, A', M)$ for $T'([x{:}A]A', M)$.

The corresponding monad structure is defined by pair of compatible terms

val val'	$A \to TA$ $\forall x{:}A.A' \to T'(A, A', val(x))$
	$val(x) = in_0(x)$ $val'(x, x') = x'$
let let'	$(A \to TB), TA \to TB$ $\forall f, x{:}(A \to TB), TA,$ $(\forall x{:}A.A' \to T'(B, B', fx)), T'(A, A', x) \to T'(B, B', let(f, x))$
	$let(f, in_0(x)) = f(x)$ $let'(f, in_0(x), f', x') = f'(x, x')$ $let(f, in_1(*)) = in_1(*)$ $let'(f, in_1(*), f', *) = *$

We write $[M]_T$ for $val(M)$, ($\text{let}\ x{\Leftarrow}M\ \text{in}\ N$) for $let([x{:}A]N, M)$, M' for $val'(M, M')$ and ($\text{let}'\ x, x'{\Leftarrow}M, M'\ \text{in}\ N'$) for $let'([x{:}A]N, M, [x{:}A, x'{:}A']N', M')$.

In defining the mixed translation of types $[\tau]$ we use the list type-constructor (L, L') acting on families of r-types.

Definition 6 (List objects for families of types). Given $x{:}A{:}c \vdash A'{:}r$ the family $x{:}LA{:}c \vdash L'([x{:}A]A', x){:}r$ is given by $LA = \Sigma n{:}N.V(n, A)$ and $L'([x{:}A]A', \langle n, v\rangle) = \forall i{:}n.A'(vi)$. We write $L'(A, A', M)$ for $L'([x{:}A]A', M)$.

3.3 Semantic View and Main Results

Given a model $\pi{:}\,\mathbf{Fam}(\mathcal{C}) \to \mathbf{Set}$ of $S2$ (see Section 2), one may compose the two translations of $\mathcal{NSC}$ in $S2$ with the interpretation of $S2$ in the model, and thus investigate the properties of the resulting interpretations. In fact, it is often easier to start from a direct interpretation of $\mathcal{NSC}$, and then work out the corresponding translation (with its low level details). In what follows we assume that $\mathcal{C}$ is a cartesian closed category with small products and small sums (as done in Section 2 to ensure that $\pi{:}\,\mathbf{Fam}(\mathcal{C}) \to \mathbf{Set}$ is a model of $S2$).

- The simple translation corresponds to an interpretation of $\mathcal{NSC}$ in the Kleisli category $\mathcal{C}_T$ for the monad $T(\_) = \_ + 1$.
- The mixed translation corresponds to an interpretation of $\mathcal{NSC}$ in the Kleisli category $\mathbf{Fam}(\mathcal{C})_T$ for the monad $T(\_) = \_ + 1$, namely $T(\langle I, c\rangle) = \langle I + 1, [c, 1]\rangle$, where $[c, 1] \in \mathcal{C}^{I+1}$ is the family s.t. $[c, 1]_{in_0(i)} = c_i$ and $[c, 1]_{in_1(*)} = 1$.

Remark. These interpretations could be parameterized w.r.t. a (strong) monad S on $\mathcal{C}$, i.e.: $TA = S(A+1)$ in $\mathcal{C}$; $T(\langle I, c\rangle) = \langle I+1, [c', 1]\rangle$ with $c'_i = S(c_i)$ in $\mathbf{Fam}(\mathcal{C})$. This generalization is interesting because it suggests a way for dealing with non-termination and other computational effects. Unfortunately, in intensional $S2$ it is not possible to mimic the definition of T in $\mathbf{Fam}(\mathcal{C})$ from S (the difficulty is in the definition of *let*). The reason is the lack of extensivity (see Proposition 3).

In order to interpret $\mathcal{NSC}$ in the Kleisli category $\mathcal{A}_T$ for a strong monad T over $\mathcal{A}$, the category $\mathcal{A}$ must have finite products, binary sums and list objects (satisfying certain additional properties). Moreover, one can always take $T(\_) = \_+1$. When $\mathcal{A}$ is cartesian closed and has countable sums, the necessary structure and properties are automatically available.

The following lemma says that we can interpret $\mathcal{NSC}$ in $\mathbf{Set}$, $\mathcal{C}$ and $\mathbf{Fam}(\mathcal{C})$ by taking $T(\_) = \_+1$ (and fixing an interpretation for the signature Σ).

Lemma 7. *The categories* $\mathbf{Set}$, $\mathcal{C}$ *and* $\mathbf{Fam}(\mathcal{C})$ *are cartesian closed and have small sums. The following are full reflections*

$$\mathbf{Set} \underset{\Delta}{\overset{\pi}{\leftrightarrows}} \mathbf{Fam}(\mathcal{C}) \underset{}{\overset{\exists}{\rightleftarrows}} \mathcal{C} \qquad (\text{with } \pi \dashv \Delta,\ \exists \dashv \mathcal{C}\hookrightarrow\mathbf{Fam}(\mathcal{C}))$$

Moreover, the functors $\Delta\colon \mathbf{Set} \to \mathbf{Fam}(\mathcal{C})$, $\pi\colon \mathbf{Fam}(\mathcal{C}) \to \mathbf{Set}$ *and* $\exists\colon \mathbf{Fam}(\mathcal{C}) \to \mathcal{C}$ *preserve finite products, exponentials and small sums.*

Proof. The relevant categorical structure in $\mathbf{Fam}(\mathcal{C})$ is defined as follows: $1 = \langle 1, 1\rangle$, $\langle I, a\rangle \times \langle J, b\rangle = \langle I \times J, c\rangle$ with $c_{i,j} = a_i \times b_j$, $\langle J, b\rangle^{\langle I, a\rangle} = \langle J^I, c\rangle$ with $c_f = \prod_{i\in I} b_{fi}^{a_i}$, $\coprod_{i\in I}\langle J_i, b_i\rangle = \langle \coprod_{i\in I} J_i, c\rangle$ with $c_{i,j} = (b_i)_j$. The adjoint functors are given by

$$I \overset{\pi}{\longleftarrow\!\!\shortmid} \langle I, a\rangle \qquad \langle I, a\rangle \overset{\exists}{\longmapsto} \coprod_{i\in I} a_i$$

$$J \underset{\Delta}{\longmapsto} \langle J, 1\rangle \qquad \langle 1, b\rangle \longleftarrow\!\!\shortmid b$$

A simple check shows that all functors preserve finite products and exponentials, and all functors except $\mathcal{C} \hookrightarrow \mathbf{Fam}(\mathcal{C})$ preserve small sums.

Theorem 8. *The following diagrams commute (up to a natural isomorphism)*

$$\begin{array}{ccccc} & & \mathcal{NSC} & & \\ & \swarrow^{\textit{compile}} & \downarrow{\scriptstyle \textit{mixed}} & \searrow^{\textit{simple}} & \\ \mathbf{Set}_T & \underset{\pi}{\longleftarrow} & \mathbf{Fam}(\mathcal{C})_T & \underset{\exists}{\longrightarrow} & \mathcal{C}_T \end{array}$$

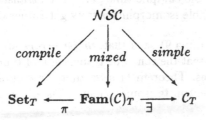

Remark. There is a proviso to the above theorem: the simple and mixed interpretation of $\mathcal{NSC}$ are related (as stated), if and only if the simple and mixed interpretation of Σ are. The syntactic counterpart of this theorem says that the following assertions are provable (in extensional $S2$):

- $\tau^* \cong \exists x\colon \tau^c.\tau^r$ and $T\tau^* \cong \exists x\colon T\tau^c.T'(\tau^c, \tau^r, x)$;
- $\overline{x}\colon\overline{\tau}^c, \overline{x}'\colon\overline{\tau}^r \vdash [\langle x_i, x'_i\rangle/\overline{x}_i]e^* = \langle e^c, e^r\rangle\colon T\tau^*$ (up to isomorphism).

The delicate step in the proof of the syntactic result is the case $[\tau]$, where one should use Proposition 2. Informally speaking, the theorem says that the mixed translation *extracts* more information than the simple translation.

Lemma 9. *If $\mathcal{C}$ is extensive and non-trivial (i.e. $0 \not\cong 1$), then $\exists\colon \mathbf{Fam}(\mathcal{C})(1, x) \to \mathcal{C}(1, \exists x)$ is injective for any x.*

Thus one can conclude (when the hypothesis of the lemma are satisfied) that the interpretation of a closed expression of $\mathcal{NSC}$ is equal to error in the simple semantics if and only if it is in the mixed semantics.

The mixed interpretation of $\mathcal{NSC}$ is rather boring when the interpretation of atomic types in Σ are *trivial*, i.e. isomorphic to the terminal object.

Theorem 10. *If the mixed interpretation of base types are trivial, then the following diagram commutes (up to a natural isomorphism)*

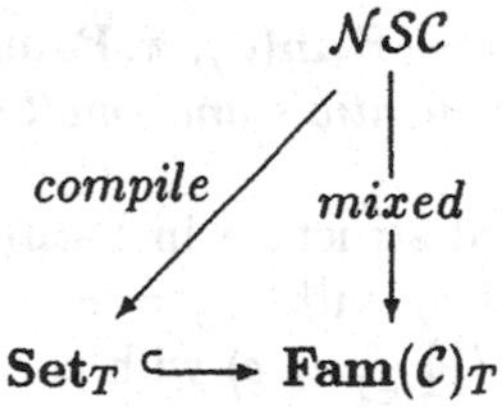

Remark. The syntactic counterpart of this theorem says that $x\colon \tau^c \vdash \tau^r \cong 1$ is provable (in extensional $S2$). Therefore, the run-time part of $S2$ is not really used.

Theorem 11. *The compile-time part compile: $\mathcal{NSC} \to \mathbf{Set}_T$ of the mixed interpretation factors through the full sub-category of countable sets.*

Remark. The syntactic counterpart of this result is much stronger, namely (in extensional $S2$) τ^c is provable isomorphic either to a finite cardinal or to the NNO. This means that the compile-time part of the translation uses very simple types (though the provable isomorphisms may get rather complex).

Theorem 8 (and Lemma 9) says that *shape errors* are detected at compile-time. Theorem 10 says that the run-time translation of types is *trivial*, when it is trivial for all base types. Theorem 11 says that the compile-time translation of types is very simple, i.e. (up to isomorphism) it is either a finite cardinal or the NNO.

4 The Simple Translation

The simple translation $\_^*$ corresponds to translate $\mathcal{NSC}$ in a simply typed lambda calculus with unit, sums, products, NNO and list objects (extended with the analogue Σ^* of the $\mathcal{NSC}$-signature Σ).

- types $\vdash_{\mathcal{NSC}} \tau$ are translated to r-types $\vdash_{S2} \tau^*: r$

τ of $\mathcal{NSC}$	$\tau^*: r$ of $S2$
1	1
N	$\exists n: N.1$
D	D
$\tau_1 \times \tau_2$	$\tau_1^* \times \tau_2^*$
$\tau_1 + \tau_2$	$\tau_1^* + \tau_2^*$
$[\tau]$	$\exists n: N.n \Rightarrow \tau^*$
arities of $\mathcal{NSC}$	
$\overline{\tau} \to \tau$	$\overline{\tau}^* \to T\tau^*$
$\overline{\sigma}, \overline{\tau} \to \tau$	$\overline{\sigma}^*, \overline{\tau}^* \to T\tau^*$

- terms $\overline{x}: \overline{\tau} \vdash_{\mathcal{NSC}} e: \tau$ are translated to terms $\overline{x}: \overline{\tau}^*: r \vdash_{S2} e^*: T\tau^*: r$

$e: \tau$ of $\mathcal{NSC}$	$e^*: T\tau^*: r$ of $S2$
x	$[x]_T$
$op(\overline{e})$	$\text{let } \overline{x} \Leftarrow \overline{e}^* \text{ in } op^*(\overline{x})$
$f(\overline{e})$	$\text{let } \overline{x} \Leftarrow \overline{e}^* \text{ in } f^*(\overline{x})$
$c(\overline{f}, \overline{e})$	$\text{let } \overline{x} \Leftarrow \overline{e}^* \text{ in } c^*(\overline{f}^*, \overline{x})$
$\lambda \overline{x}: \overline{\tau}.e$	$\lambda \overline{x}: \overline{\tau}^*.e^*$

  when a term-constructor c cannot raise an error (i.e. is not marked by † in Figure 1), we translate $c(\overline{f}, \overline{e})$ to $\text{let } \overline{x} \Leftarrow \overline{e}^* \text{ in } [c^*(\overline{f}^*, \overline{x})]_T$.

- term-constructors $c: \overline{\sigma}, \overline{\tau} \to \tau$ are translated to terms $\vdash_{S2} c^*: \overline{\sigma}^*, \overline{\tau}^* \to T\tau^*: r$, or to $\vdash_{S2} c^*: \overline{\sigma}^*, \overline{\tau}^* \to \tau^*: r$ when c cannot raise an error.

Figure 2 gives c^* for the term-constructors c of $\mathcal{NSC}$ which can raise an error, the reader could figure out for himself the definition of c^* for the other term-constructors. In Figure 2 we use ML-style notation for function definitions and other auxiliary notation for $S2$, which is defined in Appendix A.

Remark. Given a $\mathcal{NSC}$-signature Σ, its analogue Σ^* in $S2$ is defined as follows:

- a constant type $D: r$ for each atomic type D in Σ;
- a constant term $\overline{x}: \overline{\tau}^*: r \vdash op^*(\overline{x}): T\tau^*: r$ for each operation $op: \overline{\tau} \to \tau$ in Σ.

5 The Mixed Translation

The mixed translation highlights phase distinction between compile-time and run-time, and exploits fully the features of $S2$ (extended with the analogue of the $\mathcal{NSC}$-signature Σ).

c^*	arity and ML-like definition of $c^*(\overline{f}, \overline{x})$
err^*	$T\tau$
	$err^* = in_1(*)$
for^*	$(N, \tau \to T\tau), \tau, N \to T\tau$
	$for^*(f, x, 0) \;= [x]_T$ $for^*(f, x, sn) = \text{let } y \Leftarrow for^*(f, x, n) \text{ in } f(n, y)$
$case^*$	$(\tau_1 \to T\tau), (\tau_2 \to T\tau), \tau_1 + \tau_2 \to T\tau$
	$case^*(f_0, f_1, in_i(x)) = f_i(x) \qquad (i = 0, 1)$
map^*	$(\tau_1 \to T\tau_2), L\tau_1 \to T(L\tau_2)$
	$map^*(f, nil) \;= [nil]_T$ $map^*(f, h::t) = \text{let } x \Leftarrow f(h) \text{ in let } l \Leftarrow map^*(f, t) \text{ in } [x::l]_T$
get^*	$L\tau \to T\tau$
	$get^*(nil) \;= err^*$ $get^*(h::t) = case^*([h]_T, err^*, eq^*(0, length^*(t)))$ where $eq^*: N, N \to 1 + 1$ is equality for the NNO
zip^*	$L\tau_1, L\tau_2 \to T(L(\tau_1 \times \tau_2))$
	$zip^*(nil, nil) \;= [nil]_T$ $zip^*(nil, h_2::t_2) \;= err^*$ $zip^*(h_1::t_1, nil) \;= err^*$ $zip^*(h_1::t_1, h_2::t_2) = \text{let } l \Leftarrow zip^*(t_1, t_2) \text{ in } [\langle h_1, h_2\rangle::l]_T$
$split^*$	$L\tau, LN \to T(L(L\tau))$
	$split^*(nil, nil) \;= [nil]_T$ $split^*(h::t, nil) \;= err^*$ $split^*(x, 0::p) \;= \text{let } l \Leftarrow split^*(x, p) \text{ in } [nil::l]_T$ $split^*(nil, sn::p) \;= err^*$ $split^*(h::t, sn::p) = \text{let } l \Leftarrow split^*(t, n::p) \text{ in } cons'(x, l)$ where $cons'(x, nil) = err^* \mid cons'(x, h::t) = [(x::h)::t]_T$

Fig. 2. Simple translation of $\mathcal{NSC}$ term-constructors

- types $\vdash_{\mathcal{NSC}} \tau$ are translated to families of r-types $x: \tau^c : c \vdash_{S2} \tau^r(x) : r$

τ of $\mathcal{NSC}$	$x: \tau^c : c$ and	$\tau^r(x) : r$ of $S2$
1	$\_ : 1$	1
N	$\_ : N$	1
D	$\_ : 1$	D
$\tau_1 \times \tau_2$	$\langle x_1, x_2\rangle : \tau_1^c \times \tau_2^c$	$\tau_1^r(x_1) \times \tau_2^r(x_2)$
$\tau_1 + \tau_2$	$in_i(x_i) : \tau_1^c + \tau_2^c$	$\tau_i^r(x_i)$
$[\tau]$	$\langle n, v\rangle : \Sigma n: N.V(n, \tau^c)$	$\forall i: n.\tau^r(vi)$
arities of $\mathcal{NSC}$		
$\overline{\tau} \to \tau$	$f: \overline{\tau}^c \to T\tau^c$	$\forall \overline{x}: \overline{\tau}^c.\overline{\tau}^r \to T'(\tau^c, \tau^r, f(\overline{x}))$
$\overline{\sigma}, \overline{\tau} \to \tau$	$F: \overline{\sigma}^c, \overline{\tau}^c \to T\tau^c$	$\forall \overline{f}, \overline{x}: \overline{\sigma}^c, \overline{\tau}^c.\overline{\sigma}^r, \overline{\tau}^r \to T'(\tau^c, \tau^r, F(\overline{f}, \overline{x}))$

- terms $\overline{x}: \overline{\tau} \vdash_{\mathcal{NSC}} e : \tau$ are translated to pairs of *compatible* terms $\overline{x}: \overline{\tau}^c : c \vdash_{S2} e^c : T\tau^c : c$ and

$\overline{x}:\overline{\tau}^c:c, \overline{x}':\overline{\tau}^r:r \vdash_{S2} e^r : T'(\tau^c,\tau^r,e^c):r$

$e:\tau$ of $\mathcal{NSC}$	$e^c:T\tau^c:c$ and	$e^r:T'(\tau^c,\tau^r,e^c):r$ of $S2$
x	$[x]_T$	x'
$f(\overline{e})$	$\text{let}\, \overline{x} \Leftarrow \overline{e}^c \text{ in } f^c(\overline{x})$	$\text{let}'\, \overline{x},\overline{x}' \Leftarrow \overline{e}^c,\overline{e}^r \text{ in } f^r(\overline{x},\overline{x}')$
$op(\overline{e})$	$\text{let}\, \overline{x} \Leftarrow \overline{e}^c \text{ in } op^c(\overline{x})$	$\text{let}'\, \overline{x},\overline{x}' \Leftarrow \overline{e}^c,\overline{e}^r \text{ in } op^r(\overline{x},\overline{x}')$
$c(\overline{f},\overline{e})$	$\text{let}\, \overline{x} \Leftarrow \overline{e}^c \text{ in } c^c(\overline{f}^c,\overline{x})$	$\text{let}'\, \overline{x},\overline{x}' \Leftarrow \overline{e}^c,\overline{e}^r \text{ in } c^r(\overline{f}^c,\overline{f}^r,\overline{x},\overline{x}')$
$\lambda\overline{x}:\overline{\tau}.e$	$\lambda\overline{x}:\overline{\tau}^c.e^c$	$\Lambda\overline{x}:\overline{\tau}^c.\lambda\overline{x}':\overline{\tau}^r.e^r$

when a term-constructor c cannot raise an error, we translate $c(\overline{f},\overline{e})$ to $\text{let}\, \overline{x} \Leftarrow \overline{e}^c \text{ in } [c^c(\overline{f}^c,\overline{x})]_T$ and $\text{let}'\, \overline{x},\overline{x}' \Leftarrow \overline{e}^c,\overline{e}^r \text{ in } c^r(\overline{f}^c,\overline{f}^r,\overline{x},\overline{x}')$.

- term-constructors $c:\overline{\sigma},\overline{\tau} \to \tau$ are translated to pairs of *compatible* terms
  $\vdash_{S2} c^c:\overline{\sigma}^c,\overline{\tau}^c \to T\tau^c:c$ and
  $\vdash_{S2} c^r:\forall \overline{f},\overline{x}:\overline{\sigma}^c,\overline{\tau}^c.\overline{\sigma}^r,\overline{\tau}^r \to T'(\tau^c,\tau^r,c^c(\overline{f},\overline{x})):r$, or to
  $\vdash_{S2} c^c:\overline{\sigma}^c,\overline{\tau}^c \to \tau^c:c$ and
  $\vdash_{S2} c^r:\forall \overline{f},\overline{x}:\overline{\sigma}^c,\overline{\tau}^c.\overline{\sigma}^r,\overline{\tau}^r \to \tau^r(c^c(\overline{f},\overline{x})):r$ when c cannot raise an error.

Figure 3 and 4 give c^c and c^r for the term-constructors c of $\mathcal{NSC}$ which can raise an error. The tables are organized as follows:

- For each term-constructor c of $\mathcal{NSC}$ we write

c^c	$\overline{f},\overline{x}:\overline{\sigma},\overline{\tau} \to T(\tau)$
c^r	$\overline{\sigma}',\overline{\tau}' \to T'(\tau,\tau',c^c(\overline{f},\overline{x}))$

  to mean that $c^c:\overline{\sigma},\overline{\tau} \to T(\tau)$ and $c^r:\forall \overline{f},\overline{x}:\overline{\sigma},\overline{\tau}.\overline{\sigma}',\overline{\tau}' \to T'(\tau,\tau',c^c(\overline{f},\overline{x}))$.
- Each case of the ML-like definition of c^c is immediately followed by the corresponding case for c^r.
- Arguments of trivial run-time type (i.e. isomorphic to 1) are omitted. This happens in the definition of *for* and *split*.

Remark. Given a $\mathcal{NSC}$-signature Σ, its analogue $\Sigma^{c,r}$ in $S2$ is defined as follows:

- a constant type $D:r$ for each atomic type D in Σ;
- a pair of compatible constant terms $\overline{x}:\overline{\tau}^c:c \vdash op^c(\overline{x}):T\tau^c:c$ and $\overline{x}:\overline{\tau}^c:c,\overline{x}':\overline{\tau}^r:r \vdash op^r(\overline{x},\overline{x}'):T'(\tau,\tau',op^c(\overline{x})):r$ for each $op:\overline{\tau} \to \tau$ in Σ.

At the semantic level this translation of Σ imposes strong restrictions. For instance, consider the translation of $op: D, D \to 1{+}1$. This is given by $op^c: 1{+}1$ and $op^r: D, D \to 1$ (when ignoring arguments of trivial type). Therefore, we cannot interpret op as equality on D, because both op^c and op^r are constant. The mixed translation can only cope with *shapely* operations (see [14]), where the shape of the result is determined uniquely by the shape of the arguments. However, in $S2$ one can give a type to non-shapely operations, for instance $op: D, D \to (\exists x: 1 + 1.1)$.

6 Conclusions and Further Research

[19] advocates the use of Martin-Löf Type Theory for program construction. This paper advocates the use of Martin-Löf Type Theory as part of an intermediate

c	arity and ML-like definition of $c^c(\overline{f},\overline{x})$ and $c^r(\overline{f},\overline{f}',\overline{x},\overline{x}')$
err^c err^r	$T\tau$ $T'(\tau,\tau',err^c)$
	$err^c = in_1(*)$ $err^r = *$
for^c for^r	$f,x,n\colon (N,\tau\to T\tau),\tau,N\to T\tau$ $(\forall n,x\colon N,\tau.\tau'\to T'(\tau,\tau',f(n,x))),\tau'(x)\to T'(\tau,\tau',for^c(f,x,n))$
	$for^c(f,x,0) = [x]_T$ $for^r(f,x,0,f',x') = x'$ $for^c(f,x,sn) = \text{let } y\Leftarrow for^c(f,x,n) \text{ in } f(n,y)$ $for^r(f,x,sn,f',x') = \text{let}'\ y,y'\Leftarrow for^c(f,x,n),for^r(f,x,n,f',x') \text{ in } f'(n,y,y')$
$case^c$ $case^r$	$f_1,f_2,in_i(x)\colon(\tau_1\to T\tau),(\tau_2\to T\tau),\tau_1+\tau_2\to T\tau$ $(\forall y\colon\tau_1.\tau_1'\to T'(\tau,\tau',f_1y)),(\forall y\colon\tau_2.\tau_2'\to T'(\tau,\tau',f_2y)),\tau_i'(x)\to$ $\to T'(\tau,\tau',case^c(f_1,f_2,x))$
	$case^c(f_0,f_1,in_i(x)) = f_i(x)\quad (i=0,1)$ $case^r(f_0,f_1,in_i(x),f_1',f_2',x') = f_i'(x,x')\quad (i=0,1)$
map^c map^r	$f,l\colon(\tau_1\to T\tau_2),L\tau_1\to T(L\tau_2)$ $\forall x\colon\tau_1.\tau_1'\to T'(\tau_2,\tau_2',fx),L'(\tau_1,\tau_1',l)\to T'([x\colon L\tau_2]L'(\tau_2,\tau_2',x),map^c(f,l))$
	$map^c(f,nil) = [nil]_T$ $map^r(f,nil,f',[]) = []$ $map^c(f,h::t) = \text{let } y\Leftarrow f(h) \text{ in let } z\Leftarrow map^c(f,t) \text{ in } [y::z]_T$ $map^r(f,h::t,f',[h',t']) = \text{let}'\ y,y'\Leftarrow f(h),f'(h,h') \text{ in}$ $\text{let}'\ z,z'\Leftarrow map^c(f,t),map^r(f,t,f',t') \text{ in } [y',z']$
get^c get^r	$l\colon L\tau\to T\tau$ $L'(\tau,\tau',l)\to T'(\tau,\tau',get^c(l))$
	$get^c(nil) = err^c$ $get^r(nil,[]) = err^r$ $get^c(h::t) = case^c([h]_T,err^c,eq^c(0,length^c(t)))$ $get^r(h::t,[h',t']) = case^r([h]_T,err^c,eq^c(0,length^c(t)),h',err^r,*)$ where $eq^c\colon N,N\to 1+1$ is equality for the NNO
zip^c zip^r	$l_1,l_2\colon L\tau_1,L\tau_2\to T(L(\tau_1\times\tau_2))$ $L'(\tau_1,\tau_1',l_1),L'(\tau_2,\tau_2',l_2)\to T'([x\colon L(\tau_1\times\tau_2)]L'(\tau_1\times\tau_2,\tau_1'\times\tau_2',x),zip^c(l_1,l_2))$
	$zip^c(nil,nil) = [nil]_T$ $zip^r(nil,nil,[],[]) = []$ $zip^c(nil,h_2::t_2) = err^c$ $zip^r(nil,h_2::t_2,\_,\_) = err^r$ $zip^c(h_1::t_1,nil) = err^c$ $zip^r(h_1::t_1,nil,\_,\_) = err^r$ $zip^c(h_1::t_1,h_2::t_2) = \text{let } y\Leftarrow zip^c(t_1,t_2) \text{ in } [\langle h_1,h_2\rangle::y]_T$ $zip^r(h_1::t_1,h_2::t_2,[h_1',t_1'],[h_2',t_2']) = \text{let}'\ y,y'\Leftarrow zip^c(t_1,t_2),zip^r(t_1,t_2,t_1',t_2') \text{ in}$ $[\langle h_1',h_2'\rangle,y']$

Fig. 3. Mixed translation of $\mathcal{NSC}$ term-constructors

$split^c$	$l, k\colon L\tau, LN \to T(L(L\tau))$	
$split^r$	$L'(\tau,\tau',l) \to T'([x\colon L(L\tau)]L'([y\colon L\tau]L'(\tau,\tau',y),x), split^c(l,k))$	
	$split^c(nil, nil)$	$= [nil]_T$
	$split^r(nil, nil, [])$	$= []$
	$split^c(h::t, nil)$	$= err^c$
	$split^r(h::t, nil, [h',t'])$	$= err^r$
	$split^c(x, 0::p)$	$=$ let $y \Leftarrow split^c(x,p)$ in $[nil::y]_T$
	$split^r(x, 0::p, x')$	$=$ let$'$ $y, y' \Leftarrow split^c(x,p), split^r(x,p,x')$ in $[[], y']$
	$split^c(nil, sn::p)$	$= err^c$
	$split^r(nil, sn::p, [])$	$= err^r$
	$split^c(h::t, sn::p)$	$=$ let $y \Leftarrow split^c(t, n::p)$ in $cons'^c(h,y)$
	$split^r(h::t, sn::p, [h',t'])$	$=$ let$'$ $y, y' \Leftarrow split^c(t,n::p), split^r(t,n::p,t')$ in $cons'^r(h,y,h',y')$
	where $cons'^c(x, nil) = err^c \mid cons'^c(x, h::t) = [(x::h)::t]_T$	
	where $cons'^r(x, nil, x', []) = err^r \mid cons'^r(x, h::t, x', [h',t']) = [[x',h'],t']$	

Fig. 4. Mixed translation of $\mathcal{NSC}$ term-constructors (cont.)

language $S2$. This avoids two major problems: a programmer does not have to deal with dependent type directly and decidability of type-checking in $S2$ does not rely on strong normalization of run-time expressions (since dependent types are confined to the compile-time part of $S2$). [16] introduces a simply typed language for vectors with an operator $\#\colon \tau \to \#\tau$ to extract shape information from terms. The type $\#\tau$ is like our τ^c, whereas the term $\#e$ *performs* the translation e^c lazily. Our approach gains in clarity and generality by separating the programming language from the intermediate language. There are many unresolved issues about $S2$ that should be addressed. The following is a partial list with some hints on how one may proceed.

$S2$ is based on intensional type theory (to ensure decidability of type-checking), however most of the semantic properties of translations rely on extensionality. It would be interesting to investigate whether some of the extensional properties considered (e.g. extensivity) could be added safely to intensional type theory.

We have not given an operational semantics for $S2$, probably this can be done relatively easy by borrowing ideas from [7, 6].

$\mathcal{NSC}$ is a simply typed language, whereas NESL has also ML-like polymorphism. This is likely to require a refinement of $S2$ by adding a sort of *shapes*. Shape theory should provide useful guidelines for such refinement. There are also obvious extensions to the run-time part of $S2$, e.g. recursive types.

One must fill the gap between $S2$ and parallel machines (or already implemented intermediate languages, like VCODE). Moreover, translations should be *efficient* in the sense of [22, 3].

We have used a modified version of $\mathcal{NSC}$ with for-loops rather than while-loops. It should be possible to incorporate run-time computational aspects in $S2$ using monads, and thus translate more realistic languages.

One cannot expect that array bound-checking can all be done at compile-time. Indeed shape analysis proposes a more pragmatic approach, where execution and analysis alternate (see [15]). Existential types should provide a clean way of expressing when shape information is available only at run-time.

In shape theory one can distinguish between arrays and lists (see [13]): the elements of an array have the same shape, those of a list may have different shapes. This suggests a different translation of $\mathcal{NSC}$ into $S2$ worth studying, namely: $[\tau]^c = N \times \tau^c$ and $[\tau]^r(\langle n, x\rangle) = n \Rightarrow \tau^r(x)$.

References

1. L. Birkedal, M. Tofte, and M. Vejlstrup. From Region Inference to von Neumann Machines via Region Representation Inference. In *Proceedings from the 23rd annual ACM SIGPLAN-SIGACT Symposium on Principles of Programming Languages*, 1996.
2. G.E. Blelloch, S. Chatterjee, J.C. Hardwick, J. Sipelstein, and M. Zagha. Implementation of a portable nested data-parallel language. *Journal of Parallel and Distributed Computing*, 21(1), April 1994.
3. G.E. Blelloch and J. Greiner. A provable time and space efficient implementation of NESL. In *ACM SIGPLAN International Conference on Functional Programming*, pages 213–225, May 1996.
4. A. Carboni, S. Lack, and R.F.C. Walters. Introduction to extensive and distributive categories. *Journal of Pure and Applied Algebra*, 84:145–158, 1993.
5. T. Coquand and C. Paulin-Mohring. Inductively defined types. volume 389, LNCS, 1989.
6. R. Davies. A temporal-logic approach to binding-time analysis. In E. Clarke, editor, *Proceedings of the Elenventh Annual Symposium on Logic in Computer Science*, New Brunswick, New Jersey, July 1996. IEEE Computer Society Press.
7. H. Goguen. *A Typed Operational Semantics for Type Theory*. PhD thesis, University of Edinburgh, 1994.
8. R. Harper, J. Mitchell, and E. Moggi. Higher-order modules and the phase distinction. In *17th POPL*. ACM, 1990.
9. R. Harper and G. Morrisett. Compiling polymorphism using intensional type analysis. In *Conference Record of POPL '95: 22nd ACM SIGPLAN-SIGACT Symposium on Principles of Programming Languages*, pages 130–141, San Francisco, California, January 1995.
10. M. Hofmann. Dependent types: Syntax, semantics, and applications. Summer School on Semantics and Logics of Computation, University of Cambridge, Newton Institute for Mathematical Sciences, September 1995.
11. B. Jacobs, E. Moggi, and T. Streicher. Relating models of impredicative type theories. In *Proceedings of the Conference on Category Theory and Computer Science, Manchester, UK, Sept. 1991*, volume 389 of *LNCS*. Springer Verlag, 1991.
12. B.P.F. Jacobs. *Categorical Type Theory*. PhD thesis, University of Nijmegen, 1991.
13. C.B. Jay. Matrices, monads and the fast fourier transform. In *Proceedings of the Massey Functional Programming Workshop 1994*, pages 71–80, 1994.
14. C.B. Jay. A semantics for shape. *Science of Computer Programming*, 25:251–283, 1995.
15. C.B. Jay. Shape in computing. *ACM Computing Surveys*, 28(2):355–357, 1996.

16. C.B. Jay and M. Sekanina. Shape checking of array programs. In *Computing: the Australasian Theory Seminar, Proceedings, 1997*, 1997. accepted for publication.
17. E. Moggi. A category-theoretic account of program modules. *Math. Struct. in Computer Science*, 1:103–139, 1991.
18. F. Nielson and H.R. Nielson. *Two-Level Functional Languages.* Number 34 in Cambridge Tracts in Theoretical Computer Science. Cambridge University Press, 1992.
19. B. Nordström, K. Petersson, and J.M. Smith. *Programming in Martin-Löf's type theory:an introduction.* Number 7 in International series of monographs on computer science. Oxford University Press, New York, 1990.
20. S. Peyton Jones. Unboxed values as first-class citizens. In *Functional Programming and Computer Architecture*, volume 523 of *LNCS*, 1991.
21. A.M. Pitts. Notes on categorical logic. University of Cambridge, Computer Laboratory, Lent Term 1989.
22. D. Suciu and V. Tannen. Efficient compilation of high-level data parallel algorithms. In *Proc. ACM Symposium on Parallel Algorithms and Architectures*, June 1994.
23. J.-P. Talpin and P. Jouvelot. The type and effect discipline. *Information and Computation*, 111(2):245–296, June 1994.
24. M. Tofte and J.-P. Talpin. Implementation of the typed call-by-value lambda-calculus using a stack of regions. In *Proceedings from the 21st annual ACM SIGPLAN-SIGACT Symposium on Principles of Programming Languages*, 1994.

A The 2-Level Intensional Type-Theory $S2$

In presenting the rules we follow [19, 10]. The elimination rules for inductive types are over sorts (since $S2$ has no universes).

A.1 Rules for c-Types

Π-types. $(\Pi)\ \dfrac{\Gamma, x{:}\,A{:}\,c \vdash B{:}\,c}{\Gamma \vdash (\Pi x{:}\,A.B){:}\,c}$ $\quad(\Pi\text{-I})\ \dfrac{\Gamma, x{:}\,A{:}\,c \vdash M{:}\,B{:}\,c}{\Gamma \vdash (\lambda x{:}\,A.M){:}\,(\Pi x{:}\,A.B){:}\,c}$

$$(\Pi\text{-E})\ \frac{\Gamma \vdash M{:}\,(\Pi x{:}\,A.B){:}\,c \quad \Gamma \vdash N{:}\,A{:}\,c}{\Gamma \vdash MN{:}\,[N/x]B{:}\,c}$$

Σ-types. $(\Sigma)\ \dfrac{\Gamma, x{:}\,A{:}\,c \vdash B{:}\,c}{\Gamma \vdash (\Sigma x{:}\,A.B){:}\,c}$ $\quad(\Sigma\text{-I})\ \dfrac{\Gamma, x{:}\,A{:}\,c \vdash B(x){:}\,c \quad \Gamma \vdash M{:}\,A{:}\,c \quad \Gamma \vdash N{:}\,B(M){:}\,c}{\Gamma \vdash \langle M, N\rangle{:}\,(\Sigma x{:}\,A.B){:}\,c}$

$$(\Sigma\text{-E})\ \frac{\Gamma, z{:}\,(\Sigma x{:}\,A.B){:}\,c \vdash C(z){:}\,\alpha \quad \Gamma, x{:}\,A{:}\,c, y{:}\,B{:}\,c \vdash M{:}\,C(\langle x,y\rangle){:}\,\alpha \quad \Gamma \vdash N{:}\,(\Sigma x{:}\,A.B){:}\,c}{\Gamma \vdash R^{\Sigma}([x{:}\,A, y{:}\,B]M, N){:}\,C(N){:}\,\alpha}$$

$$(\Sigma\text{-E-}\alpha)\ \frac{\Gamma, x{:}\,A{:}\,c, y{:}\,B{:}\,c \vdash C{:}\,\alpha \quad \Gamma \vdash N{:}\,(\Sigma x{:}\,A.B){:}\,c}{\Gamma \vdash R^{\Sigma}([x{:}\,A, y{:}\,B]C, N){:}\,\alpha}$$

Sums. (+) $\dfrac{\Gamma \vdash A_i : c \quad (i=0,1)}{\Gamma \vdash A_0 + A_1 : c}$ (+-I) $\dfrac{\Gamma \vdash A_i : c \quad (i=0,1) \qquad \Gamma \vdash M : A_i : c}{\Gamma \vdash in_i(M) : A_0 + A_1 : c}$

(+-E) $\dfrac{\Gamma, z : A_0 + A_1 : c \vdash C(z) : \alpha \qquad \Gamma, x : A_i : c \vdash M_i : C(in_i(x)) : \alpha \quad (i=0,1) \qquad \Gamma \vdash N : A_0 + A_1 : c}{\Gamma \vdash R^+([x : A_0]M_0, [x : A_1]M_1, N) : C(N) : \alpha}$

(+-E-α) $\dfrac{\Gamma, x : A_i : c \vdash C_i : \alpha \quad (i=0,1) \qquad \Gamma \vdash N : A_0 + A_1 : c}{\Gamma \vdash R^+([x : A_0]C_0, [x : A_1]C_1, N) : \alpha}$

Unit. (1) $\dfrac{\Gamma \vdash}{\Gamma \vdash 1 : c}$ (1-I) $\dfrac{\Gamma \vdash}{\Gamma \vdash * : 1 : c}$ (1-E) $\dfrac{\Gamma, z : 1 : c \vdash C(z) : \alpha \qquad \Gamma \vdash M : C(*) : \alpha \qquad \Gamma \vdash N : 1 : c}{\Gamma \vdash R^1(M, N) : C(N) : \alpha}$

NNO. (N) $\dfrac{\Gamma \vdash}{\Gamma \vdash N : c}$ (N-0) $\dfrac{\Gamma \vdash}{\Gamma \vdash 0 : N : c}$ (N-s) $\dfrac{\Gamma \vdash M : N : c}{\Gamma \vdash s(M) : N : c}$

(N-E) $\dfrac{\Gamma, n : N : c \vdash A(n) : \alpha \qquad \Gamma \vdash M_0 : A(0) : \alpha \qquad \Gamma, n : N : c, x : A(n) : \alpha \vdash M_s : A(sn) : \alpha \qquad \Gamma \vdash m : N : c}{\Gamma \vdash R^N(M_0, [n : N, x : A(n)]M_s, m) : A(m) : \alpha}$

Finite cardinals. (n) $\dfrac{\Gamma \vdash n : N : c}{\Gamma \vdash n : c}$ (s-0) $\dfrac{\Gamma \vdash n : N : c}{\Gamma \vdash 0 : sn : c}$ (s-s) $\dfrac{\Gamma \vdash n : N : c \qquad \Gamma \vdash M : n : c}{\Gamma \vdash sM : sn : c}$

(0-E) $\dfrac{\Gamma, i : 0 : c \vdash C(i) : \alpha \qquad \Gamma \vdash M : 0 : c}{\Gamma \vdash R^0(M) : C(M) : \alpha}$ (s-E) $\dfrac{\Gamma \vdash n : N : c \qquad \Gamma, i : sn : c \vdash C(i) : \alpha \qquad \Gamma \vdash M_0 : C(0) : \alpha \qquad \Gamma, i : n : c \vdash M_s : C(si) : \alpha \qquad \Gamma \vdash P : sn : c}{\Gamma \vdash R^s(M_0, [i : n]M_s, P) : C(P) : \alpha}$

Arrays. (V) $\dfrac{\Gamma \vdash n : N : c \qquad \Gamma \vdash A : c}{\Gamma \vdash V(n, A) : c}$ (V_0-I) $\dfrac{\Gamma \vdash A : c}{\Gamma \vdash [] : V(0, A) : c}$

(V_s-I) $\dfrac{\Gamma \vdash n : N : c \qquad \Gamma \vdash M : A : c \qquad \Gamma \vdash N : V(n, A) : c}{\Gamma \vdash [M, N] : V(sn, A) : c}$ (V_0-E) $\dfrac{\Gamma, z : V(0, A) : c \vdash C(z) : \alpha \qquad \Gamma \vdash M : C([]) : \alpha \qquad \Gamma \vdash N : V(0, A) : c}{\Gamma \vdash R^{V_0}(M, N) : C(N) : \alpha}$

$$(V_s\text{-E})\ \frac{\begin{array}{l}\Gamma \vdash n{:}\, N{:}\, c\\ \Gamma, z{:}\, V(sn, A){:}\, c \vdash C(z){:}\, \alpha\\ \Gamma, x{:}\, A{:}\, c, y{:}\, V(n, A){:}\, c \vdash M{:}\, C([x, y]){:}\, \alpha\\ \Gamma \vdash N{:}\, V(sn, A){:}\, c\end{array}}{\Gamma \vdash R^{V_s}([x{:}\, A, y{:}\, V(n, A)]M, N){:}\, C(N){:}\, \alpha}$$

Remark. In a stronger version of *intensional* $S2$, where sort c and r are universes (i.e. types of some bigger sort), one could have defined finite cardinals and arrays by induction on the natural numbers

$$F(0) = 0 \mid F(sn) = 1 + F(n) \qquad V(0, A) = 1 \mid V(sn, A) = A \times V(n, A)$$

and derived the corresponding introduction and elimination rules. We have not used the stronger version of intensional $S2$, because its categorical semantics is more involved. On the other hand, the categorical models of $S2$ are *extensional* (see Section 2), and the interpretation of finite cardinals and arrays is defined in terms of the NNO by exploiting extensionality.

A.2 Rules for r-Types

$\forall$-types. $(\forall)\ \dfrac{\Gamma, x{:}\, A{:}\, c \vdash B{:}\, r}{\Gamma \vdash (\forall x{:}\, A.B)}$ $\quad (\forall\text{-I})\ \dfrac{\Gamma, x{:}\, A{:}\, c \vdash M{:}\, B{:}\, r}{\Gamma \vdash (\Lambda x{:}\, A.M){:}\, (\forall x{:}\, A.B){:}\, r}$

$$(\forall\text{-E})\ \frac{\Gamma \vdash M{:}\, (\forall x{:}\, A.B){:}\, r \quad \Gamma \vdash N{:}\, A{:}\, c}{\Gamma \vdash MN{:}\, [N/x]B{:}\, r}$$

$\exists$-types. $(\exists)\ \dfrac{\Gamma, x{:}\, A{:}\, c \vdash B{:}\, r}{\Gamma \vdash (\exists x{:}\, A.B){:}\, r}$ $\quad (\exists\text{-I})\ \dfrac{\begin{array}{l}\Gamma, x{:}\, A{:}\, c \vdash B(x){:}\, r\\ \Gamma \vdash M{:}\, A{:}\, c\\ \Gamma \vdash N{:}\, B(M){:}\, r\end{array}}{\Gamma \vdash (\langle M, N\rangle){:}\, (\exists x{:}\, A.B){:}\, r}$

$$(\exists\text{-E})\ \frac{\begin{array}{l}\Gamma \vdash C{:}\, r\\ \Gamma, x{:}\, A{:}\, c, y{:}\, B{:}\, r \vdash N{:}\, C{:}\, r\\ \Gamma \vdash M{:}\, (\exists x{:}\, A.B){:}\, r\end{array}}{\Gamma \vdash R^{\exists}([x{:}\, A, y{:}\, B]M, N){:}\, C{:}\, r}$$

$\times$-types. $(\times)\ \dfrac{\Gamma \vdash A_i{:}\, r \quad (i = 0, 1)}{\Gamma \vdash (A \times B){:}\, r}$ $\quad (\times\text{-I})\ \dfrac{\Gamma \vdash M_i{:}\, A_i{:}\, r \quad (i = 0, 1)}{\Gamma \vdash \langle M_0, M_1\rangle{:}\, A_0 \times A_1{:}\, r}$

$$(\times\text{-E})\ \frac{\Gamma \vdash M{:}\, A_0 \times A_1{:}\, r}{\Gamma \vdash \pi_i(M){:}\, A_i{:}\, r}$$

Sums. $(+)\ \dfrac{\Gamma \vdash A_i{:}\, r \quad (i = 0, 1)}{\Gamma \vdash A_0 + A_1{:}\, r}$ $\quad (+\text{-I})\ \dfrac{\begin{array}{l}\Gamma \vdash A_i{:}\, r \quad (i = 0, 1)\\ \Gamma \vdash M{:}\, A_i{:}\, r\end{array}}{\Gamma \vdash in_i(M){:}\, A_0 + A_1{:}\, r}$

$$(+\text{-E})\ \frac{\begin{array}{l}\Gamma, x{:}\, A_i{:}\, c \vdash M_i{:}\, C{:}\, r \quad (i = 0, 1)\\ \Gamma \vdash N{:}\, A_0 + A_1{:}\, r\end{array}}{\Gamma \vdash R^{+}([x{:}\, A_0]M_0, [x{:}\, A_1]M_1, N){:}\, C{:}\, r}$$

→-types. (→) $\dfrac{\Gamma \vdash A, B{:}\, r}{\Gamma \vdash (A \to B){:}\, r}$ (→-I) $\dfrac{\Gamma, x{:}\, A{:}\, r \vdash M{:}\, B{:}\, r}{\Gamma \vdash (\lambda x{:}\, A.M){:}\, (A \to B){:}\, r}$

(→-E) $\dfrac{\Gamma \vdash M{:}\,(A \to B){:}\, r \quad \Gamma \vdash N{:}\, A{:}\, r}{\Gamma \vdash MN{:}\, B{:}\, r}$

Unit. (1) $\dfrac{\Gamma \vdash}{\Gamma \vdash 1{:}\, r}$ (1-I) $\dfrac{\Gamma \vdash}{\Gamma \vdash *{:}\, 1{:}\, r}$

Remark. Sums, products and unit types of sort r could have been defined in terms of finite cardinals, universal and existential types.

The computation rules on raw terms are the usual ones and we omit them.

A.3 Auxiliary Notation and Notational Conventions

This section introduces auxiliary notations and notational conventions for $S2$.

Auxiliary notation for types.

- $A \Rightarrow B$ stands for $\forall \_{:}\, A.B$
- $A \to B$ stands for $\Pi \_{:}\, A.B$ when A and B have sort c
- $A \times B$ stands for $\Sigma \_{:}\, A.B$ when A and B have sort c
- A^n stands for $V(n, A)$
- N^r stands for $\exists n{:}\, N.1$ (the NNO of sort r)
- $L(A)$ stands for $\Sigma n{:}\, N.V(n, A)$ (the list object of sort c)
- $V^r(n, [i{:}\, n]A)$ with $n{:}\, N{:}\, c$ and $i{:}\, n{:}\, c \vdash A{:}\, r$ stands for $\forall i{:}\, n.A(i)$, i.e. the r-type of heterogeneous arrays of size n
- $L^r(A)$ stands for $\exists n{:}\, N.n \Rightarrow A$ (the list object of sort r)

When there is no ambiguity with sorts the superscript r is omitted.

ML-style notation for function definitions.

- $f(\langle x, y\rangle) = M(x, y)$ stands for $f(z) = R^{\Sigma}([x{:}\, A, y{:}\, B]M, z)$ or $R^{\exists}([x{:}\, A, y{:}\, B]M, z)$ or $M(\pi_0(z), \pi_1(z))$ depending on the domain of f
- $f(in_i(x)) = M_i(x)$ $(i = 0, 1)$ stands for $f(z) = R^{+}([x{:}\, A_0]M_0, [x{:}\, A_1]M_1, z)$
- $f(*) = M$ stands for $f(z) = R^1(M, z)$
- $\begin{cases} f(0) = M_0 \\ f(sn) = M_s(n, f(n)) \end{cases}$ stands for $f(z) = R^N(M_0, [n{:}\, N, x{:}\, A(n)]M_s, z)$
- $(M_0, [x{:}\, n]M_s)$ stands for $f(z{:}\, sn) = R^s(M_0, [x{:}\, n]M_s, z)$ and $()$ stands for R^0
- $f([]) = M$ stands for $f(z) = R^{V_0}(M, z)$
  $f([x, v]) = M$ stands for $f(z) = R^{V_s}([x{:}\, A, v{:}\, V(n, A)]M, z)$ when n is clear from the context.
- array selection $select{:}\, \Pi n{:}\, N.V(n, A) \to n \to A$ is given by
  $select(0, []) = ()$
  $select(sn, [a, v]) = (a, select(n, v))$
  and we write MN for $select(n, M, N)$ when $M{:}\, V(n, A){:}\, c$ and $N{:}\, n{:}\, c$.

Auxiliary notation for derived universal objects.

- NNO N^r of sort r

0^r	N^r
	$0^r = \langle 0, * \rangle$
s^r	$N^r \to N^r$
	$s^r(\langle n, * \rangle) = \langle sn, * \rangle$
R^{N^r}	$A, (N^r, A \to A), N^r \to A$
	$R^{N^r}(x, f, \langle 0, * \rangle) \;\; = x$ $R^{N^r}(x, f, \langle sn, * \rangle) = f(n, R^{N^r}(x, f, \langle n, * \rangle))$

- r-type of heterogeneous arrays $V^r(n, [i{:}\,n]A)$ with $n{:}\,N{:}\,c$ and $i{:}\,n{:}\,c \vdash A{:}\,r$

$[]^r$	$V^r(0, [i{:}\,0]A)$
	$[]^r = ()$
$[\_,\_]^r$	$A(0), V^r(n, [i{:}\,n]A(si)) \to V^r(sn, [i{:}\,sn]A)$
	$[a, v]) = (a, v)$
$R^{V_0^r}$	$B, V^r(0, [i{:}\,0]A) \to B$
	$R^{V_0^r}(b, \_) = b$
$R^{V_s^r}$	$(A(0), V^r(n, i{:}\,nA(si)) \to B), V^r(sn, [i{:}\,sn]A) \to B$
	$R^{V_s^r}(f, v) = f(v_0, v_s)$ where $v_0 = v(0)$ and $v_s = \Lambda i{:}\,n.v(si)$

- list object $L(A)$ of sort c

nil	$L(A)$
	$0 = \langle 0, [] \rangle$
$cons$	$A, L(A) \to L(A)$
	$cons(a, \langle n, v \rangle) = \langle sn, [a, v] \rangle$
R^L	$B([]), (l{:}\,L(A), a{:}\,A, B(l) \to B([a, l])), l{:}\,L(A) \to B(l)$
	$R^L(b, f, \langle 0, [] \rangle) \quad\quad = b$ $R^L(b, f, \langle sn, [a, v] \rangle) = f(\langle n, v \rangle, a, R^L(b, f, \langle n, v \rangle))$

we may write $M{:}\,N$ for $cons(M, N)$

- list object $L^r(A)$ of sort r

nil^r	$L^r(A)$
	$0^r = \langle 0, () \rangle$
$cons^r$	$A, L^r(A) \to L^r(A)$
	$cons^r(a, \langle n, l \rangle) = \langle sn, (a, l) \rangle$
R^{L^r}	$B, (L^r(A), A, B \to B), L^r(A) \to B$
	$R^{L^r}(b, f, \langle 0, \_ \rangle) \;\; = b$ $R^{L^r}(b, f, \langle sn, l \rangle) = f(\langle n, l_s \rangle, l_0, R^{L^r}(b, f, \langle n, l_s \rangle))$ where $l_0 = l(0)$ and $l_s = \Lambda i{:}\,n.l(si)$

we may write $M{::}\,N$ for $cons^r(M, N)$

ML-style notation for function definitions will be used also for these derived types. When there is no ambiguity, we may drop the superscript r.

Minimum Information Code in a Pure Functional Language with Data Types

S. Berardi[1] and L. Boerio[2]

[1] Dipartimento di Informatica, Universita' di Torino
Corso Svizzera 185, 10149 Torino, Italy
stefano@di.unito.it

[2] Dipartimento di Informatica, Universita' di Torino
Corso Svizzera 185, 10149 Torino, Italy
lucab@di.unito.it

Abstract. In this paper we study programs written in a purely functional language with Data Types. We introduce a class of redundant code, the minimum information code, consisting either of "dead" code (code we may avoid evaluating), or of code whose "useful part" is constant. Both kinds of code may be removed, speeding-up the evaluation in this way: we introduce an algorithm for doing it. We prove the correctness of the method and we characterize the code we remove.

1 Introduction

We often write programs by combining and instantiating pieces of code we already have, that is, by a massive re-use of older code. In this way, it may happen that some pieces of code we use compute information which is now useless, while some others may be turned into constant functions. The first kind of code is called dead code, the second kind of code, constant code. The problem of discovering which pieces of code are constant, and thus may be evaluated at compile time, is called Binding Time Analysis (see for example [11]). Both the removal of dead code and constant code are essential to achieve efficiency. None of the two problems may be solved completely, because there is no algorithm able to decide in general if a piece of code returns a useless or constant answer under all possible inputs. Yet interesting subproblems and related problems of both issues have been solved (e.g. in [11] and in [2]).

The topic of this paper will be a curious problem, related to, but not perfectly coinciding with Binding Time Analysis for typed λ-calculus with data types (an ML-like language). Rather than trying to detect, say, the pieces of code returning a constant integer value $0, 1, 2, 3 \ldots$, we will try to detect the pieces of codes returning (in symbol form) either no useful information, or a constant and a minimum amount of useful information. We call such code *minimum information code* for short. Since integers in data types have the notation 0 or $S(t)$, a piece of code returns a constant and minimum amount of useful symbols if it always returns 0 or $S(t)$, where 0, S are useful in the rest of the program, while t is useless (it is dead code). To put it in other words, a piece of code is minimum information if we use only the first symbol of its output, and this first symbol is always the same.

A few remarks in order to make the concept precise. Pieces of codes returning a constant and useful output 1, 2, 3 ... are not minimum informative, since 1, 2, 3 ... have notation $S(0)$, $S(S(0))$, $S(S(S(0)))$, ... consisting of more than one symbol. Thus not all constant code is minimum information code. A piece of code always returning 0, *true*, *false*, *nil*, ... is both constant and minimum information code. A trivial example of minimum information code which is not constant is $S(x)$ in $t = it(S(x), a, \lambda\_.b)$. The term t evaluates to $(\lambda\_b)it(x, \ldots)$, hence to b; thus the value of x is of no use (x is dead code), and $S(x)$ is minimum informative. If we know it in advance, we may replace t with b, without having to evaluate x in its environment.

Through examples (see section 6), we may convince ourselves that minimum information code arises naturally if we program in a functional language using data types. In this paper we address the question of detecting and removing it. Again, the general problem is undecidable, but we have found an interesting decidable subproblem: to find the pieces of code which we may *re*type by a singleton type (even if the actual type is *not* a singleton). The singleton types we use are Ω (used to type useless code) and singleton *data* types. The only inhabitant of any singleton data type D has a one-symbol notation, hence any piece of code u we may retype with D is a minimum information code. The method we have used in order to check whether we may assign the type D to u is built over a method introduced in [13], [4], [7] and [5] in order to discover dead code. We first remove subterms which are dead code (performing in this way a first optimization of the term), and all types corresponding to types assigned to removed subterms. In this way we may remove some subtypes of the original type T of u. Suppose T is turned into T'. The idea is now of checking whether T' is a singleton data type: if it is, we may replace u with the only inhabitant of T'. We are forced to delay an example of this technique until after we have introduced the syntax for data types.

The contributions of this paper will be

- a formal definition, through chains of simplifications, of a technique for removing dead and minimum informative code, and a proof that it is correct;
- a proof of the fact that all possible chains of the simplifications converge to the same term; this allow us to chose any strategy we like for simplifying a term, without affecting the final result.

By building over an algorithm for removing dead code we introduced in [8], it is easy to compute a simplification chain requiring at most quadratic time in the term we want to simplify. We conjecture the existence of an algorithm solving the same problem in linear time.

This is the plan of the paper. In section 2 we introduce the typed λ-calculus with data types we will deal with in this paper. In section 3 we informally describe the method for removing dead code introduced in [8], and a method for removing minimum information code built over it. In section 4 we give the corresponding formal definitions, and a proof that our simplification technique is correct. In section 5 we prove the convergence of all simplification chains to a unique simplification; in section 6 some examples, and in section 7 the conclusion.

We often omit details: for a complete version of this paper we refer to [6].

2 The System $\lambda \rightarrow_{DT}$

In this section we briefly illustrate the pure functional programming language we will deal with in this paper: simply typed lambda calculus, extended with algebraic reduction rules and inductively defined data types. We call such system $\lambda \rightarrow_{\mathrm{DT}}$: essentially it corresponds to the pure functional core of ML (see e.g. [10]). For a more detailed description of $\lambda \rightarrow_{\mathrm{DT}}$ the reader can see [8], chs. 2,3 and [6], sec. 2.

We start by defining the types of $\lambda \rightarrow_{\mathrm{DT}}$, denoting the sets of objects our programs will deal with. We define types, data types and arities of $\lambda \rightarrow_{\mathrm{DT}}$ by mutual induction, as follows:

Definition 1.

i) *Types* are Ω (a singleton type), any data type D, and $A \rightarrow B$ if A, B are types. As usual, we will write $A_1, \ldots, A_n \rightarrow B$ as a shorthand for $A_1 \rightarrow (A_2 \rightarrow (\ldots \rightarrow (A_n \rightarrow B) \ldots))$.

ii) *Data types* are all expressions of the form

$$\delta X.C_1[X]; \ldots; C_n[X] \ \ (n \geq 0)$$

with $C_1[X], \ldots, C_n[X]$ arities. $C_1[X], \ldots, C_n[X]$ are the arities defining D, and X is a local name bound by δ.

iii) *Arities* are all expressions of the form

$$C[X] = \Omega \text{ or } C[X] = T_1, \ldots, T_n \rightarrow X$$

where each T_i is either a (previously defined) type or X itself. If $C[X] = T_1, \ldots, T_n \rightarrow X$ is any arity defining D, then any $T_i \neq X$ is a *parameter* of D.

We postpone the explanation of the notation $\delta X.C_1[X]; \ldots; C_n[X]$. However, for those who know polymorphism, (see [9] and [14]) we anticipate that the meaning of such a notation is roughly the same as $\forall X.C_1[X], \ldots, C_n[X] \rightarrow X$.

We denote by $\mathcal{T}_{\mathrm{DT}}$ the set of types of $\lambda \rightarrow_{\mathrm{DT}}$. We define the occurrence relation (of a type in another one) as usual. Remark that an arity $C[X] \neq \Omega$ is not a type, but just a formal notation defining a data type. If A is any type and $C[A]$ denotes the result of replacing each X in $C[X]$ by A, then $C[A]$ is a type.

We will now informally explain how to associate to each type T a set interpreting it, this set we will denote by $[\![T]\!]$.

Ω denotes the singleton set $\{\omega\}$. Intuitively ω will be a useless term, the prototype of dead code, and Ω the type of the dead code.

The arrow constructor denotes the ordinary function space.

If D is a data type defined as $D =_{\text{def}} \delta X.C_1[X]; \ldots; C_n[X]$ then we mean that D is a data type with n constructors $c_1, \ldots, c_n$ having types $C_1[D], \ldots, C_n[D]$ respectively. Such constructors are used to build the values of the set denoted by D.

For instance, if $N =_{\text{def}} \delta X.X;\ X \to X$, then according to our definition N is a data type having two constructors with types $C_1[N] = N$ and $C_2[N] = N \to N$. If we give to the constructors their usual names 0 and S, N denotes the set of all formal expressions 0, $S(0)$, $S(S(0)), \ldots$ representing natural numbers. By reasoning along the same lines, we may define the set *Bool* of truth values by $\mathit{Bool} =_{\text{def}} \delta X.X;\ X$. For definitions of lists, trees, sums and products of types, etc. we refer again to [8], chs. 2,3 and [6], sec. 2.

For each data type D, we may define the type of the subset of D consisting of all elements not requiring the constructors $c_{i_1}, \ldots, c_{i_h}$ in their construction. We only have to replace the arities $C_{i_1}[X], \ldots, C_{i_h}[X]$ of D by the dummy arity Ω. We obtain in this way a new data type D', denoting this subset. For instance, $\delta X.X;\ \Omega$ denotes the set of elements of $[\![N]\!]$ not requiring the constructor S, that is, the subset $\{0\}$ of $[\![N]\!]$. The data type $\delta X.\Omega;\ X \to X$ denotes instead the set of elements of $[\![N]\!]$ not requiring the constructor 0, that is, the empty subset of $[\![N]\!]$. Let $\mathit{Void}_n =_{\text{def}} \delta X.\Omega; \ldots; \Omega$ ($n \geq 0$ times Ω). Then Void_n is the set of elements of a data type requiring no constructors: again, the empty set. Two more examples are $\delta X.X;\ \Omega$ and $\delta X.\Omega;\ X$, denoting the subsets of the set of truth values $\{\mathit{true}\}$ and $\{\mathit{false}\}$. These are examples of minimal types.

Definition 2. We denote by *Minimal types* the data types of the form:

$$D = \delta X.\Omega; \ldots; \Omega; (\Omega, \ldots, \Omega \to X); \Omega; \ldots; \Omega$$

From a semantical point of view, minimal types are singleton data types. The only element of a minimal type M has the form $m = c_i(\omega, \ldots, \omega)$ and, since $\omega, \ldots, \omega$ are intended to be "useless" terms (dead code), only the first symbol of m may be "useful" information. Thus, m is minimum information code, in a sense, the prototype of the code of this kind.

Now we can define the programs or terms of $\lambda \to_{\text{DT}}$.

We type $\lambda \to_{\text{DT}}$ à la Church, that is variables of $\lambda \to_{\text{DT}}$ are pairs x^A of a variable name x and a type A. We suppose having infinitely many variable names. A context is any finite set of variables with pairwise different names. We denote contexts by $\Gamma, \Delta, \Theta, \ldots$

We suppose we have a fixed set of algebraic constants, to introduce as primitive some maps of frequent use in programming, like $+, *$ over natural numbers, $\mathit{cons}_T, \mathit{car}_T, \mathit{cdr}_T$ over lists of objects of type T, etc.

Definition 3. The language of terms of $\lambda \rightarrow_{\mathrm{DT}}$ is defined by the following syntax:

$$t ::= \omega \mid f(t_1,\ldots,t_n) \mid c_{i,D}(t_1,\ldots,t_n) \mid it_{D,T}(t;t_1,\ldots,t_n) \mid dummy_M(\Gamma) \mid x^A \mid \lambda x^A.t \mid t(t)$$

where x^A is any variable, f any algebraic constant, $c_{i,D}$ the i-th constructor of the data type D, $it_{D,T}$ the iterator over the data type D and resulting type T, M any *minimal* data type, Γ any context, D any data type, A, T any types, n any integer ≥ 0. We will often drop the type indexes whenever no ambiguity arises, writing $dummy(\Gamma)$, $c_i(t_1,\ldots,t_n)$, $it(t;t_1,\ldots,t_n)$, x, $\lambda x.t$.

We will use the expressions $\lambda x_1^{A_1},\ldots,x_n^{A_n}.b$ and $f(a_1,\ldots,a_n)$ as syntactic sugar for $\lambda x_1^{A_1}.\ldots\lambda x_n^{A_n}.b$ and $f(a_1)\ldots(a_n)$. We call such a set of terms Λ_{DT}.

As we said, ω denotes a useless term and it is the prototype of the dead code. It has type Ω. The terms $dummy_M$ (with M minimal type) are constants denoting the minimum information code $c_i(\omega,\ldots,\omega)$ of type M.

The set of free variables of a term t is denoted by $\mathrm{FV}(t)$, and it is defined as usual. We just extend such a definition with the clause $\mathrm{FV}(dummy(\Gamma)) = \Gamma$.

The typing relation $t : A$, to be read: "t has type A", is defined by induction on t, according to the following rules:

- (unit) $\omega : \Omega$;
- (alg) if the algebraic constant f is associated to $D_1,\ldots,D_n \rightarrow D$, and $t_1 : D_1, \ldots, t_n : D_n$, then $f(t_1,\ldots,t_n) : D$;
- (var) $x^A : A$;
- (λ) if $t : B$ and in $\mathrm{FV}(t)$ no variable $\neq x^A$ has name x, then $\lambda x^A.t : A \rightarrow B$;
- (appl) if $t : A \rightarrow B$ and $u : A$ then t(u) : B;
- (constr) if $C_{i,D}[X] \neq \Omega$ is the i-th arity defining D, and $C_{i,D}[D] = T'_1,\ldots,T'_n \rightarrow D$, $t_1 : T'_1,\ldots,t_n : T'_n$, then $c_{i,D}(t_1,\ldots,t_n) : D$;
- (iter) if $t : D, t_1 : C_{1,D}[A],\ldots,t_n : C_{n,D}[A]$, then $it_{D,A}(t;t_1,\ldots,t_n) : A$;
- (dummy) $dummy_M(\Gamma) : M$ (where Γ is any context).

We say that t *has context* Γ if $\mathrm{FV}(t)$ is included in Γ. By "$\Gamma \vdash t : A$" we abbreviate "t has context Γ and type A". t is *well-typed* if $\Gamma \vdash t : A$ for some Γ, A. From now on, by "term" we will in fact mean "well-typed term", and by "Γ-term", "well-typed term with context Γ".

In order to complete our description of $\lambda \rightarrow_{\mathrm{DT}}$, we have only to describe how to compute the value denoted by t, introducing some reduction rules.

We assume the standard definitions and notations about the substitutions. The only special clause concerns $dummy_M(\Gamma)$: if σ is a substitution, we define $\sigma(\Gamma)$ as the *context* union of all $\mathrm{FV}(\sigma(x^A))$ for $x^A \in \Gamma$, and $\sigma(dummy_M(\Gamma))$ as $dummy_M(\sigma(\Gamma))$.

We can now introduce the reduction rules, motivated by the intended interpretation of terms.

Definition 4.

i) The elementary contraction rules of the system $\lambda \rightarrow_{\mathrm{DT}}$ are:
- (β) $(\lambda x^A.b)(a) \rightarrow_\beta b[x := a]$
- (iter) If $c_{i,D}$ is a constructor corresponding to $C_i[X] = T_1, \ldots, T_n \rightarrow X$, then

$$it_{D,A}(c_{i,D}(d_1, \ldots, d_n); f_1, \ldots, f_m) \rightarrow_{it} f_i(d'_1, \ldots, d'_n)$$

with $d'_i = it_{D,A}(d_i; f_1, \ldots, f_m)$ if $T_i = X$ and $d'_i = d_i$ otherwise
- (alg) Suppose f denotes $\phi : [\![D_1]\!], \ldots, [\![D_n]\!] \rightarrow [\![D]\!]$. Then for all normal closed $d_1 : D_1, \ldots, d_n : D_n$ denoting $\delta_1 \in [\![D_1]\!], \ldots, \delta_n \in [\![D_n]\!]$, if d of type D is the only closed normal denotation of $\phi(\delta_1, \ldots, \delta_n) \in [\![D]\!]$, then

$$f(d_1, \ldots, d_n) \rightarrow_{\mathrm{alg}} d$$

- (dummy) If M is any minimal data type, and C_i the only arity of M different from Ω, then

$$dummy_M(\Gamma) \rightarrow_{\mathrm{d}} c_{i,M}(\omega, \ldots, \omega)$$

ii) We use $\rightarrow_1$ to denote the contextual closure of the union $\rightarrow_\beta \cup \rightarrow_{it} \cup \rightarrow_{\mathrm{d}} \cup \rightarrow_{\mathrm{alg}}$ and we use $\rightarrow^*$ to indicate the transitive closure of $\rightarrow_1$. We use $=_{\mathrm{conv}}$ for the convertibility. We assume the standard definition of redex and normal form.

No reduction rule has been introduced for ω, corresponding to the idea that ω is "useless". We note that the rule (*dummy*), in particular, is justified by the fact that both $dummy_M(\Gamma)$ and $c_{i,M}(\omega, \ldots, \omega)$ denote the only element of M. We just state some important properties of $\lambda \rightarrow_{\mathrm{DT}}$.

Proposition 5. *$\rightarrow^*$ is type preserving, Church-Rosser and strongly normalizing.*

At this point we can introduce a notion of program equivalence, observational equivalence.

Definition 6.

i) $C[z^A] : Bool\ (= \delta X.X;\ X)$ is a *Γ-observation* if z^A (not in Γ) is its only free variable, and for any Γ-term t, if we literally replace z^A by t (allowing capture of free variables), then $C[t]$ is closed.

ii) Let t, u be two terms of type A and context Γ. We say that t, u are observationally equal in Γ, A, and we write $\Gamma \vdash t =_{obs} u : A$, if for all Γ-observations $C[z^A]$, $C[t] =_{\mathrm{conv}} C[u]$. When Γ is empty, we write $(t =_{obs} u : A)$ for $(\vdash t =_{obs} u : A)$.

Observational equivalence formally translates the intuitive idea of "having the same input/output behaviour under all possible values for the parameters.". We refer to [3], secs. 2,3 and [8], ch. 2 for a justification of this claim. In this paper, we will use observational equality only to check the correctness of replacing a program p with a more efficient program q: if p, q are observationally equivalent, then the replacing is correct, because q will return the same output as p for the same input.

Let A be any type. With $\mathrm{card}(A)$ we denote the cardinality of A, namely the number of closed terms in normal form having type A.

We call A a *void type* if $\mathrm{card}(A) = 0$, and a *singleton type* if $\mathrm{card}(A) = 1$. For example $Void_n$ for each n is a void type, while each minimal type M is a singleton type. We say that A is *void-free* (*singleton-free*) if all types occurring in A are non-void (non-singleton). Notice that the types really used in pure functional programs are void- and singleton-free; as we already said, we introduced singleton types only in order to retype useless subterms, while void types arose for technical reasons we shall make precise later.

3 Dead Code and Minimum Information Code

In this section we will outline the general ideas behind our optimization technique. We will sketch the "pruning" technique for removing dead code introduced in [8] and [6], and define out of it a technique for removing minimum information code from programs of $\lambda \rightarrow_{\mathrm{DT}}$.

3.1 The "Pruning" Technique for Removing Dead Code

In general, finding all dead code in a typed λ-term is an undecidable problem, but we found in [8] an interesting decidable subproblem: to find all dead code we may remove without losing the typability of the term. Through examples (see section 6 of this paper), we found that this kind of dead code arises often when we program by reusing code.

The first problem we considered was to formally describe this class of dead code. To this aim, we introduced a simplification relation $\leq_{\mathrm{pr}}$ we called "pruning". Given two well-formed terms t and u of $\lambda \rightarrow_{\mathrm{DT}}$, we wrote $t \leq_{\mathrm{pr}} u$ (to be read "t is a pruning of u") if t may be obtained out of u by replacing some subterms of u by ω, and some subtypes of u with Ω. The intuition behind it is that if we may replace a subterm with the useless ω, without altering the consistency of typing, then the subterm is useless (it is dead code); and that if we may replace a subtype with the type Ω of dead code (again without altering the consistency of typing), then the subtype is assigned to a useless subterm. We have that if $t \leq_{\mathrm{pr}} u$ and t, u have the same type and context, then $t =_{obs} u$, that is, we may replace u with t without altering the input/output behaviour of the term. This means that if $t \leq_{\mathrm{pr}} u$ and t, u have the same type and context, then the subterms we remove when passing from u to t are indeed useless for computing the output of u (are indeed dead code); and consequently, that the subtypes we replaced by Ω are assigned to subterms we replaced by ω, that is, to useless subterms.

After having formally defined a class of simplifications of u obtained by removing dead code from u, we considered the problem of selecting a particular simplification and computing it. The choice of a particular simplification of u was an easy thing to do. We proved that the set CLE(u) of terms $t \leq_{\text{pr}} u$ having the same type and context as u has a complete lattice structure (w.r.t. $\leq_{\text{pr}}$). Hence there exists a simplification $v \in \text{CLE}(u)$ of u which is a simplification of any $t \in \text{CLE}(u)$. Such a v correspond to the maximum of subterms we may erase using $\leq_{\text{pr}}$, thus it is the best possible choice for a simplification of u. We have studied the behaviour of v under several evaluation strategies, proving that the computation of v requires in general less time and space, and in the worst case the same time and space.

In order to compute v, we have developed in [8], ch. 3, an algorithm Fl which is a kind of abstract version of a Data Flow Algorithm used in Compiler Theory. It proceeds from the output of a program, it follows the computation flow backward, always with the help of types, until it reaches the inputs. During this process, Fl marks all subterms and subtypes of the program that may be useful to produce the output. At the end of the analysis, unmarked parts are exactly the dead code we remove when forming the simplification v of u, that is, exactly *all* dead code we may replace by ω without altering the consistency of typing in u

At this point our goal is to obtain similar results about removal of minimum information code.

3.2 Removing Minimum Information Code

We have already pointed out that all terms having a minimal type M are minimum information code. In fact, the only value they may return has the form $dummy_M =_{\text{conv}} c(\omega, \ldots, \omega)$, and since ω represents an object of no use, $dummy_M$ contains only one piece of useful information, c, which is always the same. We may define a function Dummy that, given a term t, replaces all the maximum subterms v of t having a minimal type M with $dummy_M$. This very simple function, however, does not erase an interesting class of minimum information code, since in real programming no subterm will have type M (M is a singleton, and no singleton type is used in pure functional programming).

As we said in the Introduction, the idea is to remove the dead code first, using the pruning technique. As a side effect, we will replace by Ω some subtypes of each type in the program. Afterwards, a subterm w whose type was, say, $T = Bool = \delta X.X; X$ may now have type $T' = \delta X.X; \Omega = \{true\}$. This may happen if the constructor $\{false\}$ is not used in the evaluation of w, and therefore it is useless (it is dead code) as far as w is concerned. In this case, the pruning technique replaces the second X in *Bool*, the one corresponding to the type of *false*, by Ω, the type of useless code, turning T into T'. Since $T' = \delta X.X; \Omega$ is a minimal type, after applying pruning we may replace w by $dummy_{\text{true}} =_{\text{conv}} true$ simply applying Dummy. Before, we could not. In other words, the pruning technique discovered that the value of w was forced to be *true*, and the Dummy map exploited this information.

Now the immediate idea is to always apply the pruning technique first, and then the Dummy map. But we may do better: in fact (see section 6) after applying the Dummy map, we may apply the pruning technique again, and discover some dead code which before eluded detection. If we remove this latter, we may discover new minimum information code using the Dummy map, and so on. Given any program t, we may consider any simplification chain t, t_1, t_2, ..., starting with any pruning $t_1 \leq_{\text{pr}} t$, then apply Dummy, continuing with a pruning, and so on. At each stage we have several choices: there are many prunings of a term, and we do not know in advance which one will allow the best application of the Dummy map. In fact, one pruning may erase more dead code than another one, but it may also generate *less* minimal types, making the next application of Dummy less useful. *A priori*, we may obtain in this way different final simplifications, and we do not know which one is the best.

A posteriori, in section 5 we will prove that all simplification chains from t end in the same u: thus, this latter represents an optimal choice. We will discuss how to compute u. Before, in section 4, we will turn the informal discussion above into a formal definition of simplification.

4 A Simplification Relation on Programs

In this section we will introduce a relation "$t \leq_1 u$" on well-typed programs t, u, to be read "t is a simplification of u", in the sense introduced in subsection 3.2: we may obtain t from u by replacing some subterms of u by ω, some subtypes of u by Ω, and some subterm of u having minimal type M by $dummy_M$.

We will then show that $t \leq_1 u$, with t, u both having type A and context Γ, implies $\Gamma \vdash t =_{obs} u : A$, in other words, that we may safely replace t by u without altering the input/output behaviour of u.

Now, we inductively define the relation $\leq_1$ over well-typed terms. We first define a relation $\leq$ over types, arities and contexts. $A \leq B$ will mean that we may obtain A out of B by replacing some types and arities in B by Ω. We will use $\leq$ to replace types of redundant subprograms by Ω.

i) Let A, B denote types, $C_1, \ldots, C_n$ arities, and T, U either types or arities. We inductively define the relation $\leq$ on types and arities by the rules:

$$(\Omega)\quad \Omega \leq T \qquad (\rightarrow)\quad \frac{A \leq B \quad T \leq U}{A \rightarrow T \leq B \rightarrow U}$$

$$(\text{Var})\quad X \leq X \quad (\text{Data})\quad \frac{C_1 \leq C_1' \quad \ldots \quad C_n \leq C_n'}{\delta X.C_1; \ldots; C_n \leq \delta X.C_1'; \ldots; C_n'}$$

ii) We extend $\leq$ to contexts by: $\Gamma \leq \Delta$ iff $(\forall x^A \in \Gamma)$ $(\exists B)$ $(x^B \in \Delta$ and $A \leq B)$.

iii) We inductively define the relation $\leq_1$ on well-typed terms by the rules:

$$\text{(unit)}\quad \omega \leq_1 t \qquad\qquad \text{(alg)}\quad \frac{d_1 \leq_1 d'_1 \quad \ldots \quad d_n \leq_1 d'_n}{f(d_1,\ldots,d_n) \leq_1 f(d'_1,\ldots,d'_n)}$$

$$\text{(var)}\quad \frac{A \leq B}{x^A \leq_1 x^B} \qquad\qquad (\mathit{dummy})\quad \frac{\Gamma \leq \Delta}{\mathit{dummy}_M(\Gamma) \leq_1 \mathit{dummy}_M(\Delta)}$$

$$\text{(app)}\quad \frac{f \leq_1 g \quad a \leq_1 b}{f(a) \leq_1 g(b)} \quad (\lambda)\quad \frac{A \leq B \quad t[z^A] \leq_1 u[z^B]}{\lambda x^A.t[x^A] \leq_1 \lambda x^B.u[x^B]}\,{}^{3}$$

$$\text{(constr)}\quad \frac{D \leq D' \quad d_1 \leq_1 d'_1 \ldots d_n \leq_1 d'_n}{c_{i,D}(d_1,\ldots,d_n) \leq_1 c_{i,D'}(d'_1,\ldots,d'_n)}$$

$$(\mathit{it})\quad \frac{D \leq D' \quad A \leq A' \quad t \leq_1 t' \quad t_1 \leq_1 t'_1 \ldots t_n \leq_1 t'_n}{\mathit{it}_{D,A}(t;t_1,\ldots,t_n) \leq_1 \mathit{it}_{D',A'}(t';t'_1,\ldots,t'_n)}$$

$$\text{(singleton)}\quad \frac{t \leq_1 u \quad t : M \quad M \text{ minimal singleton}}{\mathit{dummy}_M(\mathrm{FV}(t)) \leq_1 u}$$

iv) We will use $t <_1 u$ to denote $t \leq_1 u$ and not t (α)-convertible to u. By $u >_1 t$, $u \geq_1 t$ we will mean $t <_1 u$, $t \leq_1 u$. We define $\leq^*$ as the transitive closure of $\leq_1$. We say that t is a minimal simplification if $t >_1 u$ for no u having the same type and context of t.

A trivial example: if $t = (\lambda x^A.\lambda y^B.x^A)(a)(b)$ and $u = (\lambda x^A.\lambda y^\Omega.x^A)(a)(\omega)$, then $t >_1 u$. The context and the type of t, u are the same, and indeed the subterm b we remove by replacing t with u is dead code.[4] We postpone non-trivial examples to the next section.

In the remaining of this subsection we only mention some properties of $\leq_1$ and of $\leq$.

Lemma 7.

i) $\leq$ *is an order relation on types and for any two types* A, B *there exists their g.l.b. w.r.t.* $\leq$, *denoted by* $\inf(A,B)$.

ii) $\leq$ *is an order relation on contexts and for any two contexts* Γ, Δ *there exists their g.l.b. w.r.t.* $\leq$*:*

$$\inf(\Gamma,\Delta) = \{x^{\inf(A,B)} \mid x^A \in \Gamma, x^B \in \Delta\}$$

[3] In expressions involving binders we have to introduce a fresh variable z in order to have the relation $\leq_1$ invariant up to (α)-conversion. Without this caution we would have, for example, $\lambda x^A.x \leq_1 \lambda x^A.x$ but not $\lambda x^A.x \leq_1 \lambda y^A.y$.

[4] A short remark. One may wonder if, by adding more dummy constants and by generalizing the singleton simplification rule to:

$$\frac{t \leq_1 u \quad t : A \quad \mathrm{card}(A) = 1}{\mathit{dummy}_M(\mathrm{FV}(t)) \leq_1 u}$$

one may get more possible simplifications. We have checked that this is not the case, but we will include no more details about this fact here.

iii) if $t \leq_1 u$, $t : A$, $u : B$, *then* $FV(t) \leq FV(u)$ *and* $A \leq B$.

iv) $\leq_1$ *is compatible with substitutions: if* $t \leq_1 u$ *and* $x_1^{A_1} \leq_1 x_1^{B_1}, \ldots, x_n^{A_n} \leq_1 x_n^{B_n}$, *and* $t_1 \leq_1 u_1, \ldots, t_n \leq_1 u_n$, *then*

$$t[x_1^{A_1} := t_1, \ldots, x_n^{A_n} := t_n] \leq_1 u[x_1^{B_1} := u_1, \ldots, x_n^{B_n} := u_n]$$

v) $\leq_1$ *is compatible with the observations: if* t *and* u *are* Γ*-terms of type* A *and* $C[z^A]$ *is a* Γ*-observation, then* $t \leq_1 u \Rightarrow C[t] \leq_1 C[u]$

vi) $<_1$ *is a well-founded relation.*

We end this section by stating one of the most important results about the simplification relation, namely that if $t \leq_1 u$ and t, u are terms with the same type A and context Γ, then $\Gamma \vdash t =_{obs} u : A$. In other words, all subprograms we may remove from u by using $\leq_1$ and without altering the type and context of u are indeed redundant (do not influence the output of u): t is a simplification of u.

Theorem 8. *Let* t, u *have context* Γ *and type* A. *Then*

$$t \leq_1 u \Rightarrow \Gamma \vdash t =_{obs} u : A$$

We have proved (in [6]) this result by interpreting $\leq_1$ by a kind of Plotkin logical relation (see [12]). For each pair of types A, A', such that $A \leq A'$, we have defined a relation $R_{A,A'}(t,u)$ on closed $t : A, u : A'$. $R_{A,A'}$ is chosen in such a way that $t \leq_1 u \Rightarrow R_{A,A'}(t,u)$ for *closed* t, u. The only base case in the definition of $R_{A,A'}$ is that $R_{\Omega,A'}$ always holds; this case validates the rule $\omega \leq_1 u$. Then $R_{A,A'}$ is extended to higher types in the usual way. Our thesis follows (essentially) from the fact that if $A = A'$ then $R_{A,A'}(t,u) \Rightarrow t =_{obs} u : A$; therefore, $t \leq_1 u \Rightarrow R_{A,A'}(t,u) \Rightarrow t =_{obs} u : A$.

5 Minimum Simplifications

In this section we are going to prove that the set

$$\mathrm{CLE}(u) = \{t \mid t \leq^* u \text{ and } t, u \text{ have the same type } A \text{ and the same context } \Gamma\}$$

has a minimum, that is, that all sequences $u >_1 u_1 >_1 u_2 \ldots$ of terms with type A and context Γ end in the same v. Such a v is the optimal simplification of u we may obtain by repeatedly using $\leq_1$. We will then discuss how to compute v from u.

5.1 Existence of a Minimum w.r.t $\leq^*$

We only sketch the main steps of the proof. For the complete proof we refer to [6].

Lemma 9. *$\leq_1$ is confluent over CLE(u), that is, if t, u, v have type A and context Γ, both void-free, then:*

$$u, v \leq_1 t \Rightarrow w \leq_1 u, v, \text{ for some } w \text{ s.t. } \Gamma \vdash w : A^5$$

Theorem 10. *Let t be any term having type and context void-free (as it is the case for any real program). Then the set CLE(t) has a minimum element w.r.t. $\leq^*$.*

Proof

$\leq^*$, being the transitive closure of $\leq_1$, is confluent as well. (Proof: see Barendregt [1]) If we now take any minimal $v \leq^* u$ (existing since $\leq_1$ is well-founded), then v is indeed a minimum, because $u \geq^* w$ implies $v, w \geq^* z$ for some z, and, by the minimality of v, $v = z$, hence, $v \leq^* w$.

5.2 Computing the Minimum w.r.t. $\leq^*$

Let $\leq_{du}$ denote the subrelation of $\leq^*$ obtained as contextual closure of the rule about singleton terms:

$$t \geq_1 u : M \text{ minimal singleton } \Rightarrow t \geq_1 \mathit{dummy}_M(\mathrm{FV}(u))$$

and let $\leq_{pr}$ be another subrelation of $\leq_1$, obtained by taking away this rule. $\leq_{pr}$ is the relation we called "pruning" in subsection 3.1. As we said there, we have defined in [8], ch. 3, a Data Flow algorithm Fl(u), computing the best simplification of u we may obtain using $\leq_{pr}$. In subsection 3.2, we also defined a map Dummy(u) computing the best simplification of u we may obtain using $\leq_{du}$. The idea is now to apply in turn Fl and Dummy, computing in this order $u \geq_1 \mathrm{Fl}(u) \geq_1 \mathrm{Fl}(\mathrm{Dummy}(u)) \geq_1 \mathrm{Fl}(\mathrm{Dummy}(\mathrm{Fl}(u))) \geq_1 \mathrm{Fl}(\mathrm{Dummy}(\mathrm{Fl}(\mathrm{Dummy}(u)))) \geq_1 \ldots$, until we find some v such that $v = \mathrm{Fl}(v) = \mathrm{Dummy}(v)$. We might wonder if v is really the minimum of $\leq_1$ in CLE(u); after all, we only get a term we cannot simplify further by $\leq_{pr}$, $\leq_{du}$, but $\leq_1$ is the merging of the two, and *a priori* could be stronger than each of them separately, and it could allow some more simplifications. But fortunately this is not the case:

Lemma 11. *If $v = Fl(v) = Dummy(v) \in CLE(u)$, then v is the minimum of $\leq_1$ in CLE(u).*

The algorithm we sketched is terminating in a number of applications of Fl and Dummy bounded by the size of u, since the size of u decreases at each step. In [8] we pointed out that Fl requires a number of steps proportional to the size of u (if we include in u the type superscripts of variables). The same clearly

---

5 We only mention that the proof of this simple statement was surprisingly difficult. The reason is that, even if we do not have void types to start with, they arise as by-product of some simplification step on t; and confluence in presence of void types is problematic.

holds for Dummy. Thus, the algorithm we sketched has a time complexity at most quadratic in u. On all the examples we could think of, at most three or four iterations of Fl are enough, so in practice the time required seems linear in the size of u.

However, we are not satisfied by the algorithm we have for computing v. As we may see in [8], each application of Fl to some t requires the building of a complex data structure, representing symbolically the computation flow in t. This work takes away most of the time; if we apply Fl first to u and then to $u_2 = \text{Dummy}(\text{Fl}(u))$, the data structure we need to compute $\text{Fl}(u_2)$ might be easily obtained from the one used to compute $\text{Fl}(u)$. Yet we are not reusing this latter, but we are building a completely new one. We conjecture we may design, by reusing the data structure of Fl, an algorithm computing v in linear time in u (and anyway quite a lot faster).

6 Examples

We consider some situations where minimum information code arises when composing and instantiating code we already have. In each case we show that it is erased (together with dead code) by repeatedly applying the maps Fl and Dummy.

(1) As the first example, we consider the term

$$t = (\lambda x^N.it(x; a, \lambda\_.b))(S(y))$$

where _ denotes a variable not in b and $N = \delta X.X; X \to X$ has constructors $0 : N$ and $S : N \to N$. The term t reduces β to $it(S(y); a, \lambda\_.b)$, then by an (iter)-reduction to $(\lambda\_.b)it(y; \ldots)$, and eventually (since _ is not in b) to b. Therefore a, the argument y of S, and the value we assign to _ are useless information for the computation, that is they are dead code. The variable x is minimum information code because we use only the first symbol S of its value $S(y)$. Thus, the first step in our algorithm, an application of Fl, replaces both a and y with ω, and their types with Ω. The result is $t_1 = \text{Fl}(t) = (\lambda x^{N_1}.it(x; \omega, \lambda\_^{\Omega}.b))(S_1(\omega))$. The new type of _ is Ω because the value we assign to _ is dead code. The new type N_1 of x is obtained from the previous one, $N = \delta X.X; X \to X$, by replacing by Ω the subtypes of N corresponding to objects useless in t. The first X after δX in N corresponds to the constructor $0 : N$, not used in t; thus it is replaced by Ω in N_1. The next X, the first one in $X \to X$, corresponds to the argument of the constructor $S : N \to N$, hence to the dead code y, and it will be replaced by Ω too. The result is $N_1 = \delta X.\Omega; \Omega \to X$. This N_1 is a minimal type. Thus the next step in our algorithm, the application of Dummy, will replace anything with type N_1 by $dummy_{N_1}$. The result is $t_2 = \text{Dummy}(t_1) = (\lambda x^{N_1}.it(dummy; \omega, \lambda\_^{\Omega}.b))(dummy)$. Eventually, we have got rid of the minimum information code x. The algorithm now stops, because there is no dead code nor minimum information code left, and therefore $t_2 = \text{Dummy}(t_2) = \text{Fl}(t_2)$; t_2 is the result, maybe a little ugly to see. If we use simple techniques (quoted in the next section) which "clean up" the result by removing all λ-abstractions, applications and iterations over singleton types, we are left with just b.

In this example (but not in general) we could have got the same result just by reducing t; this example was only intended to show the simplification mechanism at work.

(2) This is an example of a minimum information code t which cannot be removed just by reducing t. We need some preliminary definitions before introducing t.

Let $D =_{\text{def}} \delta X.X; X; X$. D is a data type with three constructors, $a_1, a_2, a_3 : D$, which are the only closed normal forms of type D. Put $\{a_1, a_2\} = \delta X.X; X; \Omega$. We can define the type $List(D) =_{\text{def}} \delta X.X; D, X \to X$ of lists of elements of type D. As usual we call its constructors $nil : List(D)$ and $cons : D, List(D) \to List(D)$. We may define a function F that checks whether a list l is a sequence of just a_1's and a_2's by: $F(l) =$ if $(l = nil)$ then *true* else if $car(l) = a_1$ or $car(l) = a_2$ then $F(cdr(l))$ else *false*. We may formally express F in our syntax by:

$$F =_{\text{def}} \lambda l.it(l; true, \lambda a, ris^{Bool}.case(a; ris, ris, false)) : List(D) \to Bool,$$

where *true* and *false* are the constructors of the type $Bool =_{\text{def}} \delta X.X; X$ and *case* denotes the iteration $it_{D,Bool}$ on D (which is in fact a case operator). For instance, if we apply F to the list $L_1 = cons(a_3, nil)$ then $(F\,L_1)$ (β)-reduces to $it(L_1; true, \lambda a, ris^{Bool}.case(a; ris, ris, false))$. Then with two steps of (iter) we obtain: $(\lambda a, ris.case(a; ris, ris, false))(a_3)(true)$. At this point by ($\beta$) and evaluating *case* we immediately obtain *false*, the expected answer for L_1.

We may now introduce t by $t =_{\text{def}} F(L(n))$ (with n integer variable), where $L(n)$ is a map that computes a sequence of alternating a_1, a_2 of length n.[6] Since $F(L(n)) =_{\text{conv}} true$, then t is a minimum information code, because it always returns the same one-symbol information, *true*. The clause *false* in the *case* operator in F never applies (there is no a_3 in $L(n)$), so it becomes dead code. We will see what happens if we apply our algorithm to t.

The first step is an application of Fl to $F(L(n))$, which removes the dead code *false*. We obtain the term $t_1 = \text{Fl}(t)$ defined as:

$$t_1 = F'(L'(n)) = (\lambda l.it(l; true, \lambda a, ris^{\{true\}}.case'(a; ris, ris, \omega)))(L'(n)).$$

---

[6] We may define $L(n) =_{\text{def}} G(n, true) : List(D)$, with $G(n, flag) =$ if $n = 0$ then *nil* else if *flag* then $cons(a_1, G(false, n-1))$ else $cons(a_2, G(true, n-1))$. If we call T the type $Bool \to List(D)$, then in our syntax we may translate G as: $G = \lambda n.it_{N,T}(n; \lambda flag^{Bool}.nil, \lambda H^T, flag.\ if(flag, cons(a_1, H(false)), cons(a_2, H(true))))$ $: N \to T$. Here *if* stays for the iterator $it_{Bool,List(D)}$ on *Bool* (which is in fact a case operator).

where L' is L with $List(D)$ replaced by $List(\{a_1, a_2\})$. The type of ris is changed from $Bool =_{\text{def}} \delta X.X; X$ to $\{true\} = \delta X.X; \Omega$. The reason is that the last X in $Bool$ corresponds to the type of the constructor *false*, which is dead code, hence it is replaced with Ω by Fl. But the type $Bool$ of ris was also the type of t, hence the new type $\{true\}$ of ris is the type of t_1. Therefore t_1 has a minimal type. Thus the next step of our algorithm, the application of the map Dummy to t_1, produces $t_2 = \text{Dummy}(t_1) = dummy_{\{true\}} =_{\text{conv}} true$. We could not obtain *true* just by reducing t, because of the presence of the free variable n.

We refer to [6] for an example of a term t where the removing of minimum information code allows the map Fl to remove more dead code. The same example shows that, for realistic programs, we may need three or more iterations of our algorithm.

7 Final Remarks

In this paper we have presented a combined technique to erase dead and minimum information code from functional programs represented by λ-terms of a simply typed λ-calculus extended with data types. Such a technique can be very useful in a variety of situations in which code is automatically generated. In such cases the code can contain a great amount of useless parts that a human programmer would not introduce in programs.

In [8], ch. 3, are considered also some minor optimizations such as η-contraction, linear β-contraction, isomorphism between types etc. that we have not described here for reasons of space. They are just syntactic optimizations that can be executed after the application of our main technique and that can be useful to put an expression in a more readable form.

The most important simplification step that we have omitted (again for space reasons) in this paper is the application of the function Isom (see again [8], ch. 3 and [6], sec. 2). The reader may have noticed that when we remove the dead and the minimum information code, we replace it with the constants ω, $dummy_M$ and Ω. Of course this is not our final goal since also such constants are useless. They are used just as an optimization tool to divide the optimization task in different subproblems. The difficult part is the search for the useless code and the replacement with the constants still maintaining consistency and the original meaning. The easy part is to remove the useless constants. The function Isom remove such useless constants returning a clean expression. This is again a syntactic optimization that we can apply as a final optimizing step and so we have omitted it.

References

1. H. Barendregt, *The Lambda Calculus its Syntax and Semantics*, Studies in Logic and the Foundation of Mathematics, North-Holland, 1984

2. P. N. Benton, ***Strictness Analysis of Lazy Functional Programs***, Ph. D. Thesis, University of Cambridge, Pembroke College, 1992, Cambridge.
3. S. Berardi, ***Extensional Equality for Simply Typed λ-calculi***, Technical Report, Turin University, 1993 (available via ftp at ftp.di.unito.it).
4. S. Berardi, ***Pruning Simply Typed λ-terms***, to appear in Journal of Logic and Computation.
5. S. Berardi, L. Boerio, ***Using Subtyping in Program Optimization***, Proceedings of TLCA '95, Eds M. Dezani and G. Plotkin, Edinburgh, April 1995, LNCS 902, pp. 63-77, Springer-Verlag.
6. S. Berardi, L. Boerio, ***Removing Redundant Code in a Pure Functional Language with Data Types***, Technical Report, Turin University, 1996 (available via ftp at ftp.di.unito.it).
7. L. Boerio, ***Extending Pruning Techniques to Polymorphic Second Order λ-Calculus***, Proceedings of ESOP '94, Edinburgh, April 1994, LNCS 788, D. Sannella (ed.), Springer-Verlag, pp. 120-134.
8. L. Boerio, ***Optimizing Programs Extracted from Proofs***, Ph. D. Thesis in Computer Science, University of Turin, February 1995, Turin (available via ftp at ftp.di.unito.it).
9. J. Y. Girard, ***Interpretation Fonctionelle et Elimination des Coupures dans l'Aritmetique d'Ordre Superieur***, Ph. D. Thesis, Université Paris VII, 1971, Paris.
10. M. J. C. Gordon, R. Milner, C. P. Wadsworth, ***Edinburgh LCF***, LNCS 78, Springer-Verlag, 1979.
11. L.S. Hunt, D. Sands, ***Binding Time Analysis: A New PERspective***, Proceedings of the ACM Symposium on Partial Evaluation and Semantics-based Program Manipulation, 1991.
12. J. C. Mitchell, ***Type Systems for Programming Languages***, In: J. van Leeuwen ed., Handbook of Theoretical Computer Science, Elsevier Science Publ., 1990, 366-458.
13. C. Paulin-Mohring, ***Extracting F_ω's Programs from Proofs in the Calculus of Constructions***, In: Sixteenth Annual ACM Symposium on Principles of Programming Languages, Austin, Texas, Association for Computing Machinery publisher, 11-13 January 1989, pp. 89-104.
14. J. C. Reynolds, ***Towards a Theory of Type Structure***, in Colloque sur la Programmation, LNCS 19, Springer-Verlag, 1974.

Matching Constraints for the Lambda Calculus of Objects

Viviana Bono
Dipartimento di Informatica, Università di Torino
C.so Svizzera 185, I-10149 Torino, Italy
e-mail: `bono@di.unito.it`

Michele Bugliesi
Dipartimento di Matematica, Università di Padova
Via Belzoni 7, I-35131 Padova, Italy
e-mail: `michele@math.unipd.it`

Abstract. We present a new type system for the *Lambda Calculus of Objects* [16], based on matching. The new system retains the property of type safety of the original system, while using implicit match-bounded quantification over type variables instead of implicit quantification over row schemes (as in [16]) to capture *Mytype* polymorphic types for methods. Type soundness for the new system is proved as a direct corollary of subject reduction. A study of the relative expressive power of the two systems is also carried out, that shows that the new system is as powerful as the original one on derivations of closed-object typing judgements. Finally, an extension of the new system is presented, that gives provision for a class-based calculus, where primitives such as creation of class instances and method update are rendered in terms of delegation.

1 Introduction

The problem of deriving safe and flexible type systems for object-oriented languages has been addressed by many theoretical studies in the last years. The interest of these studies has initially been focused on *class-based* languages, while, more recently, type systems have also been proposed for *object-based* (or *delegation-based*) languages. Clearly, the work on the latter has been strongly influenced by the study of the former. For example, the notion of *row-variables* introduced by [18] to type extensible records was refined in [16, 7, 17, 5] to type extensible objects. Similarly, *recursive object types* have first been used to provide functional models of class-based languages [11, 9, 12, 14, 13], and then applied to the case of an object calculus supporting method override in presence of subsumption [3]. A further notion that has been studied extensively in class-based languages (as well as in the record calculus of [10]) is that of *(F-)bounded quantification* as a tool for modeling the subclass relation. Following this line of research, in this paper we investigate the role of *matching* in the design of a type system for the delegation-based *Lambda Calculus of Objects* [16].

Matching is a relation over object types that has first been introduced by [8] as an alternative to F-bounded subtyping [13] in modeling the subclass relation in class-based languages, and then as a complement to subtyping to model method inheritance between classes [1, 2].

The *Lambda Calculus of Objects* [16] is a delegation-based calculus where method addition and override, as well as method inheritance, all take place at the object level rather than at the class level. In [16] a type system for this calculus is defined, that provides for static detection of errors, such as *message not understood*, while at the same time allowing types of methods to be specialized to the type of the inheriting objects. This mechanism, that is commonly referred to as *Mytype* specialization, is rendered in the type system in terms of a form of higher-order polymorphism which, in turn, uses implicit quantification over *row schemes* to capture the underling notion of *protocol extension*.

The system we present in this paper takes a different approach to the rendering of *Mytype* specialization, while retaining the property of type safety of the original system. Technically, the new solution is based on implicit match-bounded quantification over type variables to characterize methods as functions with polymorphic types, and to enfore correct instantiation of these types as methods are inherited.

A similar solution for the polymorphic typing of methods in Lambda Calculus of Objects is proposed in [6]. The key difference, with respect to the system of this paper, is that [6] uses subtyping, instead of matching, and (implicit) subtype-bounded quantification. There appear to be fundamental tradeoffs between the two solutions: in fact, while subtyping has the advantage of allowing object subsumption, matching appears to be superior to subtyping in the rendering of the desired typing of methods. The reason is explained, briefly, as follows: in order to ensure safe uses of subsumption, the system of [6] allows type promotion for an object-type only when the methods in the promoted type do not reference any of the methods of the original type. As in [7], labeled types are used to encode the cross references among methods in the methods' types. In [6], however, labeled types involve some additional limitations over [7] for subsumption, and require a rather more complex bookkeeping that affects the typing of methods as well.

Relying on matching, instead, has the advantage of isolating the typing of methods from subtyping, thus allowing a rather elegant and simple rendering of the method polymorphism. The simplicity of the resulting system also allows us to draw a formal analysis of the relationship between the original system and ours. In particular, we show that every closed-object typing judgement derivable in [16] is also derivable in our system. We then show that the new system may naturally be extended to give provision for a class-based calculus, where subclassing, and primitives such as creation of class instances and method update are rendered in terms of delegation.

The rest of the paper is organized as follows. In Section 2, we review the untyped calculus of [16]. In Section 3, we present the new typing rules for objects and we prove type soundness. In Section 4, we present the encoding of the type system of [16] into the new system. In Section 5, we present the extended system that gives provision for classes, and then we conclude in Section 6 with some final remarks.

2 The Untyped Calculus

An expression of the untyped calculus can be any of the following:

$$e ::= x \mid c \mid \lambda x.e \mid e_1\, e_2 \mid \langle\rangle \mid \langle e_1 \leftarrow\!\!+\, m{=}e_2\rangle \mid \langle e_1 \leftarrow m{=}e_2\rangle \mid e \Leftarrow m$$

where x is a variable, c a constant and m a method name. The reading of the object-related forms is as follows:

$\langle\rangle$	is the empty object,
$\langle e_1 \leftarrow\!\!+\, m{=}e_2\rangle$	extends object e_1 with a new method m having body e_2,
$\langle e_1 \leftarrow m{=}e_2\rangle$	replaces e_1's method body for m with e_2,
$e \Leftarrow m$	sends message m to object e,

The expression $\langle e_1 \leftarrow\!\!+\, m{=}e_2\rangle$ is defined only when e_1 denotes an object that does not have an m method, whereas $\langle e_1 \leftarrow m{=}e_2\rangle$ is defined only when e_1 denotes an object that *does* contain an m method. As in [16], both these conditions are enforced statically by the type system.

The other main object operation is method invocation, which comprises two separate actions, *search* and *self-application*: evaluating the message $e \Leftarrow m$ requires a search, within e, of the body of the m method which is then applied to e itself. To formalize this behavior, we introduce a subsidiary object expression, $e \hookleftarrow m$, whose intuitive semantics is as follows: evaluating $e \hookleftarrow m$ results into a recursive traversal of the "sub-objects" of e, that succeeds upon reaching the right-most addition or override of the method in question.

The operational semantics $\xrightarrow{eval}\!\!\!\!\rightarrow$ of the untyped calculus can be thus defined as the reflexive, transitive and contextual closure of the reduction relation defined below ($\leftarrow\!\circ$ stands for both $\leftarrow\!\!+$ and $\leftarrow$)

(β)	$(\lambda x.e_1)\, e_2$	$\xrightarrow{eval}$	$[e_2/x]\, e_1$
$(\Leftarrow)$	$e \Leftarrow m$	$\xrightarrow{eval}$	$(e \hookleftarrow m)\, e$
$(\hookleftarrow$ *succ*$)$	$\langle e_1 \leftarrow\!\circ\, m{=}e_2\rangle \hookleftarrow m$	$\xrightarrow{eval}$	e_2
$(\hookleftarrow$ *next*$)$	$\langle e_1 \leftarrow\!\circ\, n{=}e_2\rangle \hookleftarrow m$	$\xrightarrow{eval}$	$e_1 \hookleftarrow m$
(*fail* $\langle\rangle$)	$\langle\rangle \hookleftarrow m$	$\xrightarrow{eval}$	*err*
(*fail abs*)	$\lambda x.e \hookleftarrow m$	$\xrightarrow{eval}$	*err*

plus a few additional reductions for error propagation (see Appendix A). The use of the search operator expressions in our calculus is inspired to [7], and it provides a more direct and concise technical device than the *bookkeeping* relation originally introduced in [16].

3 Object Types and Matching

Object types have the same structure as in [16]: an object type has the form

$$\texttt{Obj}\, t.\langle\!\langle m_1{:}\tau_1, \ldots, m_k{:}\tau_k \rangle\!\rangle$$

where the m_i's are method names, whereas the τ_i's are type expressions.

The row $\langle\!\langle m_1{:}\tau_1, \ldots, m_k{:}\tau_k \rangle\!\rangle$ defines the interface or *protocol* of the objects of that type, i.e. the list of the methods (with their types) that may be invoked on these objects. The binder Obj scopes over the row, and the bound variable t may occur free within the scope of the binder, with every free occurrence referring to the object type itself. Obj-types are thus a form of recursively-defined types, even though Obj is not to be understood as a fixed-point operator: as in [16], the self-referential nature of these types is axiomatized directly by the typing rules, rather than defined in terms of an explicit unfolding rule.

3.1 Types and Rows

Type expressions include type-constants, type variables, function types and object types. The sets of rows and types are defined recursively as follows:

$$\begin{array}{ll} \text{Rows} & R ::= \langle\!\langle\rangle\!\rangle \mid \langle\!\langle R \mid m{:}\tau \rangle\!\rangle \\ \text{Types} & \tau ::= \mathtt{b} \mid v \mid \tau \to \tau \mid \texttt{Obj}\, t.R. \end{array}$$

The symbol b denotes type constants, t, u, and v denote type variables, whereas τ, σ, ρ, ... range over types; all symbols may appear indexed.

Row expressions that differ only for the name of the bound variable, or for the order of the component $m{:}\tau$ pairs are considered identical. More formally, α-conversion of type variables bound by Obj, as well as applications of the principle:

$$\langle\!\langle\langle\!\langle R \mid n{:}\tau_1 \rangle\!\rangle \mid m{:}\tau_2 \rangle\!\rangle = \langle\!\langle\langle\!\langle R \mid m{:}\tau_2 \rangle\!\rangle \mid n{:}\tau_1 \rangle\!\rangle,$$

are taken as syntactic conventions rather than as explicit rules. The structure of valid contexts (see Appendix B) is defined as follows:

$$\Gamma ::= \varepsilon \mid \Gamma, x : \tau \mid \Gamma, u \lessdot\!\!\# \, \tau,$$

Correspondingly, the judgements are $\Gamma \vdash *$, $\Gamma \vdash e : \tau$, $\Gamma \vdash \tau_1 \lessdot\!\!\# \, \tau_2$, where $\Gamma \vdash *$ stands for "Γ is a well-formed context", and the reading of the other judgements is standard.

As an important remark, we note that, as in [6] and in contrast to [16], rows in our system are formed only as "ground" collections of pairs "method-name:type". One advantage of this choice is a simplified notion of well-formedness for rows: instead of the kinding judgements of [16], in our system this notion is axiomatized, syntactically, as follows. Let $\mathcal{M}(R)$ denote the set of method names of the row R defined inductively as follows:

$$\mathcal{M}(\langle\!\langle\rangle\!\rangle) = \{\}, \quad \text{and} \quad \mathcal{M}(\langle\!\langle R \mid m{:}\tau \rangle\!\rangle) = \mathcal{M}(R) \cup \{m\}.$$

Then, we say that a row is well formed if and only if it is (i) either the empty row $\langle\!\langle\rangle\!\rangle$, or (ii) a row of the form $\langle\!\langle R \mid m{:}\tau \rangle\!\rangle$ with R well formed and $m \notin \mathcal{M}(R)$.

3.2 Matching

Matching is the only relation over types that we assume in the type system; it is a reflexive and transitive relations over all types, while for Obj-types it also formalizes the notion of ***protocol extension*** needed in the typing of inheritance. The relation we use here is a specialization of the original matching relation [8], defined by the following rule ($\overline{m{:}\tau}$ is short for $m_1{:}\tau_1, \ldots, m_k{:}\tau_k$):

$$\frac{\Gamma \vdash * \qquad \langle \overline{m{:}\tau}, \overline{n{:}\sigma} \rangle \textit{ well formed}}{\Gamma \vdash \texttt{Obj}\, t.\langle \overline{m{:}\tau}, \overline{n{:}\sigma} \rangle \mathrel{<\!\!\#} \texttt{Obj}\, t.\langle \overline{m{:}\tau} \rangle} \quad (\mathrel{<\!\!\#})$$

Unlike the original definition [8], that allows the component types of a Obj-type to be promoted by subtyping, our definition requires that these types coincide with the component types of every Obj-type placed higher-up in the $\mathrel{<\!\!\#}$ -hierarchy. Like the original relation, on the other hand, our relation has the peculiarity that it is ***not*** used in conjunction with a subsumption rule. As noted in [2], this restriction is crucial to prevent unsound uses of type promotion in the presence of method override.

3.3 Typing Rules for Objects

For the most part, the type system is routine. The object-related rules are discussed below. The first defines the type of the empty object (whose type is the top of object-types in the $\mathrel{<\!\!\#}$ -hierarchy):

$$\frac{\Gamma \vdash *}{\Gamma \vdash \langle\rangle : \texttt{Obj}\, t.\langle\rangle} \quad (\textit{empty object}).$$

The typing rule for method invocation has the following format:

$$\frac{\Gamma \vdash e : \sigma \quad \Gamma \vdash \sigma \mathrel{<\!\!\#} \texttt{Obj}\, t.\langle n{:}\tau \rangle}{\Gamma \vdash e \Leftarrow n : [\sigma/t]\tau} \quad (\textit{send}).$$

As in [16], the substitution for t in τ reflects the recursive nature of object types. In order for a call to an n method on an object e to be typed, we require that e has any type containing the method name n. An interesting aspect of the above rule is that the type σ may either be a Obj-type matching Obj $t.\langle n{:}\tau \rangle$, or else an unknown type (i.e. a type variable) occurring (match-bounded) in the context Γ. Rules like (***send***) are sometimes referred to as ***structural rules*** [1], and their use is critical for an adequate rendering of ***Mytype*** polymorphism: it is the ability to refer to possibly unknown types in the type rules, in fact, that allows methods to act parametrically over any $u \mathrel{<\!\!\#} A$, where u is the type of ***self***, and A is a given Obj-type.

The next rule defines the typing of an object-extension with a new method:

$$\frac{\begin{array}{l}\Gamma \vdash e_1 : \texttt{Obj}\, t.\langle R \mid \overline{m{:}\rho} \rangle \\ \Gamma, u \mathrel{<\!\!\#} \texttt{Obj}\, t.\langle \overline{m{:}\rho}, n{:}\tau \rangle \vdash e_2 : [u/t](t \to \tau) \qquad n \notin \mathcal{M}(\langle R \mid \overline{m{:}\tau} \rangle)\end{array}}{\Gamma \vdash \langle e_1 \longleftrightarrow\!\!+\; n{=}e_2 \rangle : \texttt{Obj}\, t.\langle R \mid \overline{m{:}\rho}, n{:}\tau \rangle} \quad (\textit{ext}).$$

Typing the method addition for n requires (i) that the n method be not in the type of the object e_1 that is being extended, and (ii) that the protocol of the resulting object contains at least the n method as well as the $\overline{m}$ methods needed to type the body e_2 of n.

The typing of a method override is defined similarly. As for the (*send*) rule, the generality that derives from the use of the type σ is needed to carry out derivations where the (*over*) rule is applied with e_1 variable (e.g. *self*).

$$\frac{\Gamma \vdash e_1 : \sigma \qquad \Gamma \vdash \sigma \lessdot\!\!\# \; \texttt{Obj}\, t.\langle\!\langle \overline{m{:}\rho}, n{:}\tau \rangle\!\rangle \qquad \Gamma, u \lessdot\!\!\# \; \texttt{Obj}\, t.\langle\!\langle \overline{m{:}\rho}, n{:}\tau \rangle\!\rangle \vdash e_2 : [u/t](t \to \tau)}{\Gamma \vdash \langle e_1 \leftarrow n = e_2 \rangle : \sigma} \;(over).$$

We conclude with the rule for typing a search expression:

$$\frac{\Gamma \vdash e : \sigma \quad \Gamma \vdash \sigma \lessdot\!\!\# \; \texttt{Obj}\, t.\langle\!\langle n{:}\tau \rangle\!\rangle \quad \Gamma \vdash \varsigma \lessdot\!\!\# \; \sigma}{\Gamma \vdash e \hookleftarrow n : [\varsigma/t](t \to \tau)} \;(search).$$

Once more we assume a possibly unknown type for e: typing a search requires, however, more generality than typing a method invocation because the search of a method encompasses a recursive inspection of the recipient object (i.e. of *self*). This explains the intuitive roles of the two types ς and σ in the above rule: ς is the type of the *self* object, to which the n message was sent and to which the body of n will be applied; σ, on the other hand, is the type of e, the sub-object of *self* where the body of n method is eventually found while searching within *self*.

3.4 An Example of Type Derivations

We conclude the description of the type system with an example that illustrates the use of the typing rules in typing derivations. The example is borrowed from [16], and shows that the type system captures the desired form of method specialization. The following object expression represents a point object with an x coordinate and a move method:

$$\texttt{pt} \equiv \langle\langle \texttt{x} = \lambda\texttt{self}.3\rangle \leftarrow\!\!+ \; \texttt{move} = \lambda\texttt{self}.\lambda\texttt{dx}.\langle \texttt{self} \leftarrow \texttt{x} = \lambda\texttt{s}.(\texttt{self} \Leftarrow \texttt{x}) + \texttt{dx}\rangle\rangle.$$

Below, we sketch the derivation of the judgement $\varepsilon \vdash \texttt{pt} : \texttt{Obj}\, t.\langle\!\langle \texttt{x} : int, \texttt{move} : int \to t \rangle\!\rangle$, using the assumption $\varepsilon \vdash \langle \texttt{x} = \lambda\texttt{self}.3\rangle : \texttt{Obj}\, t.\langle\!\langle \texttt{x} : int \rangle\!\rangle$.

Consider then defining cp as a new point, obtained from pt with the addition of a color method, namely: $\texttt{cp} \equiv \langle \texttt{pt} \leftarrow\!\!+ \; \texttt{color} = \lambda\texttt{self}.\texttt{blue}\rangle$. With a similar derivation, we may now prove that cp has type $\texttt{Obj}\, t.\langle\!\langle \texttt{x} : int, \texttt{move} : int \to t, \texttt{color} : colors \rangle\!\rangle$, thus showing how the type of move gets specialized as the method is inherited from a pt to a cp.

Contexts

$\Gamma_1 \;=\; u \not\ll\!\# \; \texttt{Obj}\, t.\langle \texttt{x} : int, \texttt{move} : int \to t\rangle, \texttt{self} : u, \texttt{dx} : int$

$\Gamma_2 \;=\; \Gamma_1, v \not\ll\!\# \; \texttt{Obj}\, t.\langle \texttt{x} : int, \texttt{move} : int \to t\rangle, \texttt{s} : v$

Derivation

1. $\Gamma_2 \vdash (\texttt{self} \Leftarrow \texttt{x}) + \texttt{dx} : int$
   by (*send*) from $\Gamma_2 \vdash \texttt{self} : u$ and $\Gamma_2 \vdash u \not\ll\!\# \; \texttt{Obj}\, t.\langle \texttt{x} : int\rangle$.
2. $\Gamma_2 - \texttt{s} \vdash \lambda \texttt{s}.(\texttt{self} \Leftarrow \texttt{x}) + \texttt{dx} : v \to int$
3. $\Gamma_1 \vdash \langle \texttt{self} \leftarrow \texttt{x} = \lambda \texttt{s}.(\texttt{self} \Leftarrow \texttt{x}) + \texttt{dx}\rangle : u$
   by (*over*) from $\Gamma_1 \vdash \texttt{self} : u$ and $\Gamma_1 \vdash u \not\ll\!\# \; \texttt{Obj}\, t.\langle \texttt{x} : int, \texttt{move} : int \to t\rangle$ and 2.
4. $\Gamma_1 - \texttt{self} - \texttt{dx} \vdash \lambda \texttt{self}.\lambda \texttt{dx}.\langle \texttt{self} \leftarrow \texttt{x} = \lambda \texttt{s}.(\texttt{self} \Leftarrow \texttt{x}) + \texttt{dx}\rangle : u \to int \to u$
5. $\varepsilon \vdash \texttt{pt} : \texttt{Obj}\, t.\langle \texttt{x} : int, \texttt{move} : int \to t\rangle$
   by (*ext*) from $\varepsilon \vdash \langle \texttt{x} = \lambda \texttt{self}.3\rangle : \texttt{Obj}\, t.\langle \texttt{x} : int\rangle$ and 4.

3.5 Type Soundness

We conclude this section with a theorem of type soundness. We first show that types are preserved by reduction.

Theorem 1 [Subject Reduction]. *If $e_1 \overset{eval}{\twoheadrightarrow} e_2$ and the judgement $\Gamma \vdash e_1 : \tau$ is derivable, then so is the judgement $\Gamma \vdash e_2 : \tau$.*

The proof is omitted here due to the lack of space, and can be found in [4]. Type soundness follows directly as a corollary of Subject Reduction, for if we may derive a type for an expression e, then e may not be reduced to *err* (which has no type). Hence we have:

Theorem 2 [Type Soundness]. *If $\varepsilon \vdash e : \tau$ is derivable, then $e \overset{eval}{\not\twoheadrightarrow}$ wrong.*

4 Relationship with the Type System of [FHM94]

The relationship between our system and the system of [16] is best illustrated by looking at the example of Section 3.4, where we showed that the move method for the pt object may be typed with the following judgement:

$$u \not\ll\!\# \; \texttt{Obj}\, t.\langle \texttt{x} : int, \texttt{move} : int \to t\rangle \vdash \lambda \texttt{self}.\lambda \texttt{x}.\langle \texttt{self} \leftarrow \texttt{x} = \lambda \texttt{s}.(\texttt{self} \Leftarrow \texttt{x}) + \texttt{dx}\rangle : u \to int \to u$$

The corresponding judgement in the original system, (cfr. [16], Section 3.4)) is:

$$r : T \to [\texttt{x}, \texttt{move}] \vdash \lambda \texttt{self}.\lambda \texttt{x}.\langle \texttt{self} \leftarrow \texttt{x} = \lambda \texttt{s}.(\texttt{self} \Leftarrow \texttt{x}) + \texttt{dx}\rangle : [\texttt{Obj}\, t.\langle rt \mid \texttt{x} : int, \texttt{move} : int \to t\rangle / t](t \to int \to t).$$

The two judgements have essentially the same structure, but the rendering of polymorphism is fundamentally different. In the original system, polymorphism is placed *inside* the row of the type $\texttt{Obj}\, t.\langle rt \mid \texttt{x}{:}int, \texttt{move}{:}int \to t\rangle$, and arises from using the (second order) row variable r: instances of the Obj-type are obtained by substituting any *well-kinded* row (i.e. one that does not contain the names x and move) for the row rt that results from the application of r to the type t. In our system, instead, we place polymorphism outside rows: the type of move is polymorphic in the type-variable u, and instances of this type are obtained by substituting any *syntactically well-formed* type that matches the bound $\texttt{Obj}\, t.\langle \texttt{x}{:}int, \texttt{move}{:}int{\rightarrow}t\rangle$ for u.

As it turns out, the correspondence between polymorphic Obj-types of the original system, and the type-variables of our systems carries over to typing derivations. As a matter of fact, the correspondence works correctly as long as the polymorphic Obj-types are used in a "disciplined", or *regular* way in typing judgements and derivations from [16]; it breaks, instead, in other cases. Consider, for instance, the following judgement:

$$r : T{\rightarrow}[m],\ e : \texttt{Obj}\, t.\langle rt\rangle \vdash \langle e \leftarrow\!\!+\ m{=}\lambda s.s\rangle : \texttt{Obj}\, t.\langle rt \mid m{:}t\rangle$$

While this judgement is derivable in [16], replacing the polymorphic Obj-types with the corresponding type variables fails to produce a derivable judgement in our system. There are several reasons for this, the most evident being that our (ext) rule requires a Obj-type (not a variable) in the type of the extended object.

In general, the correspondence between row and type variables breaks whenever the same row-variable occurs in different polymorphic Obj-types of a derivation. Fortunately, however, it can be shown that such undesired uses of polymorphic Obj-type may always be dispensed with in derivations for closed-object typing judgements of the form $\varepsilon \vdash e : \tau$. The proof of this fact is rather complex, and requires a number of technical lemmas establishing some useful properties of the typing derivations of [16]. Due to the lack of space, below we only state the main result, referring the reader to [4] for full details.

Definition 3. [4] Let $\varepsilon \vdash e : \tau$ be a judgement from [16], and let Ξ be a derivation for this judgement. We say that Ξ is *regular* if and only if every every row-variable of Ξ occurs always within the same polymorphic Obj-type in Ξ.

Theorem 4. [4] *Every judgement of the form $\varepsilon \vdash e : \tau$ has a derivation in [16] iff it has a regular derivation.*

Using this result, we may then prove that every typing judgement of the form $\varepsilon \vdash e : \tau$ derivable in [16] is derivable also in our system. We do this, in the next subsection, introducing an encoding function that translates every regular derivation from [16] into a corresponding derivation in our system. Restricting to regular derivation is convenient in that it greatly simplifies the definition of the encoding; on the other hand, given the result established in Theorem 4, it involves no loss of generality for the judgements of interest.

4.1 Encoding the *Fisher-Honsell-Mitchell* Type System

Throughout this subsection, types, contexts, judgements and derivations from [16] will be referred to as, respectively row–types, row-contexts, row-judgements. Also we will implicitly assume that they occur within regular derivations.

Encoding of Types, Contexts and Judgements. The encoding of a row-type τ is the type τ^* that results from replacing every occurrence of a polymorphic Obj-type with a corresponding type-variable. The correspondence between the polymorphic Obj-types of τ and the type-variables of τ^* may be established using any injective map from row-variables to corresponding type variables (this is because in every regular row-derivation there is a one-to-one correspondence between row-variables and polymorphic Obj-types). To ease the presentation, we will thus assume that the row-variables of regular derivations are all chosen from a given set $\mathcal{V}_r$, and that an injective map $\xi : \mathcal{V}_r \mapsto \mathcal{V}_t$ is given, where $\mathcal{V}_t$ is a corresponding set of type variables.

Definition 5 [ENCODING OF TYPES]. Let τ be a row-type. The encoding of τ, denoted by τ^*, is defined as follows:

- $\tau^* = \tau$, for every type variable or type constant τ;
- $(\tau_1 \to \tau_2)^* = \tau_1^* \to \tau_2^*$
- $(\mathtt{Obj}\, t.\langle m_1{:}\tau_1, \ldots, m_k{:}\tau_k \rangle)^* = \mathtt{Obj}\, t.\langle m_1{:}\tau_1^*, \ldots, m_k{:}\tau_k^* \rangle$
- $(\mathtt{Obj}\, t.\langle rt \mid m_1{:}\tau_1, \ldots, m_k{:}\tau_k \rangle)^* = \xi(r)$

It follows from the definition that the encoding of a "ground" type, i.e. one that does not contain any occurrence of row-variables, leaves the type unchanged: in other words, τ^* is equal to τ, whenever τ is ground in the sense we just explained.

The encoding of a row-context Γ is more elaborate, and it is given with respect to the regular derivation where the context occurs.

Definition 6 [ENCODING OF CONTEXTS]. Let Γ be a row-context of a regular derivation Ξ, and let $\mathtt{Obj}\, t.\langle rt \mid \overline{m{:}\tau} \rangle$ denote *the* polymorphic Obj-type of Ξ where r occurs. The encoding of Γ, denoted by Γ_Ξ^*, is defined as follows:

- $\varepsilon_\Xi^* = \varepsilon$
- $(\Gamma, t : T)_\Xi^* = \Gamma_\Xi^*$;
- $(\Gamma, r : \kappa)_\Xi^* = \Gamma_\Xi^*,\ \xi(r) \mathrel{<\!\!\#} \mathtt{Obj}\, t.\langle \overline{m{:}\tau^*} \rangle$
- $(\Gamma, x : \tau)_\Xi^* = \Gamma_\Xi^*,\ x : \tau^*$.

The definition is well-posed as every row-variable occurs in just one polymorphic Obj-type, Ξ being regular. Also note that the encoding of every valid row-context Γ is a valid context, since every row-variable r may occur at most once in Γ.

The encoding of row-judgements is also given with respect to the regular derivation where they occur.

Definition 7 [ENCODING OF JUDGEMENTS]. Let $\Gamma \vdash e : \tau$ be a row-judgement of a regular derivation Ξ. Then the encoding of this row-judgement is $\Gamma_\Xi^* \vdash e : \tau^*$.

Note that if $\varepsilon \vdash e : \tau$ is a derivable judgement, then its encoding is the judgement $\varepsilon \vdash e : \tau$ itself. This is easily seen as the type τ must be ground, the judgement being derivable.

Encoding of Type Rules. The row-portion of the type system from [16] has no counterpart in our system, as we axiomatize well-formedness for rows syntactically. The encoding of a type rule for terms, instead, is the result of encoding the row-judgements in the conclusion and in the premises of the rule, with the exception of the rules (*send*), (*ext*) and (*over*).

Definition 8 [Encoding of (*send*)].

$$\left(\frac{\Gamma \vdash e : \texttt{Obj}\, t.\langle R \mid \overline{m{:}\tau}\rangle}{\Gamma \vdash e \Leftarrow m : [\texttt{Obj}\, t.\langle R \mid \overline{m{:}\tau}\rangle / t]\tau}\right)^*_{\Xi} = \frac{\begin{array}{c}\Gamma^*_{\Xi} \vdash e : (\texttt{Obj}\, t.\langle R \mid \overline{m{:}\tau}\rangle)^* \\ \Gamma^*_{\Xi} \vdash (\texttt{Obj}\, t.\langle R \mid \overline{m{:}\tau}\rangle)^* \lll\!\!\# \; \texttt{Obj}\, t.\langle m{:}\tau^*\rangle\end{array}}{\Gamma^*_{\Xi} \vdash e \Leftarrow m : ([\texttt{Obj}\, t.\langle R \mid \overline{m{:}\tau}\rangle / t]\tau)^*}$$

where Ξ is the regular derivation where the rule occurs.

The type $(\texttt{Obj}\, t.\langle R \mid \overline{m{:}\tau}\rangle)^*$ may either be a a Obj-type, if R is a list of $m{:}\tau$ pairs, or the type variable $u = \xi(r)$ if $R = \langle rt \mid \ldots\rangle$. The two cases correspond respectively to messages to an object, and messages to *self*. As in [16], they are treated uniformly in our type system.

Definition 9 [Encoding of (*ext*)].

$$\left(\frac{\begin{array}{ll}\Gamma \vdash e_1 : \texttt{Obj}\, t.\langle R \mid \overline{m{:}\tau}\rangle & \\ \Gamma, r : T \to [\overline{m}, n] \vdash & r \text{ not in } \tau \\ \quad e_2 : [\texttt{Obj}\, t.\langle rt \mid \overline{m{:}\tau}, n{:}\tau\rangle / t](t \to \tau) & \end{array}}{\Gamma \vdash \langle e_1 \leftarrow\!\!+\, n{=}e_2\rangle : \texttt{Obj}\, t.\langle R \mid \overline{m{:}\tau}, n{:}\tau\rangle}\right)^*_{\Xi} =$$

$$= \frac{\begin{array}{c}\Gamma^*_{\Xi} \vdash e_1 : \texttt{Obj}\, t.\langle R^* \mid \overline{m{:}\tau^*}\rangle \;\; n \notin \mathcal{M}(R*) \cup \{\overline{m}\} \\ \Gamma^*_{\Xi}, u \lll\!\!\# \; \texttt{Obj}\, t.\langle \overline{m{:}\tau}^*, n{:}\tau^*\rangle \vdash e_2 : [u/t](t \to \tau)^*\end{array}}{\Gamma^*_{\Xi} \vdash \langle e_1 \leftarrow\!\!+\, n{=}e_2\rangle : \texttt{Obj}\, t.\langle R^* \mid \overline{m{:}\tau}^*, n{:}\tau^*\rangle}$$

where $u = \xi(r)$, and Ξ is the regular derivation where the rule occurs.

Note that the context $\Gamma^*_{\Xi}, u \lll\!\!\# \; \texttt{Obj}\, t.\langle\overline{m{:}\tau^*}\rangle$ is just the encoding, in Ξ, of the row-context $\Gamma, r : T \to [\overline{m}]$. The notation R^* is consistent because, as we show in [4], for every instance of the (*ext*) rule occurring in a regular derivation it must be the case that R is a ground row of the form $\overline{p{:}\sigma}$ for some methods $\overline{p}$ and types $\overline{\sigma}$. The encoding of (*over*) is defined similarly to the (*ext*) case.

Theorem 10 [Completeness]. *Let $\varepsilon \vdash e : \tau$ be a row-judgement derivable in [16]. Then $\varepsilon \vdash e : \tau$ is derivable in our system.*

Proof. Since $\varepsilon \vdash e : \tau$ is derivable, the encoding of this judgement coincides with the judgement itself. Let then Ξ be a regular derivation for $\varepsilon \vdash e : \tau$ (the existence of such a derivation follows from Theorem 4): to prove the claim, it is enough to show that the encoding (in Ξ) of every other judgement of Ξ is derivable in our system.

Let then $\Gamma' \vdash e' : \tau'$ be a judgement of Ξ, and let Ξ' be the sub-derivation of Ξ rooted at $\Gamma' \vdash e' : \tau'$. The proof is by induction on Ξ'. The basis of induction, the (*projection*) case, follows immediately, and most of the inductive cases follow easily by induction.

The only slightly more elaborate cases is when Ξ' ends up with (*send*). In this case the last rule of Ξ', ans its encoding are, respectively:

$$\left(\frac{\Gamma \vdash e : \texttt{Obj}\, t.\langle R \mid \overline{m{:}\tau}\rangle}{\Gamma \vdash e \Leftarrow m : [\texttt{Obj}\, t.\langle R \mid \overline{m{:}\tau}\rangle / t]\tau}\right)^*_{\Xi} \quad\text{and}\quad \frac{\Gamma^*_{\Xi} \vdash e : (\texttt{Obj}\, t.\langle R \mid \overline{m{:}\tau}\rangle)^* \qquad \Gamma^*_{\Xi} \vdash (\texttt{Obj}\, t.\langle R \mid \overline{m{:}\tau}\rangle)^* \lessdot\!\!\# \;\texttt{Obj}\, t.\langle m{:}\tau^*\rangle}{\Gamma^*_{\Xi} \vdash e \Leftarrow m : ([\texttt{Obj}\, t.\langle R \mid \overline{m{:}\tau}\rangle / t]\tau)^*}$$

That $\Gamma^*_{\Xi} \vdash e : (\texttt{Obj}\, t.\langle R \mid \overline{m{:}\tau}\rangle)^*$ is derivable follows from the induction hypothesis. For the other judgement in the premise, instead, we distinguish the two possible sub-cases. If $(\texttt{Obj}\, t.\langle R \mid \overline{m{:}\tau}\rangle)^* = \texttt{Obj}\, t.\langle R^* \mid \overline{m{:}\tau^*}\rangle$ then the judgement is question is derivable directly by ($\lessdot\!\!\#$). If, instead, $(\texttt{Obj}\, t.\langle R \mid \overline{m{:}\tau}\rangle)^*$ is a type variable, say $\xi(r)$, then it must be the case that $R = \langle rt \mid \overline{p{:}\sigma}\rangle$ for some $\overline{p{:}\sigma}$. But then $r \in Dom(\Gamma)$ and $\texttt{Obj}\, t.\langle rt \mid \overline{p{:}\sigma}, \overline{m{:}\tau}\rangle$ is the polymorphic Obj-type for r in Ξ. Hence, from Definition 6, we have that $\xi(r) \lessdot\!\!\# \;\texttt{Obj}\, t.\langle \overline{p{:}\sigma}, \overline{m{:}\tau}\rangle \in \Gamma^*_{\Xi}$, and then $\Gamma^*_{\Xi} \vdash (\texttt{Obj}\, t.\langle R \mid \overline{m{:}\tau}\rangle)^* \lessdot\!\!\# \;\texttt{Obj}\, t.\langle m{:}\tau^*\rangle$ is derivable by ($\lessdot\!\!\#$ *proj*) and ($\lessdot\!\!\#$ *trans*). □

5 Classes

Introducing classes relies on the idea of distinguishing two kinds of types, `Class` types whose elements are classes, and `Obj` types whose elements are instances created by the classes. This distinction is inspired by [17], but we use it here in a completely different way and with orthogonal purposes.

5.1 Class Types and Object Types

Object types have the usual form $\texttt{Obj}\, t.\langle m_1^\iota{:}\tau_1, \ldots, m_k^\iota{:}\tau_k\rangle$, with the difference that we now require that the component methods be ***instance-methods*** annotated by the superscript $^\iota$. Class types, instead, have the form:

$$\texttt{Class}\, t.\langle m_1^c{:}\tau_1, \ldots, m_k^c{:}\tau_k, m_1^\iota{:}\sigma_1, \ldots, m_l^\iota{:}\sigma_l\rangle$$

with the superscript ^{c} distinguishing ***class-methods*** from the remaining instance-methods. The intention is to have each class define a set of class-methods for exclusive use of the class and its sub-classes, as well as a set of instance-methods for the instances of the class (and of the class' sub-classes). Instance-methods may only invoke or override other instance-methods, so that classes are protected from updates caused by their instances; class-methods, instead, are not subject to this restriction.

Every class defines (or inherits from its super-classes) at least one class-method – *new* – for creating new instances of that class. As in [17], instances of a class are created by ***packaging*** a class, an operation that does nothing but "sealing" the class, so that no methods may be added. Sealing a class changes its type into a corresponding Obj type, that results from hiding all of the class-methods of the class. As for subclassing, new classes may be derived from a class by adding new methods or redefining existing methods of that class.

To account for the above features, the object-related forms of the untyped calculus are extended as follows:

$$e ::= \mid topclass \mid \langle e_1 \overset{o}{\longleftarrow\!\!+} m{=}e_2 \rangle \mid pack(e) \mid \langle e_1 \leftarrow m{=}e_2 \rangle \mid e \Leftarrow m \mid e \hookleftarrow m$$

The operational meaning of the operations of override, send, and search is exactly as in Section 2. The meaning of the remaining expressions is given next.

topclass is a pre-defined constant representing the empty class and defined as follows: $topclass \equiv \langle new^c{=}\lambda s.pack(s) \rangle$. This class defines class-method *new* for creating instances: the result of a call to *new* on a class is the instance of that class that results from packaging the class.

New classes may be derived from the empty class by a sequence of method overrides and extensions. The symbol $\overset{o}{\longleftarrow\!\!+}$ above denotes two different operators, $\overset{c}{\longleftarrow\!\!+}$ and $\overset{\iota}{\longleftarrow\!\!+}$, denoting class-extension with, respectively, class-methods, and instance methods.

The operational semantics of the *pack* operator is defined by few additional cases of the reduction relation; the effect of packaging on types, in turn, is rendered in terms of a corresponding type operator, denoted by **pack**. The new cases of reduction for *pack* and the equational rules for packaged types are defined below:

$$\begin{array}{lll} pack(c) & \xrightarrow{eval} c \\ pack(\lambda x.e) & \xrightarrow{eval} \lambda x.e \\ pack(e) \Leftarrow m & \xrightarrow{eval} (e \hookleftarrow m)\, pack(e) \\ pack(pack(e)) & \xrightarrow{eval} pack(e) \end{array} \qquad \begin{array}{ll} \mathbf{pack}(b) & = b \\ \mathbf{pack}(\tau_1 \to \tau_2) & = \tau_1 \to \tau_2 \\ \mathbf{pack}(\mathtt{Class}\, t.\langle\!\langle \overline{m^c{:}\rho}, \overline{p^\iota{:}\tau} \rangle\!\rangle) & = \mathtt{Obj}\, t.\langle\!\langle \overline{p^\iota{:}\tau} \rangle\!\rangle \\ \mathbf{pack}(\mathtt{Obj}\, t.R) & = \mathtt{Obj}\, t.R \\ \mathbf{pack}(\mathbf{pack}(\tau)) & = \mathbf{pack}(\tau) \end{array}$$

Packaging a class produces the corresponding Obj type that results from hiding all of the class-methods of the class. Packaging any other expression, instead, has no effect on the type of the expression. The introduction of the **pack** operator induces a new, and richer, notion of type equality that now allows two types to be identified if they are equal modulo the equational rules for the **pack** operator.

The definition of matching is readily extended to the newly introduced types. Matching over Class-types is exactly as matching over Obj-types, namely:

$$\frac{\Gamma \vdash * \qquad \langle\!\langle \overline{m^o{:}\tau}, \overline{n^o{:}\sigma} \rangle\!\rangle \ well\ formed}{\Gamma \vdash \mathtt{Class}\, t.\langle\!\langle \overline{m^o{:}\tau}, \overline{n^o{:}\sigma} \rangle\!\rangle \lessdot\!\# \ \mathtt{Class}\, t.\langle\!\langle \overline{m^o{:}\tau} \rangle\!\rangle} \ (\lessdot\!\#\ class),$$

where ^{o} may either be ^{c} or $^\iota$. As for matching between Class and Obj types, we have the following rule:

$$\frac{\Gamma \vdash * \qquad \langle\!\langle \overline{m^c{:}\tau}, \overline{n^\iota{:}\sigma} \rangle\!\rangle \ well\ formed}{\Gamma \vdash \mathtt{Class}\, t.\langle\!\langle \overline{m^c{:}\tau}, \overline{n^\iota{:}\sigma} \rangle\!\rangle \lessdot\!\# \ \mathtt{Obj}\, t.\langle\!\langle \overline{n^\iota{:}\sigma} \rangle\!\rangle} \ (\lessdot\!\#\ obj),$$

that is, every Class type matches its corresponding Obj type.

5.2 Typing Rules

The new types require few additional changes in the object-related portion of the type system, which are described next. The following two rules define the type of the empty class and of packaged expressions.

$$\frac{\Gamma \vdash *}{\Gamma \vdash \mathit{topclass} : \mathtt{Class}\, t.\langle\!\langle \mathit{new}^c : \mathtt{pack}(t) \rangle\!\rangle} \quad (\mathit{topclass}).$$

$$\frac{\Gamma \vdash e : \tau}{\Gamma \vdash \mathit{pack}(e) : \mathtt{pack}(\tau)} \quad (\mathit{pack}).$$

For the remaining object-expression, we need different rules for distinguishing the different behavior of class-methods and instance-methods.

Class Methods. **The typing of class-method invocation is as follows:**

$$\frac{\Gamma \vdash e : \sigma \quad \Gamma \vdash \sigma \lessdot\!\!\# \ \mathtt{Class}\, t.\langle\!\langle n^c{:}\tau \rangle\!\rangle}{\Gamma \vdash e \Leftarrow n : [\sigma/t]\tau} \quad (\mathit{c\text{-}send}),$$

If n is a class-method (as indicated by the annotation ^{c}), then e must have a Class-type matching $\mathtt{Class}\, t.\langle\!\langle n^c{:}\tau \rangle\!\rangle$. Therefore, only class-methods may be invoked on a class; class-methods may, instead, perform (internal) invocations or overrides also on instance-methods of that class. The following rule for class-method addition allows this behaviour.

$$\frac{\begin{array}{c}\Gamma \vdash e_1 : \mathtt{Class}\, t.\langle\!\langle R \mid \overline{m^o{:}\rho} \rangle\!\rangle \quad n \notin \mathcal{M}(R) \cup \{\overline{m}\} \\ \Gamma,\, u \lessdot\!\!\# \ \mathtt{Class}\, t.\langle\!\langle \overline{m^o{:}\rho}, n^c{:}\tau \rangle\!\rangle \vdash e_2 : [u/t](t \to \tau)\end{array}}{\Gamma \vdash \langle e_1 \overset{c}{\longleftarrow\!\!+} n{=}e_2 \rangle : \mathtt{Class}\, t.\langle\!\langle R \mid \overline{m^o{:}\rho}, n^c{:}\tau \rangle\!\rangle} \quad (\mathit{c\text{-}ext}).$$

As usual, method addition, is subject to the constraint that the n method be not already in the type of class. The typing rule for method override is similar to ($\mathit{c\text{-}ext}$), while the typing of search expressions follows the same idea as ($\mathit{c\text{-}send}$) above.

Instance Methods. **As we said, instance-method may only invoke other instance-methods of the *self* object. The following typing rule enforces this constraint:**

$$\frac{\begin{array}{c}\Gamma \vdash e_1 : \mathtt{Class}\, t.\langle\!\langle R \mid \overline{m^\iota{:}\rho} \rangle\!\rangle \qquad n \notin \mathcal{M}(R) \cup \{\overline{m}\} \\ \Gamma,\, u \lessdot\!\!\# \ \mathtt{Obj}\, t.\langle\!\langle \overline{m^\iota{:}\rho}, n^\iota{:}\tau \rangle\!\rangle \vdash e_2 : [\mathtt{pack}(u)/t](t \to \tau)\end{array}}{\Gamma \vdash \langle e_1 \overset{\iota}{\longleftarrow\!\!+} n{=}e_2 \rangle : \mathtt{Class}\, t.\langle\!\langle R \mid \overline{m^\iota{:}\rho}, n^\iota{:}\tau \rangle\!\rangle} \quad (\iota\text{-}\mathit{ext}).$$

Note that the bound for u is so defined as to allow the n method to be applied on objects of every type matching the Obj type corresponding to the Class type (as in ($\lessdot\!\!\#$ *obj*) rule). This way, instance-methods may safely be inherited by

the instances of the class. Note, furthermore, that the polymorphic type of e_2 is so defined as to ensure that e_2 may only be applied on elements of packaged types (i.e., objects).

A corresponding constraint is imposed on the typing of instance-method invocation.

$$\frac{\Gamma \vdash e : \mathtt{pack}(\sigma) \quad \Gamma \vdash \sigma \lll\!\# \; \mathtt{Obj}\, t.\langle\!\langle n^{\iota}{:}\tau \rangle\!\rangle}{\Gamma \vdash e \Leftarrow n : [\mathtt{pack}(\sigma)/t]\tau} \; (\iota\text{-}send),$$

The type of the receiver must be a packaged type consistent with the polymorphic type that is associated with the body of the method. The rules for method search and override may be defined following this idea.

5.3 A Simple Example

Using the above type rules, it is now possible to define an object expression representing the class of points with an x coordinate and, say, a *set* method for updating the position of the points in the class. Assuming that *create* is a class method, and that x and *set* are instance methods, the class of points may be defined as follows:

$$\begin{aligned} \mathtt{ptClass} \equiv \langle\, & new &&= \lambda s.pack(s); \\ & create &&= \lambda s.\lambda v.((s \Leftarrow new) \Leftarrow set\, v); \\ & x &&= \lambda s.0; \\ & set &&= \lambda s.\lambda v.\langle s \leftarrow x = \lambda self.v \rangle \,\rangle \end{aligned}$$

where the value of the x field is defaulted to 0 in the class definition. The following type may then be derived:

$$\mathtt{ptClass} : \mathtt{Class}\, t.\langle\!\langle new^{c}{:}\mathtt{pack}(t),\; create^{c}{:}\mathtt{pack}(t),\; x^{\iota}{:}int,\; set^{\iota}{:}int \rightarrow \mathtt{pack}(t) \rangle\!\rangle$$

Instances of this class may then be created with a *create* message to `ptclass`. For instance, the expression ($\mathtt{ptclass} \Leftarrow create\, 3$) creates a point instance of type $\mathtt{Obj}\, t.\langle\!\langle x^{\iota}{:}int,\; set^{\iota}{:}int \rightarrow \mathtt{pack}(t) \rangle\!\rangle$, with x coordinate valued 3 and a *set* method.

6 Conclusions

We have presented a new type system for the *Lambda Calculus of Objects*. The main difference with respect to the original proposal, is that in [16] method polymorphism is rendered in terms of quantification over row-schemas, whereas in our system it is captured by means of match-bounded quantification and matching. A formal analysis of the relative expressive power of the two systems shows that they coincide on derivations for closed-object typing judgements. On the other hand, there are some fundamental tradeoffs between the two approaches, both in terms of complexity and of their logical rendering. In fact, while the new solution appears to reduce the complexity of the system, freeing it from the calculus of rows of the original system, on the other hand the auxiliary judgements and side-conditions needed for matching are not costless, and may have undesired consequences in the encoding of the new system in Logical Frameworks [15].

We have then presented an extension of the new system that gives provision for classes in an object-based setting. The extended calculus may, in some respects, be seen as a functional counterpart of the *Imperative Object Calculus* of [1]. Besides the different operational setting (i.e. imperative versus functional), that proposal differs from ours in that, while using matching to model method inheritance between classes, it relies on subtyping for the treatment of self types. Instead, our system requires no subtyping, since it relies on matching as the only relation over class and object types.

Acknowledgements. We would like to thank Mariangiola Dezani-Ciancaglini for inspiring this work and for endless discussions on earlier drafts. Suggestions from the anonymous referees helped to improve the presentation substantially.

References

1. M. Abadi and L. Cardelli. An Imperative Objects Calculus. In P.D. Mosses, M. Nielsen, and M.I. Schwartzbach, editors, *Proceedings of TAPSOFT'95: Theory and Practice of Software Development*, volume 915 of *LNCS*, pages 471–485. Springer–Verlag, May 1995.
2. M. Abadi and L. Cardelli. On Subtyping and Matching. In *Proceedings of ECOOP'95: European Conference on Object-Oriented Programming*, volume 952 of *LNCS*, pages 145–167. Springer–Verlag, August 1995.
3. M. Abadi and L. Cardelli. A Theory of Primitive Objects. *Information and Computation*, 125(2):78–102, March 1996.
4. V. Bono and M. Bugliesi. Matching Lambda Calculus of Objects. Submitted for publication, 1996.
5. V. Bono, M. Bugliesi, and L. Liquori. A Lambda Calculus of Incomplete Objects. In *Proc. of MFCS*, volume 1113 of *Lecture Notes in Computer Science*, pages 218–229. Springer-Verlag, 1996.
6. V. Bono, M. Bugliesi, M.Dezani, and L. Liquori. Subtyping Constraints for Incomplete Objects. In *Proc. of CAAP*, Lecture Notes in Computer Science. Springer-Verlag, 1997. To appear.
7. V. Bono and L. Liquori. A Subtyping for the Fisher-Honsell-Mitchell Lambda Calculus of Objects. In *Proc. of CSL*, volume 933 of *Lecture Notes in Computer Science*, pages 16–30. Springer-Verlag, 1995.
8. K.B. Bruce. A Paradigmatic Object-Oriented Programming Language: Design, Static Typing and Semantcs. *Journal of Functional Programming*, 1(4):127–206, 1994.
9. L. Cardelli. A Semantics of Multiple Inheritance. *Information and Computation*, 76:138–164, 1988.
10. L. Cardelli and J.C. Mitchell. Operations on Records. *Mathematical Structures in Computer Sciences*, 1(1):3–48, 1991.
11. L. Cardelli and P. Wegner. On Understanding Types, Data Abstraction and Polymorphism. *Computing Surveys*, 17(4):471–522, 1985.

12. W. Cook. A Self-ish Model of Inheritance. Manuscript, 1987.
13. W. Cook, W. Hill, and P. Canning. Inheritance is not Subtyping. In *Proc. of ACM Symp. POPL*, pages 125–135. ACM Press, 1990.
14. W.R. Cook. *A Denotational Semantics of Inheritance*. PhD thesis, Brown University, 1989.
15. Harper R. Honsell F. and Plotkin G. A Framework for Defining Logics. *J.ACM*, 40(1):143–184, 1993.
16. K. Fisher, F. Honsell, and J. C. Mitchell. A Lambda Calculus of Objects and Method Specialization. *Nordic Journal of Computing*, 1(1):3–37, 1994.
17. K. Fisher and J. C. Mitchell. A Delegation-based Object Calculus with Subtyping. In *Proc. of FCT*, volume 965 of *Lecture Notes in Computer Science*, pages 42–61. Springer-Verlag, 1995.
18. M. Wand. Complete Type Inference for Simple Objects. In *Proc. of IEEE Symp. LICS*, pages 37–44. Silver Spring, 1987.

A Operational Semantics

(β)	$(\lambda x.e_1)\, e_2$	$\xrightarrow{eval}$	$[e_2/x]\, e_1$
$(\Leftarrow)$	$e \Leftarrow m$	$\xrightarrow{eval}$	$(e \hookleftarrow m)\, e$
$(\hookleftarrow\ succ)$	$\langle e_1 \leftarrow\!\!\circ\ m{=}e_2\rangle \hookleftarrow m$	$\xrightarrow{eval}$	e_2
$(\hookleftarrow\ next)$	$\langle e_1 \leftarrow\!\!\circ\ n{=}e_2\rangle \hookleftarrow m$	$\xrightarrow{eval}$	$e_1 \hookleftarrow m$
$(fail\ \langle\rangle)$	$\langle\rangle \hookleftarrow m$	$\xrightarrow{eval}$	err
$(fail\ abs)$	$\lambda x.e \hookleftarrow m$	$\xrightarrow{eval}$	err
$(err \leftarrow\!\!\circ\ )$	$\langle err \leftarrow\!\!\circ\ m = e\rangle$	$\xrightarrow{eval}$	err
$(err\ abs)$	$\lambda x.err$	$\xrightarrow{eval}$	err
$(err\ app)$	$err\ e$	$\xrightarrow{eval}$	err
$(err \hookleftarrow)$	$err \hookleftarrow n$	$\xrightarrow{eval}$	err

B Typing Rules

General Rules: $\Gamma \vdash A$ stands for any derivable judgement in the system.

(start)
$$\frac{}{\varepsilon \vdash *}$$

(projection)
$$\frac{\Gamma \vdash * \quad x : \tau \in \Gamma}{\Gamma \vdash x : \tau}$$

(weakening)
$$\frac{\Gamma \vdash A \quad \Gamma, \Gamma' \vdash *}{\Gamma, \Gamma' \vdash A}$$

Rules for Terms

(*var*) $$\frac{\Gamma \vdash * \quad x \notin Dom(\Gamma)}{\Gamma, x{:}\tau \vdash *}$$

(*abs*) $$\frac{\Gamma, x{:}\tau_1 \vdash e{:}\tau_2}{\Gamma \vdash \lambda x.e{:}\tau_1 \rightarrow \tau_2}$$

(*app*) $$\frac{\Gamma \vdash e_1{:}\tau_1 \rightarrow \tau_2 \quad \Gamma \vdash e_2{:}\tau_1}{\Gamma \vdash e_1 e_2{:}\tau_2}$$

(*empty*) $$\frac{\Gamma \vdash *}{\Gamma \vdash \langle\rangle : \mathtt{Obj}\, t.\langle\!\langle\rangle\!\rangle}$$

(*ext*) $$\frac{\begin{array}{l}\Gamma \vdash e_1 : \mathtt{Obj}\, t.\langle\!\langle R \mid \overline{m{:}\rho} \rangle\!\rangle \\ \Gamma, u \lessdot\!\!\# \; \mathtt{Obj}\, t.\langle\!\langle \overline{m{:}\rho}, n{:}\tau \rangle\!\rangle \vdash e_2 : [u/t](t \rightarrow \tau) \qquad n \notin \mathcal{M}(\langle\!\langle R \mid \overline{m{:}\tau} \rangle\!\rangle)\end{array}}{\Gamma \vdash \langle e_1 \longleftrightarrow\!\!+\; n{=}e_2 \rangle : \mathtt{Obj}\, t.\langle\!\langle R \mid \overline{m{:}\rho}, n{:}\tau \rangle\!\rangle}$$

(*over*) $$\frac{\begin{array}{l}\Gamma \vdash e_1 : \sigma \qquad \Gamma \vdash \sigma \lessdot\!\!\# \; \mathtt{Obj}\, t.\langle\!\langle \overline{m{:}\rho}, n{:}\tau \rangle\!\rangle \\ \Gamma, u \lessdot\!\!\# \; \mathtt{Obj}\, t.\langle\!\langle \overline{m{:}\rho}, n{:}\tau \rangle\!\rangle \vdash e_2 : [u/t](t \rightarrow \tau)\end{array}}{\Gamma \vdash \langle e_1 \leftarrow n{=}e_2 \rangle : \sigma}$$

(*send*) $$\frac{\Gamma \vdash e : \sigma \quad \Gamma \vdash \sigma \lessdot\!\!\# \; \mathtt{Obj}\, t.\langle\!\langle n{:}\tau \rangle\!\rangle}{\Gamma \vdash e \Leftarrow n : [\sigma/t]\tau}$$

(*search*) $$\frac{\Gamma \vdash e : \sigma \quad \Gamma \vdash \sigma \lessdot\!\!\# \; \mathtt{Obj}\, t.\langle\!\langle n{:}\tau \rangle\!\rangle \quad \Gamma \vdash \varsigma \lessdot\!\!\# \; \sigma}{\Gamma \vdash e \hookleftarrow n : [\varsigma/t](t \rightarrow \tau)}$$

Rules for Matching

($\lessdot\!\!\#$ *var*) $$\frac{\Gamma \vdash * \quad u \notin \Gamma \quad u \notin \tau}{\Gamma, u \lessdot\!\!\# \; \tau \vdash *}$$

($\lessdot\!\!\#$ *proj*) $$\frac{\Gamma \vdash * \quad u \lessdot\!\!\# \; \tau \in \Gamma}{\Gamma \vdash u \lessdot\!\!\# \; \tau}$$

($\lessdot\!\!\#$ *refl*) $$\frac{\Gamma \vdash *}{\Gamma \vdash \tau \lessdot\!\!\# \; \tau}$$

($\lessdot\!\!\#$ *trans*) $$\frac{\Gamma \vdash \sigma \lessdot\!\!\# \; \tau \quad \Gamma \vdash \tau \lessdot\!\!\# \; \rho}{\Gamma \vdash \sigma \lessdot\!\!\# \; \rho}$$

($\lessdot\!\!\#$) $$\frac{\Gamma \vdash * \qquad \langle\!\langle \overline{m{:}\tau}, \overline{n{:}\sigma} \rangle\!\rangle \textit{ well formed}}{\Gamma \vdash \mathtt{Obj}\, t.\langle\!\langle \overline{m{:}\tau}, \overline{n{:}\sigma} \rangle\!\rangle \lessdot\!\!\# \; \mathtt{Obj}\, t.\langle\!\langle \overline{m{:}\tau} \rangle\!\rangle}$$

Coinductive Axiomatization of Recursive Type Equality and Subtyping*

Michael Brandt and Fritz Henglein

DIKU, University of Copenhagen, Universitetsparken 1, DK-2100 Copenhagen East, Denmark, Email: {mick,henglein}@diku.dk

Abstract. We present new sound and complete axiomatizations of type equality and subtype inequality for a first-order type language with regular recursive types. The rules are motivated by coinductive characterizations of type containment and type equality via simulation and bisimulation, respectively. The main novelty of the axiomatization is the *fixpoint rule* (or *coinduction principle*), which has the form

$$\frac{A, P \vdash P}{A \vdash P}$$

where P is either a type equality $\tau = \tau'$ or type containment $\tau \leq \tau'$. We define what it means for a proof (formal derivation) to be formally *contractive* and show that the fixpoint rule is *sound* if the proof of the premise $A, P \vdash P$ is contractive. (A proof of $A, P \vdash P$ using the assumption axiom is, of course, *not* contractive.) The fixpoint rule thus allows us to capture a coinductive relation in the fundamentally inductive framework of inference systems.

The new axiomatizations are "leaner" than previous axiomatizations, particularly so for type containment since no separate axiomatization of type equality is required, as in Amadio and Cardelli's axiomatization. They give rise to a natural operational interpretation of proofs as *coercions*. In particular, the fixpoint rule corresponds to *definition by recursion*. Finally, the axiomatization is closely related to (known) efficient algorithms for deciding type equality and type containment. These can be modified to not only *decide* type equality and type containment, but also construct proofs in our axiomatizations efficiently. In connection with the operational interpretation of proofs as coercions this gives *efficient* ($O(n^2)$ time) algorithms for constructing *efficient* coercions from a type to any of its supertypes or isomorphic types.

1 Introduction

The simply typed λ-calculus is paradigmatic for both type inference for programming languages and the Curry-Howard isomorphism. Whereas adding recursive

* This research was partially supported by the Danish Research Council, Project *DART*.

types destroys its strong normalization property and its logical soundness under the Curry-Howard interpretation, recursive types preserve and extend the "well-typed programs don't go wrong" interpretation of λ-terms. To use recursive types it is necessary to add the rule

$$\frac{A \vdash e : \tau \qquad \vdash \tau = \tau'}{A \vdash e : \tau'} \quad (\text{EQUAL})$$

for simple typing with recursive types [CC91] or

$$\frac{A \vdash e : \tau \qquad \vdash \tau \leq \tau'}{A \vdash e : \tau'} \quad (\text{SUBTYPE})$$

for simple subtyping with recursive types [AC91, AC93]. The question, now, is when two recursive types are equal or in the subtyping relation. This is what we study in this paper.

1.1 Recursive Types

Definition 1. The *recursive types (in canonical form)* μTp are generated by the grammar

$$\tau \equiv \bot \mid \top \mid \alpha \mid \tau_1 \to \tau_2 \mid \mu\alpha.(\tau_1 \to \tau_2)$$

where α ranges over an infinite set TVar of *type variables*, μ binds its type variable, α-congruent recursive types are identified, and in every $\mu\alpha.\tau$ the bound variable α occurs freely in τ.

Intuitively, $\mu\alpha.\tau$ denotes the recursive type defined by the type equation $\alpha = \tau$ (note that α occurs in τ), $\bot$ is contained in all other types, and $\top$ contains all other types. Our results extend to other types and type constructors such as product and sum. We let τ, σ range over recursive types and write $\text{fv}(\tau)$ for the set of free type variables in τ.

1.2 Regular Trees

We define $\text{Tree}(\tau)$ to be the regular (possibly infinite) tree obtained by completely unfolding all occurrences of $\mu\alpha.\tau$ to $\tau[\mu\alpha.\tau]$. (For a precise definition of $\text{Tree}(\tau)$, regular trees and their properties see [Cou83, CC91, AC93].)

Henceforth we shall assume that all trees are over the ranked alphabet $\{\bot^0, \to^2, \top^0\} \cup \{\alpha^0 : \alpha \in \text{TVar}\}$ of *labels*, which are ordered by the reflexive-transitive closure of $\bot < \underset{\alpha}{\to} < \top$.

We can define *depth-k lower and upper approximations* $T|_k$ and $T|^k$ of a tree T as follows:

$$\begin{aligned} T|_0 &= \bot & T|^0 &= \top \\ (T' \to T'')|_{k+1} &= T'|^k \to T''|_k & (T' \to T'')|^{k+1} &= T'|_k \to T''|^k \\ \bot|_{k+1} &= \bot & \bot|^{k+1} &= \bot \\ \top|_{k+1} &= \top & \top|^{k+1} &= \top \\ \alpha|_{k+1} &= \alpha & \alpha|^{k+1} &= \alpha \end{aligned}$$

For tree T, $\mathcal{L}(T)$, the *label* of T, is the label of the root node of T. The label of a recursive type τ is the label of the tree it denotes: $\mathcal{L}(\tau) = \mathcal{L}(\text{Tree}(\tau))$.

1.3 Recursive Type Equality

Cardone and Coppo [CC91] show the following results about recursive type equality (for types without $\top$):

- Interpreting recursive types as ideals in a universal domain, τ, τ' are *semantically equivalent* (denote the same ideal) if and only if $\mathrm{Tree}(\tau) = \mathrm{Tree}(\tau')$.
- *Weak type equality*, the congruence generated by axiom FOLD/UNFOLD in Figure 1, is properly weaker than semantic type equivalence.
- The principal typing property in the sense of Hindley [Hin69] and Ben-Yelles [BY79] extends to simple typing with recursive types if type equality in Rule (EQUAL) is taken to be semantic type equivalence, yet it breaks if it is defined as weak equality.

Let us write $\tau \approx \tau'$ if $\mathrm{Tree}(\tau) = \mathrm{Tree}(\tau')$. Axiomatizations of $\approx$ are given by Amadio/Cardelli [AC91] and Ariola/Klop [AK95]. It is clear, however, that this kind of axiomatization has been known for a long time; see for example Salomaa [Sal66] and Milner [Mil84].

All these axiomatizations are a variant of the inference system presented in Figure 1.[2] (In Rule CONTRACT, recursive type τ is contractive in type variable α if α occurs in τ only under $\rightarrow$, if at all.)

$$\vdash \tau = \tau \qquad \frac{\vdash \tau = \tau' \quad \vdash \tau' = \tau''}{\vdash \tau = \tau''} \qquad \frac{\vdash \tau' = \tau}{\vdash \tau = \tau'}$$

$$\frac{\vdash \tau = \tau' \quad \vdash \sigma = \sigma'}{\vdash \tau \rightarrow \sigma = \tau' \rightarrow \sigma'} \qquad \frac{\vdash \tau = \tau'}{\vdash \mu\alpha.\tau = \mu\alpha.\tau'} \ (\mu\text{-COMPAT})$$

$$\vdash \mu\alpha.\tau = \tau[\mu\alpha.\tau/\alpha] \qquad (\text{FOLD/UNFOLD})$$

$$\frac{\vdash \tau_1 = \tau[\tau_1/\alpha] \quad \vdash \tau_2 = \tau[\tau_2/\alpha]}{\vdash \tau_1 = \tau_2} \ (\tau \text{ contractive in } \alpha) \quad (\text{CONTRACT})$$

Fig. 1. Classical axiomatization of recursive type equality

[2] Instead of Rule CONTRACT Ariola and Klop [AK95] use the equivalent rule

$$\frac{\vdash \tau_1 = \tau[\tau_1/\alpha]}{\mu\alpha.\tau = \tau_1} \quad \tau \text{ contractive in } \alpha.$$

$$\Gamma \vdash \bot \leq \tau \qquad\qquad \Gamma \vdash \tau \leq \top$$

$$\Gamma \vdash \tau \leq \tau \qquad\qquad \frac{\Gamma \vdash \tau \leq \tau' \quad \Gamma \vdash \tau' \leq \tau''}{\Gamma \vdash \tau \leq \tau''}$$

$$\Gamma, \tau \leq \tau', \Gamma' \vdash \tau \leq \tau' \qquad\qquad \frac{\Gamma \vdash \tau = \tau'}{\Gamma \vdash \tau \leq \tau'}$$

$$\frac{\Gamma \vdash \tau' \leq \tau \quad \Gamma \vdash \sigma \leq \sigma'}{\Gamma \vdash \tau \to \sigma \leq \tau' \to \sigma'} \qquad \frac{\Gamma, \alpha \leq \beta \vdash \tau \leq \sigma}{\Gamma \vdash \mu\alpha.\tau \leq \mu\beta.\sigma} \ (\alpha, \beta \text{ not free in } \sigma, \tau)$$

Fig. 2. Amadio/Cardelli axiomatization of subtyping for recursive types

1.4 Recursive Subtyping

Amadio and Cardelli [AC93] extend the standard contravariant structural subtyping relation on μ-free types (to be thought of as finite trees) defined by

$$\bot \leq_{\text{fin}} T \ (\bot_{\text{fin}}) \qquad T \leq_{\text{fin}} \top \ (\top_{\text{fin}})$$

$$T \leq_{\text{fin}} T \ (\text{REF}_{\text{fin}}) \qquad \frac{S_1 \leq_{\text{fin}} T_1 \quad T_2 \leq_{\text{fin}} S_2}{T_1 \to T_2 \leq_{\text{fin}} S_1 \to S_2} \ (\text{ARROW}_{\text{fin}})$$

in a natural fashion to infinite trees.

Definition 2 (Amadio/Cardelli subtype relation). Let τ, σ be recursive types. Define $\tau \leq_{AC} \sigma$ if $\text{Tree}(\tau)|_k \leq_{\text{fin}} \text{Tree}(\sigma)|_k$ for all $k \in \mathbb{N}_0$.

In the definition we could replace the lower approximations by upper approximations since both induce the same subtyping relation:

Proposition 3. *$T|_k \leq_{\text{fin}} T'|_k$ if and only if $T|^k \leq_{\text{fin}} T'|^k$.*

Amadio and Cardelli build on the axiomatization of type equality in Figure 1 and give a sound and complete axiomatization of $\leq_{AC}$ in [AC93], shown in Figure 2.

1.5 The New Axiomatizations

In this paper we show that the Amadio/Cardelli subtype relation can be directly axiomatized by the inference system in Figure 3.

The most noteworthy aspect of the system is rule ARROW/FIX for proving inequality between function types. It can be understood as the composition of the two separate rules

$$\frac{A \vdash \sigma_1 \leq \tau_1 \quad A \vdash \tau_2 \leq \sigma_2}{A \vdash \tau_1 \to \tau_2 \leq \sigma_1 \to \sigma_2} \ (\text{ARROW}) \qquad \frac{A, \tau \leq \tau' \vdash \tau \leq \tau'}{A \vdash \tau \leq \tau'} \ (\text{FIX})$$

$$A \vdash \bot \leq \tau \quad (\bot) \qquad\qquad A \vdash \tau \leq \top \quad (\top)$$

$$A \vdash \tau \leq \tau \quad (\textsc{Ref}) \qquad\qquad \frac{A \vdash \tau \leq \delta \quad A \vdash \delta \leq \sigma}{A \vdash \tau \leq \sigma} \quad (\textsc{Trans})$$

$$A \vdash \mu\alpha.\tau \leq \tau[\mu\alpha.\tau/\alpha] \quad (\textsc{Unfold}) \qquad\qquad A \vdash \tau[\mu\alpha.\tau/\alpha] \leq \mu\alpha.\tau \quad (\textsc{Fold})$$

$$A, \tau \leq \sigma, A' \vdash \tau \leq \sigma \quad (\textsc{Hyp})$$

$$\frac{A, \tau_1 \to \tau_2 \leq \sigma_1 \to \sigma_2 \vdash \sigma_1 \leq \tau_1 \quad A, \tau_1 \to \tau_2 \leq \sigma_1 \to \sigma_2 \vdash \tau_2 \leq \sigma_2}{A \vdash \tau_1 \to \tau_2 \leq \sigma_1 \to \sigma_2} \quad (\textsc{Arrow/Fix})$$

Fig. 3. Coinductive axiomatization of Amadio/Cardelli subtyping

where the premise of obviously "dangerous" Rule FIX must be proved by Rule ARROW. Rule FIX says that we may actually use as a hypothesis what we want to prove when trying to prove it. We are just not allowed to use it "right away"!

The system is is *sound and complete* for Amadio/Cardelli subtyping: $\vdash \tau \leq \tau'$ if and only if $\tau \leq_{AC} \tau'$. Because of Rule ARROW/FIX, more specifically the part corresponding to Rule FIX, soundness is actually a tricky issue. The proof is accomplished by giving sequents a level stratified interpretation. Completeness is shown by exhibiting an algorithm that builds a derivation for given (τ, τ') and succeeds whenever $\tau \leq_{AC} \tau'$. The crucial part here is showing that the algorithm terminates. The algorithm not only *decides* whether $\vdash \tau \leq \tau'$, but also returns an explicit proof. (It is relatively easy to see that the algorithm can be implemented in time $O(n^2)$ where n is the number of symbols in the two input types, but this is not elaborated in this paper. See [Car93, KPS93, KPS95] for efficient algorithms for deciding recursive subtyping.)

Given our "innate" axiomatization of $\leq_{AC}$ by $\leq$ in Figure 3 the semantic type equivalence $\approx$ can now be *defined* in terms of subtyping since $\tau \approx \tau'$ if and only $\tau \leq_{AC} \tau'$ and $\tau' \leq_{AC} \tau$ if and only if $\vdash \tau \leq \tau'$ and $\vdash \tau' \leq \tau$. Alternatively, we can provide a direct axiomatization of $\approx$, see Figure 4. Note that it requires neither Rule CONTRACT nor Rule μ-COMPAT, which are difficult to interpret denotationally and operationally.

The above results for subtyping are presented in Section 2. The corresponding results for type equality are analogous; they are omitted for space reasons.

1.6 Proofs as Programs: Subtyping/Type Equality Proofs as Coercions

The point of our coinductive axiomatizations is not only to support direct coinductive reasoning, but also to provide a natural foundation for a proof theory

$$A, \tau = \tau', A' \vdash \tau = \tau' \qquad A \vdash \tau = \tau$$

$$\frac{A \vdash \tau = \tau'}{A \vdash \tau' = \tau} \qquad \frac{A \vdash \tau = \tau' \quad A \vdash \tau' = \tau''}{A \vdash \tau = \tau''}$$

$$A \vdash \mu\alpha.\tau = \tau[\mu\alpha.\tau/\alpha]$$

$$\frac{A, \tau \to \tau' = \sigma \to \sigma' \vdash \tau = \sigma \quad A, \tau \to \tau' = \sigma \to \sigma' \vdash \tau' = \sigma'}{A \vdash \tau \to \tau' = \sigma \to \sigma'}$$

Fig. 4. Coinductive axiomatization of recursive type equality

and operational interpretation of proofs. In Section 3 we briefly introduce the term language of *coercions* for proofs in our subtyping axiomatization. Each rule corresponds to a natural construction on coercions; in particular, the fixpoint rule corresponds to definition by recursion.

Finally, Section 4 concludes with a brief summary and possible future work.

2 Recursive Types: Subtyping

2.1 Simulations on Recursive Types

We give a characterization of $\leq_{AC}$ that highlights the coinductive nature of $\leq_{AC}$. Its advantages are that it is intrinsically in terms of recursive types, without referring to infinite trees, and it directly reflects the characteristic closure properties of $\leq_{AC}$. It will be used in the proof of completeness for our axiomatization of $\leq_{AC}$.

Definition 4 (Simulation on recursive types). A *simulation (on recursive types)* is a binary relation $\mathcal{R}$ on recursive types satisfying:

(i) $(\tau_1 \to \tau_2)\,\mathcal{R}\,(\sigma_1 \to \sigma_2) \;\Rightarrow\; \sigma_1\,\mathcal{R}\,\tau_1$ and $\tau_2\,\mathcal{R}\,\sigma_2$
(ii) $\mu\alpha.\tau\,\mathcal{R}\,\sigma \;\Rightarrow\; \tau[\mu\alpha.\tau/\alpha]\,\mathcal{R}\,\sigma$
(iii) $\tau\,\mathcal{R}\,\mu\beta.\sigma \;\Rightarrow\; \tau\,\mathcal{R}\,\sigma[\mu\beta.\sigma/\beta]$
(iv) $\tau\,\mathcal{R}\,\sigma \;\Rightarrow\; \mathcal{L}(\tau) \leq \mathcal{L}(\sigma)$

Lemma 5. *$\leq_{AC}$ is a simulation.*

Proof. We prove the four properties of Definition 4.

(i) Assume $\tau_1 \to \tau_2 \leq_{AC} \sigma_1 \to \sigma_2$ and let $k \in \mathbb{N}_0$. By Definition 2 we have $\mathrm{Tree}(\tau_1 \to \tau_2)\,|\,(k+1) \leq_{\mathrm{fin}} \mathrm{Tree}(\sigma_1 \to \sigma_2)\,|\,(k+1)$. By definition of $\mathrm{Tree}(\cdot)$

and of the k'th lower approximation we find

$$\left(\text{Tree}(\tau_1)|^k \to \text{Tree}(\tau_2)|_k\right) \leq_{\text{fin}} \left(\text{Tree}(\sigma_1)|^k \to \text{Tree}(\sigma_2)|_k\right)$$

and thus $\text{Tree}(\sigma_1) \mid^k \leq_{\text{fin}} \text{Tree}(\tau_1) \mid^k$ and $\text{Tree}(\tau_2) \mid_k \leq_{\text{fin}} \text{Tree}(\sigma_2) \mid_k$. Since k was chosen arbitrary and $\text{Tree}(\sigma_1)|^k \leq_{\text{fin}} \text{Tree}(\tau_1)|^k$ if and only if $\text{Tree}(\sigma_1)|_k \leq_{\text{fin}} \text{Tree}(\tau_1)|_k$ (Proposition 3) we finally obtain $\sigma_1 \leq_{\text{AC}} \tau_1$ and $\tau_2 \leq_{\text{AC}} \sigma_2$ as desired.

(*ii*) Consider $\mu\alpha.\tau \leq_{\text{AC}} \sigma$. By definition of Tree($\cdot$) we have $\text{Tree}(\mu\alpha.\tau) = \text{Tree}(\tau[\mu\alpha.\tau/\alpha])$ and thereby $\tau[\mu\alpha.\tau/\alpha] \leq_{\text{AC}} \sigma$.

(*iii*) Exactly as (*ii*).

(*iv*) Let $\tau \leq_{\text{AC}} \sigma$ and thus $\text{Tree}(\tau) \mid_k \leq_{\text{fin}} \text{Tree}(\sigma) \mid_k$ for all $k \in \mathbb{N}_0$. By inspection of $\leq_{\text{fin}}$ we get $\mathcal{L}(\text{Tree}(\tau) \mid_k) \leq \mathcal{L}(\text{Tree}(\sigma) \mid_k)$. For $k > 0$ we obviously have $\mathcal{L}(\text{Tree}(\tau)|_k) = \mathcal{L}(\text{Tree}(\tau)) = \mathcal{L}(\tau)$ and hence $\mathcal{L}(\tau) \leq \mathcal{L}(\sigma)$.

Lemma 6. *If $\mathcal{R}$ is a simulation then $\tau\mathcal{R}\sigma \Rightarrow \tau \leq_{\text{AC}} \sigma$ for all $\tau, \sigma \in \mu Tp$.*

Proof. We prove $\forall k \in \mathbb{N}_0.\, \forall \tau, \sigma \in \mu\text{Tp}.\, (\tau\mathcal{R}\sigma \Rightarrow \text{Tree}(\tau)|_k \leq_{\text{fin}} \text{Tree}(\sigma)|_k)$ by induction on k.

Case $k = 0$: Trivial, since $\bot \leq_{\text{fin}} \bot$.

Case $k > 0$: Let τ, σ be given such that $\tau\mathcal{R}\sigma$. We perform a case analysis on the syntactic forms of τ, σ where $\mathcal{L}(\tau) \leq \mathcal{L}(\sigma)$.

Case $\tau = \bot$ or $\sigma = \top$ or $\tau = \alpha, \sigma = \alpha$: Trivial.

Case $\tau = \mu\alpha.\tau'$, $\sigma = \sigma_1 \to \sigma_2$: τ is canonical so $\tau'[\tau/\alpha] = \tau_1 \to \tau_2$ for some τ_1, τ_2. By Definition 4 (*ii*) we have $(\tau_1 \to \tau_2)\mathcal{R}(\sigma_1 \to \sigma_2)$ and thus by (*i*) $\sigma_1\mathcal{R}\tau_1$, $\tau_2\mathcal{R}\sigma_2$. Our induction hypothesis yields

$$\text{Tree}(\sigma_1)|_{(k-1)} \leq_{\text{fin}} \text{Tree}(\tau_1)|_{(k-1)} \text{ and } \text{Tree}(\tau_2)|_{(k-1)} \leq_{\text{fin}} \text{Tree}(\sigma_2)|_{(k-1)}$$

By Proposition 3 and Rule $\text{ARROW}_{\text{fin}}$ we conclude

$$\text{Tree}(\tau_1)|^{(k-1)} \to \text{Tree}(\tau_2)|_{(k-1)} \leq_{\text{fin}} \text{Tree}(\sigma_1)|^{(k-1)} \to \text{Tree}(\sigma_2)|_{(k-1)}$$

which by definition of Tree(.) and $|_k$ implies

$$\text{Tree}(\tau) = \text{Tree}(\tau_1 \to \tau_2)|_k \leq_{\text{fin}} \text{Tree}(\sigma_1 \to \sigma_2)|_k = \text{Tree}(\sigma).$$

The remaining cases follow the same schema as the previous one, since they all have $\to$ labels.

Theorem 7 (Characterization of Amadio/Cardelli subtyping). $\tau \leq_{\text{AC}} \sigma$ *if and only if there exists a simulation $\mathcal{R}$ such that $\tau\mathcal{R}\sigma$.*

Proof. Follows by Lemma 5 and Lemma 6.

2.2 Soundness

We might want to interpret a sequent $\sigma_{11} \leq \sigma_{11}, \ldots, \sigma_{n1} \leq \sigma_{n2} \vdash \tau \leq \tau'$ conventionally as "if $\sigma_{11} \leq_{AC} \sigma_{11}, \ldots, \sigma_{n1} \leq_{AC} \sigma_{n2}$ then $\tau \leq_{AC} \tau'$" and prove every inference rule sound under this interpretation.

The problem is that Rule ARROW/FIX — more specifically the part that corresponds to Rule FIX — is ***unsound*** under this interpretation! To see this, consider for example $(\bot \to \top) \leq (\top \to \bot) \vdash \top \leq \bot$. Since $\bot \to \top \not\leq_{AC} \top \to \bot$ it is vacuously valid under the conventional interpretation. Application of Rule ARROW/FIX lets us deduce $\vdash \bot \to \top \leq \top \to \bot$, which is, however, *not* valid.

This does not mean that our inference system is unsound. The problem is that the interpretation of sequents is *too strong* (in the sense of "too many sequents are valid"): the premise $(\bot \to \top) \leq (\top \to \bot) \vdash \top \leq \bot$, which is obviously not derivable anyway, should not be valid. As suggested by Martín Abadi [Aba96] we give sequents a *level stratified* interpretation, under which all inference rules are sound.

Definition 8 (Stratified sequent interpretation). Let k range over the non-negative integers. Define:

1. $\models_k \tau \leq \tau'$ if $\text{Tree}(\tau)|_k \leq \text{Tree}(\tau')|_k$.
2. $\models_k A$ if $\models_k \tau \leq \tau'$ for all $\tau \leq \tau' \in A$.
3. $A \models_k \tau \leq \tau'$ if $\models_k A$ implies $\models_k \tau \leq \tau'$.
4. $A \models \tau \leq \tau'$ if $A \models_k \tau \leq \tau'$ for all $k \in \mathbb{N}_0$.

Note that $\bot \to \top \leq \top \to \bot \vdash \top \leq \bot$ does *not* hold under this interpretation; that is, $(\bot \to \top) \leq (\top \to \bot) \not\models \top \leq \bot$. To wit, we have $(\bot \to \top) \leq (\top \to \bot) \models_0 \top \leq \bot$ and $(\bot \to \top) \leq (\top \to \bot) \models_k \top \leq \bot$ for all $k \geq 2$, but $(\bot \to \top) \leq (\top \to \bot) \not\models_1 \top \leq \bot$ since $\text{Tree}(\bot \to \top)|_1 = \top \to \bot = \text{Tree}(\top \to \bot)|_1$, yet $\text{Tree}(\top)|_1 = \top \not\leq \bot = \text{Tree}(\bot)|_1$. Intuitively, a sequent $A \vdash \tau \leq \tau'$ that holds vacuously (because the assumptions are false) under the conventional interpretation holds under the stratified interpretation only if $\tau \leq \tau'$ is not wrong "earlier" than an assumption in A when descending into the trees in A and $\tau \leq \tau'$ in lockstep.

Lemma 9 (Soundness of inference rules). *If $A \vdash \tau \leq \tau'$ then $A \models \tau \leq \tau'$.*

Proof. The proof is by rule induction on the inference rules in Figure 3. For all rules but ARROW/FIX it is easy to prove $A \models_k \tau \leq \tau'$ for arbitrary k and then generalize over k. For Rule ARROW/FIX we require induction on k.

Recall Rule ARROW/FIX:

$$\frac{A, \tau_1 \to \tau_2 \leq \sigma_1 \to \sigma_2 \vdash \sigma_1 \leq \tau_1 \quad A, \tau_1 \to \tau_2 \leq \sigma_1 \to \sigma_2 \vdash \tau_2 \leq \sigma_2}{A \vdash \tau_1 \to \tau_2 \leq \sigma_1 \to \sigma_2}$$

Our major induction hypothesis IH1 is $A, \tau_1 \to \tau_2 \leq \sigma_1 \to \sigma_2 \models \sigma_1 \leq \tau_1$ and $A, \tau_1 \to \tau_2 \leq \sigma_1 \to \sigma_2 \models \tau_2 \leq \sigma_2$. We now prove $\forall k \in \mathbb{N}_0.\ A \models_k \tau_1 \to \tau_2 \leq \sigma_1 \to \sigma_2$ by induction on k.

Base case: $k = 0$. Trivial since $\mathrm{Tree}(\tau)|_0 = \bot$ for all τ.

Inductive case: $k > 0$. Assume $A \models_{(k-1)} \tau_1 \to \tau_2 \leq \sigma_1 \to \sigma_2$ (minor induction hypothesis IH2). We need to show $A \models_k \tau_1 \to \tau_2 \leq \sigma_1 \to \sigma_2$; that is, $\models_k A$ implies $\models_k \tau_1 \to \tau_2 \leq \sigma_1 \to \sigma_2$.
Assume $\models_k A$. This implies that $\models_{(k-1)} A$. Since $A \models_{(k-1)} \tau_1 \to \tau_2 \leq \sigma_1 \to \sigma_2$ by IH2 we obtain $\models_{(k-1)} \tau_1 \to \tau_2 \leq \sigma_1 \to \sigma_2$ and thus $\models_{(k-1)} A, \tau_1 \to \tau_2 \leq \sigma_1 \to \sigma_2$. Invoking IH1 we derive $\models_{(k-1)} \sigma_1 \leq \tau_1$ and $\models_{(k-1)} \tau_2 \leq \sigma_2$, which together are equivalent to $\models_k \tau_1 \to \tau_2 \leq \sigma_1 \to \sigma_2$, and we are done.

Theorem 10 (Soundness). *If* $\vdash \tau \leq \tau'$ *then* $\tau \leq_{\mathrm{AC}} \tau'$.

Proof. Follows immediately from Lemma 9 and the observation that $\models \tau \leq \tau'$ if and only if $\tau \leq_{\mathrm{AC}} \tau'$.

2.3 Completeness

This section is concerned with the completeness of the inference system in Figure 3 with respect to $\leq_{\mathrm{AC}}$. The proof is divided into three parts; 1) an algorithm **S** that produces derivations, 2) a termination proof and finally 3) a correctness proof for **S**.

Algorithm S Consider Algorithm **S** in Figure 5. The first clause in **S** that matches a particular argument tuple is executed. The only cases requiring remarks are those concerning function types. A pair of function types may have been encountered earlier in the computation and is therefore stored in the assumption set. If that is the case, rule Hyp is applied and otherwise rule Arrow/Fix. It is of vital importance that assumptions are checked before applying the Arrow/Fix rule, since otherwise we would never be able to use them.

Termination of S

Syntactic subterms We first introduce the concept of syntactic subterms and prove a crucial property about them: every recursive type has only a finite number of syntactic subterms. This is not entirely obvious since recursive types may have syntactic subterms that are larger than themselves. We require a number of preliminary technical results.

Definition 11. A recursive type τ' is a *syntactic subterm* (or just *subterm*) of τ if $\tau' \sqsubseteq \tau$, where $\sqsubseteq$ is defined by the following rules:

$$\tau \sqsubseteq \tau \quad (\textsc{Ref}) \qquad\qquad \frac{\tau \sqsubseteq \sigma[\mu\alpha.\sigma/\alpha]}{\tau \sqsubseteq \mu\alpha.\sigma} \quad (\textsc{Unfold})$$

$$\frac{\tau \sqsubseteq \sigma_1}{\tau \sqsubseteq \sigma_1 \to \sigma_2} \quad (\textsc{Arrow}_L) \qquad\qquad \frac{\tau \sqsubseteq \sigma_2}{\tau \sqsubseteq \sigma_1 \to \sigma_2} \quad (\textsc{Arrow}_R)$$

```
S(A, μα.τ, σ) =
    let
        D1 = UNFOLD
        D2 = S(A, τ[μα.τ/α], σ)
    in
        TRANS(D1, D2)
    end
S(A, τ, μβ.σ) =
    let
        D1 = S(A, τ, σ[μβ.σ/β])
        D2 = FOLD
    in
        TRANS(D1, D2)
    end
S((A, τ ≤ σ, A'), τ, σ) = HYP
S(A, τ1 → τ2, σ1 → σ2) =
    let
        A' = A ∪ {τ1 → τ2 ≤ σ1 → σ2}
        D1 = S(A', σ1, τ1)
        D2 = S(A', τ2, σ2)
    in
        ARROW/FIX(D1, D2)
    end
S(A, α, α) = REF
S(A, ⊥, τ) = ⊥
S(A, τ, ⊤) = ⊤
S(A, τ, σ) = exception
```

Fig. 5. Algorithm S

Lemma 12. *The subterm relation is transitive, i.e. if* $\tau \sqsubseteq \delta$, $\delta \sqsubseteq \sigma$ *then* $\tau \sqsubseteq \sigma$.

Proof. Induction on the derivation of $\delta \sqsubseteq \sigma$.

We define a subterm closure operation on recursive types. As we shall see the subterm closure contains all subterms of a recursive type.

Definition 13. The *subterm closure* τ^* of τ is the set of recursive types defined by

$$\begin{aligned} \bot^* &= \{\bot\} & (\tau_1 \to \tau_2)^* &= \{\tau_1 \to \tau_2\} \cup \tau_1^* \cup \tau_2^* \\ \top^* &= \{\top\} & (\mu\alpha.\tau_1)^* &= \{\mu\alpha.\tau_1\} \cup \tau_1^*[\mu\alpha.\tau_1/\alpha] \\ \alpha^* &= \{\alpha\} \end{aligned}$$

Obviously, the subterm closure is finite; indeed $|\tau^*| = O(|\tau|)$.

Proposition 14. $|\tau^*| < \infty$.

An important technical property of the closure operation is its commutation with substitution:

Lemma 15. $(\tau'[\tau/\beta])^* = (\tau')^*[\tau/\beta] \cup \tau^*$ *if* $\beta \in fv(\tau')$.

Proof. Induction on the structure of τ'.

Using this property we can show that τ^* contains all syntactic subterms of τ:

Lemma 16. *If* $\tau \sqsubseteq \sigma$ *then* $\tau \in \sigma^*$.

Proof. Induction on the derivation of $\tau \sqsubseteq \sigma$. The only interesting case is UNFOLD.

Case UNFOLD : So $\sigma = \mu\alpha.\tau'$ and $\tau \sqsubseteq \tau'[\mu\alpha.\tau'/\alpha]$. By IH we get $\tau \in (\tau'[\mu\alpha.\tau'/\alpha])^*$. By Lemma 15 substitution and closure commute and we can conclude

$$\tau \in (\tau')^*[\mu\alpha.\tau'/\alpha] \cup (\mu\alpha.\tau')^* = (\mu\alpha.\tau')^*$$

since $(\tau')^*[\mu\alpha.\tau'/\alpha] \subseteq (\mu\alpha.\tau')^*$ by definition of $(\mu\alpha.\tau')^*$.

Lemma 16 and Proposition 14 together finally give us the desired property:

Theorem 17. *For recursive type τ the set $\{\tau' \mid \tau' \sqsubseteq \tau\}$ is finite.*

Algorithm execution We now study the computations performed by **S**. The main result is that all recursive types encountered in calls to **S** during the computation are syntactic subterms of the initial recursive types. Combined with Theorem 17 we can prove that **S** terminates. To reason about the steps performed by **S** we define the notions of call tree and call path.

Definition 18. The *call tree* of $\mathbf{S}(A_0, \tau_0, \sigma_0)$ is defined to be a root node labeled $\mathbf{S}(A_0, \tau_0, \sigma_0)$ whose subtrees are the call trees of all the recursive calls $\mathbf{S}(A_i, \tau_i, \sigma_i)$ (finitely many) occurring in the first clause in **S** that matches $\mathbf{S}(A_0, \tau_0, \sigma_0)$.

A *call path* in $\mathbf{S}(A_0, \tau_0, \sigma_0)$ is a path in the call tree of $\mathbf{S}(A_0, \tau_0, \sigma_0)$, starting at its root.

Theorem 19. *Let τ_0, σ_0 be recursive types and A_0 an assumption set. For all nodes $\mathbf{S}(A_i, \tau_i, \sigma_i)$ in the call tree of $\mathbf{S}(A_0, \tau_0, \sigma_0)$ we have that τ_i, σ_i are syntactic subterms of either τ_0 or σ_0.*

Proof. Induction on the depth d of nodes.

Case $d = 0$: Root node $\mathbf{S}(A_0, \tau_0, \sigma_0)$. Trivial from reflexivity of $\sqsubseteq$.

Case $d > 0$: Case analysis of nodes at depth $d - 1$.

Case $\mathbf{S}(A, \mu\alpha.\tau, \sigma)$: The unique child of $\mathbf{S}(A, \mu\alpha.\tau, \sigma)$ at depth d is then $\mathbf{S}(A, \tau[\mu\alpha.\tau/\alpha], \sigma)$. By induction hypothesis we know that $\mu\alpha.\tau$ and σ are in canonical form and furthermore that

$$(\mu\alpha.\tau \sqsubseteq \tau_0 \land \sigma \sqsubseteq \sigma_0) \quad \text{or} \quad (\mu\alpha.\tau \sqsubseteq \sigma_0 \land \sigma_i \sqsubseteq \tau_0)$$

It is easily seen that $\tau[\mu\alpha.\tau/\alpha]$ is in canonical form. Assume that $\mu\alpha.\tau \sqsubseteq \tau_0$ (second case similar). By Rule UNFOLD we have $\tau[\mu\alpha.\tau/\alpha] \sqsubseteq \mu\alpha.\tau$ and thus by transitivity (Lemma 12) $\tau[\mu\alpha.\tau/\alpha] \sqsubseteq \tau_0$.

Case $\mathbf{S}(A, \tau, \mu\beta.\sigma)$: Analogous to the above case.

Case $\mathbf{S}(A, \tau_1 \to \tau_2, \sigma_1 \to \sigma_2)$: There are two child nodes at depth d:

1. $\mathbf{S}(A', \sigma_1, \tau_1)$. By IH we know that $\tau_1 \to \tau_2 \sqsubseteq \tau_0$ and $\sigma_1 \to \sigma_2 \sqsubseteq \sigma_0$, or $\tau_1 \to \tau_2 \sqsubseteq \sigma_0$ and $\sigma_1 \to \sigma_2 \sqsubseteq \tau_0$. But then the result follows directly from transitivity (Lemma 12) since $\tau_1 \sqsubseteq \tau_1 \to \tau_2$ and $\sigma_1 \sqsubseteq \sigma_1 \to \sigma_2$ by Rule ARROW$_L$.

2. $\mathbf{S}(A', \tau_2, \sigma_2)$ Exactly as previous case.

Lemma 20. *If* $\mathbf{S}(A_0, \tau_0, \sigma_0), \ldots, \mathbf{S}(A_i, \tau_i, \sigma_i), \ldots$ *is a call path of* $\mathbf{S}(A_0, \tau_0, \sigma_0)$ *then* $A_0 \subseteq A_1 \subseteq \ldots \subseteq A_i \subseteq \ldots$

Proof. By inspection of Algorithm **S** we see that, for every clause $\mathbf{S}(A, \tau, \sigma)$ and every recursive call $\mathbf{S}(A', \tau', \sigma')$ occurring in it, we have $A \subseteq A'$.

Lemma 21. *If* $\mathbf{S}(A_0, \tau_0, \sigma_0), \ldots, \mathbf{S}(A_n, \tau_n, \sigma_n), \ldots$ *is a call path of* $\mathbf{S}(A_0, \tau_0, \sigma_0)$ *then:* $\exists N \forall i : (\tau_i, \sigma_i) \in \left\{(\tau_j, \sigma_j) \mid 0 \leq j \leq N\right\}$.

The lemma states that every path has only finitely many different type arguments.

Proof. The statement is proved by contradiction. Assume that

$$\forall N \exists i : (\tau_i, \sigma_i) \notin \left\{(\tau_j, \sigma_j) \mid 0 \leq j \leq N\right\}$$

This fact directly implies that $\{(\tau_j, \sigma_j) \mid j \in \mathbb{N}_0\}$ is an infinite set. Theorem 19 states that all terms in a call tree are subterms of the initial two terms. We thus have

$$\{(\tau_j, \sigma_j) \mid j \in N_0\} \subseteq (\{\tau_j \mid j \in N_0\}) \times (\{\sigma_j \mid j \in \mathbb{N}_0\}) \subseteq \{\tau \mid \tau \sqsubseteq \tau_0\} \times \{\sigma \mid \sigma \sqsubseteq \sigma_0\}$$

Theorem 17, however, implies that

$$\left|\{(\tau_j, \sigma_j) \mid j \in \mathbb{N}_0\}\right| \leq \left|\{\tau \mid \tau \sqsubseteq \tau_0\} \cup \{\sigma \mid \sigma \sqsubseteq \sigma_0\}\right|^2 < \infty$$

which contradicts our assumption that $\{(\tau_j, \sigma_j) \mid j \in \mathbb{N}_0\}$ is infinite.

The above results enable us to prove termination of **S**.

Theorem 22 (Termination of S). *If* τ, σ *are canonical and* A *an assumption set then* $\mathbf{S}(A, \tau, \sigma)$ *terminates.*

Proof. The proof is once again by contradiction. Assume that $\mathbf{S}(A, \tau, \sigma)$ does not terminate, *i.e.* there exists an infinite call path p in the call tree of $\mathbf{S}(A, \tau, \sigma)$. Let N be determined by Lemma 21 such that

$$\forall i : (\tau_i, \sigma_i) \in \left\{(\tau_j, \sigma_j) \mid 0 \leq j \leq N\right\} \quad (1)$$

Let us consider the calls (τ_i, σ_i) of p where $i > N$. There must exist a call (τ_n, σ_n) with $n > N$ where $\tau_n = \tau_1 \rightarrow \tau_2$ and $\sigma_n = \sigma_1 \rightarrow \sigma_2$, because otherwise all calls would be unfoldings, which is not possible since the terms are in canonical form (Theorem 19). From (1) we conclude that $(\tau_1 \rightarrow \tau_2, \sigma_1 \rightarrow \sigma_2) \in \{(\tau_j, \sigma_j) \mid 0 \leq j \leq N\}$ which implies that there exists $m \leq N < n$ such that $(\tau_n, \sigma_n) = (\tau_m, \sigma_m)$. The assumption set associated with call n inherits all assumptions from its ancestors in p (by Lemma 20), but then call n must be an application of Hyp, which corresponds to a leaf in the call tree. Path p is therefore not infinite and the assumption is false.

Correctness of S Finally we show that, whenever $\tau \leq_{AC} \sigma$, $\mathbf{S}(A, \tau, \sigma)$ does not fail (it does not raise an exception) and it returns a proof of $A \vdash \tau \leq \sigma$.

Lemma 23. *Let τ, σ be recursive types in canonical form and A an assumption set. If $\tau \leq_{AC} \sigma$ then $\mathbf{S}(A, \tau, \sigma)$ returns a derivation of $A \vdash \tau \leq \sigma$.*

Proof. The termination theorem (Theorem 22) gives that $\mathbf{S}(A, \tau, \sigma)$ terminates with, say, n recursive calls. Correctness is proved by induction on n. In each case we verify the derivation returned by **S**.

Case $n = 0$ **:** No recursive calls performed at all. Case analysis on the clauses in **S** with no recursive calls.

Case $\mathbf{S}((A, \tau \leq \sigma, A'), \tau, \sigma)$, $\mathbf{S}(A, \alpha, \alpha)$, $\mathbf{S}(A, \bot, \tau)$ **or** $\mathbf{S}(A, \tau, \top)$**:** Obvious.

Case $\mathbf{S}(A, \tau, \sigma)$ **:** If this clause is reached, it means that $\mathcal{L}(\tau) \neq \bot$, $\mathcal{L}(\sigma) \neq \top$ and $\mathcal{L}(\tau) \neq \mathcal{L}(\sigma)$, *i.e.* $\mathcal{L}(\tau) \not\leq \mathcal{L}(\sigma)$. Since $\leq_{AC}$ is a simulation and $\tau \leq_{AC} \sigma$ it must hold that $\mathcal{L}(\tau) \leq \mathcal{L}(\sigma)$, which contradicts $\mathcal{L}(\tau) \not\leq \mathcal{L}(\sigma)$. Thus this clause is never reached!

Case $n > 0$ **:** Induction hypothesis: Computations $\mathbf{S}(A', \tau', \sigma')$ with fewer than n recursive calls, where $\tau' \leq_{AC} \sigma'$, produces a correct derivation of $A' \vdash \tau' \leq \sigma'$. Case analysis of rules containing recursive calls.

Case $\mathbf{S}(A, \mu\alpha.\tau, \sigma)$**:** Since $\leq_{AC}$ is a simulation (Lemma 5) it follows that $\tau[\mu\alpha.\tau/\alpha] \leq_{AC} \sigma$. The induction hypothesis does thus apply and gives that $\mathbf{S}(A, \tau[\mu\alpha.\tau/\alpha], \sigma)$ returns a derivation of $A \vdash \tau[\mu\alpha.\tau/\alpha] \leq \sigma$. By Unfold and Trans we then get a proof of $A \vdash \mu\alpha.\tau \leq \sigma$, which is exactly what $\mathbf{S}(A, \mu\alpha.\tau, \sigma)$ returns.

Case $\mathbf{S}(A, \tau, \mu\beta.\sigma)$**:** Since $\leq_{AC}$ is a simulation (Lemma 5) we get $\tau \leq_{AC} \sigma[\mu\beta.\sigma/\beta]$. Thus the induction hypothesis is applicable: $\mathbf{S}(A, \tau, \sigma[\mu\beta.\sigma/\beta])$ returns a proof of $A \vdash \tau \leq \sigma[\mu\beta.\sigma/\beta]$. We conclude

$$\dfrac{\dfrac{(\mathrm{IH})}{A \vdash \tau \leq \sigma[\mu\beta.\sigma/\beta]} \qquad \dfrac{(\textsc{Fold})}{A \vdash \sigma[\mu\beta.\sigma/\beta] \leq \mu\beta.\sigma}}{A \vdash \tau \leq \mu\beta.\sigma}\,(\textsc{Trans})$$

which is the result of $\mathbf{S}(A, \tau, \mu\beta.\sigma)$.

Case $\mathbf{S}(A, \tau_1 \to \tau_2, \sigma_1 \to \sigma_2)$**:** Let $A' = A \cup \{\tau_1 \to \tau_2 \leq \sigma_1 \to \sigma_2\}$. Two recursive calls are issued from this rule.

1. $\mathbf{S}(A', \sigma_1, \tau_1)$. From the simulation property of $\leq_{AC}$ and IH we get that the call returns a proof of $A' \vdash \sigma_1 \leq \tau_1$.
2. $\mathbf{S}(A', \tau_2, \sigma_2)$. As above, the call returns a proof of $A' \vdash \tau_2 \leq \sigma_2$.

$\mathbf{S}(A, \tau_1 \to \tau_2, \sigma_1 \to \sigma_2)$ returns Rule Arrow/Fix applied to the two subproofs above, which is a proof of $A \vdash \tau_1 \to \tau_2 \leq \sigma_1 \to \sigma_2$.

Theorem 24 (Completeness). *If $\tau \leq_{AC} \sigma$ then $\vdash \tau \leq \sigma$*

Proof. Follows from Lemma 23 for $A = \emptyset$.

3 Proofs as Coercions

In this section we present a somewhat generalized axiomatization of recursive subtyping in which rules ARROW and FIX are separated instead of being melded into a single rule as in Figure 3. This is made possible by using an explicit term representation of proofs as *coercions*. Sound application of Rule FIX is then guaranteed by requiring the coercion in the premise to be formally *contractive* in a sense to be defined. The coercion constructions for the rules in the axiomatization can be interpreted as natural functional programming constructs. Notably, the fixpoint rule corresponds to definition by recursion. Coercions can then be used as a basis for *proof theory* as well as *operational interpretation* of proofs in the sense of the Curry-Howard isomorphism.

The latter is important where coercions are not only (constructive) evidence of some subsumption relation, but have semantic significance; that is, they denote functions that map an element of one type to an element of its supertype. Furthermore, coercions may have *operational significance*; that is, different proofs of the *same* subtyping statement may yield coercions with different operational characteristics. For example, folding and unfolding to and from a recursive type (corresponding to the axioms $A \vdash \tau[\mu\alpha.\tau/\alpha] \leq \mu\alpha.\tau$ and $A \vdash \mu\alpha.\tau \leq \tau[\mu\alpha.\tau/\alpha]$, respectively) may operationally require execution of a referencing (heap allocation) and (pointer) dereferencing step, respectively. In this case it is important to replace their composition by the identity coercion (corresponding to the axiom of reflexivity), since the latter is obviously operationally more efficient than the former.

In a separate paper we shall explore the semantics and operational interpretation of coercions. Here we shall only briefly describe their operational interpretation in order to demonstrate, in intuitive and nontechnical terms, how each rule corresponds to a natural program construct.

3.1 Coercions and their functional interpretation

Coercions are defined by the grammar

$$C := \iota_\tau \mid f \mid \mathbf{fix}\, f : \tau \leq \tau'.c \mid c; c \mid c \to c \mid \mathrm{fold}_{\mu\alpha.\tau} \mid \mathrm{unfold}_{\mu\alpha.\tau} \mid \mathrm{abort}_\tau \mid \mathrm{discard}_\tau$$

Each coercion can be interpreted as a function:

- ι_τ denotes the identity on type τ.
- $\mathbf{fix}\, f : \tau \leq \tau'.c$ denotes the function f recursively defined by the equation $f = c$ (note that f may occur in c).
- $c; c'$ denotes the composition of c' with c.
- $c \to c'$ denotes the functional F defined by $F f x = c'(f(cx))$.
- The pair $\mathrm{fold}_{\mu\alpha.\tau}$ and $\mathrm{unfold}_{\mu\alpha.\tau}$ denotes the isomorphism between $\tau[\mu\alpha.\tau/\alpha]$ and $\mu\alpha.\tau$.
- abort_τ maps any argument to $\bot$; that is, operationally it enters an infinite loop.
- $\mathrm{discard}_\tau$ discards its argument and returns (), the only defined element of type $\top$.

3.2 Well-typed coercions

Definition 25 (Contractiveness). A coercion c is *(formally) contractive* in coercion variable f if: f does not occur in c; or $c \equiv c_1 \to c_2$; or $c \equiv c_1; c_2$ and both c_1 and c_2 are contractive in f; or $c \equiv \mathbf{fix}\, g : \tau \leq \sigma.\, c_1$ and c_1 is contractive in f.

Definition 26 (Well-typed coercions, canonical coercions). A coercion c is *well-typed* if $E \vdash c : \tau \leq \sigma$ is derivable for some E, τ, σ in the inference system of Figure 6.

A well-typed coercion c is *canonical* if every **fix**-coercion occurring in c has the form $\mathbf{fix}\, f : \tau' \leq \sigma'.\, c_1 \to c_2$.

$$E \vdash \iota_\tau : \tau \leq \tau$$

$$E \vdash \mathrm{abort}_\tau : \bot \leq \tau \qquad\qquad E \vdash \mathrm{discard}_\tau : \tau \leq \top$$

$$E \vdash \mathrm{unfold}_{\mu\alpha.\tau} : \mu\alpha.\tau \leq \tau[\mu\alpha.\tau/\alpha] \qquad E \vdash \mathrm{fold}_{\mu\alpha.\tau} : \tau[\mu\alpha.\tau/\alpha] \leq \mu\alpha.\tau$$

$$\frac{E \vdash c : \tau \leq \delta \quad E \vdash d : \delta \leq \sigma}{E \vdash c; d : \tau \leq \sigma} \qquad \frac{E \vdash c : \tau \leq \tau' \quad , \quad d : \sigma \leq \sigma'}{E \vdash (c \to d) : (\tau' \to \sigma) \leq (\tau \to \sigma')}$$

$$E, f : \tau \leq \sigma, E' \vdash f : \tau \leq \sigma \qquad \frac{E, f : \tau \leq \sigma \vdash c_f : \tau \leq \sigma \quad c_f \text{ contr. in } f}{E \vdash \mathbf{fix}\, f : \tau \leq \sigma.\, c_f : \tau \leq \sigma}$$

Fig. 6. Coercion typing rules

Thinking of coercions as a special class of functions, $\leq$ can be understood as a special (coercion) type constructor in Figure 6. Assumptions in E are of the form $f : \tau \leq \sigma$ where coercion variable f occurs at most once in E. Note that the subscripting of coercions guarantees that there is exactly one proof for every derivable $E \vdash c : \tau \leq \sigma$. Let us write $\bar{E}$ for the subtyping assumptions we get from E by erasing all coercion variables in it.

Well-typed coercions are a term interpretation of the axiomatization in Figure 3 in the sense that for every E and every proof of $\bar{E} \vdash \tau \leq \tau'$ there exists a unique canonical coercion c such that $E \vdash c : \tau \leq \tau'$, where every coercion of the form $c_1 \to c_2$ is the body of some **fix**-coercion. Conversely, it is easily seen that every canonical coercion of this form corresponds to a proof using the inference rules of Figure 3.

Theorem 27. $\vdash \tau \leq \tau'$ *if and only if there exists a canonical coercion* c *such that* $\vdash c : \tau \leq \tau'$.

Proof. "Only if" is obvious. Let $\vdash c : \tau \leq \tau'$. "If" follows from the observation that every coercion occurrence of the form $c_1 \rightarrow c_2$ can be "wrapped" with $\mathbf{fix}\, f : \sigma \leq \sigma'$ (f fresh) for suitable recursive types σ, σ' if it is not already the body of a **fix**-coercion. Once this is done, the transformed coercion corresponds directly to a derivation in Figure 3.

This theorem holds not only for *canonical* coercions, but also for the larger class of *well-typed* coercions; that is, well-typed coercions give more *proofs*, but not more *theorems* than canonical coercions. Since this requires a rather lengthy and involved proof, however, we omit it here.

4 Conclusion

4.1 Summary

We have given sound and complete axiomatizations of type equality and type containment using a novel fixpoint rule, which represents a coinduction principle. We have argued that this gives rise to a natural interpretation of proofs as coercions where the fixpoint rule corresponds to definition by recursion.

4.2 Future work

Proof theory, semantics and operational interpretation of coercions In continuation of the work reported here we have formulated an equational theory of coercions that is complete in the sense that two coercions are provably equal if and only if they have identical type signatures. Interestingly, this theory is coinductive, too, as it is based on the fixpoint rule, though for coercion equalities instead of type equalities or subtypings. We can show that the equational theory is verified in a number of functional (cpo-based) interpretations of coercions. This shows that, extensionally, any two coercions with the same type signature are equivalent. Conversely, the equational theory codifies the requirements on a semantics of coercions if we demand that any two coercions be extensionally equivalent. The equational theory can be used as a starting point for optimization of coercions by rewriting: even though coercions of equal type signature are extensionally equivalent, they are not necessarily equally *efficient*!

On the theoretical side, we would like to extend the equational theory for coercions to the typed lambda calculus with embedded coercions in the style of [BTCGS91, CG90, Hen94, Reh95] in order to obtain a general coherence characterization for simply typed lambda-calculus with recursive subtyping. This should give another approach to comparing the semantics of FPC under type equality on the one hand and under type isomorphism on the other hand [AF96]. Furthermore, the interrelation of our fixpoint rule and the co-induction principle of Pitts [Pit94] needs to be illuminated; see also [Pit96].

On the more practical side, coercion reduction by rewriting appears to be useful in representation optimization (such as *boxing* analysis [Jør95]) for recursively defined types; this may, however, require admitting more powerful transformations to be useful, for example the isomorphism $S \times (T + U) \approx (S \times T) + (S \times U)$.

Coinduction principles in other formal systems Observational congruence [Mor68] of programs is intuitively a coinductive notion since it states that, if it is impossible to provide finitary evidence that two expressions behave differently, then they are observationally congruent. Indeed for many programming languages observational congruence can be characterized by a notion of bisimulation. (Since the literature on this topic is voluminous we make no attempt at completeness; see e.g. [Mil77, Abr90].) It is thus not surprising that coinduction principles play an important role in proving program properties and in particular equivalences [MT91, Gor95, HL95, Len96]. Relatively little, however, seems to have been done on incorporating coinduction principles in formal proof systems. Coquand [Coq93] formulates a *guarded induction principle* for reasoning about infinite objects within Type Theory, to which our fixpoint rule and its contractiveness requirement is a close pendant.

We are interested in applying our coinduction principle (fixpoint rule) to other coinductive notions such as program equivalence. For example, we hope to formulate a λ-theory that is strong enough to capture regular Böhm tree equality. More abstractly, it seems to be possible to characterize when coinductively defined relations can be completely axiomatized using the fixpoint rule.

Extensions to richer type languages and systems We have studied recursive types and subtyping within a type language of simple types. It would be interesting to extend this study to richer type disciplines with polymorphism (predicative and impredicative), intersection types or object typing.

Acknowledgments

We would like to thank Martín Abadi, Luca Cardelli, Andrew Gordon, Furio Honsell, Jakob Rehof, Simona Ronchi della Rocca, Dave Sands, Mads Tofte and the anonymous referees for helpful discussions, feedback, corrections, and pointers to related or interesting literature. Martín Abadi's help has been particularly valuable in several respects: he came up with the notion of a level stratified interpretation of judgements which is the basis of the soundness proof presented in this paper, replacing our original soundness proof; he found a number of mistakes in our submission; and he has provided interesting pointers to relevant topics and literature. Dave Sands pointed out the coinductive nature of recursive type equality in the early stages of this work and provided valuable help during a number of discussions over a period of two years.

References

[Aba96] Martín Abadi. Personal communication, September 1996. ACM State of the Art Summer School on Functional and Object-Oriented Programming in Sobotka, Poland, September 8-14, 1996.

[Abr90] Samson Abramsky. The lazy lambda calculus. In D. Turner, editor, *Research Topics in Functional Programming*, pages 65–117. Addison-Wesley, 1990. Also available by anonymous ftp from theory.doc.ic.ac.uk.

[AC91] R. Amadio and L. Cardelli. Subtyping recursive types. In *Proc. 18th Annual ACM Symposium on Principles of Programming Languages (POPL), Orlando, Florida*, pages 104–118. ACM Press, January 1991.

[AC93] Roberto M. Amadio and Luca Cardelli. Subtyping recursive types. *ACM Transactions on Programming Languages and Systems (TOPLAS)*, 15(4):575–631, September 1993.

[AF96] Martín Abadi and Marcelo P. Fiore. Syntactic considerations on recursive types. In *Proc. 1996 IEEE 11th Annual Symp. on Logic in Computer Science (LICS), New Brunswick, New Jersey*. IEEE Computer Society Press, June 1996.

[AK95] Zena M. Ariola and Jan Willem Klop. Equational term graph rewriting. Technical report, University of Oregon, 1995. To appear in Acta Informatica.

[BTCGS91] V. Breazu-Tannen, T. Coquand, C. Gunter, and A. Scedrov. Inheritance as implicit coercion. *Information and Computation*, 93(1):172–221, July 1991. Presented at LICS '89.

[BY79] Ch. Ben-Yelles. Type assignment in the lambda-calculus: Syntax and semantics. Technical report, Department of Pure Mathematics, University College of Swansea, September 1979. Author's current address: Univérsite des Sciènces et dé la Technologie Houari Boumediene, Institut D'Informatique, El-Alia B.P. No. 32, Alger, Algeria.

[Car93] Luca Cardelli. Algorithm for subtyping recursive types (in Modula-3). http://www.research.digital.com/SRC/personal/Luca_Cardelli/Notes/RecSub.txt, 1993. Originally implemented in Quest and released in 1990.

[CC91] F. Cardone and M. Coppo. Type inference with recursive types: Syntax and semantics. *Information and Computation*, 92(1):48–80, May 1991.

[CG90] P. Curien and G. Ghelli. Coherence of subsumption. In A. Arnold, editor, *Proc. 15th Coll. on Trees in Algebra and Programming, Copenhagen, Denmark*, pages 132–146. Springer, May 1990.

[Coq93] Thierry Coquand. Infinite objects in type theory. In *Proc. Int'l Workshop on Types for Proofs and Programs*, Lecture Notes in Computer Science (LNCS), pages 62–78. Springer-Verlag, 1993.

[Cou83] B. Courcelle. Fundamental properties of infinite trees. *Theoretical Computer Science*, 25:95–169, 1983.

[Gor95] Andrew Gordon. Bisimilarity as a theory of functional programming. In *Proceedings of the Eleventh Conference on the Mathematical Foundations of Programming Semantics (MFPS), New Orleans, March 29 to April 1, 1995, Elsevier Electronic Notes in Theoretical Computer Science, volume 1*, 1995.

[Hen94] Fritz Henglein. Dynamic typing: Syntax and proof theory. *Science of Computer Programming (SCP)*, 22(3):197–230, 1994.

[Hin69] R. Hindley. The principal type-scheme of an object in combinatory logic. *Trans. Amer. Math. Soc.*, 146:29–60, December 1969.

[HL95] Furio Honsell and Marina Lenisa. Final semantics for untyped lambda-calculus. In *Proc. Int'l Conf. on Typed Lambda Calculi and Applications (TLCA)*, volume 902 of *Lecture Notes in Computer Science (LNCS)*, pages 249–265. Springer-Verlag, 1995.

[Jør95] Jesper Jørgensen. *A Calculus for Boxing Analysis of Polymorphically Typed Languages*. PhD thesis, DIKU, University of Copenhagen, October 1995.

[KPS93] Dexter Kozen, Jens Palsberg, and Michael Schwartzbach. Efficient recursive subtyping. In *Proc. 20th Annual ACM SIGPLAN-SIGACT Symp. on Principles of Programming Languages*, pages 419–428. ACM, ACM Press, January 1993.

[KPS95] Dexter Kozen, Jens Palsberg, and Michael Schwartzbach. Efficient recursive subtyping. *Mathematical Structures in Computer Science (MSCS)*, 5(1), 1995.

[Len96] Marina Lenisa. Final semantics for a higher order concurrent language. In H. Kirchner, editor, *Proc. Coll. on Trees in Algebra and Programming (CAAP)*, volume 1059 of *Lecture Notes in Computer Science (LNCS)*, pages 102–118. Springer-Verlag, 1996.

[Mil77] Robin Milner. Fully abstract models of typed *lambda*-calculi. *Theoretical Computer Science (TCS)*, 4(1):1–22, February 1977.

[Mil84] Robin Milner. A complete inference system for a class of regular behaviours. *Journal of Computer and System Sciences (JCSS)*, 28:439–466, 1984.

[Mor68] J. Morris. *Lambda-Calculus Models of Programming Languages*. PhD thesis, MIT, 1968.

[MT91] Robin Milner and Mads Tofte. Co-induction in relational semantics. *Theoretical Computer Science*, 87(1):209–220, 1991. Note.

[Pit94] Andrew M. Pitts. A co-inducion principle for recursively defined domains. *Theoretical Computer Science (TCS)*, 124:195–219, 1994.

[Pit96] Andrew M. Pitts. Relational properties of domains. *Information and Computation*, 127(2):66–90, June 1996.

[Reh95] J. Rehof. Polymorphic dynamic typing — aspects of proof theory and inference. Master's thesis, DIKU, University of Copenhagen, March 1995.

[Sal66] A. Salomaa. Two complete axiom systems for the algebra of regular events. *Journal of the Association for Computing Machinery (JACM)*, 13(1):158–169, 1966.

A Simple Adequate Categorical Model for PCF

Torben Braüner

BRICS*
Department of Computer Science
University of Aarhus
Ny Munkegade
DK-8000 Aarhus C, Denmark
Internet: tor@brics.dk

Abstract. Usually types of PCF are interpreted as cpos and terms as continuous functions. It is then the case that non-termination of a closed term of ground type corresponds to the interpretation being bottom; we say that the semantics is adequate. We shall here present an axiomatic approach to adequacy for PCF in the sense that we will introduce categorical axioms enabling an adequate semantics to be given. We assume the presence of certain "bottom" maps with the role of being the interpretation of non-terminating terms, but the order-structure is left out. This is different from previous approaches where some kind of order-theoretic structure has been considered as part of an adequate categorical model for PCF. We take the point of view that partiality is the fundamental notion from which order-structure should be derived, which is corroborated by the observation that our categorical model induces an order-theoretic model for PCF in a canonical way.

1 Introduction

The programming language PCF (an acronym for *P*rogramming language for *C*omputable *F*unctions) was in 1969 introduced by Scott in the until recently unpublished paper [Sco93]. The formal system of PCF consists of the λ-calculus augmented with booleans and numerals together with recursion where the usual reduction rules are replaced by an operational semantics. Scott's pioneering idea was to interpret types as cpos and terms as continuous functions. The possibility of non-termination of programs (that is, closed terms) is then reflected in the presence of bottom elements in cpos. It was shown by Plotkin [Plo77] that non-termination of a program of ground type actually corresponds to the interpretation being the bottom element of the appropriate cpo. A semantics with that property is called adequate; the adequacy result thus expresses that the syntactic notion of non-termination coincides with the semantic notion of undefinedness.

---

* Basic Research in Computer Science,
Centre of the Danish National Research Foundation.

We shall here present an axiomatic approach to adequacy for PCF in the sense that we will introduce categorical axioms enabling an adequate semantics to be given. We follow Scott's idea in assuming the presence of certain "bottom" maps with the role of being the interpretation of non-terminating programs, but the order-structure is left out motivated by the point of view that partiality is the fundamental notion from which order-structure should be derived (which is in accordance with [Fio94b]). This is different from previous approaches where some kind of order-theoretic structure has been considered as part of an adequate categorical model for PCF. For example, in [BCL86] **cpo**-enrichment is taken as part of a categorical model, which in [AJM96] is replaced by rationality, that is, **poset**-enrichment such that only certain increasing chains are assumed to have least upper bounds. Our motivation for adhering to the point of view that partiality is the fundamental notion from which order-structure should be derived is that no order-theoretic notions are used in the formal definition of PCF, but an appropriate order-structure is indeed derivable, namely the observational preorder on terms. Moreover, the point of view is corroborated by the observation that our categorical model induces a rational category; this is with respect to the observational preorders on hom-sets quotiented down to posets. Historically, the axioms of our categorical model were discovered essentially by extracting the properties of the order of a rational category which are actually used to obtain an adequacy result. The axioms consist of some equational (that is, first order) ones together with one non-equational, namely the axiom of *rational openness* of a fixpoint operator. Having fixed an appropriate categorical model it is proved that the categorical semantics is *adequate* in the sense that we have $[\![t]\!] \neq \perp \;\Rightarrow\; t \Downarrow$ for every program t (note that this is stronger than the original notion of adequacy which only takes ground types into account). If the program t is of numerals type, then we have a converse to adequacy; a type where this happens to be the case is called *observable*. It is possible to restrict the axiom of rational openness such that it is not only sufficient, but also necessary for the interpretation to be adequate. The "necessary" direction of this result relies on the *unwinding theorem*, which we prove using a concrete instance of the adequacy result.

We consider two concrete categories as we introduce our categorical model, namely the category of cpos and continuous functions, **cpo**, and the category of dI domains and continuous stable functions, **dI**. They are both well known from the literature; the first category is for example described in the textbooks[Win93, Gun92], and the latter category in the article [Win87].

Very recently, there has been considerable progress in axiomatising sufficient conditions for obtaining full abstraction for PCF, [Abr96]. The intuitions leading to this work stem from the theory of games rather than from traditional order-theoretic domain theory, but rational openness is, however, recognised[2] as the essential notion for an axiomatic approach to adequacy.

Now, the numerals type is observable, but we cannot expect that to be the case for exponential types because the bottom elements here do not correspond

[2] With due reference to a preliminary version of this paper.

to non-termination in general; for example, the terminating program $\lambda x.\Omega$ is interpreted as bottom. We stress that what we do in this paper is consider the question of adequacy in the situation where exponential types are interpreted as categorical exponentials in the model. But it should be mentioned that it is possible to dodge the impossibility of obtaining a converse to adequacy at exponential types by introducing a monad with appropriate properties and then interpret types in a way that reflects the evaluation strategy following the ideas put forward in [Mog89]. This is the approach taken in [Gun92, Win93] where PCF-like languages are interpreted in the category **cpo** using the usual "lift" monad. The use of a monad to give an adequate interpretation with the property that the adequacy result has a converse at all types is implicit in [FP94, Fio94a] where a categorical semantics is given for FPC, which is the λ-calculus augmented with recursive types and equipped with an operational semantics. The starting point is a categorical notion of partial maps put forward in [Mog86, Ros86] together with **cpo**-enrichment. A comparison with our approach shows that they obtain a stronger result, namely an adequacy result which has a converse at all types, at the expense of having to use a more technically involved categorical notion of partial maps; and moreover, they take partiality as well as order as primitive notions, whereas we only take a notion of partiality as primitive.

In Section 2 appropriate categorical machinery is introduced and it is shown that our non-order-theoretic categorical axioms induce a rational category. The programming language PCF is introduced in Section 3 and a sound categorical interpretation is given in Section 4. In Section 5 additional axioms are imposed on the categorical model such that an adequacy result can be proved. Section 6 deals with observable types and in Section 7 we prove the unwinding theorem using adequacy; this enables us to show that a restricted version of rational openness is not only sufficient, but also necessary for the interpretation to be adequate. A more detailed presentation of the results given in this article can be found in [Bra96].

2 PCF - Semantic Issues

2.1 Undefinedness

In this subsection we introduce a notion of undefinedness by assuming the existence of "undefined" maps $\bot_A \colon 1 \to A$ which are supposed to be the interpretation of non-terminating programs. So all non-terminating programs of a given type are identified by our interpretation.

Definition 1. A category with a terminal object is *pointed* iff for each object A a map $\bot_A \colon 1 \to A$ is given. A map $f : A \to B$ in a pointed category is *strict* iff $\bot_A; f = \bot_B$. A map $f : A \times B \to C$ in a pointed cartesian category is *left-strict* iff $\langle \bot_A, h\rangle; f = \bot_C$ for any map $h : 1 \to B$.

Observation: Strictness of all maps does not go well together with cartesian closure; it can simply be shown that a pointed cartesian closed category where all maps are strict is equivalent to the one-object-one-map category.

2.2 Numerals

The following definition is essentially as given in [HO96]. Note that booleans are represented by numerals.

Definition 2. An *object of numerals* in a cartesian category is an object N equipped with maps $zero : 1 \to N$ and $succ, pred : N \to N$ such that $\tilde{0}; pred = \tilde{0}$ and $\widetilde{n+1}; pred = \tilde{n}$ for any number n where the maps $\tilde{p} : 1 \to N$ are defined in the obvious way from the *zero* and *succ* maps. Furthermore, a map $cond_A : N \times (A \times A) \to A$ is given for each object A such that $\langle \tilde{0}, \langle g, h \rangle \rangle; cond = g$ and $\langle \widetilde{n+1}, \langle g, h \rangle \rangle; cond = h$ for any maps $g, h : 1 \to A$ and any number n. An object of numerals in a pointed cartesian category is *standard* iff for any map $h : 1 \to N$ we have $h = \perp$ or $h = \tilde{n}$ for some number n.

A notable feature of an object of numerals in a pointed cartesian category is that for each number n it is possible to define a map $\phi_n : N \to N$ with the property that $\tilde{n}; \phi_n = \tilde{0}$ and $\tilde{p}; \phi_n = \perp$ whenever $p \neq n$. The ϕ_n maps are defined inductively by

$$\phi_0 = \langle id, (\langle\rangle; \langle \tilde{0}, \perp \rangle) \rangle; cond$$
$$\phi_{n+1} = \langle id, \langle (\langle\rangle; \perp), (pred; \phi_n) \rangle \rangle; cond$$

If the map *cond* is left-strict then each map ϕ_n has the property of being strict.

Now we will give a couple of objects of numerals in the concrete categories **cpo** and **dI**; note that the finite products of **dI** coincide with the ones of **cpo**. The obvious choice of a standard object of numerals in these categories is to take the object N to be equal to ω with a bottom element adjoined, and equipped with the appropriate maps. In [Plo77] another option is suggested; the object N is taken to be equal to ω with a bottom element adjoined together with an element ∞ unrelated to any other element but bottom, and equipped with appropriate extensions of the maps from the standard object of numerals.

2.3 Fixpoints

The main concern of this subsection is fixpoints as introduced in [Law69].

Definition 3. Let C be a category with a terminal object. A map $h : 1 \to B$ is a *fixpoint* of a map $f : B \to B$ iff $h; f = f$. A *fixpoint operator* for an object B is an operation on maps $(-)^{\sharp}_B : C(B, B) \longrightarrow C(1, B)$ such that $f^{\sharp}$ is a fixpoint of f for any map $f : B \to B$.

Given an endomap g we denote n iterations of g by g^n. In the categories **cpo** and **dI** the obvious choice of fixpoint operators defines the fixpoint $f^{\sharp}$ of a map $f : B \to B$ as $f^{\sharp} = \sqcup_{n \in \omega} f^n(\perp)$.

Definition 4. A fixpoint operator for an object B in a pointed category is *rationally open* with respect to an object P iff for all maps $f : B \to B$ and $g : B \to P$ it is the case that

$$f^{\sharp}; g \neq \perp \;\Rightarrow\; \exists n \in \omega.\ \perp; f^n; g \neq \perp .$$

Consider a map $g : B \to P$ living **cpo** or in **dI**. In the category **cpo** the set $\{x \in B \mid g(x) \neq \perp\}$ is a Scott-open subset of B, and in the category **dI**, it is a union of sets of the form $\{x \in B \mid x \geq b\}$ where b is a finite element of B, so it is straightforward to see that the fixpoint operators in the categories **cpo** and **dI** are rationally open with respect to any object P. Under appropriate circumstances rational openness with respect to an object of numerals N can be restated to a different form:

Proposition 5. *Let C be a pointed cartesian category equipped with a standard object of numerals N such that the map cond is left strict. Assume that a fixpoint operator is given for an object B. Then the fixpoint operator is rationally open with respect to N iff for all maps $f : B \to B$ and $g : B \to N$ and numbers p it is the case that*

$$f^{\sharp}; g = \tilde{p} \;\Rightarrow\; \exists n \in \omega.\ \perp; f^{n}; g = \tilde{p}.$$

Proof. Use the ϕ_p maps defined in the remark following Definition 2. □

The role of rational openness in the adequacy result for PCF in Section 3 is analogous to the role of the *absoluteness* condition in [Fio94b]. Note, however, that our notion of rational openness does not assume the presence of any order-structure; this is not the case with the notion of absoluteness. We can internalise the notion of a fixpoint operator when we deal with a cartesian closed category.

Definition 6. An *internal fixpoint operator* for an object B in a cartesian closed category is a map $Y_B : B \Rightarrow B \to B$ such that the diagram

$$\begin{array}{ccc} B \Rightarrow B & \xrightarrow{\Delta} & (B \Rightarrow B) \times (B \Rightarrow B) \\ {\scriptstyle Y}\downarrow & & \downarrow{\scriptstyle id \times Y} \\ B & \xleftarrow{eval} & (B \Rightarrow B) \times B \end{array}$$

commutes.

Proposition 7. *Let B be an object in a cartesian closed category. A fixpoint operator for $(B \Rightarrow B) \Rightarrow B$ induces an internal fixpoint operator for B.*

Proof. Firstly, there is an straightforward way to define a map

$$(B \Rightarrow B) \Rightarrow B \xrightarrow{H_B} (B \Rightarrow B) \Rightarrow B$$

with the property that for every map $h : B \Rightarrow B \to B$ we have $\ulcorner h \urcorner; H = \ulcorner \Delta; (id \times h); eval \urcorner$. We then have that $\ulcorner h \urcorner$ is a fixpoint for H iff $h = \Delta; (id \times h); eval$. We now define the internal fixpoint operator $Y : B \Rightarrow B \to B$ by the equation $\ulcorner Y \urcorner = H^{\sharp}$; hence $Y = \Delta; (id \times Y); eval$. □

2.4 The Observational Preorder

The goal of this subsection is to introduce the observational preorder. We need some definitions and results from [HO96].

Definition 8. Let $(\mathcal{C}, I, \otimes, \multimap)$ be a symmetric monoidal closed category. A *notion of observables* $\mathcal{O}$ associates to each object A a set $\mathcal{O}_A$ of subsets of $\mathcal{C}(I, A)$ with the property that if $S \in \mathcal{O}_A$ then $g^*S \in \mathcal{O}_B$ for any map $g : B \to A$ where $g^*S = \{h \in \mathcal{C}(I, B) \mid h; g \in S\}$.

Definition 9. Let $\mathcal{C}$ be a symmetric monoidal closed category with a notion of observables $\mathcal{O}$. The *observational preorder* $\lesssim_{A,B}$ on each hom-set $\mathcal{C}(A, B)$ is defined as

$$f \lesssim g \text{ iff } \forall R \in \mathcal{O}_{A \multimap B}.\ \ulcorner f \urcorner \in R \ \Rightarrow\ \ulcorner g \urcorner \in R$$

for any maps $f, g : A \to B$.

The observational preorder behaves properly with respect to points:

Proposition 10. *If $\mathcal{C}$ is a symmetric monoidal closed category with a notion of observables $\mathcal{O}$, then for any maps $f, g : I \to A$ it is the case that*

$$f \lesssim g \text{ iff } \forall R \in \mathcal{O}_A.\ f \in R \ \Rightarrow\ g \in R$$

Proof. Straightforward. □

All the given structure enriches with respect to the observational preorder:

Proposition 11. *Let $\mathcal{C}$ be a symmetric monoidal closed category with a notion of observables and consider the observational preorder. The symmetric monoidal closed category is* **preorder**-*enriched. If the category is actually cartesian closed then it is* **preorder**-*enriched as such.*

Proof. [HO96]. □

2.5 Rationality

In this subsection we will show that a rationally open fixpoint operator under appropriate circumstances induces a rational category, which in [AJM96] is taken to be an appropriate (order-theoretic) notion of a categorical model for PCF.

Definition 12. A *rational category* is a pointed cartesian closed category which is **poset**-enriched as a cartesian closed category such that for every object A it is the case that

- the map $\bot_A$ is least,
- for every map $f : A \to A$ the increasing chain $\{\bot; f^n\}_{n \in \omega}$ has a least upper bound $f^\natural$ with the property that for any map $g : A \to D$ the map $f^\natural; g$ is a least upper bound for the increasing chain $\{\bot; f^n; g\}_{n \in \omega}$.

We will now consider a notion of observables for a pointed cartesian closed category that assumes the presence of a distinguished object P. The object P should be thought of as the interpretation of a distinguished type. In [HO96] the following notion of observables is called the termination notion of observables when a map $f : 1 \to P$ is taken to be a *value* iff $f \neq \perp$.

Definition 13. Let $\mathcal{C}$ be a pointed cartesian closed category with a distinguished object P. The *termination notion of observables* is defined by to any object A assigning the set

$$\mathcal{O}_A = \{\mathcal{O}_h \mid h \in \mathcal{C}(A, P)\}$$

where

$$\mathcal{O}_h = \{f \in \mathcal{C}(1, A) \mid f; h \neq \perp\}.$$

This is a notion of observables because $g^*\mathcal{O}_h = \mathcal{O}_{g;h}$ for any maps $g : B \to A$ and $h : A \to P$. The observational preorder induced by the termination notion of observables can be stated explicitly as

$$f \lesssim g \text{ iff } \forall h \in \mathcal{C}(A \Rightarrow B, P).\ \ulcorner f \urcorner; h \neq \perp \ \Rightarrow\ \ulcorner g \urcorner; h \neq \perp$$

for any maps $f, g : A \to B$. We will refer to this preorder as the *termination preorder*. If we are dealing with a pointed cartesian closed category equipped with an object of numerals, then we take the object P to be N unless otherwise is stated.

Theorem 14. *Let $\mathcal{C}$ be a pointed cartesian closed category equipped with a standard object of numerals such that the map cond is left-strict. Assume that a fixpoint operator is given for each object. Then for any object A it is the case that*

- *the map $\perp_A$ is least and for any maps $f : A \to A$ and $g : A \to D$ the map $f^\sharp; g$ is an upper bound for the increasing chain $\{\perp; f^n; g\}_{n\in\omega}$,*
- *the fixpoint operator for the object A is rationally open with respect to N iff for any maps $f : A \to A$ and $g : A \to D$ the map $f^\sharp; g$ is a least upper bound for the increasing chain $\{\perp; f^n; g\}_{n\in\omega}$,*

where we consider the termination preorder.

Proof. Without loss of generality we will assume that $\tilde{0} \neq \perp$ because $\tilde{0} = \perp$ entails that every hom-set $\mathcal{C}(1, A)$ has exactly one element.

We first show that the map $\perp_A$ is least[3]. We will use the $\phi_n : N \to N$ maps defined in the remark following Definition 2. Take any map $h : 1 \to A$. We will then show that $\perp \lesssim h$. Assume $\perp; f \neq \perp$ for some map $f : A \to N$, that is, $\perp; f = \tilde{n}$ for some number n; we then have to show that $h; f \neq \perp$. Now, define a map $k : N \to A$ as

$$k = \langle id, (\langle\rangle; \langle h, \perp\rangle)\rangle; cond$$

[3] Here we essentially make use of an argument suggested by Alex Simpson.

and consider the fixpoint $(k; f; \phi_n)^\sharp : 1 \to N$ of the composition

$$N \xrightarrow{k} A \xrightarrow{f} N \xrightarrow{\phi_n} N$$

which has the property that $(k; f; \phi_n)^\sharp = (k; f; \phi_n)^\sharp; k; f; \phi_n$. We cannot have $(k; f; \phi_n)^\sharp = \perp$ because this contradicts $\perp; f; \phi_n = \tilde{0}$, so we have $(k; f; \phi_n)^\sharp = \tilde{0}$ which entails that $h; f; \phi_n = \tilde{0}$. We conclude that $h; f = \tilde{n}$, and thus $h; f \neq \perp$.

For any maps $f : A \to A$ and $g : A \to D$ it is straightforward to check by induction that the chain $\{\perp; f^n\}_{n\in\omega}$ is increasing and $f^\sharp$ is an upper bound, so the chain $\{\perp; f^n; g\}_{n\in\omega}$ is increasing and $f^\sharp; g$ is an upper bound.

Assume that the fixpoint operator for the object A is rationally open with respect to N, and consider any maps $f : A \to A$ and $g : A \to D$. Let $k : 1 \to D$ be an arbitrary upper bound for the increasing chain $\{\perp; f^n; g\}_{n\in\omega}$. If $f^\sharp; g; h \neq \perp$ for some map $h : D \to N$ then there exists a number p such that $\perp; f^p; g; h \neq \perp$ because the fixpoint operator is assumed to be rationally open with respect to N, which entails that $k; h \neq \perp$. We conclude that $f^\sharp; g \lesssim k$.

Consider any maps $f : A \to A$ and $g : A \to N$, and assume that the map $f^\sharp; g$ is a least upper bound for the increasing chain $\{\perp; f^n; g\}_{n\in\omega}$. Then $\perp; f^n; g = \perp$ for every number n entails that $\perp$ is an upper bound for the increasing chain $\{\perp; f^n; g\}_{n\in\omega}$, but $f^\sharp; g$ is a least upper bound, so $f^\sharp; g \lesssim \perp$ and thus $f^\sharp; g = \perp$. We conclude that the fixpoint operator is rationally open with respect to N. □

Note that in the context of Theorem 14 none of the maps in the hom-set $\mathcal{C}(1, N)$ are equivalent with respect to the equivalence relation induced by the termination preorder.

Corollary 15. *Let $\mathcal{C}$ be a pointed cartesian closed category equipped with a standard object of numerals such that the map cond is left-strict. For each object assume that a fixpoint operator which is rationally open with respect to N is given. Then the quotient category $\widehat{\mathcal{C}}$ is a rational category when $\mathcal{C}$ is considered as* **preorder***-enriched with respect to the termination preorder.*

Proof. Follows from Theorem 14 and the observation that the fixpoint operator is a congruence with respect to the equivalence relation induced by the termination preorder. □

3 Syntax and Operational Semantics

The programming language PCF has exponential types together with product and sum types, and it has one ground type, namely a type for numerals. Booleans are then represented by numerals in the traditional way. The units for product and sum types are of limited computational interest, but we have included them for the sake of completeness. Types of PCF are given by the grammar

$$s \ ::= \ N \mid 1 \mid s \times s \mid s \Rightarrow s \mid 0 \mid s + s$$

and terms are given by the grammar

$$
\begin{aligned}
t \; ::= \; & x \mid \\
& \mathtt{zero} \mid \mathtt{succ}(t) \mid \mathtt{pred}(t) \mid \mathtt{if}\; t \;\mathtt{then}\; t \;\mathtt{else}\; t \mid \\
& \mathtt{true} \mid (t,t) \mid \mathtt{fst}(t) \mid \mathtt{snd}(t) \mid \lambda x^A.t \mid tt \mid \\
& \mathtt{false}^C(t) \mid \mathtt{inl}^{A+B}(t) \mid \mathtt{inr}^{A+B}(t) \mid \mathtt{case}\; t \;\mathtt{of}\; \mathtt{inl}(x).t \mid \mathtt{inr}(y).t \mid \\
& \Omega_A \mid \mathrm{Y}_A
\end{aligned}
$$

where x is a variable. Rules for assignment of types to terms are given in Fig. 1.

Fig. 1. Type Assignment Rules for PCF

$$\frac{}{x_1 : A_1, \ldots, x_n : A_n \vdash x_p : A_p}$$

$$\frac{}{\Gamma \vdash \mathtt{zero} : N} \qquad \frac{\Gamma \vdash u : N}{\Gamma \vdash \mathtt{succ}(u) : N} \qquad \frac{\Gamma \vdash u : N}{\Gamma \vdash \mathtt{pred}(u) : N}$$

$$\frac{\Gamma \vdash u : N \quad \Gamma \vdash v : A \quad \Gamma \vdash w : A}{\Gamma \vdash \mathtt{if}\; u \;\mathtt{then}\; v \;\mathtt{else}\; w : A}$$

$$\frac{}{\Gamma \vdash \mathtt{true} : 1} \qquad \frac{\Gamma \vdash u : A \quad \Gamma \vdash v : B}{\Gamma \vdash (u,v) : A \times B} \qquad \frac{\Gamma \vdash u : A \times B}{\Gamma \vdash \mathtt{fst}(u) : A} \qquad \frac{\Gamma \vdash u : A \times B}{\Gamma \vdash \mathtt{snd}(u) : B}$$

$$\frac{\Gamma, x : A \vdash u : B}{\Gamma \vdash \lambda x^A.u : A \Rightarrow B} \qquad \frac{\Gamma \vdash f : A \Rightarrow B \quad \Gamma \vdash u : A}{\Gamma \vdash fu : B}$$

$$\frac{\Gamma \vdash w : 0}{\Gamma \vdash \mathtt{false}^C(w) : C} \qquad \frac{\Gamma \vdash u : A}{\Gamma \vdash \mathtt{inl}^{A+B}(u) : A + B} \qquad \frac{\Gamma \vdash u : B}{\Gamma \vdash \mathtt{inr}^{A+B}(u) : A + B}$$

$$\frac{\Gamma \vdash w : A + B \quad \Gamma, x : A \vdash u : C \quad \Gamma, y : B \vdash v : C}{\Gamma \vdash \mathtt{case}\; w \;\mathtt{of}\; \mathtt{inl}(x).u \mid \mathtt{inr}(y).v : C}$$

$$\frac{}{\Gamma \vdash \Omega_A : A} \qquad \frac{}{\Gamma \vdash \mathrm{Y}_A : (A \Rightarrow A) \Rightarrow A}$$

We will now give a lazy operational semantics of PCF. A *program* is an arbitrary closed term and a *value* is a closed term of one of the forms

$$\mathtt{succ}^n(\mathtt{zero}) \qquad \mathtt{true} \qquad (v,w) \qquad \lambda x.u \qquad \mathtt{inl}(u) \qquad \mathtt{inr}(u)$$

where n is a number and a term $\mathtt{succ}^n(\mathtt{zero})$ is defined in the obvious way. Let T be the set of programs and C the set of values. The evaluation rules in Fig. 2 induce a relation $\Downarrow \subset T \times C$ called the *evaluation relation*. Given a program u, we will write $u \Downarrow$ iff there exists a term c such that $u \Downarrow c$.

Fig. 2. Operational Semantics of PCF

$$\frac{}{\mathtt{zero} \Downarrow \mathtt{zero}} \qquad \frac{u \Downarrow c}{\mathtt{succ}(u) \Downarrow \mathtt{succ}(c)} \qquad \frac{u \Downarrow \mathtt{zero}}{\mathtt{pred}(u) \Downarrow \mathtt{zero}} \qquad \frac{u \Downarrow \mathtt{succ}(c)}{\mathtt{pred}(u) \Downarrow c}$$

$$\frac{u \Downarrow \mathtt{zero} \quad v \Downarrow d}{\mathtt{if}\ u\ \mathtt{then}\ v\ \mathtt{else}\ w \Downarrow d} \qquad \frac{u \Downarrow \mathtt{succ}(c) \quad w \Downarrow e}{\mathtt{if}\ u\ \mathtt{then}\ v\ \mathtt{else}\ w \Downarrow e}$$

$$\frac{}{\mathtt{true} \Downarrow \mathtt{true}} \qquad \frac{}{(u,v) \Downarrow (u,v)} \qquad \frac{u \Downarrow (v,w) \quad v \Downarrow c}{\mathtt{fst}(u) \Downarrow c} \qquad \frac{u \Downarrow (v,w) \quad w \Downarrow d}{\mathtt{snd}(u) \Downarrow d}$$

$$\frac{}{\lambda x.u \Downarrow \lambda x.u} \qquad \frac{f \Downarrow \lambda x.v \quad v[u/x] \Downarrow c}{fu \Downarrow c}$$

$$\frac{}{\mathtt{inl}(u) \Downarrow \mathtt{inl}(u)} \qquad \frac{}{\mathtt{inr}(u) \Downarrow \mathtt{inr}(u)}$$

$$\frac{w \Downarrow \mathtt{inl}(t) \quad u[t/x] \Downarrow c}{\mathtt{case}\ w\ \mathtt{of}\ \mathtt{inl}(x).u \mid \mathtt{inr}(y).v \Downarrow c} \qquad \frac{w \Downarrow \mathtt{inr}(t) \quad v[t/x] \Downarrow c}{\mathtt{case}\ w\ \mathtt{of}\ \mathtt{inl}(x).u \mid \mathtt{inr}(y).v \Downarrow c}$$

$$\frac{}{\mathtt{Y} \Downarrow \lambda f.f(\mathtt{Y}f)}$$

4 Categorical Semantics

In this section we will give a sound categorical semantics of PCF. In Section 5 additional assumptions on our category are imposed with the aim of proving an adequacy result. We will interpret the product and function types using a cartesian closed structure. It is not so obvious how to interpret the sum type. If we assume the presence of finite sums, then the "case" construct can be interpreted using the operation on maps

$$\frac{A \times \Gamma \xrightarrow{u} C \qquad B \times \Gamma \xrightarrow{u} C}{(A+B) \times \Gamma \xrightarrow{\lambda^{-1}([\lambda(u),\lambda(v)])} C}$$

which can be shown to be natural in Γ. This would be equivalent to using the natural isomorphism

$$(A+B) \times \Gamma \cong (A \times \Gamma) + (B \times \Gamma)$$

given by the observation that the functor $(-)\times\Gamma$ has a right adjoint and thus preserves sums. But in [HP90] a cartesian closed category with finite sums together with a fixpoint operator is shown to be *inconsistent*, that is, it is equivalent to the category consisting of one object and one map. So we have to be content with a less demanding assumption than finite sums:

Definition 16. A *categorical premodel for PCF* is a pointed cartesian closed category with

- an object of numerals,
- a fixpoint operator for each object,
- weak finite sums such that the diagram

$$\begin{array}{ccc} A+B & \xrightarrow{[f,g]} & \Gamma \Rightarrow C \\ & \searrow^{[(f;(h\Rightarrow C)),(g;(h\Rightarrow C))]} & \downarrow{\scriptstyle h\Rightarrow C} \\ & & \Delta \Rightarrow C \end{array}$$

commutes for any maps $f : A \to \Gamma \Rightarrow C$, $g : B \to \Gamma \Rightarrow C$ and $h : \Delta \to \Gamma$.

Then the above mentioned operation can still be defined, and commutativity of the diagram is equivalent to the operation being natural in Γ, as can be shown by some equational manipulation. Given a premodel for PCF, we can interpret types as objects, typing rules as natural operations on maps, and derivations of type assignments as maps. A derivable sequent

$$x_1 : A_1, ..., x_n : A_n \vdash u : B$$

is interpreted as a map

$$[\![A_1]\!] \times ... \times [\![A_n]\!] \xrightarrow{[\![u]\!]} [\![B]\!]$$

by induction on its derivation using the appropriate operations on maps induced by the categorical assumptions. Note that it follows from Proposition 7 that a cartesian closed category with a fixpoint operator for each object also has an internal fixpoint operator for each object; this is used to interpret the fixpoint constant of PCF.

Lemma 17. *(Substitution) If the sequents $\Gamma \vdash u : A$ and $\Gamma, x : A, \Lambda \vdash v : B$ are derivable, then the sequent $\Gamma, \Lambda \vdash v[u/x] : B$ is also derivable and it has the interpretation*

$$\Gamma \times \Lambda \xrightarrow{\Delta \times \Lambda} \Gamma \times \Gamma \times \Lambda \xrightarrow{\Gamma \times [\![u]\!] \times \Lambda} \Gamma \times A \times \Lambda \xrightarrow{[\![v]\!]} B$$

Proof. Induction on the derivation of $\Gamma, x : A, \Lambda \vdash v : B$. We use the observation that the operations on maps induced by the typing rules are natural in the interpretation of the unchanged components of the contexts of the sequents. □

The interpretation is preserved by evaluation:

Theorem 18. *(Soundness) Given a program u such that $u \Downarrow c$, then $[\![u]\!] = [\![c]\!]$.*

Proof. Induction on the derivation of $u \Downarrow c$ where we use Lemma 17. Note that the evaluation rule for the fixpoint constant preserves the interpretation because it is essentially a syntactic restatement of the defining diagram for an internal fixpoint operator, Definition 6. □

5 Adequacy

In our adequacy proof for PCF we will use the now standard technique of logical relations introduced in [Plo73]. The original adequacy proof in [Plo77] made use of a unary logical relation $Comp_A \subseteq T_A$ called the *computability predicate.* Recall that T_A is the set of programs of type A. The logical relation $Comp_A$ is defined by induction on the type A such that at ground type N it coincides with adequacy, that is, $Comp_N(t)$ is defined to hold iff $[\![t]\!] \neq \perp$ entails $t \Downarrow$. It is then proved that every program satisfies the computability predicate. Another method for proving adequacy is given in [Gun92, Win93]; here a binary logical relation $\preceq_A \subseteq [\![A]\!] \times T_A$ is used, where $d \preceq_A t$ expresses that the element $d \in A$ "approximates" the term t. The logical relation $\preceq_A$ is defined by induction on the type A such that at ground type $d \preceq_N t$ is defined to hold iff $d \neq \perp$ entails $t \Downarrow \mathtt{succ}^q(\mathtt{zero})$ for some number q such that $d = \tilde{q}$. One then proves that $[\![t]\!] \preceq_A t$ for every program t of type A. Here we will also use the binary logical relation $\preceq_A$, but in a different way adapted to our categorical setting. We need some extra assumptions on our premodel:

Definition 19. A *categorical model for PCF* is a categorical premodel such that

- the maps π_1 and π_2 are strict and the map *eval* is left-strict,
- the maps *succ* and *pred* are strict and the map *cond* is left-strict,
- the map $[] : 0 \to C$ is strict for any object C and the map $[f, g] : A+B \to C$ is strict for any maps $f : A \to C$ and $g : B \to C$.

Note that the adequacy result below is stated solely in terms of the categorical premodel for PCF, but the assumptions added in Definition 19 are indeed used in the proof.

Definition 20. For each type A the binary logical relation

$$\preceq_A \subseteq \mathcal{C}(1, A) \times T_A$$

is defined by induction on the structure of A as

$$f \preceq_N t \text{ iff } f \neq \perp \Rightarrow \exists n \in \omega.\ t \Downarrow \mathtt{succ}^n(\mathtt{zero}) \wedge f = \tilde{n}$$

$$f \preceq_1 t \text{ iff } true$$

$$\begin{aligned} f \preceq_{B\times C} t \text{ iff } & (f \neq \perp \Rightarrow t \Downarrow) \wedge \\ & (f;\pi_1 \preceq_B \mathtt{fst}(t) \wedge f;\pi_2 \preceq_C \mathtt{snd}(t)) \end{aligned}$$

$$\begin{aligned} f \preceq_{B\Rightarrow C} t \text{ iff } & (f \neq \perp \Rightarrow t \Downarrow) \wedge \\ & (\forall g \in \mathcal{C}(1, B). \forall v \in T_B.\ g \preceq_B v \Rightarrow \langle f, g\rangle; eval \preceq_C tv) \end{aligned}$$

$$f \preceq_0 t \text{ iff } f = \perp$$

$$\begin{aligned} f \preceq_{B+C} t \text{ iff } & f \neq \perp \Rightarrow \\ & (\exists h \in \mathcal{C}(1, B). \exists v \in T_B.\ f = h; in_1 \wedge t \Downarrow \mathtt{inl}(v) \wedge h \preceq_B v) \vee \\ & (\exists h \in \mathcal{C}(1, C). \exists v \in T_C.\ f = h; in_2 \wedge t \Downarrow \mathtt{inr}(v) \wedge h \preceq_C v) \end{aligned}$$

Lemma 21. *For each program t of type A we have* $\bot \preceq_A t$.

Proof. Induction on the structure of A. □

In the following lemma an essential connection between the logical relation $\preceq$ and the evaluation relation $\Downarrow$ is shown:

Lemma 22. *If* $f \preceq_A t$ *and* $u \Downarrow c$ *whenever* $t \Downarrow c$, *then* $f \preceq_A u$.

Proof. Induction on the structure of A. □

Note that in the following lemma the term u is assumed not to contain any occurrences of the fixpoint constant. Also note that such a restriction is not imposed on the t_i terms.

Lemma 23. *Consider a derivable sequent*

$$x_1 : A_1, ..., x_n : A_n \vdash u : C$$

such that the term u does not contain any occurrences of the fixpoint constant. Assume that for each $i \in \{1, ..., n\}$ *we have a map* $f_i : 1 \to A_i$ *and a program* t_i *of type* A_i *such that* $f_i \preceq_{A_i} t_i$. *We then have*

$$\langle f_1, ..., f_n \rangle; [\![u]\!] \preceq_C u[t_1, ..., t_n/x_1, ..., x_n].$$

Proof. Induction on the derivation of $x_1 : A_1, ..., x_n : A_n \vdash u : C$. □

The following lemma says that the (formal) finite approximants to an internal fixpoint operator are $\preceq$-related to the corresponding fixpoint constant:

Lemma 24. *For every type A and number n we have* $\bot; H_A^n \preceq_{(A \Rightarrow A) \Rightarrow A} \mathtt{Y}_A$.

Proof. Recall that the map H_A is given in the proof of Proposition 7. We proceed by induction on n. The assertion is true in case $n = 0$ according to Lemma 21. Now assume that the assertion is true for an arbitrary number n; we then have to show that $\bot; H^{n+1} \preceq_{(A \Rightarrow A) \Rightarrow A} \mathtt{Y}$. First, observe that $\mathtt{Y} \Downarrow$. Assume that we have a map h and a term v such that $h \preceq_{A \Rightarrow A} v$; we then have to show that

$$\langle (\bot; H^{n+1}), h \rangle; eval \preceq_A \mathtt{Y}v. \tag{1}$$

We have

$$\bot; H^{n+1} = \ulcorner \Delta; (id \times \ulcorner \bot; H^n \urcorner^{-1}); eval \urcorner$$

cf. the proof of Proposition 7, so we get

$$\begin{aligned} \langle (\bot; H^{n+1}), h \rangle; eval &= h; \Delta; (id \times \ulcorner \bot; H^n \urcorner^{-1}); eval \\ &= \langle h, (\langle (\bot; H^n), h \rangle; eval) \rangle; eval \end{aligned}$$

Now

$$\langle (\bot; H^n), h \rangle; eval \preceq_A \mathtt{Y}v$$

cf. the induction hypothesis, which entails that

$$\langle h, (\langle (\bot; H^n), h \rangle; eval) \rangle; eval \preceq_A v(\mathtt{Y}v)$$

and $v(\mathtt{Y}v) \Downarrow c$ for some value c entails that $\mathtt{Y}v \Downarrow c$ so we conclude (1) according to Lemma 22. □

We are finally able to state a result expressing that the categorical interpretation is adequate with respect to the operational semantics. Note how we "extract" the fixpoint constants of a term.

Theorem 25. *(Adequacy) Let u be a program of type B. If the fixpoint operator is rationally open with respect to B, then $[\![u]\!] \neq \perp$ entails that $u \Downarrow$.*

Proof. It is straightforward to show by induction that there exists a derivable sequent

$$z_1 : (A_1 \Rightarrow A_1) \Rightarrow A_1, \; \dots, z_n : (A_n \Rightarrow A_n) \Rightarrow A_n \vdash u' : B$$

such that

$$u = u'[\mathrm{Y}_{A_1}, \dots, \mathrm{Y}_{A_n}/z_1, \dots, z_n]$$

and such that the term u' does not contain any occurrences of the fixpoint constant. But then

$$[\![u'[\mathrm{Y}_{A_1}, \dots, \mathrm{Y}_{A_n}/z_1, \dots, z_n]]\!] = \langle [\![\mathrm{Y}_{A_1}]\!], \dots, [\![\mathrm{Y}_{A_n}]\!] \rangle ; [\![u']\!]$$

and for each $i \in \{1, \dots, n\}$ we have $[\![\mathrm{Y}_{A_i}]\!] = H^{\sharp}_{A_i}$ so there exist numbers $p_1, \dots, p_n$ such that

$$\langle (\perp; H^{p_1}_{A_1}), \; \dots, (\perp; H^{p_n}_{A_n}) \rangle ; [\![u']\!] \neq \perp$$

because the fixpoint operator is rationally open with respect to B and because $[\![u]\!] \neq \perp$. But for each $i \in \{1, \dots, n\}$ it is the case that

$$\perp; H^{p_i}_{A_i} \preceq_{(A_i \Rightarrow A_i) \Rightarrow A_i} \mathrm{Y}_{A_i}$$

according to Lemma 24, which entails that

$$\langle (\perp; H^{p_1}_{A_1}), \dots, (\perp; H_{A_n})^{p_n} \rangle ; [\![u']\!] \preceq_B u'[\mathrm{Y}_{A_1}, \dots, \mathrm{Y}_{A_n}/z_1, \dots, z_n]$$

according to Lemma 23. We conclude that $u \Downarrow$ cf. the definition of $\preceq_B$. □

Note that the preceding theorem reveals information about how the semantics is related to termination/non-termination behaviour at any type, that is, $[\![u]\!] \neq \perp$ entails that $u \Downarrow$ whichever type the program u has.

6 Observable Types

Types where we have a converse to adequacy, that is, to Theorem 25, will be called *observable*. To be explicit, a type B is observable iff $u \Downarrow$ entails that $[\![u]\!] \neq \perp$ for any program u of type B when we assume that $\tilde{0} \neq \tilde{1}$.

Now, the ground type N is observable according to soundness and the observation that if $\tilde{n} = \perp$ for some number n then $\tilde{0} = \tilde{1}$. It is actually possible to obtain a more informative result, which is a categorical generalisation of the traditional notion of adequacy for PCF:

Corollary 26. *Let u be a program of type N. If $\tilde{0} \neq \tilde{1}$ and the fixpoint operator is rationally open with respect to N, then $[\![u]\!] = \tilde{q}$ iff $u \Downarrow \mathtt{succ}^q(\mathtt{zero})$.*

Proof. The result follows from Theorem 18 and Theorem 25 together with the observation that if $\tilde{n} = \perp$ for some number n, or if $\tilde{p} = \tilde{q}$ for different numbers p and q, then $\tilde{0} = \tilde{1}$. □

Sum types are observable too; at binary sum types $A + B$ it follows from soundness and the observation that if in_1 or in_2 factors through $\perp : 1 \to A + B$, then $\tilde{0} = \tilde{1}$, and at unary sum type 0 it follows from the observation that we cannot have $t \Downarrow$, where t is a program of type 0, as there are no a values of type 0. It actually turns out that a result analogous to Corollary 26 can be obtained at binary sum types:

Corollary 27. *Let u be a program of type $A + B$. If $\tilde{0} \neq \tilde{1}$ and the fixpoint operator is rationally open with respect to $A + B$, then the map $[\![u]\!]$ factors through in_1 iff $u \Downarrow \mathtt{inl}(t)$ for some term t, and, analogously, it factors through in_2 iff $u \Downarrow \mathtt{inr}(v)$ for some term v.*

Proof. The result follows from Theorem 18 and Theorem 25 together with the observation that if in_1 or in_2 factors through $\perp : 1 \to A + B$, or if in_1 and in_2 both factor through an arbitrary map $f : 1 \to A + B$, then $\tilde{0} = \tilde{1}$. □

Product and exponential types are not observable as the following examples show: The programs `true` and (Ω, Ω) of product type are values, but are interpreted as $\perp$, and, similarly, the program $\lambda x.\Omega$ of exponential type is canonical, but is interpreted as $\perp$.

7 Unwinding

The adequacy result can be used to prove non-trivial syntactic properties of PCF in the presence of an appropriate model. Here we shall give the *unwinding theorem* which establishes a relation between the fixpoint constant Y and its finite approximants. Purely syntactic proofs of similar results can be found in [Gun92, Pit95]. We first need a convention: For any type A we inductively define a program $\mathtt{Y}^n_A$ of type $(A \Rightarrow A) \Rightarrow A$ for every number n by

$$\mathtt{Y}^0 = \Omega$$
$$\mathtt{Y}^{n+1} = \lambda f. f(\mathtt{Y}^n f)$$

The terms $\mathtt{Y}^n$ are called *finite approximants* to the fixpoint constant Y. By a small induction proof it can be shown that $[\![\mathtt{Y}^n]\!] = \perp ; H^n$ where the map H is given in the proof of Proposition 7.

Theorem 28. *(Unwinding) If t is a term of observable type with one free variable z of type $(A \Rightarrow A) \Rightarrow A$, then*

$$t[\mathtt{Y}/z] \Downarrow \;\Leftrightarrow\; \exists n \in \omega.\ t[\mathtt{Y}^n/z] \Downarrow .$$

Proof. We will interpret PCF in the concrete category **cpo**; this is a model for PCF such that the fixpoint operator is rationally open with respect to the interpretation of any observable type. We then have $t[\mathtt{Y}/z] \Downarrow$ iff $[\![t[\mathtt{Y}/z]]\!] \neq \perp$ cf. adequacy and its converse. But $[\![t[\mathtt{Y}/z]]\!]$ is a least upper bound for the increasing chain $\{[\![t[\mathtt{Y}^n/z]]\!]\}_{n\in\omega}$ because $[\![t[\mathtt{Y}/z]]\!] = H^{\sharp}; [\![t]\!]$ and $[\![t[\mathtt{Y}^n/z]]\!] = \perp; H^n; [\![t]\!]$ according to the remark above, so $[\![t[\mathtt{Y}/z]]\!] \neq \perp$ iff there exists a number n such that $[\![t[\mathtt{Y}^n/z]]\!] \neq \perp$, that is, iff there exists a number n such that $t[\mathtt{Y}^n/z] \Downarrow$ cf. adequacy and its converse. □

At ground type N it is possible to obtain a more informative result: If t is a term of ground type N with one free variable z of type $(A \Rightarrow A) \Rightarrow A$, then $t[\mathtt{Y}/z] \Downarrow c$ iff there exists a number n such that $t[\mathtt{Y}^n/z] \Downarrow c$. The proof of the result is similar to the proof of Theorem 28. An analogous result can be obtained at binary sum types.

In the previous section we showed how the assumption of rational openness on a fixpoint operator entails that the categorical interpretation is adequate. It is possible to weaken this assumption such that it is not only sufficient, but also necessary for the interpretation to be adequate[4]. Essentially, what we do is we restrict rational openness to maps definable in PCF. Note that the "upwards" direction of the following theorem relies on the unwinding theorem.

Theorem 29. *Assume that we have a category that is a model for PCF in the sense of Definition 19 such that $\tilde{0} \neq \tilde{1}$. Let an observable type B be given, then the assertions*

- *for all types A and programs t of type B with one free variable z of type $(A \Rightarrow A) \Rightarrow A$ it is the case that*
$$H^{\sharp}; [\![t]\!] \neq \perp \;\Rightarrow\; \exists n \in \omega.\ \perp; H^n; [\![t]\!] \neq \perp,$$
- *for every program u of type B we have $[\![u]\!] \neq \perp$ entails that $u \Downarrow$,*

are equivalent.

Proof. The "downwards" direction comes from the observation that the assumption made here is sufficient to prove Theorem 25 in the relevant case. The proof of the "upwards" direction goes as follows: Assume that the assumption made here holds, and consider a term t of type B with one free variable z of type $(A \Rightarrow A) \Rightarrow A$. If $H^{\sharp}; [\![t]\!] \neq \perp$ then $t[\mathtt{Y}/z] \Downarrow$ according to the assumption because $H^{\sharp}; [\![t]\!] = [\![t[\mathtt{Y}/z]]\!]$. But then there exists a number n such that $t[\mathtt{Y}^n/z] \Downarrow$ cf. the unwinding theorem, which entails the existence of a number n such that $\perp; H^n; [\![t]\!] \neq \perp$ because the type B is observable and $[\![t[\mathtt{Y}^n/z]]\!] = \perp; H^n; [\![t]\!]$. □

Acknowledgements. I am grateful to my supervisor, Glynn Winskel, for guidance and support. The work presented here has benefitted from discussions with Marcelo Fiore, Gordon Plotkin and Alex Simpson. Also thanks to Guy McCusker for a discussion on the topic of this paper. The diagrams and proof-rules are produced using Paul Taylor's macros.

[4] This result was conjectured by Gordon Plotkin while the author visited Edinburgh in March '96.

References

[Abr96] S. Abramsky. Axioms for full abstraction and full completeness. Manuscript, 1996.

[AJM96] S. Abramsky, R. Jagadeesan, and P. Malacaria. Full abstraction for PCF. Submitted for publication, 1996.

[BCL86] G. Berry, P.-L. Curien, and J.-J. Levy. Full abstraction for sequential languages: the state of the art. In *Algebraic Semantics*. Cambridge University Press, 1986.

[Bra96] T. Braüner. *An Axiomatic Approach to Adequacy*. PhD thesis, Department of Computer Science, University of Aarhus, 1996. 168 pages. Published as Technical Report BRICS-DS-96-4.

[Fio94a] M. P. Fiore. *Axiomatic Domain Theory in Categories of Partial Maps*. PhD thesis, University of Edinburgh, 1994.

[Fio94b] M. P. Fiore. First steps on the representation of domains. Manuscript, 1994.

[FP94] M. P. Fiore and G. D. Plotkin. An axiomatisation of computationally adequate domain theoretic models of FPC. In *9th LICS Conference*. IEEE, 1994.

[Gun92] C. A. Gunter. *Semantics of Programming Languages: Structures and Techniques*. The MIT Press, 1992.

[HO96] J. M. E. Hyland and C.-H. L. Ong. On full abstraction for PCF. Submitted for publication, 1996.

[HP90] H. Huwig and A. Poigne. A note on inconsistencies caused by fixpoints in a cartesian closed category. *Theoretical Computer Science*, 73, 1990.

[Law69] F. W. Lawvere. Diagonal arguments and cartesian closed categories. In *Category Theory, Homology Theory and their Applications II, LNM*, volume 92. Springer-Verlag, 1969.

[Mog86] E. Moggi. Categories of partial morphisms and the partial lambda-calculus. In *Proceedings Workshop on Category Theory and Computer Programming, Guildford 1985, LNCS*, volume 240. Springer-Verlag, 1986.

[Mog89] E. Moggi. Computational lambda-calculus and monads. In *4th LICS Conference*. IEEE, 1989.

[Pit95] A. M. Pitts. Operationally-based theories of program equivalence. Notes to accompany lectures given at the Summer School *Semantics and Logics of Computation*, Isaac Newton Institute for Mathematical Sciences, University of Cambridge, 1995.

[Plo73] G. D. Plotkin. Lambda-definability and logical relations. Memorandum SAI-RM-4, University of Edinburgh, 1973.

[Plo77] G. D. Plotkin. LCF considered as a programming language. *Theoretical Computer Science*, 5, 1977.

[Ros86] G. Rosolini. *Continuity and Effectiveness in Topoi*. PhD thesis, University of Oxford, 1986.

[Sco93] D. S. Scott. A type theoretical alternative to CUCH, ISWIM, OWHY. In *Böhm Festscrift, Theoretical Computer Science*, volume 121. Elsevier, 1993.

[Win87] G. Winskel. Event structures. In *LNCS*, volume 255. Springer-Verlag, 1987.

[Win93] G. Winskel. *The Formal Semantics of Programming Languages*. The MIT Press, 1993.

Logical Reconstruction of Bi-domains

Antonio Bucciarelli

Dipartimento di Scienze dell'Informazione,
Università di Roma "La Sapienza", via Salaria 113, 00198 Roma

Abstract. We introduce a technique based on logical relations, which, given two models M and N of a simply typed lambda-calculus L, allows us to construct a model M/N whose L-theory is a superset of both Th(M) and Th(N).
Keywords: simply typed λ-calculi, logical relations, PCF, Scott-continuous model, stable model.

1 Introduction

Two phrases of a programming language are said to be operationally equivalent if, when plugged into any context, they both produce the same result, if any.

Proving operational equivalences by syntactic means is usually difficult. Any model $\mathcal{M}$ provides a theory ($M =_{\mathcal{M}} N$ if $[\![M]\!]^{\mathcal{M}} = [\![N]\!]^{\mathcal{M}}$) which approximates, as a relation, the operational equivalence. From this point of view the natural way of comparing different models of a given language is to compare their theories: the closer the theory is to the operational theory, the better the model is.

On the other hand if we want to use models in order to prove operational equivalences, they should be effective, in the sense that, at least for finitary languages, their theories should be decidable.

An example of this trade-off is provided by the state of the art in the study of the finitary fragment of Scott's PCF ([18], see section 2.1): there exist fully abstract models[1] ([1, 12, 16]), but it has been recently shown that their theory is undecidable [14]. On the other hand there exist several models whose theory is decidable ([18, 4, 3, 8, 19], the list is not exhaustive see [9] for a survey), but each of them gives us only an approximation of the operational equivalence.

In order to obtain better approximations of the operational theory of PCF, several authors have proposed techniques for "refining" two given models $\mathcal{M}$ and $\mathcal{N}$, which roughly consist in cutting down the size of their function spaces by taking into account only functions which are both $\mathcal{M}$-functions and $\mathcal{N}$-functions. These approaches are usually referred to as "bidomain constructions" ([2, 7, 19]).

The ultimate goal of these constructions is to define models whose theory is a superset of both $\mathrm{Th}(\mathcal{M})$ and $\mathrm{Th}(\mathcal{N})$.

In this paper we show how this goal can be systematically achieved: given two models of PCF, $\mathcal{M}$ and $\mathcal{N}$, we start by defining their *extensional collapse*, roughly the logical relation which is the identity at ground types.

[1] A model is fully abstract when its theory coincides with the operational one.

Extensional collapses of the same kind have already been used in [5] in order to find relatively simple sufficient conditions which allow one to prove that $\mathrm{Th}(\mathcal{M}) \subseteq \mathrm{Th}(\mathcal{N})$. Applications may be found in [5] and [10].

Here we show that the extensional collapse of $\mathcal{M}$ and $\mathcal{N}$ induces a family of partial equivalence relations on $\mathcal{M}$, denoted by $\{\mathcal{Q}(\mathcal{M},\mathcal{N})^\sigma\}_{\sigma \in Types}$. Then we show that the quotienting of $\mathcal{M}$ by $\mathcal{Q}(\mathcal{M},\mathcal{N})$ yields a model, $\mathcal{M}/\mathcal{N}$, whose theory includes both the theory of $\mathcal{M}$ and that of $\mathcal{N}$.

The paper is structured as follows: in section 2 we introduce applied λ-calculi and their models, showing as examples the language PCF and its continuous model. In section 3 we introduce logical relations, and extensional collapses. Section 4 presents the construction of quotient models, and their properties. It should be noticed that, even though we focus on PCF here, the technique used allows one to define quotient models for any applied calculus, provided that the extensional collapse defined in section 3 is a logical relation for that calculus.

2 Applied typed λ-calculi

Following [15] we adopt the term *applied λ-calculus* to designate a simply typed λ-calculus endowed with a set of ground types, a set of typed constants and a set of rules (often called δ-rules) stipulating the operational behaviour of the constants. Constants may be first-order (e.g. arithmetical operations or conditional statements) or higher-order (e.g. fixpoint operators).

Models of applied λ-calculi can be presented in several way. Here we use the notion of *environment models*, which is general enough for our purposes, and we refer to [15] for a comprehensive discussion on the relations between this and other approaches (combinatorial, categorical).

2.1 Syntax

Given a set K of ground types, the set of types of the corresponding applied calculus is defined by the following BNF-expression:

$$T ::= K \mid T \rightarrow T$$

Given a family $\{C_\sigma\}_{\sigma \in T}$ of typed constants, and for any type σ a denumerable set Var^σ of variables, the set of terms of the corresponding applied calculus is defined by the following rules:

$$\frac{x \in Var^\sigma}{x : \sigma} \qquad \frac{c \in C_\sigma}{c : \sigma}$$

$$\frac{M : \sigma \rightarrow \tau \quad N : \sigma}{MN : \tau} \qquad \frac{x \in Var^\sigma \quad M : \tau}{\lambda x\, M : \sigma \rightarrow \tau}$$

The operational semantics of the language is provided by the usual β rule:

$$(\lambda x\, M)N \rightarrow^\beta [N/x]M$$

combined with the δ-rules specific to the constants.

A *program* P is a closed term of ground type, and it *converges* to a ground constant c if there exists a reduction path $P \to^* c$.

The language PCF As an example of applied λ-calculus, we introduce Scott's PCF ([18]). PCF has two ground types, **B** and **N**.

The ground constants are

$$C_{\mathbf{B}} = \{\text{true}, \text{false}\}$$

and

$$C_{\mathbf{N}} = \{\underline{0}, \underline{1}, \ldots, \underline{n}, \ldots\}$$

Other constants are succ, pred : $\mathbf{N} \to \mathbf{N}$, $Z : \mathbf{N} \to \mathbf{B}$ (the "test for zero"), $\mathtt{if} : \mathbf{B} \to (\mathbf{B} \to (\mathbf{B} \to \mathbf{B}))$ (boolean conditional statement) $\mathtt{intif} : \mathbf{B} \to (\mathbf{N} \to (\mathbf{N} \to \mathbf{N}))$ (integer conditional statement), and, for any type σ, $Y_\sigma : (\sigma \to \sigma) \to \sigma$ (fixpoint operators). The δ-rules concerning these constants are the usual ones, for instance $Y_\sigma M \to^\delta M(Y_\sigma M)$, and

$$\mathtt{if}\ \text{true}\ M\ N \to^\delta M \qquad\qquad \mathtt{if}\ \text{false}\ M\ N \to^\delta N$$

Two programs are *operationally equivalent* if they converge to the same ground constant, and two terms $M, N : \alpha$ are operationally equivalent if for any program context $P[\ ]$, $P[M]$ and $P[N]$ are.

Finitary PCF (FPCF) has only **B** as ground type. The ground constants are $\{\bot, \text{true}, \text{false}\}$ and the only non-ground constant is $\mathtt{if}$.

2.2 Models

A *typed applicative structure* $\mathcal{M}$ for an applied calculus L is a tuple

$$< \{\mathcal{M}^\sigma\}, \{\mathbf{App}^{\sigma,\tau}\}, \mathit{Const} >$$

such that:

- $\mathcal{M}^\sigma$ is a set.
- $\mathbf{App}^{\sigma,\tau}$ is a set-theoretic map $\mathbf{App}^{\sigma,\tau} : \mathcal{M}^{\sigma\to\tau} \to (\mathcal{M}^\sigma \to \mathcal{M}^\tau)$.
- *Const* maps each constant of L onto an element of the appropriate $\mathcal{M}^\sigma$.

If $\mathcal{M}^{\sigma\to\tau}$ is some set of functions from $\mathcal{M}^\sigma$ to $\mathcal{M}^\tau$ then $\mathbf{App}^{\sigma,\tau}$ will simply be function application, more generally if an applicative structure is built out of a cartesian closed category and $\mathcal{M}^{\sigma\to\tau}$ is the object $(\mathcal{M}^\tau)^{\mathcal{M}^\sigma}$, then $\mathbf{App}^{\sigma,\tau}$ is categorical application.

An applicative structure $\mathcal{M}$ is *extensional* if for all $\sigma, \tau \in T$, $f, g \in \mathcal{M}^{\sigma\to\tau}$, if for all $x \in \mathcal{M}^\sigma$ $\mathbf{App}^{\sigma,\tau} fx = \mathbf{App}^{\sigma,\tau} gx$ then $f = g$.

An *environment* ρ for an applicative structure $\mathcal{M}$ is a type-respecting mapping from variables of L to the union of $\mathcal{M}^\sigma$.

An *environment model* is an extensional applicative structure such that the clauses below define a total meaning function $[\![.]\!]$ which maps terms and environments to elements of the appropriate $\mathcal{M}^\sigma$. The interpretations of constants are required to validate the δ-rules (if $c \in C_{\sigma\to\tau}$ is such that, for a given $M : \sigma$, $cM \to^\delta N$, then for any ρ $\mathbf{App}^{\sigma,\tau}[\![c]\!]\rho[\![M]\!]\rho = [\![N]\!]\rho$).

- $[\![x]\!]\rho = \rho(x)$,
- $[\![c]\!]\rho = Const(c)$,
- $[\![MN]\!]\rho = \mathbf{App}^{\sigma,\tau}[\![M]\!]\rho[\![N]\!]\rho$ for the appropriate σ and τ,
- $[\![\lambda x\ M]\!]\rho = f \in \mathcal{M}^{\sigma\to\tau}$ such that for all $d \in \mathcal{M}^\sigma$ $\mathbf{App}^{\sigma,\tau} fd = [\![M]\!]\rho[d/x]$, for the appropriate σ and τ.

The extensionality of $\mathcal{M}$ guarantees that the function f defined in the last clause is unique.

It is easy to prove by structural induction that if M is closed then for all ρ, ρ', $[\![M]\!]\rho = [\![M]\!]\rho'$. Hence the interpretation of a closed term M will be called just $[\![M]\!]$.

The *theory* of an environment model $\mathcal{M}$, $\mathrm{Th}(\mathcal{M})$, is the set of closed equations $M = N$ such that $[\![M]\!] = [\![N]\!]$. We will assume that the models we are considering are *adequate*, that their theories are included in the operational equivalence.

The continuous model We define the well-known continuous model of PCF. Instead of getting it out of the general definition of the cartesian closed category of complete partial orders and Scott-continuous functions, we content ourselves with an inductive definition on PCF types: let $\mathcal{C}^{\mathbf{B}}$, $\mathcal{C}^{\mathbf{N}}$ be the flat domains of boolean values and natural numbers respectively:

$$\mathcal{C}^{\mathbf{B}} =< \{\bot, \mathtt{tt}, \mathtt{ff}\}, x \le y \text{ if and only if } x = \bot \text{ or } x = y >$$

$$\mathcal{C}^{\mathbf{N}} =< \{\bot, 0, 1, \ldots, n, \ldots\}, x \le y \text{ if and only if } x = \bot \text{ or } x = y >$$

and $\mathcal{C}^{\sigma\to\tau}$ be the cpo of Scott-continuous functions from $\mathcal{C}^\sigma$ to $\mathcal{C}^\tau$ ordered pointwise:

$$\mathcal{C}^{\sigma\to\tau} =< \{f : \mathcal{C}^\sigma \to \mathcal{C}^\tau \mid f \text{ continuous }\}, f \le g \text{ iff for all } d \in \mathcal{C}^\sigma\ f(d) \le g(d) >$$

In $\mathcal{C}$ $\mathbf{App}$ is simply functional application, that is $\mathbf{App}^{\sigma,\tau} fd = f(d)$.

The ground constants and first-order constants of PCF are interpreted by the corresponding ground values and first-order continuous functions, for instance:

$$[\![\mathtt{if}]\!]^{\mathcal{C}} xyz = \begin{cases} y & \text{if } x = \mathtt{tt} \\ z & \text{if } x = \mathtt{ff} \\ \bot & \text{otherwise} \end{cases}$$

and fixpoint constants are interpreted as domain-theoretic least-fixpoints operators: $[\![Y_\sigma]\!]^{\mathcal{C}} = \underline{\lambda} f \bigvee_{n\in\omega} f^n(\bot_\sigma)$.

3 Logical Relations and the Extensional Collapse

Logical relations have been extensively used in the study of both syntactic and semantic properties of typed λ-calculi (see [15] for a survey).

Here we just need a particular kind of binary logical relations, establishing a correspondence between two given models of PCF. Throughout this section the application $\mathbf{App}^{\sigma,\tau} fx$ in an applicative structure will be abbreviate by $f(x)$.

Definition 1. Let $\mathcal{M}$ and $\mathcal{N}$ be models of PCF; a *logical relation* between $\mathcal{M}$ and $\mathcal{N}$ is a family $\{R^\sigma\}_{\sigma \in T}$ such that:

- $\forall \sigma \in T \; R^\sigma \subseteq \mathcal{M}^\sigma \times \mathcal{N}^\sigma$.
- $\forall \sigma, \tau \in T \; \forall f \in \mathcal{M}^{\sigma \to \tau} \; \forall g \in \mathcal{N}^{\sigma \to \tau}$:

$$f R^{\sigma \to \tau} g \text{ if and only if } \forall x \in \mathcal{M}^\sigma \; \forall y \in \mathcal{N}^\sigma \; [\, x R^\sigma y \Rightarrow f(x) R^\tau g(y) \,]$$

- for any constant $c \in C_\sigma$ of PCF, $[\![c]\!]^{\mathcal{M}} R^\sigma [\![c]\!]^{\mathcal{N}}$.

We will denote the fact that $\{R^\sigma\}_{\sigma \in T}$ is a logical relation between $\mathcal{M}$ and $\mathcal{N}$ by $R \subseteq \mathcal{M} \times \mathcal{N}$.

The following is the basic lemma of logical relations:

Lemma 2. *If $R \subseteq \mathcal{M} \times \mathcal{N}$ then for any closed term $M : \sigma$ of PCF: $[\![M]\!]^{\mathcal{M}} R^\sigma [\![M]\!]^{\mathcal{N}}$.*

As shown by Friedman in [11], logical relations with certain properties are particularly well suited for studying λ-theories:

Definition 3. A logical relation $R \subseteq \mathcal{M} \times \mathcal{N}$ is a *logical partial function* if

$$\forall \sigma \in T \; \forall x \in \mathcal{M}^\sigma \; \forall y, y' \in \mathcal{N}^\sigma \; (x R^\sigma y \text{ and } x R^\sigma y') \Rightarrow y = y'$$

Lemma 4. *If $R \subseteq \mathcal{M} \times \mathcal{N}$ is a logical partial function, then $\mathrm{Th}(\mathcal{M}) \subseteq \mathrm{Th}(\mathcal{N})$.*

Proof. Let $M, N : \sigma$ be closed terms such that $[\![M]\!]^{\mathcal{M}} = [\![N]\!]^{\mathcal{M}}$. By the basic lemma we have that $[\![M]\!]^{\mathcal{M}} R^\sigma [\![M]\!]^{\mathcal{N}}$ and $[\![N]\!]^{\mathcal{M}} R^\sigma [\![N]\!]^{\mathcal{N}}$. Hence, since R is a logical partial function, we conclude that $[\![M]\!]^{\mathcal{N}} = [\![N]\!]^{\mathcal{N}}$

The following lemma provides sufficient conditions for a logical relation to be a logical partial function:

Definition 5. A logical relation $R \subseteq \mathcal{M} \times \mathcal{N}$ is *surjective* if

$$\forall \sigma \in T \forall y \in \mathcal{N}^\sigma \exists x \in \mathcal{M}^\sigma \; x R^\sigma y$$

Lemma 6. *If $R \subseteq \mathcal{M} \times \mathcal{N}$ is surjective, and if for any ground type k R^k is a logical partial function, then R is a logical partial function.*

Proof. By structural induction on types. By hypothesis R is a logical partial function at ground types. Let us suppose that R^σ and R^τ are logical partial functions and prove that $R^{\sigma \to \tau}$ also is. Let $f \in \mathcal{M}^{\sigma \to \tau}$, $g, g' \in \mathcal{N}^{\sigma \to \tau}$ be such that $f R^{\sigma \to \tau} g$ and $f R^{\sigma \to \tau} g'$. Let $y \in \mathcal{N}^\sigma$. By surjectivity of R there exists $x \in \mathcal{M}^\sigma$ such that $x R^\sigma y$; hence $f(x) R^\tau g(y)$ and $f(x) R^\tau g'(y)$. Since R^τ is a logical partial function we conclude that $g(y) = g'(y)$. Hence for all $y \in \mathcal{N}^\sigma$ $g(y) = g'(y)$, and, by extensionality of $\mathcal{N}$, $g = g'$.

3.1 The extensional collapse

The extensional collapse of $\mathcal{M}$ and $\mathcal{N}$ is the relation which, at ground types, is the intersection of all logical relations between $\mathcal{M}$ and $\mathcal{N}$.

Definition 7. Given two models $\mathcal{M}$, $\mathcal{N}$ of PCF, we define for any ground type k of PCF

$$\mathcal{E}(\mathcal{M},\mathcal{N})^k = \bigcap_{R \subseteq \mathcal{M}\times\mathcal{N}} R^k$$

The family of relations $\{\mathcal{E}(\mathcal{M},\mathcal{N})^\sigma\}_{\sigma\in T}$ (when σ is a function type $\mathcal{E}(\mathcal{M},\mathcal{N})^\sigma$ is obtained as in definition 1) is the *extensional collapse* of $\mathcal{M}$ and $\mathcal{N}$, denoted by $\mathcal{E}(\mathcal{M},\mathcal{N})$.

Given two models $\mathcal{M}$ and $\mathcal{N}$ of PCF, it is easy to see that the relation defined by

$$R^{\mathbf{B}} = \{(\llbracket \mathrm{true} \rrbracket^{\mathcal{M}}, \llbracket \mathrm{true} \rrbracket^{\mathcal{N}}), (\llbracket \mathrm{false} \rrbracket^{\mathcal{M}}, \llbracket \mathrm{false} \rrbracket^{\mathcal{N}})\}$$

$$R^{\mathbf{N}} = \{\llbracket \underline{0} \rrbracket^{\mathcal{M}}, \llbracket \underline{0} \rrbracket^{\mathcal{N}}), \ldots, (\llbracket \underline{n} \rrbracket^{\mathcal{M}}, \llbracket \underline{n} \rrbracket^{\mathcal{N}}), \ldots\}$$

is not logical, since, for instance, it does not relate the interpretations in $\mathcal{M}$ and $\mathcal{N}$ of the closed PCF-term $Y_{\mathbf{B}}(\lambda x\ x)$. Quite naturally, it turns out that by adding to $R^{\mathbf{B}}$ and $R^{\mathbf{N}}$ the pairs composed by the fixed points of the identity, we get the extensional collapse for PCF. In order to prove this, we have to make some "minimal" assumptions on $\mathcal{M}$ and $\mathcal{N}$:

- $\mathcal{M}$ and $\mathcal{N}$ are least fixpoint models, i.e. the $\mathcal{M}^\sigma$'s and $\mathcal{N}^\sigma$s are domains, and Y_σ is interpreted in both models by $\underline{\lambda} f \bigvee_{n\in\omega} f^n(\bot_\sigma)$.
- The order of higher-type domains is included in the pointwise order.
- The pointwise least upper bound of a directed family of functions is again in the model.
- $\mathcal{M}^{\mathbf{B}}$, $\mathcal{M}^{\mathbf{N}}$, $\mathcal{N}^{\mathbf{B}}$, $\mathcal{N}^{\mathbf{N}}$ do not contain infinite increasing chains.

Proposition 8. *Given two models $\mathcal{M}$ and $\mathcal{N}$ of PCF, the relation defined by*

$$R^{\mathbf{B}} = \{(\llbracket Y_{\mathbf{B}}(\lambda x\ x) \rrbracket^{\mathcal{M}}, \llbracket Y_{\mathbf{B}}(\lambda x\ x) \rrbracket^{\mathcal{N}}), (\llbracket \mathrm{true} \rrbracket^{\mathcal{M}}, \llbracket \mathrm{true} \rrbracket^{\mathcal{N}}), (\llbracket \mathrm{false} \rrbracket^{\mathcal{M}}, \llbracket \mathrm{false} \rrbracket^{\mathcal{N}})\}$$

$$R^{\mathbf{N}} = \{(\llbracket Y_{\mathbf{N}}(\lambda x\ x) \rrbracket^{\mathcal{M}}, \llbracket Y_{\mathbf{N}}(\lambda x\ x) \rrbracket^{\mathcal{N}}), \llbracket \underline{0} \rrbracket^{\mathcal{M}}, \llbracket \underline{0} \rrbracket^{\mathcal{N}}), \ldots, (\llbracket \underline{n} \rrbracket^{\mathcal{M}}, \llbracket \underline{n} \rrbracket^{\mathcal{N}}), \ldots\}$$

is logical, and it is $\mathcal{E}_{PCF}(\mathcal{M},\mathcal{N})$.

Proof. R does relate the interpretations of ground constants by definition. Easy by-cases arguments prove that the interpretations of first-order constants are also related. It remains to prove that, for any type σ, $[\![Y_\sigma]\!]^{\mathcal{M}} R^{(\sigma\to\sigma)\to\sigma} [\![Y_\sigma]\!]^{\mathcal{N}}$. It is enough to prove that R is *directed complete* (see [15]), i.e. that for any σ, if $D \subseteq R^\sigma$ is directed in $\mathcal{M}^\sigma \times \mathcal{N}^\sigma$, then $\bigvee D \in R^\sigma$. We prove this by induction on types: if σ is ground and $D \subseteq R^\sigma$ is directed then $\bigvee D \in D$, by our fourth assumption, and we are done. Let $D \subseteq R^{\sigma\to\tau}$ be directed, and let $\bigvee D = (f, g)$. If $(x, y) \in R^\sigma$ then by the second assumption $D' = \{(f'(x), g'(y)) \mid (f', g') \in D\}$ is a directed set included in R^τ, hence by hypothesis $\bigvee D' \in R^\tau$. By the second and third assumptions above, $\bigvee D' = (f(x), g(y))$, hence $f R^{\sigma\to\tau} g$ and we are done.

In order to show that $R = \mathcal{E}(\mathcal{M},\mathcal{N})$ it is enough to remark that for any logical relation $S \subseteq \mathcal{M} \times \mathcal{N}$, $R^{\mathbf{B}} \subseteq S^{\mathbf{B}}$ and $R^{\mathbf{N}} \subseteq S^{\mathbf{N}}$.

Moreover, the extensional collapse is a logical partial function at ground types:

Fact 1: Given two models of PCF $\mathcal{M}$ and $\mathcal{N}$, if σ is ground then $\mathcal{E}(\mathcal{M},\mathcal{N})^\sigma$ is a logical partial function. ∎

Proof. We have seen that, at a ground type σ, $\mathcal{E}(\mathcal{M},\mathcal{N})$ relates the pairs (x, y) formed by the interpretations of ground constants or by the interpretations of the divergent term $Y_\sigma(\lambda x\, x)$. Then the result follows by the adequacy of $\mathcal{M}$ and $\mathcal{N}$.

4 Quotient Models

In this section, we show how the extensional collapse of $\mathcal{M}$ and $\mathcal{N}$ can be used to define a quotient model $\mathcal{M}/\mathcal{N}$ whose theory is a superset of both $\mathtt{Th}(\mathcal{M})$ and $\mathtt{Th}(\mathcal{N})$. Throughout this section "model" will stand for "model of PCF". Moreover "$\mathcal{E}(\mathcal{M},\mathcal{N})$" will often be abbreviated by "$\mathcal{E}$" and the "σ" in "$\mathcal{E}(\mathcal{M},\mathcal{N})^\sigma$", "$\mathcal{Q}(\mathcal{M},\mathcal{N})^\sigma$" will sometimes be omitted.

Definition 9. Given two models $\mathcal{M}$ and $\mathcal{N}$, let $\mathcal{Q}(\mathcal{M},\mathcal{N}) = \{\mathcal{Q}(\mathcal{M},\mathcal{N})^\sigma\}_{\sigma\in T}$ be the family of relations defined by:

- for any σ, $\mathcal{Q}(\mathcal{M},\mathcal{N})^\sigma \subseteq \mathcal{M}^\sigma \times \mathcal{M}^\sigma$
- for any σ and $x, x' \in \mathcal{M}^\sigma$, $x\mathcal{Q}(\mathcal{M},\mathcal{N})^\sigma x'$ if and only if $\exists y \in \mathcal{N}^\sigma$ such that $x\mathcal{E}(\mathcal{M},\mathcal{N})^\sigma y$ and $x'\mathcal{E}(\mathcal{M},\mathcal{N})^\sigma y$

In order to prove that $\mathcal{Q}(\mathcal{M},\mathcal{N})^\sigma$ is a partial equivalence relation for any σ, we need the following simple lemma:

Lemma 10. *Given two models $\mathcal{M}$ and $\mathcal{N}$ and a type σ, let $x, x' \in \mathcal{M}^\sigma$, $y, y' \in \mathcal{N}^\sigma$ be such that $x\mathcal{E}y$, $x'\mathcal{E}y$, $x'\mathcal{E}y'$. Then $x\mathcal{E}y'$.*

Proof. By induction on types: if σ is ground, then by adequacy of $\mathcal{M}$ and $\mathcal{N}$, $\mathcal{E}^\sigma$ is a (partial) function, hence $y = y'$ and we are done. If $\sigma = \sigma_1 \to \sigma_2$ then let $z \in \mathcal{M}^{\sigma_1}$, $t \in \mathcal{N}^{\sigma_1}$ be such that $z\mathcal{E}t$. Then by hypothesis $x(z)\mathcal{E}y(t)$, $x'(z)\mathcal{E}y(t)$ and $x'(z)\mathcal{E}y'(t)$. Using the inductive hypothesis on σ_2 we get $x(z)\mathcal{E}y'(t)$ and we are done.

Proposition 11. *Given two models $\mathcal{M}$ and $\mathcal{N}$, for all types σ $\mathcal{Q}(\mathcal{M},\mathcal{N})^\sigma$ is a partial equivalence relation on $\mathcal{M}^\sigma$.*

Proof. $\mathcal{Q}(\mathcal{M},\mathcal{N})^\sigma$ is clearly symmetric. Let $x, x', x'' \in \mathcal{M}^\sigma$ be such that $x\mathcal{Q}(\mathcal{M},\mathcal{N})^\sigma x'$ and $x'\mathcal{Q}(\mathcal{M},\mathcal{N})^\sigma x''$. Then $x\mathcal{Q}(\mathcal{M},\mathcal{N})^\sigma x''$ follows easily from the previous lemma.

If for every σ $\mathcal{Q}(\mathcal{M},\mathcal{N})^\sigma$ is reflexive, and hence $\mathcal{Q}(\mathcal{M},\mathcal{N})$ is the identity, then the theory of $\mathcal{M}$ is bigger than the one of $\mathcal{N}$:

Proposition 12. *Given two models $\mathcal{M}$ and $\mathcal{N}$, if for every σ and for every $x \in \mathcal{M}^\sigma$ $x\mathcal{Q}(\mathcal{M},\mathcal{N})^\sigma x$ then $\mathcal{Q}(\mathcal{M},\mathcal{N})$ is the identity relation at all types, and moreover* $\mathrm{Th}(\mathcal{N}) \subseteq \mathrm{Th}(\mathcal{M})$.

Proof. If $\mathcal{Q}(\mathcal{M},\mathcal{N})$ is reflexive, then $\mathcal{E}(\mathcal{N},\mathcal{M})$ is surjective. Hence the result follows from Fact 1 and lemmas 6 and 4.

In general there exist elements of $\mathcal{M}$ which are not related to themselves by $\mathcal{Q}(\mathcal{M},\mathcal{N})$: for instance if $\mathcal{L}$ and $\mathcal{C}$ are the lattice-theoretic and continuous models respectively, $\top\neg\mathcal{Q}(\mathcal{L},\mathcal{C})^{\mathbf{B}}\top$, and, if $\mathcal{S}$ is the stable model (see [3]), and *por* the continuous parallel-or function, $por\neg\mathcal{Q}(\mathcal{C},\mathcal{S})^{\mathbf{B}\to(\mathbf{B}\to\mathbf{B})}por$. It turns out that whenever $\mathcal{Q}(\mathcal{M},\mathcal{N})$ is not the identity at all types, it is not a logical relation. A model $\mathcal{M}$ of PCF is *standard* if the interpretations of **B** and **N** in $\mathcal{M}$ are the flat domains of boolean values and natural numbers, respectively:

Proposition 13. *Let $\mathcal{M}$ be a standard model and $\mathcal{N}$ a model. Then $\mathcal{Q}(\mathcal{M},\mathcal{N})$ is a logical relation if and only if it is the identity relation at all types.*

Proof. Since $\mathcal{M}$ is extensional, the identity at all types is clearly a logical relation. Conversely, let σ be a minimal type such that there exists $f \in \mathcal{M}^\sigma$, $f\neg\mathcal{Q}^\sigma f$. Since $\mathcal{M}$ is standard, $\sigma = \sigma_1 \to \sigma_2$. If we show that for all $x, x' \in \mathcal{M}^{\sigma_1}$, if $x\mathcal{Q}x'$ then $f(x)\mathcal{Q}f(x')$ we are done. This is clearly the case if $\mathcal{Q}^{\sigma_1}$ is the identity. Since $\mathcal{M}$ is standard, $\mathcal{Q}$ is the identity at ground types, and by minimality of σ we know that $\mathcal{E}(\mathcal{N},\mathcal{M})$ is surjective at every subtype of σ_1. It is easy to rephrase the proof of lemma 6 in order to prove that under these hypothesis $\mathcal{Q}^{\sigma_1}$ is the identity relation.

Given $\mathcal{M}$ and $\mathcal{N}$ the idea for defining a quotiented structure is to take as elements equivalence classes with respect to $\mathcal{Q}(\mathcal{M},\mathcal{N})$, and to define the application by $\mathbf{App}^{\sigma,\tau}[f][x] = [f(x)]$. Since $\mathcal{Q}(\mathcal{M},\mathcal{N})$ is not a logical relation in general, we have to check that application is well defined in the quotiented structure:

Proposition 14. *If* $f\mathcal{Q}(\mathcal{M},\mathcal{N})^{\sigma\to\tau}f'$ *and* $x\mathcal{Q}(\mathcal{M},\mathcal{N})^{\sigma}x'$ *then* $f(x)\mathcal{Q}(\mathcal{M},\mathcal{N})^{\tau}f'(x')$.

Proof. By hypothesis there exist $g \in \mathcal{N}^{\sigma\to\tau}$, $y \in \mathcal{N}^{\sigma}$ such that $f\mathcal{E}(\mathcal{M},\mathcal{N})g$, $f'\mathcal{E}(\mathcal{M},\mathcal{N})g$, $x\mathcal{E}(\mathcal{M},\mathcal{N})y$ and $x'\mathcal{E}(\mathcal{M},\mathcal{N})y$. Hence $f(x)\mathcal{E}(\mathcal{M},\mathcal{N})g(y)$ and $f'(x')\mathcal{E}(\mathcal{M},\mathcal{N})g(y)$ and we are done.

Definition 15. Given two models $\mathcal{M},\mathcal{N}$ let $\mathcal{M}/\mathcal{N}$ be the applicative structure defined by

$$\mathcal{M}/\mathcal{N} =< \{\mathcal{M}^{\sigma}/\mathcal{Q}(\mathcal{M},\mathcal{N})^{\sigma}\}_{\sigma\in T}, \mathbf{App}_{\mathcal{M}/\mathcal{N}}, Const_{\mathcal{M}/\mathcal{N}} >$$

where $\mathbf{App}^{\sigma,\tau}_{\mathcal{M}/\mathcal{N}}[f][x] = [\mathbf{App}^{\sigma,\tau}_{\mathcal{M}} fx]$, and $Const_{\mathcal{M}/\mathcal{N}}(c) = [Const_{\mathcal{M}}(c)]$.

Proposition 16. *If* $\mathcal{M},\mathcal{N}$ *are models, then* $\mathcal{M}/\mathcal{N}$ *is a model and for any closed term* M, $[\![M]\!]^{\mathcal{M}/\mathcal{N}} = [[\![M]\!]^{\mathcal{M}}]$.

Proof. $\mathcal{M}/\mathcal{N}$ is an applicative structure.

Let us prove that $\mathcal{M}/\mathcal{N}$ is extensional: if $[f],[g] \in \mathcal{M}/\mathcal{N}^{\sigma\to\tau}$ are such that for all $[x] \in \mathcal{M}/\mathcal{N}^{\sigma}$ $[f(x)] = [g(x)]$, we have to prove that $[f] = [g]$. Since there exists $h \in \mathcal{N}^{\sigma\to\tau}$ such that $g\mathcal{E}h$, it is enough to prove that $f\mathcal{E}h$. Let $x \in \mathcal{M}^{\sigma}$, $y \in \mathcal{N}^{\sigma}$ be such that $x\mathcal{E}y$. Since $[f(x)] = [g(x)]$, we know that there exists $z \in \mathcal{N}^{\tau}$ such that $f(x)\mathcal{E}z$ and $g(x)\mathcal{E}z$. Moreover $g(x)\mathcal{E}h(y)$, and by applying lemma 10, we get $f(x)\mathcal{E}h(y)$, and we are done.

The facts that, for any term M and any $\mathcal{M}/\mathcal{N}$-environment ρ, $[\![M]\!]^{\mathcal{M}/\mathcal{N}}\rho$ as specified in subsection 2.2 does exist, and that for closed M, $[\![M]\!]^{\mathcal{M}/\mathcal{N}} = [[\![M]\!]^{\mathcal{M}}]$ are proved by a single induction on open terms: given a $\mathcal{M}$-environment ρ' such that, for all variables x, $\rho(x) = [\rho'(x)]$, we show that $[\![M]\!]^{\mathcal{M}/\mathcal{N}}\rho = [[\![M]\!]^{\mathcal{M}}\rho']$. If M is a variable, a constant or an application this is trivially the case. If $M = \lambda x\ N : \sigma \to \tau$ we have to prove that, for all $d \in \mathcal{M}/\mathcal{N}^{\sigma}$, $[\![N]\!]^{\mathcal{M}/\mathcal{N}}\rho[d/x] = [[\![\lambda x\ N]\!]^{\mathcal{M}}\rho'](d)$. Let $a \in \mathcal{M}^{\sigma}$ be such that $[a] = d$; we have that

$$[[\![\lambda x\ N]\!]^{\mathcal{M}}\rho'](d) = [[\![\lambda x\ N]\!]^{\mathcal{M}}\rho']([a]) = [[\![\lambda x\ N]\!]^{\mathcal{M}}\rho'(a)] = [[\![N]\!]^{\mathcal{M}}\rho'[a/x]]$$

and, using the inductive hypothesis, we are done.

The model $\mathcal{M}/\mathcal{N}$ is a refinement of $\mathcal{M}$ with respect to $\mathcal{N}$. Of course, given $\mathcal{E}(\mathcal{M},\mathcal{N})$, we can symmetrically construct the model $\mathcal{N}/\mathcal{M}$. The following proposition shows that, as far as PCF-theories are concerned, the two choices are equivalent, and that the theory of the refined model is indeed a superset of those of $\mathcal{M}$ and $\mathcal{N}$.

Proposition 17. *Given two models* $\mathcal{M},\mathcal{N}$, *we have that:*

i) $\mathrm{Th}(\mathcal{M}) \bigcup \mathrm{Th}(\mathcal{N}) \subseteq \mathrm{Th}(\mathcal{M}/\mathcal{N})$.
ii) $\mathrm{Th}(\mathcal{M}/\mathcal{N}) = \mathrm{Th}(\mathcal{N}/\mathcal{M})$.

Proof. The first part of this proposition follows from the second one, and the fact that $\text{Th}(\mathcal{M}) \subseteq \text{Th}(\mathcal{M}/\mathcal{N})$, which is a consequence of the previous proposition.

As for $\text{Th}(\mathcal{M}/\mathcal{N}) = \text{Th}(\mathcal{N}/\mathcal{M})$, let us just prove that $\text{Th}(\mathcal{M}/\mathcal{N}) \subseteq \text{Th}(\mathcal{N}/\mathcal{M})$ (the symmetric inclusion can be proved in the same way): let $M, N : \sigma$ be terms such that $[\![M]\!]^{\mathcal{M}/\mathcal{N}} = [\![N]\!]^{\mathcal{M}/\mathcal{N}}$. Hence there exists $g \in \mathcal{N}^\sigma$ such that $[\![M]\!]^{\mathcal{M}}\mathcal{E}(\mathcal{M},\mathcal{N})g$ and $[\![N]\!]^{\mathcal{M}}\mathcal{E}(\mathcal{M},\mathcal{N})g$. Moreover by the basic lemma of logical relations we know that $[\![M]\!]^{\mathcal{M}}\mathcal{E}(\mathcal{M},\mathcal{N})[\![M]\!]^{\mathcal{N}}$ and $[\![N]\!]^{\mathcal{M}}\mathcal{E}(\mathcal{M},\mathcal{N})[\![N]\!]^{\mathcal{N}}$. Hence $[\![M]\!]^{\mathcal{N}}\mathcal{Q}(\mathcal{N},\mathcal{M})g$ and $[\![N]\!]^{\mathcal{N}}\mathcal{Q}(\mathcal{N},\mathcal{M})g$, and by transitivity of $\mathcal{Q}(\mathcal{N},\mathcal{M})$, we get $[\![M]\!]^{\mathcal{N}}\mathcal{Q}(\mathcal{N},\mathcal{M})[\![N]\!]^{\mathcal{N}}$, i.e. $[\![M]\!]^{\mathcal{N}/\mathcal{M}} = [\![N]\!]^{\mathcal{N}/\mathcal{M}}$.

In order to show that in general $\text{Th}(\mathcal{M}/\mathcal{N})$ is not included in $\text{Th}(\mathcal{M}) \bigcup \text{Th}(\mathcal{N})$ let us consider the following example:

Example 1. Let $\mathcal{C}$ and $\mathcal{S}$ be the continuous and stable (see [3]) models of PCF respectively. We start by defining two pairs of closed terms, P_1, P_2 and Q_1, Q_2 such that

$$[\![P_1]\!]^{\mathcal{C}} \neq [\![P_2]\!]^{\mathcal{C}} \text{ and } [\![P_1]\!]^{\mathcal{S}} = [\![P_2]\!]^{\mathcal{S}}$$

$$[\![Q_1]\!]^{\mathcal{C}} = [\![Q_2]\!]^{\mathcal{C}} \text{ and } [\![Q_1]\!]^{\mathcal{S}} \neq [\![Q_2]\!]^{\mathcal{S}}$$

and then we combine them constructing two closed terms T_1, T_2 such that

$$[\![T_1]\!]^{\mathcal{C}} \neq [\![T_2]\!]^{\mathcal{C}}\ ,\ [\![T_1]\!]^{\mathcal{S}} \neq [\![T_2]\!]^{\mathcal{S}} \text{ and } [\![T_1]\!]^{\mathcal{C}/\mathcal{S}} = [\![T_2]\!]^{\mathcal{C}/\mathcal{S}}$$

As for P_1, P_2 we choose Plotkin's terms showing that $\mathcal{C}$ is not fully abstract ([17]): $P_1, P_2 : (\mathbf{B} \to (\mathbf{B} \to \mathbf{B})) \to \mathbf{B}$

$$P_1 = \lambda f\ \texttt{if}\ (f\ \text{true}\ \Omega)\ (\texttt{if}\ (f\ \Omega\ \text{true})\ (\texttt{if}\ (f\ \text{false false})\ \Omega\ \text{true})\ \Omega)\ \Omega$$

$$P_2 = \lambda f\ \texttt{if}\ (f\ \text{true}\ \Omega)\ (\texttt{if}\ (f\ \Omega\ \text{true})\ (\texttt{if}\ (f\ \text{false false})\ \Omega\ \text{false})\ \Omega)\ \Omega$$

Since $[\![P_1]\!]^{\mathcal{C}} por = \texttt{tt}$ and $[\![P_2]\!]^{\mathcal{C}} por = \texttt{ff}$ (*por* being the continuous "parallel-or" function), we have that $[\![P_1]\!]^{\mathcal{C}} \neq [\![P_2]\!]^{\mathcal{C}}$, whereas it is easy to see that

$$[\![P_1]\!]^{\mathcal{S}} = [\![P_2]\!]^{\mathcal{S}} = \underline{\lambda} f\ \bot$$

For defining Q_1, Q_2 we use the fact that $\mathcal{S}$ does have "or-testers": consider the terms $\text{l}-\text{or}, \text{r}-\text{or} : \mathbf{B} \to (\mathbf{B} \to \mathbf{B})$ defined by:

$$\text{l}-\text{or} = \lambda x \lambda y\ \texttt{if}\ x\ \text{true}\ y$$

$$\text{r}-\text{or} = \lambda x \lambda y\ \texttt{if}\ y\ \text{true}\ x$$

Let $Q_1, Q_2 : ((\mathbf{B} \to (\mathbf{B} \to \mathbf{B})) \to \mathbf{B}) \to \mathbf{B}$ be the following terms:

$$Q_1 = \lambda F\ \texttt{if}\ (F\ \text{l}-\text{or})\ (\texttt{if}\ (F\ \text{r}-\text{or})\ \Omega\ \mathit{true})\ \Omega$$

$$Q_2 = \lambda F\ \texttt{if}\ (F\ \text{l}-\text{or})\ (\texttt{if}\ (F\ \text{l}-\text{or})\ \Omega\ \mathit{false})\ \Omega$$

The stable model contains an "or-tester", i.e. a functional $F \in \mathcal{S}^{(\mathbf{B}\to(\mathbf{B}\to\mathbf{B}))\to\mathbf{B}}$ such that $F([\![\mathrm{l-or}]\!]^{\mathcal{S}}) = \mathtt{tt}$ and $F([\![\mathrm{r-or}]\!]^{\mathcal{S}}) = \mathtt{ff}$, hence

$$[\![Q_1]\!]^{\mathcal{S}} F = \mathtt{tt} \text{ and } [\![Q_2]\!]^{\mathcal{S}} F = \mathtt{ff}$$

whereas no continuous functionals can yield unbounded values on $[\![\mathrm{l-or}]\!]^{\mathcal{C}}$ and $[\![\mathrm{r-or}]\!]^{\mathcal{C}}$, since these two functions are bounded. Hence

$$[\![Q_1]\!]^{\mathcal{C}} = [\![Q_2]\!]^{\mathcal{C}} = \underline{\lambda} F\ \bot$$

We can now define $T_1, T_2 : \sigma \to (\tau \to (\mathbf{B} \to \mathbf{B}))$ where $\sigma = \mathbf{B} \to (\mathbf{B} \to \mathbf{B})$ and $\tau = \sigma \to \mathbf{B}$, as follows:

$$T_i = \lambda f\ \lambda F\ \lambda x\ \mathtt{if}\ x\ (P_i\ f)\ (Q_i\ F)$$

It is trivial that $[\![T_1]\!]^{\mathcal{C}} \neq [\![T_2]\!]^{\mathcal{C}}$ and $[\![T_1]\!]^{\mathcal{S}} \neq [\![T_2]\!]^{\mathcal{S}}$; hence we are left with the proof of $[\![T_1]\!]^{\mathcal{C}/\mathcal{S}} = [\![T_2]\!]^{\mathcal{C}/\mathcal{S}}$.

In order to prove it we have to show a stable functional $G \in \mathcal{S}^{\sigma\to(\tau\to(\mathbf{B}\to\mathbf{B}))}$ such that

$$[\![T_1]\!]^{\mathcal{C}} \mathcal{E}(\mathcal{C},\mathcal{S}) G \text{ and } [\![T_2]\!]^{\mathcal{C}} \mathcal{E}(\mathcal{C},\mathcal{S}) G$$

We show that $G = \underline{\lambda} f \underline{\lambda} F \underline{\lambda} x\ \bot$ is such a functional. Let $b \in \mathcal{C}^{\mathbf{B}}$, $b' \in \mathcal{S}^{\mathbf{B}}$, $f \in \mathcal{C}^{\sigma}$, $f' \in \mathcal{S}^{\sigma}$, $F \in \mathcal{C}^{\tau}$, $F' \in \mathcal{S}^{\tau}$, be such that $b\mathcal{E}(\mathcal{C},\mathcal{S})b'$, $f\mathcal{E}(\mathcal{C},\mathcal{S})f'$ and $F\mathcal{E}(\mathcal{C},\mathcal{S})F'$. Then $f \neq \mathit{por}$, since no stable function is $\mathcal{E}$-related to *por*, hence we conclude that

$$[\![T_1]\!]^{\mathcal{C}}(f)(F)(b) = [\![T_2]\!]^{\mathcal{C}}(f)(F)(b) = \bot\ \mathcal{E}(\mathcal{C},\mathcal{S}) G(f')(F')(b')$$

4.1 Extensional collapses for Finitary PCF

In order to show that the extensional collapse $\mathcal{E}(\mathcal{M},\mathcal{N})$ is a logical relation for the full PCF, we introduced in section 3.1 some conditions that $\mathcal{M}$ and $\mathcal{N}$ have to satisfy. Those conditions are meant to ensure that the interpretations of fixpoint constants in $\mathcal{M}$ and $\mathcal{N}$ are $\mathcal{E}$-related.

For the finitary fragment FPCF , defined in section 2.1, we can show that $\mathcal{E}(\mathcal{M},\mathcal{N})$ is logical without any assumption on $\mathcal{M}$, $\mathcal{N}$:

Proposition 18. *Let $\mathcal{M}$ and $\mathcal{N}$ be models of FPCF . The relation defined by*

$$R^{\mathbf{B}} = \{([\![\bot]\!]^{\mathcal{M}}, [\![\bot]\!]^{\mathcal{N}}), ([\![\mathrm{true}]\!]^{\mathcal{M}}, [\![\mathrm{true}]\!]^{\mathcal{N}}), ([\![\mathrm{false}]\!]^{\mathcal{M}}, [\![\mathrm{false}]\!]^{\mathcal{N}})\}$$

is logical, and it is $\mathcal{E}_{\ FPCF}(\mathcal{M},\mathcal{N})$.

Proof. We have just to prove that for any constant $c : \sigma$ of FPCF , $[\![c]\!]^{\mathcal{M}} R^{\sigma} [\![c]\!]^{\mathcal{N}}$. For ground constants the statement above is true by definition of R, and an easy argument by cases prove that if the interpretations of ground constants in $\mathcal{M}$ and $\mathcal{N}$ are related, so are the interpretations of the $\mathtt{if}$ constant.

In order to show that $R = \mathcal{E}_{\ \mathrm{FPCF}}(\mathcal{M},\mathcal{N})$ it is enough to remark that for any logical relation $S \subseteq \mathcal{M} \times \mathcal{N}$, $R^{\mathbf{B}} \subseteq S^{\mathbf{B}}$.

Let us call *finitary* a model of FPCF whose theory is decidable and such that at any type σ $\mathcal{M}^\sigma$ is finite.

It is worth noticing that whenever $\mathcal{M}$ and $\mathcal{N}$ are finitary, also $\mathcal{Q}(\mathcal{M},\mathcal{N})$ is finitary.

By iterating our technique, given a finite set $\mathcal{M}_1,\ldots,\mathcal{M}_m$ of finitary models of FPCF we can construct a finitary model $\mathcal{M}$ such that $\mathtt{Th}(\mathcal{M}_i) \subseteq \mathtt{Th}(\mathcal{M})$ for all i, hence we can effectively approximate the operational theory of FPCF , which is undecidable [14].

5 Conclusion

The advantage of the *ad hoc* bidomain constructions of [2, 7, 19] with respect to our systematic approach based on logical relations, is that they give more insight into the internal structure of bidomains.

The drawback is that inclusion of theories is more difficult to achieve by *ad hoc* techniques. For instance the theory of Berry's bidomain model, which was meant to refine the continuous and stable models, does not include the continuous theory ([13, 7]).

Concerning the model of *extensional embeddings*, $\mathcal{EE}$, presented in [7], which refines the strongly stable model $\mathcal{SS}$ [6] with respect to the continuous one, it has been proved that $\mathtt{Th}(\mathcal{C}) \subset \mathtt{Th}(\mathcal{EE})$.

Similarly Winskel's model of *extended bistructures*, $\mathcal{EB}$ [19], refining $\mathcal{C}$ and $\mathcal{S}$ is such that $\mathtt{Th}(\mathcal{C}) \subset \mathtt{Th}(\mathcal{EB})$.

Nevertheless it turns out that $\mathtt{Th}(\mathcal{SS}) \not\subseteq \mathtt{Th}(\mathcal{EE})$, and we conjecture that $\mathtt{Th}(\mathcal{S}) \not\subseteq \mathtt{Th}(\mathcal{EE})$.

References

1. S. Abramsky, R. Jagadeesan, P. Malacaria. *Full abstraction for PCF (Extended Abstract)*. Proc. of TACS 94, Lecture Notes in Computer Science 789, 1-15, Springer, 1994.
2. G. Berry. *Modèles complètement adéquats et stables des λ-calculs typés*, Thèse de Doctorat d'Etat, Université Paris VII, 1979.
3. G. Berry. *Stable models of typed lambda-calculi.* Proc. 5th Int. Coll. on Automata, Languages and Programming, Lecture Notes in Computer Science 62, 72-89, Springer, 1978.
4. G. Berry and P.L. Curien. *Sequential algorithms on concrete data structures.* Theoretical Computer Science 20, 265-231, 1982.
5. A. Bucciarelli. *Logical relations and λ-theories.* Proc. Imperial College Theory and Formal Methods Section Workshop, to appear, 1996.
6. A. Bucciarelli, T. Ehrhard. *Sequentiality and Strong Stability.* Proc. 6th Int. Symp. on Logic in Computer Science, 138-145, IEEE Computer Society Press, 1991.
7. A. Bucciarelli, T. Ehrhard. *Extensional embedding of a strongly stable model of PCF.* Proc. 18th Int. Coll. on Automata, Languages and Programming, Lecture Notes in Computer Science 510, 35-44, Springer, 1991.

8. A. Bucciarelli, T. Ehrhard. *Sequentiality in an extensional framework.* Information and Computation, Volume 110, Number 2, 265-296, 1994.
9. P.L. Curien. *Categorical Combinators, Sequential Algorithms and Functional Programming.* Revised edition, Birkhäuser, 1993.
10. T. Ehrhard. *A relative definability result for strongly stable functions and some corollaries.*
Submitted paper, available at http://ida.dcs.qmw.ac.uk/authors/E/EhrhardT/, 1996.
11. H. Friedman. *Equality between functionals.* Proc. Logic Colloquium, Lecture Notes in Mathematics 453, 22-37, Springer, 1973.
12. J.M.E. Hyland, L. Ong. *On full abstraction for PCF: I, II and III.* 133pp, submitted paper, 1994.
13. T. Jim, A. Meyer. *Full Abstraction and the Context Lemma.* Proc. Theoretical Aspects of Comp. Sci. 1991, Lecture Notes in Computer Science 526, 131-151, Springer, 1991.
14. R. Loader. *Finitary PCF is not decidable.* Unpublished notes, available at http://info.ox.ac.uk/ loader. 1996.
15. J.C. Mitchell. *Type Systems for Programming Languages* Handbook of Theoretical Computer Science, Volume B, edited by J. van Leeuwen, 365-458, Elsevier, 1990.
16. P. O'Hearn, J. Riecke. *Kripke Logical Relations and PCF.* To appear in Information and Computation.
17. G. Plotkin. *LCF considered as a programming language.* Theoretical Computer Science 5, 223-256, 1977.
18. D. Scott. *A type theoretic alternative to CUCH, ISWIM, OWHY.* Theoretical Computer Science 121, 411-440, 1993. Manuscript circulated since 1969.
19. G. Winskel. *Stable Bistructure Models of PCF.* BRICS Report Series 94-13, Departement of Computer Science, University of Aarhus, Denmark, 1994.

A Module Calculus for Pure Type Systems*

Judicaël Courant

LIP
46, allée d'Italie
69364 Lyon cedex 07
FRANCE

Abstract. Several proof-assistants rely on the very formal basis of Pure Type Systems (PTS). However, some practical issues raised by the development of large proofs lead to add other features to actual implementations for handling namespace management, for developing reusable proof libraries and for separate verification of distinct parts of large proofs. Unfortunately, few theoretical basis are given for these features. In this paper we propose an extension of Pure Type Systems with a module calculus adapted from SML-like module systems for programming pratiqueslanguages. Our module calculus gives a theoretical framework addressing the need for these features. We show that our module extension is conservative, and that type inference in the module extension of a given PTS is decidable under some hypotheses over this PTS.

1 Introduction

The notion of Pure Type Systems [Bar91] has been first introduced by Terlouw and Berardi. These systems are well-suited for expressing specifications and proofs and they are the basis of several proof assistants [CCF+95,Pol94,MN94,HHP93]. However, there is actually a gap between PTS and the extensions needed within proof assistants. Indeed, PTS are well-suited to type-theoretic study, but lack some features that a proof-assistant needs.

A first practical expectation when specifying and proving in a proof assistant is for definitions. Making a non-trivial proof or even a non-trivial specification in a proof assistant is often a long run task that would be impossible if one could not bind some terms to a name. The meta-theoretical study of definitions and their unfolding, although not very difficult is far from being obvious; it has been achieved for instance in [SP94].

Another highly expectable feature when developing large proofs is for a practical namespace management. Indeed, it is often difficult to find a new significant name for each theorem. In proof-assistants where proofs can be split across several files, a partial solution is to represent names as prefixed by the name of the file in which they are defined. Then, the user may either refer to a theorem by its long name, or give only the suffix part which refers to the last loaded theorem with this suffix.

Another one is the ability to parameterize a whole theory with some axioms. For instance, when defining and proving sorting algorithms, it is very convenient to have the

* This research was partially supported by the ESPRIT Basic Research Action Types and by the GDR Programmation cofinanced by MRE-PRC and CNRS.

whole theory parameterized with a set A, a function $ord : A \to A \to bool$, and three axioms stating that ord is reflexive, antisymmetric, transitive, total and decidable. This feature is implemented in the Coq proof-assistant through the sectioning mechanism [CCF+95]. In a given section, one may declare axioms or variables and use them. When the section is closed, these axioms and variables are discharged. That is, every theorem is parameterized by these hypothesis and variables. Thus, one does not have to explicitly parameterize every theorem by these hypothesis and variables.

However, this sectioning mechanism is not a definite answer. Indeed, it does not allow to instantiate a parameterized theory. For instance, once the theory of sorting algorithms has been proved, if one wants to use this theory for a given set and an ordering, one has to give the five parameters describing the ordering each time one needs to use any of the results. In order to have a more convenient way to refer to these results, we have to imagine a mechanism allowing the instantiation of several results at once.

Finally, proof assistants also raise the problem of separate verification. Thus, in proof-assistants such as Coq, the verification of standard proof-libraries can take several hours. For the user, this is annoying if the proof-assistant needs to check them each time the user references them. Therefore, a feature allows to save and restore the global state of the proof-assistant on disk ; thus, standard libraries are checked once, then the corresponding state is saved, and users start their sessions with this state. But it is not possible to save all available libraries in a given state, because they would require too much memory. Rather, one would like to have a way to load only required libraries, but at a reasonable speed. Recently, the Lego and the Coq proof-assistants allowed to put theories they check into a compiled form. Such compiled forms can be loaded very fast — several seconds instead of several minutes or hours.

But the possibility of saving proofs in compiled forms is not a true separate verification facility. In fact, we lack a notion of *specification* of a proof. Such a notion is desirable for three reasons. First, it would provide a convenient way to describe what is proved in a given proof development. Moreover, the user may like to give only a specification of a theory he needs to make a proof, in order to make his main proof first, then prove the specification he needed. And finally, that would help in making proofs robust with respect to changes: indeed, it is sometimes difficult to predict whether a change in a proof will break proofs depending on it, since there is no clear notion of the specification exported by a given file.

Some theorem provers already address some of these issues. Thus IMPS [FGT95] implements Bourbaki's notion of structures and theories [Bou70], allowing to instantiate a general theory on a given structure at once, getting every instantiations of theorems. Unfortunately, this notion is well-suited in a set-theoretic framework but less in a type-theoretic one.

The Standard ML programming language has a very powerful module system [Mac85] that allows for the definition of parametric modules and their composition, although it does not support true separate compilation. This module system was adapted to the Elf implementation of LF [HP92]. However, only the part of the SML module system that was well-understood from the semantic and pragmatic point of view was adapted, hence leaving out significant power of SML. For instance, the `sharing` construct of SML had to be ruled out. This is annoying since this construct allows to express that two structures share some components. For instance, it may be useful to make a theory over groups and monoids that share the same base set.[1]

Recent works on module systems however brought hope: Leroy [Ler94,Ler95], Harper and Lillibridge [HL94] presented "cleaner" variants of the SML module system, allowing true separate compilation since only the knowledge of the type of a module is needed in order to typecheck modules using it. Unfortunately, no proof of correctness was given for any of these system, thus preventing us to be sure their adaptation to a proof system would not lead to inconsistency. We gave one in a variant of these systems in [Cou96].

However adaptation of these module systems to Pure Type Systems raises the problem of dealing with β-equivalence that appears in the conversion rule of PTS. In this paper, we give an adaptation of the system of [Cou96] to Pure Type Systems. This system applies to the LF logical framework, the Calculus of Constructions [CH88], the Calculus of Constructions extended with universes [Luo89]. We do not deal with the problem of adding inductive types to these systems, but the addition of inductive types as first-class objects should not raise any problem as our proposal is quite "orthogonal" to the base language: as few properties of β-reduction were needed to prove our results, they should also be true in a framework with inductive types and the associated ι-reduction.

The remaining of this paper is organized as follows: we give in section 2 an informal presentation of the desired features for a module system. Then, in section 3, we expose formally our system. In section 4 we give its meta-theory. We compare our system with other approaches in section 5. Finally, we give possible directions for future work and conclusions in section 6.

2 Informal Presentation

In order to solve the problem of namespace management, we add to PTS the notion of *structure*, that is, package of definitions. An environment may now contain structures declarations. These structures can even contain *sub-structures*, which may help in structuring the environment. In fact, many mathematical structures own sub-structures. Thus, the polynomial ring $A[X]$ over a ring A may be defined as a structure having A as a component; a monoid homomorphism may be defined as a structure having the domain and the range monoids as components; ....

In order to address the issue of robustness of proofs with respect to changes, we introduce a notion of *specification*. We require every module definition be given together with a specification. A specification for a structure is a declaration of the objects the module should export, together with their types, and possibly their definitions. The specification of a structure is called a *signature* of this structure. Then, the only thing the type-checker knows about a module in a given environment is its specification. The correction of a development is ensured as soon as for every specification, a module matching this specification is given.

[1] The mathematical structure of rings is defined as the data of a group and a monoid that share the same base set, and verify some other conditions (distributivity).

Let us consider an example. Assume we want to work in the Calculus of Constructions, extended with an equality defined on any set A, $=_A$. Assuming $\diamond$ is any given term of type `Set`, we can define a monoid structure on $\diamond \to \diamond$ in the following way:

$$
\begin{array}{lll}
\texttt{module}\, M : & \texttt{sig} & \\
& E & : \texttt{Set} = \diamond \to \diamond \\
& e & : E \\
& op & : E \to E \to E \\
& assoc & : \forall x, y, z : E.(op\ (op\ x\ y)\ y) =_E (op\ x\ (op\ y\ z)) \\
& left\_neutral & : \forall x : E.(op\ e\ x) =_E x \\
& right\_neutral & : \forall x : E.(op\ x\ e) =_E x \\
& \texttt{end} & \\
= & \texttt{struct} & \\
& base & = \diamond \\
& E & = base \to base \\
& e & = \lambda x : base.x \\
& op & = \lambda f, g : base \to base.\lambda x : base.(f\ (g\ x)) \\
& assoc & = \ldots \\
& left\_neutral & = \ldots \\
& right\_neutral & = \ldots \\
& \texttt{end} &
\end{array}
$$

This definition adds to the environment a module M of the given signature. Signatures are introduced by the keyword `sig`, structures by `struct`. Both are ended by the keyword `end`.

From inside the definition, components are referred to as E, e, op ; from outside, they must be referred to as $M.E$, $M.e$, $M.op$, ... Notice that *base* is not visible outside the definition of M since it is not declared in the signature. Only the definition of $M.E$ is known outside the module definition, so that for instance no one can take advantage of a particular implementation of *op*. The declaration E : `Set` $= \diamond \to \diamond$ is said to be *manifest* since it gives the definition of E.

The naming convention $M.S.c$ might become heavy when working on a given module. Therefore, in the SML module system, there is an `open` construct such that after an `open` M, any component c of M can be referred to as c instead of $M.c$. However, this is only syntactic sugar, so we will not consider it in our theoretical study.

Since we wish to handle parameterized theories, we extend the module language in order to allow parameterized modules. Then, one can develop for instance a general theory T of monoids parameterized by a generic monoid structure, then define the module T_M of the theory of the monoid M. Parameterized modules are built through the `functor` keyword, that is the equivalent of a λ-abstraction at the module level, and of a ∀-quantification at the module type level:

module T
: functor($\mathcal{M}$: <<*monoid signature*>>)
  sig
  $left\_neutral\_unicity : \forall x : \mathcal{M}.E.(\forall y : \mathcal{M}.E.(\mathcal{M}.op\ x\ y) =_{\mathcal{M}.E} y) \rightarrow (x =_{\mathcal{M}.E} \mathcal{M}.e)$

  ⋮

  end
= functor($\mathcal{M}$: <<*monoid signature*>>)
  struct
  $left\_neutral\_unicity = \ldots$

  ⋮

  end

Then one can instantiate the general theory on a given module as follows:

module T_M
: sig
  $left\_neutral\_unicity : \forall x : M.E.(\forall y : M.E.(M.op\ x\ y) =_{M.E} y) \rightarrow (x =_{M.E} M.e)$

  ⋮

  end
= $(T\ M)$

Functors are also interesting for the construction of mathematical structures. For instance, the product monoid of two generic monoids can be defined easily through a functor, then instantiated on actual monoids.

Finally, before we give a formal definition of our system, it should be noticed that a name conflict can appear when instantiating a functor: as in λ-calculus, $(\lambda y.x\ z)\{x \leftarrow y\}$ is not $(\lambda y.y\ z)$, if

$$f : \texttt{functor}(x : \ldots)\texttt{sig}\ y = \ldots\ z = x.n\ \texttt{end}$$

then $(f\ y)$ is not of type

$$\texttt{sig}\ y = \ldots\ z = y.n\ \texttt{end}$$

The usual solution in λ-calculus is capture-avoiding substitutions that rename binders if necessary. Here, a field of a structure can not be renamed since we want to be able to access components of a structure by their names. In fact, the problem is a confusion between the notion of component name and binder. Therefore, we modify the syntax of declarations and specifications: declarations and specifications shall be of the form $x \triangleright y = \ldots$ (or $x \triangleright y : \ldots$ or $x \triangleright y : \ldots = \ldots$), the first identifier being the name of the component and the second one its binder. This syntax has been proposed by Harper and Lillibridge in [HL94]. They suggested pronouncing "▷" as "as". From inside a structure or signature, the component is referred by its binder, and from outside, it is referred by its name. Then, we avoid name clashes by capture-avoiding substitutions. For instance, the monoid previously defined could be written:

```
module M : sig
```

E	$\triangleright E'$	: Set = $\diamond \rightarrow \diamond$
e	$\triangleright e'$	: E'
op	$\triangleright op'$	: $E' \rightarrow E' \rightarrow E'$
$assoc$	$\triangleright assoc'$	: $\forall x, y, z : E'.(op'\,(op'\,x\,y)\,y)$
		$=_{E'} (op'\,x\,(op'\,y\,z))$
$left\_neutral$	$\triangleright left\_neutral'$	: $\forall x : E'.(op'\,e'\,x) =_{E'} x$
$right\_neutral$	$\triangleright right\_neutral'$	: $\forall x : E'.(op'\,x\,e') =_{E'} x$

```
        end
    = ...
```

Of course, we shall allow $x : t$ as a syntactic sugar for $x \triangleright x : t$ (similarly for $x = t$).

Another solution is using De Bruijn indices : from inside a structure or signature, a component is referred to by a De Bruijn indices, and from outside, it is referred to by its name ; then, the declaration of a component only needs to be given a name. Assuming the use of De Bruijn indices, we shall not address α-conversion problems in the remaining of this paper.

3 A Module Calculus

We now formalize our previous remarks in a module calculus derived from the propositions of [Ler94,Ler95,HL94,Cou96].

3.1 Syntax

Terms :

$e ::= v$		identifier
$\mid m.v$		access to a value field of a structure
$\mid (e_1\,e_2)$		application
$\mid \lambda v{:}e_1.e_2$		λ-abstraction
$\mid \forall v{:}e_1.e_2$		universal quantification

Module expressions :

$m ::= x$		identifier
$\mid m.x$		module field of a structure
$\mid \texttt{struct}\ s\ \texttt{end}$		structure construction
$\mid \texttt{functor}(x{:}M)m$		functor
$\mid (m_1\,m_2)$		application of a module

Structure body :

$s ::= \epsilon \mid d\ ;\ s$

Structure component :

$d ::= \texttt{term}\ v_1 \triangleright v_2 = e$		term definition
$\mid \texttt{module}\ x_1 \triangleright x_2 : M = m$		module definition

Module type :

$M ::= \texttt{sig}\ S\ \texttt{end}$ — signature type

$\quad |\ \texttt{functor}(x : M_1) M_2$ — functor type

Signature body :

$S ::= \epsilon\ |\ D\ ;\ S$

Signature component :

$D ::= \texttt{term}\ v_1 \triangleright v_2\ :\ e$ — term declaration

$\quad |\ \texttt{term}\ v_1 \triangleright v_2\ :\ e_1 = e_2$ — manifest term declaration

$\quad |\ \texttt{module}\ x_1 \triangleright x_2\ :\ M$ — module declaration

Environments :

$E ::= \epsilon$ — empty environment

$\quad |\ v\ : e$ — term declaration

$\quad |\ v\ : e = e'$ — term definition

$\quad |\ x\ : M$ — module declaration

Notice that this syntax is an extension of the syntax of pre-terms in PTS, and that this extension is quite orthogonal to the syntax of these pre-terms. Since we intend to study the reductions of the module calculus, we shall distinguish β-reductions at the level of the base-language calculus and at the level of the module calculus. Therefore we call μ-reduction the β-reduction at the level of module system. That is, μ-reduction is the least context-stable relation on the syntax such that

$$((\texttt{functor}\ (x : M)\ m_1)\ m_2) \rightarrow_\mu m_1\{x_i \leftarrow m_2\}$$

We define μ-equivalence as the least equivalence relation including the μ-reduction.

3.2 Typing Rules

Let $\mathcal{S}$ a set of constants called the sorts, $\mathcal{A}$, a set of pair (c, σ) where c is a constant and $\sigma \in \mathcal{S}$, and $\mathcal{R}$ a set of triples of elements of $\mathcal{S}$. The Pure Type System (PTS) determined by the specification $(\mathcal{S}, \mathcal{A}, \mathcal{R})$ is defined in figure 1. Three kinds of judgments are defined: a given environment is well-formed, a given term is of a given type, and two given terms are convertible. In order to build a module system over this PTS, we add rules given in figures 2 and 3, that define the following new judgments:

$E \vdash M$ **modtype**	module type M is well-formed
$E \vdash m : M$	module expression m has type M
$E \vdash M_1 <: M_2$	module type M_1 is a subtype of M_2

In these rules we make use of the following definitions. The first one helps in introducing a field of a module in the environment, the second one gives the set of fields defined in a structure body and the third one gives the set of couples (names,identifier) appearing in a given structure:

$$\overline{\texttt{term}\ v \triangleright w : e} = w : e$$

$$\overline{\texttt{term}\ v \triangleright w : e = e'} = w : e = e'$$

Context rules ($E \vdash \mathsf{ok}$):

$$\epsilon \vdash \mathsf{ok} \qquad \frac{E \vdash e : \sigma \quad \sigma \in \mathcal{S} \quad v \notin E}{E; v : e \vdash \mathsf{ok}}$$

Typing rules ($E \vdash e : e'$):

$$\frac{E; v : e; E' \vdash \mathsf{ok}}{E; v : e; E' \vdash v : e} \qquad \frac{E \vdash e : \sigma_1 \quad E; v : e \vdash e' : \sigma_2 \quad (\sigma_1, \sigma_2, \sigma_3) \in \mathcal{R}}{E \vdash \forall v : e.e' : \sigma_3}$$

$$\frac{E \vdash \mathsf{ok} \quad (c, \sigma) \in \mathcal{A}}{E \vdash c : \sigma} \qquad \frac{E \vdash e_1 : \forall v : e.e' \quad E \vdash e_2 : e}{E \vdash (e_1\, e_2) : e'\{v \leftarrow e_2\}}$$

$$\frac{E; v : e \vdash e' : e'' \quad E \vdash \forall v : e.e'' : \sigma \quad \sigma \in \mathcal{S}}{E \vdash \lambda v : e.e' : \forall v : e.e''}$$

$$\frac{E \vdash e : e' \quad E \vdash e'' : \sigma \quad \sigma \in \mathcal{S} \quad E \vdash e' \approx e''}{E \vdash e : e''}$$

Term equivalence ($E \vdash e \approx e'$):

$$\frac{e =_\beta e' \quad E \vdash \mathsf{ok}}{E \vdash e \approx e'}$$

(congruence rules omitted)

Fig. 1. PTS rules

$$\overline{\texttt{module}\, x \triangleright y : M = y : M}$$

$$N(\texttt{term}\, v \triangleright w = e; s) = \{v\} \cup N(s)$$
$$N(\texttt{module}\, x \triangleright y : M = m; s) = \{x\} \cup N(s)$$
$$N(\epsilon) = \emptyset$$

$$BV(\epsilon) = \emptyset$$
$$BV(\texttt{term}\, v \triangleright w : e[= e']; s) = \{(v, w)\} \cup BV(s)$$
$$BV(\texttt{module}\, x \triangleright y : M; s) = \{(x, y)\} \cup BV(s)$$
$$BV(E; v : e[= e']) = \{v\} \cup BV(E)$$
$$BV(E; \texttt{module}\, x : M) = \{x\} \cup BV(E)$$

Following [Ler94,Ler95], one typing rule for modules makes use of the strengthening M/m of a module type M by a module expression m: this rule is a way to express the "self" rule saying that even if the component v of a module m is declared as abstract, one knows that this component is equal to $m.v$, and may add this information to the type of m. The strengthening operation is defined as follows:

$$(\texttt{sig}\, S\, \texttt{end})/m = \texttt{sig}\, S/m\, \texttt{end}$$
$$(\texttt{functor}\,(x : M_1)\, M_2)/m = \texttt{functor}\,(x : M_1)\,(M_2/m(x))$$
$$\epsilon/m = \epsilon$$

Context formation ($E \vdash$ **ok**):

$$\frac{E \vdash M\ \textbf{modtype}\quad x \notin BV(E)}{E; \texttt{module}\ x : M \vdash \textbf{ok}} \qquad \frac{E \vdash e : e' \quad w \notin BV(E)}{E; w : e' = e \vdash \textbf{ok}}$$

Module type and signature body formation ($E \vdash M$ **modtype**):

$$\frac{E \vdash \textbf{ok}}{E \vdash \epsilon\ \textbf{modtype}} \qquad \frac{E; \texttt{module}\ x : M \vdash S\ \textbf{modtype} \quad y \notin N(S)}{E \vdash \texttt{module}\ y \triangleright x : M; S\ \textbf{modtype}}$$

$$\frac{E; v : e \vdash S\ \textbf{modtype} \quad w \notin N(S)}{E \vdash \texttt{term}\ w \triangleright v : e; S\ \textbf{modtype}} \qquad \frac{E; v : e = e' \vdash S\ \textbf{modtype} \quad w \notin N(S)}{E \vdash \texttt{term}\ w \triangleright v : e = e'; S\ \textbf{modtype}}$$

$$\frac{E \vdash S\ \textbf{modtype}}{E \vdash \texttt{sig}\ S\ \texttt{end}\ \textbf{modtype}}$$

$$\frac{E \vdash M\ \textbf{modtype} \quad x \notin BV(E) \quad E; \texttt{module}\ x : M \vdash M'\ \textbf{modtype}}{E \vdash \texttt{functor}(x : M)M'\ \textbf{modtype}}$$

Module expressions ($E \vdash m : M$) and structures ($E \vdash s : S$):

$$\frac{E; x : M; E' \vdash \textbf{ok}}{E; x : M; E' \vdash x : M} \qquad \frac{E \vdash m : \texttt{sig}\ S_1; \texttt{module}\ x \triangleright y : M; S_2\ \texttt{end}}{E \vdash m.x : M\{n \leftarrow m.n' \mid (n', n) \in BV(S_1)\}}$$

$$\frac{E; x : M \vdash m : M' \quad E \vdash \texttt{functor}(x : M)M'\ \textbf{modtype}}{E \vdash \texttt{functor}(x : M)m : \texttt{functor}(x : M)M'}$$

$$\frac{E \vdash m_1 : \texttt{functor}(x : M)M' \quad E \vdash m_2 : M}{E \vdash (m_1\ m_2) : M'\{x \leftarrow m_2\}} \qquad \frac{E \vdash m : M' \quad E \vdash M' <: M}{E \vdash m : M}$$

$$\frac{E \vdash m : M}{E \vdash m : M/m} \qquad \frac{E \vdash s : S}{E \vdash (\texttt{struct}\ s\ \texttt{end}) : (\texttt{sig}\ S\ \texttt{end})} \qquad \frac{E \vdash \textbf{ok}}{E \vdash \epsilon : \epsilon}$$

$$\frac{E \vdash e : e' \quad v \notin BV(E) \quad E; v : e' = e \vdash s : S \quad w \notin N(s)}{E \vdash (\texttt{term}\ w \triangleright v = e; s) : (\texttt{term}\ w \triangleright v : e = e'; S)}$$

$$\frac{E \vdash m : M \quad x \notin BV(E) \quad E; x : M \vdash s : S \quad y \notin N(s)}{E \vdash (\texttt{module}\ y \triangleright x : M = m; s) : (\texttt{module}\ y \triangleright x : M; S)}$$

Fig. 2. Typing rules for the Module Calculus

$$(D; S)/m = D/m; (S/m)$$
$$(\texttt{term}\ v \triangleright w : e)/m = \texttt{term}\ v \triangleright w : e = m.v$$
$$(\texttt{term}\ v \triangleright w : e_1 = e_2)/m = \texttt{term}\ v \triangleright w : e_1 = e_2$$
$$(\texttt{module}\ x \triangleright y : M)/m = \texttt{module}\ x \triangleright y : (M/m.x)$$

4 Meta-Theory

We now give our main theoretical results about our module extension: this extension is sound since it is conservative, and if type inference is possible in a PTS, it is possible in its module extension.

Module types subtyping ($E \vdash M_1 <: M_2$):

$$\frac{E \vdash \mathtt{sig}\ D'_1; \ldots; D'_m\ \mathtt{end}\ \textbf{modtype} \quad E \vdash \mathtt{sig}\ D_1; \ldots; D_n\ \mathtt{end}\ \textbf{modtype} \quad \sigma : \{1, \ldots, m\} \rightarrow \{1, \ldots, n\}\ \forall i \in \{1, \ldots, m\}\ E; \overline{D}_1; \ldots; \overline{D}_n \vdash D_{\sigma(i)} <: D'_i}{E \vdash \mathtt{sig}\ D_1; \ldots; D_n\ \mathtt{end} <: \mathtt{sig}\ D'_1; \ldots; D'_m\ \mathtt{end}}$$

$$\frac{E \vdash M_2 <: M_1 \quad E; x : M_2 \vdash M'_1 <: M'_2}{E \vdash \mathtt{functor}\,(x : M_1)\,M'_1 <: \mathtt{functor}\,(x : M_2)\,M'_2}$$

$$\frac{E \vdash M <: M'}{E \vdash \mathtt{module}\ x \triangleright y : M <: \mathtt{module}\ x \triangleright y' : M'}$$

$$\frac{E \vdash e \approx e'}{E \vdash \mathtt{term}\ v \triangleright w : e[= e''] <: \mathtt{term}\ v \triangleright w' : e'}$$

$$\frac{E \vdash e_1 \approx e'_1 \quad E \vdash w \approx e'_2}{E \vdash \mathtt{term}\ v \triangleright w : e_1[= e_2] <: \mathtt{term}\ v \triangleright w' : e'_1 = e'_2}$$

Term equivalence ($E \vdash e \approx e'$):

$$\frac{m =_{\mu\beta} m'}{E \vdash m.t \approx m'.t}$$

$$\frac{E_1; w : e = e'; E_2 \vdash \textbf{ok}}{E_1; w : e = e'; E_2 \vdash w \approx e'} \qquad \frac{E \vdash m : \mathtt{sig}\ S_1; \mathtt{term}\ v \triangleright w = e; S_2\ \mathtt{end}}{E \vdash m.v \approx e\{n \leftarrow m.n' \mid (n', n) \in BV(S_1)\}}$$

Fig. 3. Typing rules for the Module Calculus (continued)

4.1 Module Reductions

We first focus on reductions in the module language.

Theorem 1 (Subject reduction for μ-reduction). *If $E \vdash m : M$, and $m \rightarrow_\mu m'$, then $E \vdash m' : M$.*

Theorem 2 (Confluence of μ-reduction). *The μ-reduction is confluent.*

Theorem 3 (Strong normalization for μ-reduction). *The μ-reduction is strongly normalizing.*

For both reduction notions, confluence properties are proved with the standard Tait and Martin-Löf's method.

Subject reduction for μ-reduction is proved as usual (substitution property and study of possible types of a functor).

Strong normalization can be proved in a modular way. We reduce the problem of normalization of the module extension to the question of normalization of terms in a simply-typed lambda-calculus with dependent records and subtyping. The normalization proof for this latter calculus can be done by reduction to the simply-typed lambda-calculus.

However, μ-reduction in itself is not very interesting. Indeed, modules expressions are very often in μ-normal form. Instead, we can study what happens when we unfold modules and terms definitions, that is, what happens when we add to μ-reduction the ρ-reduction defined as the least context-stable relation such that

$$\begin{array}{l}\texttt{struct}\ s_1;\texttt{term}\ v \triangleright w : e = e'; s_2\ \mathsf{end}.v \\ \rightarrow_\rho e\{n \leftarrow \texttt{struct}\ s_1;\texttt{term}\ v \triangleright w : e = e'; s_2\ \mathsf{end}.n' \mid (n', n) \in BV(s_1)\} \\ \texttt{struct}\ s_1;\texttt{module}\ x \triangleright y : M = m; s_2\ \mathsf{end}.x \\ \rightarrow_\rho m\{n \leftarrow \texttt{struct}\ s_1;\texttt{module}\ x \triangleright y : M = m; s_2\ \mathsf{end}.n' \mid (n', n) \in BV(s_1)\}\end{array}$$

In an empty environment, a $\mu\rho$-normalizing expression `struct` s end.*result* normalizes to a term where no module construct appears: $\mu\rho$-normalization is a way to transform any expression of a Pure Type System extended with modules into a term of the corresponding Pure Type System.

We have the following results:

Theorem 4 (Subject reduction for $\mu\rho$ reduction). *If $E \vdash m : M$, and $m \rightarrow_{\mu\rho} m'$, then $E \vdash m' : M$.*

Theorem 5 (Confluence of $\mu\rho$-reduction). *The $\mu\rho$-reduction is confluent.*

Theorem 6 (Strong normalization for $\mu\rho$-reduction). *The $\mu\rho$-reduction is strongly normalizing.*

As a consequence of theorem 6, we have:

Theorem 7 (Conservativity of the module extension). *In the empty environment, a type T of a PTS is inhabited if and only if T is inhabited in this PTS extended with modules.*

4.2 Type Inference

In this subsection, we intend to give a type inference algorithm for our module extension. Provided some decidability properties on $\mathcal{A}$ and $\mathcal{R}$ hold, type inference is decidable for functional semi-full PTS where the β-reduction is normalizing [Pol94]. In what follows, we shall consider only PTS that enjoy these properties.

In order to obtain a type inference algorithm, we provide in figures 4 and 5 an inference system which runs in a deterministic way for a given module expression except for term comparison $\approx$ (where a main rule plus reflexivity, symmetry, transitivity and context stability may filter the same terms).

We first show that this system gives the most general type of a given module expression if this expression is well-typed. Then we state that the $\approx$ comparison relation can be decided through a suitable notion of normalization. Finally, we state that this algorithm stops even if the given module is ill-typed.

The inference system is obtained from the one given in figures 2 and 3 in the usual way by moving subsumption and strengthening rules in the application rule, and the notion of δ-reduction of a type is added in order to orient the equality between a field of structure and the corresponding declaration in its signature.

Context formation ($E \vdash_A$ **ok**):

$$\frac{E \vdash_A M \textbf{ modtype} \quad x \notin BV(E)}{E; \texttt{module } x : M \vdash_A \textbf{ok}} \qquad \frac{E \vdash_A e : e' \quad w \notin BV(E)}{E; w : e' = e \vdash_A \textbf{ok}}$$

Module type and signature body formation ($E \vdash_A M$ **modtype**):

$$\frac{E \vdash_A \textbf{ok}}{E \vdash_A \epsilon \textbf{ modtype}} \qquad \frac{E; \texttt{module } x : M \vdash_A S \textbf{ modtype} \quad y \notin N(S)}{E \vdash_A \texttt{module } y \triangleright x : M; S \textbf{ modtype}}$$

$$\frac{E; v : e \vdash_A S \textbf{ modtype} \quad w \notin N(S)}{E \vdash_A \texttt{term } w \triangleright v : e; S \textbf{ modtype}} \qquad \frac{E; v : e = e' \vdash_A S \textbf{ modtype} \quad w \notin N(S)}{E \vdash_A \texttt{term } w \triangleright v : e = e'; S \textbf{ modtype}}$$

$$\frac{E \vdash_A S \textbf{ modtype}}{E \vdash_A \texttt{sig } S \texttt{ end} \textbf{ modtype}}$$

$$\frac{E \vdash_A M \textbf{ modtype} \quad x \notin BV(E) \quad E; \texttt{module } x : M \vdash_A M' \textbf{ modtype}}{E \vdash_A \texttt{functor}(x : M) M' \textbf{ modtype}}$$

Module expressions ($E \vdash_A m : M$) and structures ($E \vdash_A s : S$):

$$\frac{E; \texttt{module } x : M; E' \vdash_A \textbf{ok}}{E; \texttt{module } x : M; E' \vdash_A x : M} \qquad \frac{E \vdash_A m : \texttt{sig } S_1; \texttt{module } x \triangleright y : M; S_2 \texttt{ end}}{E \vdash_A m.x : M\{n \leftarrow m.n' \mid (n', n) \in BV(S_1)\}}$$

$$\frac{E; \texttt{module } x : M \vdash_A m : M' \quad E \vdash \texttt{functor}(x : M) M' \textbf{ modtype}}{E \vdash_A \texttt{functor}(x : M) m : \texttt{functor}(x : M) M'}$$

$$\frac{E \vdash_A s : S}{E \vdash_A (\texttt{struct } s \texttt{ end}) : (\texttt{sig } S \texttt{ end})} \qquad \frac{E \vdash_A \textbf{ok}}{E \vdash_A \epsilon : \epsilon}$$

$$\frac{E \vdash_A m_1 : \texttt{functor}(x : M) M' \quad E \vdash_A m_2 : M'' \quad E \vdash_A M''/m_2 <: M}{E \vdash_A (m_1\ m_2) : M'\{x \leftarrow m_2\}}$$

$$\frac{E \vdash_A e : e' \quad v \notin BV(E) \quad E; v : e' = e \vdash_A s : S \quad w \notin N(s)}{E \vdash_A (\texttt{term } w \triangleright v = e; s) : (\texttt{term } w \triangleright v : e = e'; S)}$$

$$\frac{E \vdash_A m : M \quad x \notin BV(E) \quad E; \texttt{module } x : M \vdash_A s : S \quad y \notin N(s)}{E \vdash_A (\texttt{module } y \triangleright x : M = m; s) : (\texttt{module } y \triangleright x : M; S)}$$

Fig. 4. Type inference system

Soundness and Completeness

Theorem 8 (Soundness). *If* $E \vdash_A m : M$ *then* $E \vdash m : M$ *(and thus* $E \vdash m : M/m$*) ; if* $E \vdash_A M <: M'$ *then* $E \vdash M <: M'$ *; if* $E \vdash_A e \approx e'$ *then* $E \vdash e \approx e'$.

Proof: Induction on the derivation.

Theorem 9 (Completeness). *If* $E \vdash m : M$*, then there exists a unique* M' *such that* $E \vdash_A m : M'$ *and* $E \vdash_A M'/m <: M$*. Thus* M'/m *is the principal type of* m*. If* $E \vdash M <: M'$ *then* $E \vdash_A M <: M'$ *; if* $E \vdash e \approx e'$ *then* $E \vdash_A e \approx e'$.

Proof: Induction on the derivation.

Term Comparison In order two compare two terms for the $\approx$ relation, we would like to compare their normal form. Indeed, the following theorem holds:

Module types subtyping ($E \vdash_A M_1 <: M_2$):

$$\frac{E \vdash_A \texttt{sig}\ D'_1; \ldots; D'_m\ \text{end}\ \textbf{modtype} \quad E \vdash_A \texttt{sig}\ D_1; \ldots; D_n\ \text{end}\ \textbf{modtype} \\ \sigma : \{1, \ldots, m\} \to \{1, \ldots, n\} \quad \forall i \in \{1, \ldots, m\}\ E; \overline{D}_1; \ldots; \overline{D}_n \vdash_A D_{\sigma(i)} <: D'_i}{E \vdash_A \texttt{sig}\ D_1; \ldots; D_n\ \text{end} <: \texttt{sig}\ D'_1; \ldots; D'_m\ \text{end}}$$

$$\frac{E \vdash_A M_2 <: M_1 \quad E; \texttt{module}\ x : M_2 \vdash_A M'_1 <: M'_2}{E \vdash_A \texttt{functor}\,(x : M_1)\, M'_1 <: \texttt{functor}\,(x : M_2)\, M'_2}$$

$$\frac{E \vdash_A M <: M'}{E \vdash_A \texttt{module}\ x \triangleright y : M <: \texttt{module}\ x \triangleright y : M'}$$

$$\frac{E \vdash_A e \approx e'}{E \vdash_A \texttt{term}\ v \triangleright w : e[= e''] <: \texttt{term}\ v \triangleright w : e'}$$

$$\frac{E \vdash_A e_1 \approx e'_1 \quad E \vdash_A w \approx e'_2}{E \vdash_A \texttt{term}\ v \triangleright w : e_1[= e_2] <: \texttt{term}\ v \triangleright w : e'_1 = e'_2}$$

Term equivalence ($E \vdash_A e \approx e'$):

$$\frac{E \vdash_A e \to_{\mu\beta\delta} e'}{E \vdash_A e \approx e''}$$ (reflexivity, symmetry, and transitivity rules omitted)

δ-reduction:

$$\frac{E_1; w : e = e'; E_2 \vdash_A \textsf{ok}}{E_1; w : e = e'; E_2 \vdash_A w \to_\delta e'}$$

$$\frac{E \vdash_A m : \texttt{sig}\ S_1; \texttt{term}\ v \triangleright w = e; S_2\ \text{end} \quad m \text{ is } \mu\text{-normal}}{E \vdash_A m.v \to_\delta e\{n \leftarrow m.n' \mid (n', n) \in BV(S_1)\}}$$

(congruence rules omitted)

Fig. 5. Type inference system (continued)

Theorem 10 (Strong normalization of $\mu\delta$-reduction). *The $\mu\delta$-reduction is strongly normalizing*

The proof of this theorem is quite similar to the proof of μ-normalization. We also believe the following proposition is true although the proof still needs to be given at the moment of writing:

Conjecture 11 (Confluence of $\mu\delta$-reduction). *The $\mu\delta$-reduction is confluent.*

The main problem for proving this proposition appears on the following example. Let us consider

$$\begin{array}{l}(\texttt{functor}\,(x : \texttt{sig}\,\texttt{term}\,t : t' = u\ \text{end})\,\texttt{struct}\ \texttt{term}\,r = x.t\ \text{end} \\ \quad \texttt{struct}\ \texttt{term}\,t = v\ \text{end})\end{array}$$

This expression μ-reduces to

$$\texttt{struct}\ \texttt{term}\,r = \texttt{struct}\ \texttt{term}\,t = v\ \text{end}.t\ \text{end}$$

but is also δ-reduces to

$$\begin{array}{l}(\texttt{functor}\,(x : \texttt{sig}\,\texttt{term}\,t : t' = u\ \text{end})\,\texttt{struct}\ \texttt{term}\,r = u\ \text{end} \\ \quad \texttt{struct}\ \texttt{term}\,t = v\ \text{end})\end{array}$$

Then the first expressions δ-reduces to `struct term` $t = v$ `end` whereas the second one μ-reduces to `struct term` $r = u$ `end`. Since the expression we considered was well-typed, we know that $E \vdash u \approx v$, but this is not enough to conclude that u and v reduces to a common term.

Termination

Theorem 12. *$\vdash_{\mathcal{A}}$ rules give a type inference algorithm, terminating on every module expression. Therefore, type inference for the module system is decidable provided the conjecture 11 holds.*

5 Comparison with Other Works

Compared to the module system of Elf [HP92], our system is much more powerful, because of manifest declarations. Moreover, we can give a proof of its consistency through the study of reductions. Finally, we are not aware of separate compilation mechanism for the module system of Elf.

Extended ML [San90] is a very interesting framework for developing SML modular (functional) programs together with their specification and the proof of their specification. However, it is not as general as provers based on PTS can be for developing mathematical theories. Moreover, we are not aware of any proof of consistency of the EML approach.

Another way to structure a development and make parameterized theories is to add dependent record types to PTS. In systems with dependent sum types such as the Extended Calculus of Constructions [Luo89], or inductive types such as the Calculus of Constructions with Inductive Types [PM93], this is quite easy, and is more or less a syntactic sugar [Sai96]. This approach have some advantages over ours.

Firstly, functors are represented by functions from a record type to another. Therefore, there is no need for specific rules for abstraction and application of modules.

Secondly, having "modules" as first-class citizens allows powerful operations since it gives the "module" language the whole power of the base language. For instance, one can define a function taking as input a natural n and a monoid structure M and giving back as output the monoid M^n. Such a function has to be recursive whereas a functor cannot be recursive in our approach.

However the module-as-record approach suffers severe disadvantages at its turn:

Firstly, the addition of records may be difficult from a theoretical point of view. Indeed, too powerful elimination schemes can make a system logically inconsistent. For instance, Russell's paradox can be formulated in the Calculus of Constructions where one can have records of type `Set` having a set as only component if strong elimination is allowed. Hence, records are mainly useful in systems with a universe hierarchy, such as the Calculus of Constructions with Inductive Types and Universes, or the Extended Calculus of Constructions. Thus, the price of the conceptual simplicity of the record approach is the complexity of universes. On the other hand, our system is orthogonal to the considered PTS, and therefore much more robust to changes in the base language from a logical point of view.

Secondly, the abstraction mechanism is very limited. Indeed, either every component of a record is known (in the case of an explicit term or of a constant) or every component is hidden (in the case of a variable or an opaque constant)[2]. For instance, the product of two vector spaces is defined only if their field component is the same. This restriction is easily expressed in our system where we can define a module as

$$\texttt{functor}(V_1 : \, << \mathit{vector\ space} >>)$$
$$\quad \texttt{functor}(V_2 : \, << \mathit{vector\ space\ with}\ K.E = V_1.K.E,\ K.{+} = V_1.K.{+}, \ldots >>)$$
$$\quad \ldots$$

But, it is very difficult to define such a functor in a record-based formalism since there is no way to express that two given field are convertible. One could of course think of defining a notion of K-vector space, but this would require the addition of one parameter for each function on vector space.

Moreover, separate compilation of non-closed code fragments is not possible. Indeed, one sometimes needs the definition of a term in order to type-check an expression e, but the only way to know a component of a record is to know the whole record, hence it has to be compiled before e is checked. On the contrary, our notion of specification allows us to give in an interface file a specification containing only the level of details needed from the outside of a module.

6 Conclusion

We proposed a module system for Pure Type Systems. This module system can be seen as a typed lambda-calculus of its own, since it enjoys the subject reduction property. This system has several desirable properties:

- it is independent of the considered PTS, hence should be robust to changes in the base type system (addition of inductive types for instance);
- it is powerful enough to handle usual mathematical operations on usual structures;
- it is strongly normalizing;
- it is conservative with respect to the considered Pure Type System, especially it does not introduce any logical inconsistency;
- type inference is decidable under reasonnable assumption, thus allowing an effective implementation;
- it allows true separate compilation of non-closed code fragments.

Our approach also brings several new issues.

Firstly, it would also be interesting to see which mechanisms are needed for helping the user search through module libraries. The work done in [Rou92,DC95] may be of great interest in this respect.

Another issue is how to integrate proof-assistant tools in our module system. Thus, it would be interesting to add tactics components to modules helping the user by constructing proof-terms in a semi-automatic way. Similar work has been done for the

[2] It should be noticed that Jones [Jon96] proposed a way to solve this problem in a programming language with records and the ability to define abstract types, but this approach applies only in system where polymorphism is implicit and where types do not depend on terms.

IMPS prover [FGT95]: each theory comes together with a set of *macetes* that are specific tactics for a proof in this theory. A similar idea can be found in the prover CiME [CM96], where the user may declare she is in a given theory in order to get associated simplification rules.

It would also be interesting to see how far the idea of independence with respect to the base language can be formalized. In order to adapt the system of [Cou96] to PTS, we had to deal with β-equivalence; is it possible to give an abstract notion of equivalence on a base language, and general conditions allowing to extend this base language with modules (one may especially think of the Calculus of Constructions with Inductive Types and the associated ι-reduction, or of the Calculus of Constructions with $\beta\eta$-equivalence rule for conversion...)?

Finally, possible extensions of our system have to be studied. Allowing signature abbreviations as structure components may seem to be a slight extension. But, as pointed out in [HL94], such an extension can lead to subtype-checking undecidability if one allows abstract signature abbreviation components in signatures. However, while one allows only manifest abbreviations, no problem arises. More generally, a challenging extension is to add type signatures variables, type signatures operators,... without losing type inference decidability. Another direction would be the addition of overloaded functors as in [AC96].

We also hope to implement soon ideas given in this paper in the Coq proof assistant.

Acknowledgements

We would like to thank Philippe Audebaud, Xavier Leroy and Randy Pollack for their comments, encouragements, and scepticisms on this work.

References

[AC96] María Virginia Aponte and Giuseppe Castagna. Programmation modulaire avec surcharge et liaison tardive. In *JournéesFrancophones des LangagesApplicatifs*, January 1996.

[Bar91] H. Barendregt. Lambda calculi with types. Technical Report 91-19, Catholic University Nijmegen, 1991. in Handbook of Logic in Computer Science, Vol II.

[Bou70] Nicolas Bourbaki. *Eléments de Mathématique; Théorie des Ensembles*, chapter IV. Hermann, Paris, 1970.

[CCF+95] C. Cornes, J. Courant, J.-C. Filliâtre, G. Huet, P. Manoury, C. Muñoz, C. Murthy, C. Parent, C. Paulin-Mohring, A. Saïbi, and B. Werner. The Coq Proof Assistant Reference Manual Version 5.10. Technical Report 0177, INRIA, July 1995.

[CH88] T. Coquand and G. Huet. The Calculus of Constructions. *Inf. Comp.*, 76:95–120, 1988.

[CM96] Evelyne Contejean and Claude Marché. CiME: Completion Modulo E. In Harald Ganzinger, editor, *7th International Conference on Rewriting Techniques and Applications*, Lecture Notes in Computer Science. Springer-Verlag, Rutgers University, NJ, USA,, July 1996.

[Cou96] Judicaël Courant. A module calculus enjoying the subject-reduction property. Research Report RR 96-30, LIP, 1996. Preliminary version.

[DC95] Roberto Di Cosmo. *Isomorphisms of types: from λ-calculus to information retrieval and language design.* Progress in Theoretical Computer Science. Birkhauser, 1995. ISBN-0-8176-3763-X.

[FGT95] William M. Farmer, Joshua D. Guttman, and F. Javier Thayer. *The IMPS User's Manual.* The MITRE Corporation, first edition, version 2 edition, 1995.

[HHP93] Robert Harper, Furio Honsell, and Gordon Plotkin. A framework for defining logics. *Journal of the ACM*, 40(1):143–184, January 1993. Preliminary version appeared in Proc. 2nd IEEE Symposium on Logic in Computer Science, 1987, 194–204.

[HL94] R. Harper and M. Lillibridge. A type-theoretic approach to higher-order modules with sharing. In *21st Symposium on Principles of Programming Languages*, pages 123–137. ACM Press, 1994.

[HP92] Robert Harper and Frank Pfenning. A module system for a programming language based on the LF logical framework. Technical Report CMU-CS-92-191, Carnegie Mellon University, Pittsburgh, Pennsylvania, september 1992.

[Jon96] Mark P. Jones. Using parameterized signatures to express modular structures. In *23rd Symposium on Principles of Programming Languages.* ACM Press, 1996. To appear.

[Ler94] Xavier Leroy. Manifest types, modules, and separate compilation. In *21st symp. Principles of Progr. Lang.*, pages 109–122. ACM Press, 1994.

[Ler95] Xavier Leroy. Applicative functors and fully transparent higher-order modules. In *22nd Symposium on Principles of Programming Languages*, pages 142–153. ACM Press, 1995.

[Luo89] Zhaohui Luo. ECC: an Extended Calculus of Constructions. In *Proc. of IEEE 4th Ann. Symp. on Logic In Computer Science*, Asilomar, California, 1989.

[Mac85] David B. MacQueen. Modules for Standard ML. *Polymorphism*, 2(2), 1985. 35 pages. An earlier version appeared in Proc. 1984 ACM Conf. on Lisp and Functional Programming.

[MN94] Lena Magnusson and Bengt Nordström. The ALF proof editor and its proof engine. In Henk Barendregt and Tobias Nipkow, editors, *Types for Proofs and Programs*, pages 213–237. Springer-Verlag LNCS 806, 1994.

[PM93] C. Paulin-Mohring. Inductive definitions in the system Coq : Rules and Properties. In M. Bezem, J.F. Groote, editor, *Proceedings of the TLCA*, 1993.

[Pol94] Robert Pollack. *The Theory of LEGO: A Proof Checker for the Extended Calculus of Constructions.* PhD thesis, University of Edinburgh, 1994.

[Rou92] François Rouaix. The Alcool 90 report. Technical report, INRIA, 1992. Included in the distribution available at `ftp.inria.fr`.

[Sai96] Amokrane Saibi, 1996. Private Communication.

[San90] Don Sannella. Formal program development in Extended ML for the working programmer. In *Proc. 3rd BCS/FACS Workshop on Refinement*, pages 99–130. Springer Workshops in Computing, 1990.

[SP94] P. Severi and E. Poll. Pure type systems with definitions. *Lecture Notes in Computer Science*, 813, 1994.

An Inference Algorithm for Strictness

Ferruccio Damiani and Paola Giannini

Università di Torino, Dipartimento di Informatica, Corso Svizzera 185
10149 Torino (Italy)
email: {damiani,giannini}@di.unito.it

Abstract. In this paper we introduce an algorithm for detecting strictness information in typed functional programs. Our algorithm is based on a type inference system which allows to exploit the type structure of the language for the investigation of program properties. The detection of strictness information can be performed in a complete way, by reducing it to the solution of a set of constraints involving type variables. A key feature of our method is that it is compositional and there is a notion of principal type scheme. Even though the language considered in the paper is the simply typed lambda-calculus with a fixpoint constructor and arithmetic constants we can generalize our approach to polymorphic languages like Lazy ML. Although focused on strictness our method can also be applied to the investigation of other program properties, like dead code and binding time.

Introduction

Types have been recognized as useful in programming languages because they provide a semantical (context dependent) analysis of programs. Such analysis can be incorporated in the compiling process, and it is used on one side to check the consistency of programs and on the other to improve the efficiency of the code produced.

In addition to preventing run-time errors, type systems can characterize run-time properties of programs. For instance, intersection types, see [6] (and also [1]), in their full generality, provide a characterization of normalization.

Type systems tailored to specific analysis, such as strictness, totality, binding time, dead code etc. have been introduced, see [16, 14, 2, 7, 11, 20, 23, 8, 5]. In this perspective types represent program properties and their inference systems are systems for reasoning formally about them.

Type based analyzers rely on an implicit representation of types, either via type inequalities, see [18], or via lazy (implicit) types, see [11]. In this paper we pursue the first approach, reducing the inference problem to the solution of a set of constraints involving type variables.

We consider a type system similar to the one introduced in [21] and in [20] Chap. 2, where, in addition to strictness, types express also totality properties, and intersection is used at the top level. Our system makes an implicit use of intersection that gives the same results.

The representation of strictness information via sets of constraints, see Sect. 3, is very useful in the design of efficient inference algorithms. In particular the use of guarded sets of constraints produces an effect similar to that achieved in [11] through the use of lazy types.

The first section of this paper introduces the language we are dealing with and its operational semantics. Section 2 presents the strictness type assignment system. In Sect. 3 we introduce an inference algorithm for strictness types. In Sect. 4 we compare our technique with the works of Solberg, [20], and of Hankin and Le Métayer, [11]. In the last section we outline possible extensions of the system.

1 A Typed Functional Language and its Semantics

In this section we introduce a typed functional language (basically the simply typed λ-calculus with a fixpoint construct and arithmetic constants) and its operational semantics. The set of types is defined assuming as basic types **int**, the set of integers, and **bool**, the set of booleans. Types are ranged over by ρ, σ, τ...

Definition 1 (Types). The language of *types* (**T**) is defined by the grammar: $\rho ::= \iota \mid \rho \to \rho$, where $\iota \in \{\mathbf{int}, \mathbf{bool}\}$.

Typed terms are defined from a set $\mathcal{V}_\Lambda$ of typed *term variables* (ranged over by $x^\rho, y^\sigma, \ldots$), and a set $\mathcal{K}_\Lambda$ of *term constants* including the integer numbers (of type **int**) and the operations $+$, $*$, ... (of type $\mathbf{int} \to \mathbf{int} \to \mathbf{int}$). Each constant C has a type $\mathbf{T}(C)$. Typed terms, ranged over by M,N,..., are defined as follows.

Definition 2 (Typed terms). We write $\vdash_{\mathbf{T}} M : \rho$, and say that M is a typed term of type ρ, if $\vdash M : \rho$ is derivable by using the rules in Fig. 1.

$$\text{(Var)} \vdash x^\rho : \rho \qquad \text{(Con)} \vdash C : \mathbf{T}(C)$$

$$(\to \text{I}) \; \frac{\vdash M : \sigma}{\vdash \lambda x^\rho . M : \rho \to \sigma} \qquad (\to \text{E}) \; \frac{\vdash M : \rho \to \sigma \quad \vdash N : \rho}{\vdash MN : \sigma}$$

$$\text{(Fix)} \; \frac{\vdash M : \rho}{\vdash \mathrm{fix}\, x^\rho . M : \rho} \qquad \text{(If)} \; \frac{\vdash N : \mathbf{bool} \quad \vdash M_1 : \rho \quad \vdash M_2 : \rho}{\vdash \mathrm{if}\, N \,\mathrm{then}\, M_1 \,\mathrm{else}\, M_2 : \rho}$$

Fig. 1. Rules for term formation

Note that with this notation we mention explicitly in M the types of all its variables. In the following we often omit to write types which are understood. The set of *free variables* of a term M, denoted by $\mathrm{FV}(M)$, is defined in the standard way.

As usual a substitution is a finite function mapping term variables to terms, denoted by $[x_1 := N_1, \ldots, x_n := N_n]$, which respects the types, i.e., each $x_i^{\rho_i}$ is substituted by a term N_i of the same type.

Let $\Lambda_{\mathbf{T}}$ be the set of typed terms, i.e., $\Lambda_{\mathbf{T}} = \{M \mid \vdash_{\mathbf{T}} M : \rho \text{ for some type } \rho\}$, and let $\Lambda^{\circ}_{\mathbf{T}}$ be the set of *closed* terms, i.e., $\Lambda^{\circ}_{\mathbf{T}} = \{M \mid M \in \Lambda_{\mathbf{T}} \text{ and } \mathrm{FV}(M) = \emptyset\}$. Following Kahn, see [15], we define the values of the terms in $\Lambda^{\circ}_{\mathbf{T}}$ via a standard operational semantics described by judgments of the form $M \Downarrow K$, where M is a closed term and K is a closed canonical term, i.e., $K \in \mathcal{K}_{\Lambda} \cup \{\lambda x^{\rho}.N \mid \lambda x^{\rho}.N \in \Lambda^{\circ}_{\mathbf{T}}\}$.

As in Solberg, see [20], the meaning of a constant of functional type C is given by a set $\mathrm{MEAN}(C)$ of pairs of constants, i.e., if $(C_1, C_2) \in \mathrm{MEAN}(C)$ then CC_1 evaluates to C_2. For example $(2, +_2) \in \mathrm{MEAN}(+)$ and $(1, 3) \in \mathrm{MEAN}(+_2)$.

Definition 3 (Value of a term). We write $M \Downarrow K$ if this judgement is derivable using the rules in Fig. 2, and we write $M \Downarrow$ if for some K, $M \Downarrow K$.

(CAN) $K \Downarrow K$

(FIX) $\dfrac{M[x := \mathrm{fix}\, x.M] \Downarrow K}{\mathrm{fix}\, x.M \Downarrow K}$

(APP$_1$) $\dfrac{M \Downarrow \lambda x.P \quad P[x := N] \Downarrow K}{MN \Downarrow K}$

(APP$_2$) $\dfrac{M \Downarrow C \quad N \Downarrow C_1}{MN \Downarrow C_2}\ (C_1, C_2) \in \mathrm{MEAN}(C)$

(IF$_1$) $\dfrac{N \Downarrow \mathrm{true} \quad M_1 \Downarrow K}{\mathrm{if}\, N\, \mathrm{then}\, M_1\, \mathrm{else}\, M_2 \Downarrow K}$

(IF$_2$) $\dfrac{N \Downarrow \mathrm{false} \quad M_2 \Downarrow K}{\mathrm{if}\, N\, \mathrm{then}\, M_1\, \mathrm{else}\, M_2 \Downarrow K}$

Fig. 2. "Natural semantics" evaluation rules

We are interested in observing the behaviour of terms at the ground level, so, as in Pitts [19], we consider the congruence on terms induced by the contextual preorder that compares the behaviour of terms just at the ground type **int**. Let $(\mathcal{C}[\ ]^{\rho})^{\sigma}$ denote a typed context of type σ with a hole of type ρ in it. Let M and N be terms of type ρ. Define $M \preceq_{\mathbf{obs}} N$ whenever, for all contexts $(\mathcal{C}[\ ]^{\rho})^{\mathbf{int}}$, if $\mathcal{C}[M]$ and $\mathcal{C}[N]$ are closed terms, then $\mathcal{C}[M] \Downarrow$ implies $\mathcal{C}[N] \Downarrow$. Let $\simeq_{\mathbf{obs}}$ be the equivalence induced by $\preceq_{\mathbf{obs}}$. (As shown in [19] such equivalence can also be defined directly as a bisimilarity.)

The *closed term model* $\mathcal{M}$ of $\Lambda_{\mathbf{T}}$ is defined by interpreting each type ρ as the set of the equivalence classes of the relation $\simeq_{\mathbf{obs}}$ on the closed terms of type ρ. Let $\mathbf{I}(\rho)$ denote the interpretation of type ρ in this model, and let $[M]$ denote the equivalence class of the term M. For each type ρ, $[\mathrm{fix}\, x^{\rho}.x]$, which represents the notion of endless computation, is the least element, w.r.t. $\preceq_{\mathbf{obs}}$, of $\mathbf{I}(\rho)$. An *environment* is a mapping $e : \mathcal{V}_{\Lambda} \to \bigcup_{\rho \in \mathbf{T}} \mathbf{I}(\rho)$ which respects types, i.e., such that, for all x^{ρ}, $e(x^{\rho}) \in \mathbf{I}(\rho)$. The interpretation of a term M in an environment e is defined in the standard way by: $[\![M]\!]_e = [M[x_1 := N_1, \ldots, x_n := N_n]]$, where $\{x_1, \ldots, x_n\} = \mathrm{FV}(M)$ and $[N_l] = e(x_l)$ $(1 \leq l \leq n)$.

2 A Type Assignment for Strictness

To represent the divergence of terms we introduce the annotations $\bot$ and $\top$ of the basic types **int** and **bool**. They represent the behaviour of terms whose evaluation will certainly diverge, or the fact that we do not know anything about termination.

Definition 4 (Strictness types). The language $\mathbf{L}$ of *strictness types* (*s-types* for short) is defined by the following grammar: $\phi ::= \bot^\iota \mid \top^\iota \mid \phi \to \phi$, where $\iota \in \{\mathbf{int}, \mathbf{bool}\}$.

Let $\epsilon(\phi)$ denote the (standard) type obtained from the s-type ϕ by replacing $\top^\iota$ and $\bot^\iota$ by ι.

Remark. Our s-types are a restriction of the *conjunctive strictness logic* introduced by Jensen, see [14], and (independently) by Benton, see [2]. In particular s-types are obtained from the formulas of the conjunctive strictness logic, as presented in [2] pag. 36, by:

1. using $\bot$ for **f** and $\top$ for **t**,
2. forbidding formulas of the shapes $\bot^\rho$ and $\top^\rho$, where ρ is a compound type, i.e., $\rho \neq \mathbf{int}$ and $\rho \neq \mathbf{bool}$, and
3. forbidding the use of the conjunction operator.

The first point is simply a syntactic variation. We do this to emphasize the fact that we are not thinking of s-types as a logical description of abstract domains, but rather as an instance of annotated types, as explained in [20], Chap. 1. The second restriction does not affect the expressiveness of the logic, in fact it is easy to see that every formula of the conjunctive strictness logic is equivalent to a formula in which each occurrence of $\mathbf{f}^\rho$ and $\mathbf{t}^\rho$ is such that ρ is a basic type. (This holds also for the conjunction free fragment of the logic.) Finally the third restriction greatly reduces the power of the logic, and it is justified by our aim of obtaining an efficient inference algorithm. To overcome the limitation of our logic we are planning to introduce a restricted use of conjunction, e.g., rank 2 (see [22]). □

Let $\rho = \rho_1 \to \cdots \to \rho_n \to \iota$, $n \geq 0$ and $\iota \in \{\mathbf{int}, \mathbf{bool}\}$, be a type: $\top(\rho)$ denotes the s-type obtained from ρ by replacing each occurrence of the basic types $\iota \in \{\mathbf{int}, \mathbf{bool}\}$ by $\top^\iota$, and $\bot(\rho)$ denotes the s-type $\top(\rho_1) \to \cdots \to \top(\rho_n) \to \bot^\iota$.

For a basic type ι, $\bot^\iota$ means divergence whereas $\top^\iota$ means absence of information. $\bot$-*s-types* and $\top$-*s-types* formalize these notions at higher types.

Definition 5 ($\top$-s-types and $\bot$-s-types). 1. The set $\mathbf{L}_\top$ of $\top$-s-types is defined by: $\phi_1 \to \cdots \to \phi_n \to \top^\iota \in \mathbf{L}_\top$, for $\iota \in \{\mathbf{int}, \mathbf{bool}\}$, $n \geq 0$ and $\phi_1, \ldots, \phi_n \in \mathbf{L}$.

2. The set $\mathbf{L}_\bot$ of $\bot$-s-types is defined by: $\phi_1 \to \cdots \to \phi_n \to \bot^\iota \in \mathbf{L}_\bot$, for $\iota \in \{\mathbf{int}, \mathbf{bool}\}$, $n \geq 0$ and $\phi_1, \ldots, \phi_n \in \mathbf{L}_\top$.

For every type ρ, $\top(\rho) \in \mathbf{L}_\top$ and $\bot(\rho) \in \mathbf{L}_\bot$.

We now introduce a notion of inclusion between s-types, denoted by $\leq$; $\phi \leq \phi'$ means that property ϕ is more informative than property ϕ'.

Definition 6 (Inclusion relation). The *inclusion relation* $\leq$ between s-types is defined by the following rules.

$$(\text{Ref})\ \phi \leq \phi \qquad (\top)\ \frac{\psi \in \mathbf{L}_\top \quad \epsilon(\phi) = \epsilon(\psi)}{\phi \leq \psi} \qquad (\to)\ \frac{\phi_1 \leq \phi_2 \quad \psi_1 \leq \psi_2}{\phi_2 \to \psi_1 \leq \phi_1 \to \psi_2}\ .$$

It is immediate to show that $\leq$ is reflexive and transitive. Moreover, for all $\phi_1, \phi_2 \in \mathbf{L}$, $\phi_1 \leq \phi_2$ implies $\epsilon(\phi_1) = \epsilon(\phi_2)$. We write $\phi \cong \psi$ if both $\phi \leq \psi$ and $\psi \leq \phi$ hold. Notice that, for all the $\top$-s-types $\phi, \psi \in \mathbf{L}_\top$, $\epsilon(\phi) = \epsilon(\psi)$ implies $\phi \cong \psi$. The same holds for $\bot$-s-types. We also have that, for all $\phi \in \mathbf{L}_\bot$, $\psi \in \mathbf{L}$, and $\chi \in \mathbf{L}_\top$, $\phi \leq \psi \leq \chi$.

Strictness types are assigned to typed λ-terms by a set of type inference rules. If x^ρ is a term variable of type ρ an assumption for x^ρ is an expression of the shape $x^\rho : \phi$, or $x : \phi$ for short, where $\epsilon(\phi) = \rho$. A basis is a set Σ of s-types assumptions for term variables. The functions $\epsilon(\cdot)$, $\top(\cdot)$ and $\bot(\cdot)$ defined above are extended to basis. More precisely: $\epsilon(\Sigma)$ is the set of term variables which occur in Σ and, for any finite set Γ of term variables, $\top(\Gamma)$ and $\bot(\Gamma)$ denote respectively the basis $\{x^\sigma : \top(\sigma) \mid x^\sigma \in \Gamma\}$ and $\{x^\sigma : \bot(\sigma) \mid x^\sigma \in \Gamma\}$. We will prove judgements of the form $\Sigma \vdash_{\mathbf{L}} M : \phi$ where M is a typed term of type $\epsilon(\phi)$ whose free variables are in $\epsilon(\Sigma)$, i.e., such that $\vdash_{\mathbf{T}} M : \epsilon(\phi)$ and $\epsilon(\Sigma) \supseteq \mathrm{FV}(M)$. For each constant C a set $\mathbf{L}(C)$ of minimal s-types is specified. Of course, for each $\phi \in \mathbf{L}(C)$, $\epsilon(\phi) = \mathbf{T}(C)$. For example, for all integers n, $\mathbf{L}(\mathrm{n}) = \{\top^{\mathbf{int}}\}$ and $\mathbf{L}(+) = \{\bot^{\mathbf{int}} \to \top^{\mathbf{int}} \to \bot^{\mathbf{int}}, \top^{\mathbf{int}} \to \bot^{\mathbf{int}} \to \bot^{\mathbf{int}}\}$. Every s-type that includes an element of $\mathbf{L}(C)$ can be assigned to C.

Definition 7 (Strictness type assignment system). An s-typing statement is an expression $\Sigma \vdash_{\mathbf{L}} M : \phi$ where Σ is a basis containing an assumption for each free variable of M. $\Sigma, x : \psi$ denotes the basis $\Sigma \cup \{x : \psi\}$ where it is assumed that x does not appear in Σ. We write $\Sigma \vdash_{\mathbf{L}} M : \phi$ to mean that $\Sigma \vdash M : \phi$ can be derived by the rules in Fig. 3.

Note that, being $\vdash_{\mathbf{L}}$ an inference system, the same terms can have different s-types.

Remark. The rule for fixpoint (Fix) is a restriction of the fixpoint rule for the conjunctive strictness logic that follows:

$$(\text{fix})\ \frac{\Sigma, x : \phi_1 \wedge \cdots \wedge \phi_n \vdash M : \phi_1 \wedge \cdots \wedge \phi_n}{\Sigma \vdash \mathbf{fix}\, x.M : \phi_1 \wedge \cdots \wedge \phi_n}\ ,$$

see [2] pag. 45. Indeed, if in our system $\Sigma \vdash_{\mathbf{L}} \mathbf{fix}\, x.M : \phi_i$, then

$$\Sigma, x : \phi_1 \vdash M : \phi_2 \quad \cdots \quad \Sigma, x : \phi_k \vdash M : \phi_1$$

in the conjunctive strictness logic. Therefore, since $\bigwedge_{1 \leq i \leq n} \phi_i \leq \phi_j$ for all $1 \leq j \leq n$, by conjunction introduction it is possible to derive

$$\Sigma, x : \phi_1 \wedge \cdots \wedge \phi_n \vdash M : \phi_1 \wedge \cdots \wedge \phi_n$$

$$\text{(Var)}\ \Sigma, x:\phi \vdash x:\phi \qquad \text{(Con)}\ \frac{\phi \in \mathbf{L}(C)}{\Sigma \vdash C:\phi} \qquad (\leq)\ \frac{\Sigma \vdash M:\phi \quad \phi \leq \psi}{\Sigma \vdash M:\psi}$$

$$(\rightarrow \mathrm{I})\ \frac{\Sigma, x:\phi \vdash M:\psi}{\Sigma \vdash \lambda x.M:\phi \rightarrow \psi} \qquad (\rightarrow \mathrm{E})\ \frac{\Sigma \vdash M:\phi \rightarrow \psi \quad \Sigma \vdash N:\phi}{\Sigma \vdash MN:\psi}$$

$$\text{(Fix)}\ \frac{\Sigma, x:\phi_1 \vdash M:\phi_2 \ \cdots\ \Sigma, x:\phi_k \vdash M:\phi_1}{\Sigma \vdash \mathrm{fix}\, x.M:\phi_i}\ (k \geq 1 \text{ and } 1 \leq i \leq k)$$

$$(\mathrm{If}_1)\ \frac{\Sigma \vdash N:\bot^{\mathbf{bool}} \quad \Sigma \vdash M_1:\top(\rho) \quad \Sigma \vdash M_2:\top(\rho)}{\Sigma \vdash \mathbf{if}\, N\, \mathbf{then}\, M_1\, \mathbf{else}\, M_2:\bot(\rho)}$$

$$(\mathrm{If}_2)\ \frac{\Sigma \vdash N:\top^{\mathbf{bool}} \quad \Sigma \vdash M_1:\phi \quad \Sigma \vdash M_2:\phi}{\Sigma \vdash \mathbf{if}\, N\, \mathbf{then}\, M_1\, \mathbf{else}\, M_2:\phi}$$

Fig. 3. Rules for s-type assignment

and from the rule (fix) and conjunction elimination $\Sigma \vdash \mathrm{fix}\, x.M : \phi_i$ for all $1 \leq i \leq n$. Our rule (Fix) introduces an implicit (and limited) use of conjunction, by allowing more than one derivation of a typing for M in its antecedent. Indeed (for strictness) this rule is equivalent to the fixpoint rule of the system of [20] Chap. 2 (where conjunction at the top level is introduced). □

Example 1. Let $\vdash_{\mathbf{T}}$ twice : ρ where $\rho = (\mathbf{int} \rightarrow \mathbf{int}) \rightarrow \mathbf{int} \rightarrow \mathbf{int}$ and twice $= \lambda f^{\mathbf{int}\rightarrow\mathbf{int}}.\ \lambda x^{\mathbf{int}}.\ f(f\,x)$. It is easy to check that, for $i \in \{1,2\}$, $\vdash_{\mathbf{L}}$ twice : ϕ_i, where $\phi_1 = (\bot^{\mathbf{int}} \rightarrow \bot^{\mathbf{int}}) \rightarrow \bot^{\mathbf{int}} \rightarrow \bot^{\mathbf{int}}$ and $\phi_2 = (\top^{\mathbf{int}} \rightarrow \bot^{\mathbf{int}}) \rightarrow \top^{\mathbf{int}} \rightarrow \bot^{\mathbf{int}}$. □

The proof of the following fact is immediate.

Fact 8. *1.* $\Sigma \vdash_{\mathbf{L}} M : \phi$ *implies* $\vdash_{\mathbf{T}} M : \epsilon(\phi)$ *and* $\epsilon(\Sigma) \supseteq \mathrm{FV}(M)$.
2. $\vdash_{\mathbf{T}} M : \rho$ *implies* $\top(\mathrm{FV}(M)) \vdash_{\mathbf{L}} M : \top(\rho)$.

We now introduce a notion of semantics for our s-type assignment system. We interpret each s-type ϕ as a subset of the interpretation of the type $\epsilon(\phi)$.

Definition 9. 1. The interpretation $[\![\phi]\!]$ of an s-type ϕ is defined by:

$$[\![\bot^\iota]\!] = \{[\mathrm{fix}\, x^\iota.x]\} \quad [\![\top^\iota]\!] = \mathbf{I}(\iota) \quad [\![\phi \rightarrow \psi]\!] = \{[M] \mid \forall [N] \in [\![\phi]\!]\ [MN] \in [\![\psi]\!]\}\ .$$

2. An environment e *respects* Σ if for all $x : \psi \in \Sigma$, $e(x) \in [\![\psi]\!]$.

The $\leq$ relation between s-types corresponds to inclusion between interpretations. In fact the following theorem holds.

Theorem 10 (Soundness and completeness of $\leq$). $\phi \leq \psi$ *if and only if* $[\![\phi]\!] \subseteq [\![\psi]\!]$.

The proof of soundness (only if) is by induction on the derivation of $\phi \leq \psi$. The proof of completeness (if) is more complex, and can be found in [4].
Consider now the $\top$-s-types. They identify the absence of knowledge about the termination of a term. Indeed if ϕ is a $\top$-s-type, then $[\![\phi]\!] = \mathbf{I}(\epsilon(\phi))$. Divergence is identified by $\bot$-s-types. In fact if ϕ is a $\bot$-s-type, then $[\![\phi]\!] = \{[\mathrm{fix}\, x^{\epsilon(\phi)}.x]\}$.

We state now the soundness of our s-type assignment w.r.t. the semantics. The proof of the theorem is by induction on terms.

Theorem 11 (Soundness of $\vdash_{\mathbf{L}}$). *Let $\Sigma \vdash_{\mathbf{L}} M : \phi$. Then $[\![M]\!]_e \in [\![\phi]\!]$ for all e respecting Σ.*

Remark. In our semantics we identify terms such as $\mathrm{fix}\, x^{\mathbf{int}\to\mathbf{int}}.x$ and $\lambda y^{\mathbf{int}}.\ \mathrm{fix}\, z^{\mathbf{int}}.z$. This choice is justified, as explained in [2] pag. 19, by the fact that in our view *programs* are closed terms of basic type (**int** or **bool**) and the only behaviours that a program can exhibit are to diverge or to converge to an integer or boolean value. So the only way in which we can observe terms of higher types is to plug them into a complete program. The semantics proposed by Solberg in [20] Chap. 2 and Chap. 3, instead, reflects the lazy preorder, $\preceq^{\mathrm{lazy}}_{\mathrm{obs}}$, defined by: $M \preceq^{\mathrm{lazy}}_{\mathrm{obs}} N$ whenever for all types σ, for all contexts $(\mathcal{C}[\ \ ]^{\rho})^{\sigma}$, if $\mathcal{C}[M]$ and $\mathcal{C}[N]$ are closed terms, then $\mathcal{C}[M] \Downarrow$ implies $\mathcal{C}[N] \Downarrow$ (see Pitts, [19]). Clearly $\mathrm{fix}\, x^{\mathbf{int}\to\mathbf{int}}.x \not\simeq^{\mathrm{lazy}}_{\mathrm{obs}} \lambda y^{\mathbf{int}}.\ \mathrm{fix}\, z^{\mathbf{int}}.z$. Since $M \simeq^{\mathrm{lazy}}_{\mathrm{obs}} N$ implies $M \simeq_{\mathrm{obs}} N$, all optimizations preserving $\simeq^{\mathrm{lazy}}_{\mathrm{obs}}$ preserve also $\simeq_{\mathrm{obs}}$. The vice versa is not true. For instance, consider the term $M = \lambda g^{\mathbf{int}\to\mathbf{int}}.\lambda y^{\mathbf{int}}.g\,y$. For us this term is strict in the first argument, since we can assign it the s-type $(\top^{\mathbf{int}} \to \bot^{\mathbf{int}}) \to \top^{\mathbf{int}} \to \bot^{\mathbf{int}}$, where $\top^{\mathbf{int}} \to \bot^{\mathbf{int}}$ is a $\bot$-s-type. So, for any term P, the application MP can be done "by value" (first evaluating P and then doing the replacement). With the analysis of [20] this is not the case, since observing MP at the type $\mathbf{int} \to \mathbf{int}$, when $P = \mathrm{fix}\, x^{\mathbf{int}\to\mathbf{int}}.x$, we have that $MP \Downarrow$ whereas the evaluation of the application MP "by value" leads to a divergent computation. So in [20] the application would still be done "by name". □

3 An Inference Algorithm for Strictness

In this section we define an algorithm for detecting strictness information based on the system defined in the previous section. The algorithm rely on a syntax directed version of the s-type assignment system (avoiding a free use of the inclusion rule) that follows.

Definition 12 (Normalized s-type assignment system). We write $\Sigma \vdash^{\star}_{\mathbf{L}} M : \phi$ to mean that $\Sigma \vdash M : \phi$ can be derived by the rules in Fig. 4.

It is worth mentioning that, in the rule $(\to \mathrm{E}^{\star})$, the condition $\phi_1 \to \psi \leq \phi_2 \to \psi$ is used instead of $\phi_2 \leq \phi_1$. This is done to take into account the fact that: if $\phi_2 \to \psi$ is a $\top$-s-type, then ϕ_1 can be any s-type such that $\epsilon(\phi_1) = \epsilon(\phi_2)$. It is easy to see that $\vdash_{\mathbf{L}}$ and its syntax directed version $\vdash^{\star}_{\mathbf{L}}$ are equivalent.
Even though, in the rule for fixpoint of Definition 12, the number k of premises

$$(\mathrm{Var}^*)\ \frac{\phi_1 \leq \phi_2}{\Sigma, x : \phi_1 \vdash x : \phi_2} \qquad (\mathrm{Con}^*)\ \frac{\phi_1 \in \mathbf{L}(C) \quad \phi_1 \leq \phi_2}{\Sigma \vdash C : \phi_2}$$

$$(\to \mathrm{I}^*)\ \frac{\Sigma, x : \phi \vdash M : \psi}{\Sigma \vdash \lambda x.M : \phi \to \psi} \qquad (\to \mathrm{E}^*)\ \frac{\Sigma \vdash M : \phi_1 \to \psi \quad \Sigma \vdash N : \phi_2 \quad \phi_1 \to \psi \leq \phi_2 \to \psi}{\Sigma \vdash MN : \psi}$$

$$(\mathrm{Fix}^*)\ \frac{\Sigma, x : \phi_1 \vdash M : \psi_1 \ \cdots\ \Sigma, x : \phi_k \vdash M : \psi_k \quad \psi_i \leq \phi_{i+1}\ (1 \leq i \leq k-1) \quad \psi_k \leq \phi_1 \quad \psi_k \leq \psi}{\Sigma \vdash \mathbf{fix}\,x.M : \psi}\ (k \geq 1)$$

$$(\mathrm{If}_1^*)\ \frac{\Sigma \vdash N : \bot^{\mathbf{bool}} \quad \Sigma \vdash M_1 : \top(\rho) \quad \Sigma \vdash M_2 : \top(\rho) \quad \epsilon(\psi) = \rho}{\Sigma \vdash \mathbf{if}\,N\,\mathbf{then}\,M_1\,\mathbf{else}\,M_2 : \psi}$$

$$(\mathrm{If}_2^*)\ \frac{\Sigma \vdash N : \top^{\mathbf{bool}} \quad \Sigma \vdash M_1 : \phi_1 \quad \Sigma \vdash M_2 : \phi_2 \quad \phi_1 \leq \psi \quad \phi_2 \leq \psi}{\Sigma \vdash \mathbf{if}\,N\,\mathbf{then}\,M_1\,\mathbf{else}\,M_2 : \psi}$$

Fig. 4. Rules for the normalized s-type assignment

allowed to give an s-type to a fixpoint is unbounded, it can be limited. Let $[\phi]_{\cong}$ denote the $\cong$-equivalence class of the s-type ϕ, i.e., $[\phi]_{\cong} = \{\phi' \mid \phi' \cong \phi\}$, and let size$(\phi)$ denote the number of equivalence classes of the s-types χ such that $\epsilon(\chi) = \epsilon(\phi)$ and $\chi \not\leq \phi \not\leq \chi$.

Proposition 13. *The system of Definition 12 is equivalent to the system obtained by replacing the side condition "$k \geq 1$" in rule* (Fix*) *with the condition "$1 \leq k \leq \max\{\text{size}(\chi) \mid \epsilon(\chi) = \epsilon(\psi)\} + 1$".*

3.1 Strictness Type Schemes

The inference algorithm is based on a technique developped in [5] for "dead code" detection. Given a term, it returns a description of all the possible s-typings of the term. To describe sets of s-types we introduce the notion of s-type scheme.

Definition 14 (Strictness type schemes). 1. Let $\mathcal{A}_V$ be the set of *annotation variables*, ranged by α, β, γ, ...
The language **P** of *s-type patterns* (*patterns* for short) is defined from the grammar of Definition 4 by replacing $\bot$ and $\top$ by annotation variables, i.e., $\theta ::= \alpha^\iota \mid \theta \to \theta$, where $\alpha \in \mathcal{A}_V$ and $\iota \in \{\mathbf{int}, \mathbf{bool}\}$. Patterns are ranged over by θ, η,...
2. Let ζ range over $\{\bot, \top\} \cup \mathcal{A}_V$. A *constraint* is a formula of one of the following shapes:
   - $\zeta_1 \leq_{\mathrm{a}} \zeta_2$
   - $\{\zeta\} \Rightarrow \mathcal{E}$, where $\mathcal{E}$ is a finite set of constraints
   - $\mathcal{E}_1 \bigvee \mathcal{E}_2$, where $\mathcal{E}_1$ and $\mathcal{E}_2$ are finite set of constraints.
3. An *s-type scheme* is a pair $\langle \theta, \mathcal{E} \rangle$ where θ is a pattern and $\mathcal{E}$ is a finite set of constraints.

An s-type scheme $\langle\theta, \mathcal{E}\rangle$ represents the set of s-types that can be obtained from the pattern θ by replacing annotation variables with annotations in such a way that the constraints in $\mathcal{E}$ are satisfied. The order relation $\leq_a$ between the annotations $\perp$ and $\top$ is defined by: $\perp \leq_a \perp$, $\perp \leq_a \top$ and $\top \leq_a \top$. Strictness types and typings can be obtained from patterns by instantiation.

Definition 15 (Renaming and instantiation). 1. A *renaming* is a one–to–one mapping $\mathbf{r} : \mathcal{A}_V \to \mathcal{A}_V$.
2. An *instantiation* is a (total) function $\mathbf{i} : \mathcal{A}_V \to \{\perp, \top\}$.

Both renaming and instantiation can be extended to annotation constants (by defining $\mathbf{i}(a) = a$ and $\mathbf{r}(a) = a$, for $a \in \{\perp, \top\}$) and to s-types and patterns (in the obvious way). For example: $\mathbf{i}(\alpha^{\mathbf{int}} \to \beta^{\mathbf{int}}) = \mathbf{i}(\alpha)^{\mathbf{int}} \to \mathbf{i}(\beta)^{\mathbf{int}}$. Of course, for any s-type $\phi \in \mathbf{L}$, $\mathbf{i}(\phi) = \phi$ and $\mathbf{r}(\phi) = \phi$.

Definition 16. An instantiation $\mathbf{i}$ satisfies a set of constraints $\mathcal{E}$ if

- $\zeta_1 \leq_a \zeta_2 \in \mathcal{E}$ implies $\mathbf{i}(\zeta_1) \leq_a \mathbf{i}(\zeta_2)$
- $\{\zeta\} \Rightarrow \mathcal{E}' \in \mathcal{E}$ implies that, if $\mathbf{i}(\zeta) = \perp$, then $\mathbf{i}$ satisfies $\mathcal{E}'$
- $\mathcal{E}_1 \bigvee \mathcal{E}_2 \in \mathcal{E}$ implies that $\mathbf{i}$ satisfies $\mathcal{E}_1$ or $\mathbf{i}$ satisfies $\mathcal{E}_2$.

The set of all the instantiations that satisfy $\mathcal{E}$ is denoted by $\mathrm{SAT}(\mathcal{E})$. A scheme $\langle\theta, \mathcal{E}\rangle$ represents all the s-types $\mathbf{i}(\theta)$, for any $\mathbf{i} \in \mathrm{SAT}(\mathcal{E})$. Instantiations are partially ordered by: $\mathbf{i}_1 \leq_a \mathbf{i}_2$ if for all $\alpha \in \mathcal{A}_V$, $\mathbf{i}_1(\alpha) \leq_a \mathbf{i}_2(\alpha)$.

A set of constraints is said *choice-free* if it does not contain constraints of the shape $\mathcal{E}_1 \bigvee \mathcal{E}_2$.

Fact 17. *1. Let $\mathcal{E}$ be a choice-free finite set of constraints. Then the set* $\mathrm{SAT}(\mathcal{E})$ *is either empty or it has a maximum element.*
2. Let $\mathcal{E}$ be a finite set of constraints. Then the set $\mathrm{SAT}(\mathcal{E})$ *has a finite number of maximal elements.*

Example 2. Consider the sets of constraints:

$$\begin{aligned}\mathcal{E} = \{\ \{\alpha_2'\} \Rightarrow \{\ &\{\alpha_2'\} \Rightarrow \{\alpha_2 \leq_a \alpha_2', \alpha_1' \leq_a \alpha_1\},\\ &\alpha_2'' \leq_a \alpha_1',\\ &\{\alpha_2''\} \Rightarrow \{\ \{\alpha_2''\} \Rightarrow \{\alpha_2 \leq_a \alpha_2'', \alpha_1'' \leq_a \alpha_1\},\ \beta' \leq_a \alpha_1'',\ \beta \leq_a \beta'\ \}\ \}\ \}\ ,\end{aligned}$$

$$\begin{aligned}\mathcal{E}_1 &= \{\ \alpha_2' \leq_a \perp, \beta \leq_a \perp,\ \alpha_2 \leq_a \perp, \alpha_1 \leq_a \perp\ \}\ ,\\ \mathcal{E}_2 &= \{\ \alpha_2' \leq_a \perp, \top \leq_a \beta,\ \alpha_2 \leq_a \perp, \top \leq_a \alpha_1\ \}\ ,\\ \mathcal{E}_3 &= \{\ \alpha_2' \leq_a \perp, \beta \leq_a \perp,\ \top \leq_a \alpha_2\ \}\ .\end{aligned}$$

To find the maximum element $\mathbf{i}_1$ of $\mathrm{SAT}(\mathcal{E} \cup \mathcal{E}_1)$ observe that from the four constraints in $\mathcal{E}_1$ we get $\mathbf{i}_1(\alpha_2') = \mathbf{i}_1(\beta) = \mathbf{i}_1(\alpha_2) = \mathbf{i}_1(\alpha_1) = \perp$. Then from $\mathcal{E}$ we get $\mathbf{i}_1(\alpha_1') = \mathbf{i}_1(\alpha_2'') = \mathbf{i}_1(\alpha_1'') = \mathbf{i}(\beta') = \perp$.
Let $\mathcal{I}_1 = \{\alpha_1, \alpha_1', \alpha_1'', \alpha_2, \alpha_2', \alpha_2'', \beta, \beta'\}$, then $\mathbf{i}_1$ defined by: $\mathbf{i}_1(\alpha) = \perp$ if $\alpha \in \mathcal{I}_1$ and $\mathbf{i}_1(\alpha) = \top$ otherwise, is the maximum element of $\mathrm{SAT}(\mathcal{E} \cup \mathcal{E}_1)$.
Similarly observe that from $\mathcal{E}_2$ we get $\mathbf{i}_2(\alpha_2') = \mathbf{i}_2(\alpha_2) = \perp$ and $\mathbf{i}_2(\beta) = \mathbf{i}_2(\alpha_1) = \top$. Let $\mathcal{I}_2 = \{\alpha_2, \alpha_2'\}$, we have that the maximum element $\mathbf{i}_2$ of $\mathrm{SAT}(\mathcal{E} \cup \mathcal{E}_2)$ is defined by: $\mathbf{i}_2(\alpha) = \perp$ if $\alpha \in \mathcal{I}_2$ and $\mathbf{i}_2(\alpha) = \top$ otherwise.

Notice that $\mathbf{i}_2$ is also the maximum element of $\mathrm{SAT}(\mathcal{E} \cup \{\alpha'_2 \leq_a \bot\})$.
Looking for the maximum element $\mathbf{i}_3$ of $\mathrm{SAT}(\mathcal{E} \cup \mathcal{E}_3)$, we get from $\mathcal{E}_3$ that $\mathbf{i}_3(\alpha'_2) = \mathbf{i}_1(\beta) = \bot$ and $\mathbf{i}_3(\alpha_2) = \top$. Then from $\mathcal{E}$ we get $\top \leq_a \bot$. This means that $\mathrm{SAT}(\mathcal{E} \cup \mathcal{E}_3)$ is empty. □

We now define an algorithm that, given a finite set of constraints $\mathcal{E}$, finds the maximal elements $\mathbf{i}_1, \ldots, \mathbf{i}_n$ ($n \geq 0$) of $\mathrm{SAT}(\mathcal{E})$. The algorithm is presented in natural semantics style using judgements $\mathcal{E} \leadsto \mathcal{S}$, where $\mathcal{S} = \{\mathcal{I}_1, \ldots, \mathcal{I}_n\}$ and each $\mathcal{I}_i$, $1 \leq i \leq n$, is the set of annotation variables that *represents* $\mathbf{i}_i$, i.e., such that $\mathbf{i}_i(\alpha) = \bot$ if $\alpha \in \mathcal{I}_i$ and $\mathbf{i}_i(\alpha) = \top$ otherwise. The idea is simply that of recognizing, following the inequalities, all the variables that are forced to represent $\bot$. All other atomic variables are then replaced by $\top$ in a maximal solution.
To describe the algorithm we need a preliminary notation. Let $\mathcal{S}_1 = \{\mathcal{I}_1, \ldots, \mathcal{I}_n\}$ and $\mathcal{S}_2 = \{\mathcal{I}'_1, \ldots, \mathcal{I}'_m\}$. Then $\mathrm{COMP}(\mathcal{S}_1, \mathcal{S}_2)$ denotes the set of the maximal elements in $\mathcal{S}_1 \cup \mathcal{S}_2$, i.e., $\mathrm{COMP}(\mathcal{S}_1, \mathcal{S}_2) =$

$$\{\mathcal{I} \mid \mathcal{I} \in \mathcal{S}_1 \cup \mathcal{S}_2 \text{ and for no } \mathcal{I}' \in \mathcal{S}_1 \cup \mathcal{S}_2, \mathcal{I}' \subset \mathcal{I}\} \ .$$

Definition 18 ("Natural semantics" rules for constraints solution). We write $\mathcal{E} \leadsto \mathcal{S}$ to mean that $\mathcal{E} \leadsto \mathcal{S}$ is derivable by the rules in Fig. 5.

$$\text{(SUCCESS)}\ \frac{}{\mathcal{E} \leadsto \{\emptyset\}}\ \text{if no other rule can be applied} \qquad \text{(FAILURE)}\ \frac{}{\mathcal{E} \cup \{\top \leq_a \bot\} \leadsto \emptyset}$$

$$\text{(BOTTOM)}\ \frac{\mathcal{E}[\alpha := \bot] \leadsto \mathcal{S}}{\mathcal{E} \cup \{\alpha \leq_a \bot\} \leadsto \{\mathcal{I} \cup \{\alpha\} \mid \mathcal{I} \in \mathcal{S}\}} \qquad \text{(TOP)}\ \frac{\mathcal{E}[\alpha := \bot] \leadsto \mathcal{S}}{\mathcal{E} \cup \{\top \leq_a \alpha\} \leadsto \mathcal{S}}$$

$$\text{(GUARD)}\ \frac{\mathcal{E} \cup \mathcal{E}'' \leadsto \mathcal{S}}{\mathcal{E} \cup \{\{\bot\} \Rightarrow \mathcal{E}''\} \leadsto \mathcal{S}} \qquad \text{(CHOICE)}\ \frac{\mathcal{E} \cup \mathcal{E}_1 \leadsto \mathcal{S}_1 \quad \mathcal{E} \cup \mathcal{E}_2 \leadsto \mathcal{S}_2}{\mathcal{E} \cup \{\mathcal{E}_1 \bigvee \mathcal{E}_2\} \leadsto \mathrm{COMP}(\mathcal{S}_1, \mathcal{S}_2)}$$

Fig. 5. "Natural semantics" rules for constraints solution

Proposition 19. *Let $\mathcal{E}$ be a finite set of constraints and $n \geq 0$. Then $\mathcal{E} \leadsto \{\mathcal{I}_1, \ldots, \mathcal{I}_n\}$ if and only if $\mathcal{I}_1, \ldots, \mathcal{I}_n$ represent the maximal elements of* $\mathrm{SAT}(\mathcal{E})$.

Note that an efficient implementation of the previous algorithm would try to delay the application of the rules (GUARD) and (CHOICE) as much as possible to first detect possible failures. Moreover, if we want to know whether a given set of constraints $\mathcal{E}$ is satisfiable, we can define an algorithm that does not return any solution but just answers the question looking for the satisfiability of $\mathcal{E}$. Such an algorithm can be presented in natural semantics style using judgements $\mathcal{E} \leadsto' b$, where $b \in \{\mathrm{TRUE}, \mathrm{FALSE}\}$, with the property that $\mathcal{E} \leadsto' \mathrm{FALSE}$ if and only if $\mathcal{E} \leadsto \emptyset$. The rules for the new judgement would be similar to the previous ones.

3.2 The Algorithm $\mathcal{W}$

To define the inference algorithm $\mathcal{W}$ we need some preliminary notations. Let ρ be a type. By $\mathbf{fresh}(\rho)$ we denote the pattern obtained from ρ by associating a fresh annotation variable to each occurrence of any basic type in ρ. For example: $\mathbf{fresh}(\mathbf{int} \to \mathbf{int}) = \alpha^{\mathbf{int}} \to \beta^{\mathbf{int}}$. For a set of term variables, Γ, $\mathbf{fresh}(\Gamma) = \{x : \mathbf{fresh}(\rho) \mid x^\rho \in \Gamma\}$.
The function $\mathbf{vars}$ maps a pattern to its finite set of annotation variables. For example: $\mathbf{vars}(\alpha^{\mathbf{int}} \to \beta^{\mathbf{int}}) = \{\alpha, \beta\}$.
The function $\mathbf{tail}$, that maps patterns and s-types to $\{\bot, \top\} \cup \mathcal{A}_V$, is inductively defined by: $\mathbf{tail}(\zeta^\iota) = \zeta$, and $\mathbf{tail}(\theta \to \eta) = \mathbf{tail}(\eta)$. For example: $\mathbf{tail}(\alpha^{\mathbf{int}} \to \beta^{\mathbf{int}}) = \beta$.
The function $\epsilon : \mathbf{L} \to \mathbf{T}$ is extended in the obvious way to patterns. For example: $\epsilon(\alpha^{\mathbf{int}} \to \beta^{\mathbf{int}}) = \mathbf{int} \to \mathbf{int}$.
Let θ_1, θ_2 be patterns or s-types such that $\epsilon(\theta_1) = \epsilon(\theta_2)$, $\mathbf{cs}_{\leq}(\theta_1, \theta_2)$ denotes the set of constraints inductively defined by

$$\mathbf{cs}_{\leq}(\zeta_1{}^\iota, \zeta_2{}^\iota) = \{\zeta_1 \leq_{\mathsf{a}} \zeta_2\}, \; \textit{if } \zeta_1, \zeta_2 \in \{\bot, \top\} \cup \mathcal{A}_V$$
$$\mathbf{cs}_{\leq}(\eta_1 \to \cdots \to \eta_n \to \zeta_1{}^\iota, \eta_1' \to \cdots \to \eta_n' \to \zeta_2{}^\iota) =$$
$$\{\{\zeta_2\} \Rightarrow (\{\zeta_1 \leq_{\mathsf{a}} \zeta_2\} \cup \textstyle\bigcup_{1 \leq i \leq n} \mathbf{cs}_{\leq}(\eta_i, \eta_i'))\}, \; \textit{where } \zeta_1, \zeta_2 \in \{\bot, \top\} \cup \mathcal{A}_V.$$

Notice that for all instantiations $\mathbf{i}$, $\mathbf{i}(\theta_1) \leq \mathbf{i}(\theta_2)$ if and only if $\mathbf{i} \in \mathrm{SAT}(\mathbf{cs}_{\leq}(\theta_1, \theta_2))$. In particular $\mathbf{cs}_{\leq}(\theta_1, \theta_2)$ consists of a single constraint of the shape $\{\mathbf{tail}(\theta_2)\} \Rightarrow \mathcal{E}'$. This constraint is trivially satisfied by all the instantiations $\mathbf{i}$ such that $\mathbf{i}(\mathbf{tail}(\theta_2)) = \top$, i.e., such that $\mathbf{i}(\theta_2) \in \mathbf{L}_\top$.
For each constant C an s-type scheme $\mathbf{sts}(C)$ is specified. For example, for any integer n, $\mathbf{sts}(\mathrm{n}) = \langle \top^{\mathbf{int}}, \emptyset \rangle$ and $\mathbf{sts}(+) = \langle \alpha_1{}^{\mathbf{int}} \to \alpha_2{}^{\mathbf{int}} \to \beta^{\mathbf{int}}, \{\{\beta\} \Rightarrow \{\{\alpha_1 \leq_{\mathsf{a}} \bot\} \bigvee \{\alpha_2 \leq_{\mathsf{a}} \bot\}\}\} \rangle$.

We can now proceed to define the s-type inference algorithm $\mathcal{W}$. This algorithm is presented in Fig. 6. Let $\vdash_{\mathbf{T}} M : \rho$ and $\mathcal{W}(M) = \langle \Theta, \theta, \mathcal{E} \rangle$, then Θ is a basis that associates a pattern to each term variable in $\mathrm{FV}(M)$, θ is a pattern, and $\mathcal{E}$ is a finite set of constraints.
Proposition 13 implies that to get a complete inference algorithm it is sufficient to build a set of constraints that codifies the applications of the rule (Fix*) for all the values of k less or equal than 1 plus the maximum of $\{\mathrm{size}(\phi) \mid \epsilon(\phi) = \rho\}$, where ρ is the type of the fixpoint expression. The set CS(1) (see Fig. 6) codifies such applications. For example let $\rho = \mathbf{int} \to \mathbf{int} \to \mathbf{int} \to \mathbf{int}$. The set $\{[\phi]_{\cong} \mid \epsilon(\phi) = \rho\}$ has 9 elements while the maximum of $\{\mathrm{size}(\phi) \mid \epsilon(\phi) = \rho\}$ is 3. So 4 is an upper bound to the number of premises that an algorithm has to consider while looking for a property of a fixpoint expression of type ρ.

Correctness and completeness of the inference are expressed by Theorem 20. As a consequence of completeness we have that, for every $M \in \Lambda_{\mathbf{T}}$, $\mathcal{W}(M)$ is a kind of principal type scheme for M, i.e., it can be used for generating all the s-types of M.

Theorem 20. *Let $\vdash_{\mathbf{T}} M : \rho$ and $\mathcal{W}(M) = \langle \Theta, \theta, \mathcal{E} \rangle$.*

1. *For all instantiations $\mathbf{i}$, if $\mathbf{i}$ satisfies $\mathcal{E}$, then $\mathbf{i}(\Theta) \vdash^\star_{\mathbf{L}} M : \mathbf{i}(\theta)$.*

$$
\begin{array}{lll}
\mathcal{W}(P) = & \text{let} & \Theta = \mathbf{fresh}(\mathrm{FV}(P)) \\
& & \text{and } \langle\theta,\mathcal{E}\rangle = \mathcal{W}_0(\Theta, P) \\
& \text{in} & \langle\Theta,\eta,\mathcal{E}\rangle \text{ end}
\end{array}
$$

$$
\begin{array}{lll}
\multicolumn{3}{l}{\mathcal{W}_0(\Theta,P) = \text{case } P \text{ of}} \\
\quad C: & & \mathbf{sts}(C) \\
\quad x: & \text{let} & \theta_1 = \Theta(x) \\
& & \text{and } \theta_2 = \mathbf{fresh}(\epsilon(\theta_1)) \\
& \text{in} & \langle\theta_2, \mathbf{cs}_{\leq}(\theta_1,\theta_2)\rangle \text{ end} \\
\quad \lambda x^{\sigma}.M: & \text{let} & \theta = \mathbf{fresh}(\sigma) \\
& & \text{and } \langle\eta,\mathcal{E}\rangle = \mathcal{W}_0(\Theta\cup\{x:\theta\}, M) \\
& \text{in} & \langle\theta\to\eta,\mathcal{E}\rangle \text{ end} \\
\quad MN: & \text{let} & \langle\theta_1\to\eta,\mathcal{E}_1\rangle = \mathcal{W}_0(\Theta, M) \\
& & \text{and } \langle\theta_2,\mathcal{E}_2\rangle = \mathcal{W}_0(\Theta, N) \\
& \text{in} & \langle\eta, \{\{\mathbf{tail}(\eta)\} \Rightarrow (\mathcal{E}_1\cup\mathcal{E}_2\cup\mathbf{cs}_{\leq}(\theta_2,\theta_1)\}\rangle \text{ end} \\
\multicolumn{3}{l}{\quad \text{if } N \text{ then } M_1 \text{ else } M_2:} \\
& \text{let} & \langle\alpha,\mathcal{E}_0\rangle = \mathcal{W}_0(\Theta, N) \\
& & \text{and } \langle\theta_1,\mathcal{E}_1\rangle = \mathcal{W}_0(\Theta, M_1) \\
& & \text{and } \langle\theta_2,\mathcal{E}_2\rangle = \mathcal{W}_0(\Theta, M_2) \\
& & \text{and } \eta = \mathbf{fresh}(\epsilon(\theta_1)) \quad (*\ \text{Notice that } \epsilon(\theta_1)=\epsilon(\theta_2)\ *) \\
& \text{in} & \langle\eta, \{(\mathcal{E}_0\cup\{\alpha\leq_{\mathrm{a}}\bot\})\bigvee(\mathcal{E}_1\cup\mathcal{E}_2\cup\mathbf{cs}_{\leq}(\theta_1,\eta)\cup\mathbf{cs}_{\leq}(\theta_2,\eta))\}\rangle \text{ end} \\
\multicolumn{3}{l}{\quad \text{fix } x^{\rho}.M:} \\
& \text{let} & \theta = \mathbf{fresh}(\rho) \\
& & \text{and } \langle\eta,\mathcal{E}\rangle = \mathcal{W}_0(\Theta\cup\{x:\theta\}, M) \\
& & \text{and } h = \mathbf{max}\{\mathbf{size}(\phi) \mid \epsilon(\phi)=\rho\}+1 \\
& & \text{and } \mathrm{CS}(l) = \text{let} \\
& & \qquad\qquad \mathbf{r}_l = \text{“a fresh renaming of the annotation variables not in } \Theta\text{”} \\
& & \qquad\qquad \text{and } \theta_l = \mathbf{r}_l(\theta) \text{ and } \eta_l = \mathbf{r}_l(\eta) \text{ and } \mathcal{E}_l = \mathbf{r}_l(\mathcal{E}) \\
& & \qquad \text{in if } l = h \\
& & \qquad\qquad \text{then } \mathcal{E}_h\cup\mathbf{cs}_{\leq}(\eta_h,\theta_1)\cup\mathbf{cs}_{\leq}(\eta_h,\eta) \\
& & \qquad\qquad \text{else } \mathcal{E}_l\cup\{(\mathbf{cs}_{\leq}(\eta_l,\theta_1)\cup\mathbf{cs}_{\leq}(\eta_l,\eta)) \\
& & \qquad\qquad\qquad \bigvee(\mathbf{cs}_{\leq}(\theta_l,\eta_{l+1})\cup\mathrm{CS}(l+1))\} \\
& & \qquad \text{end} \\
& \text{in} & \langle\eta,\mathrm{CS}(1)\rangle \text{ end}
\end{array}
$$

Fig. 6. Algorithm $\mathcal{W}$

2. *For all s-type assignment statements* $\Sigma \vdash^{\star}_{\mathbf{L}} M : \phi$ *such that* $\epsilon(\Sigma) = \mathrm{FV}(M)$, *there exists* $\mathbf{i} \in \mathrm{SAT}(\mathcal{E})$ *such that* $\mathbf{i}(\Theta) = \Sigma$ *and* $\mathbf{i}(\theta) = \phi$.

The output of the algorithm $\mathcal{W}$ can be used either to check if a function has a given s-type, or to detect the set of arguments in which the function can be proved to be strict. Such a set can be found by examining the s-types of the shape $\phi_1 \to \cdots \to \phi_n \to \bot^{\iota}$, where $n \geq 1$, $\iota \in \{\mathbf{int}, \mathbf{bool}\}$ and there is an index i, $1 \leq i \leq n$, such that $\phi_i \in \mathbf{L}_{\bot}$ but, for all j, $1 \leq j \neq i \leq n$, $\phi_j \in \mathbf{L}_{\top}$. We call *first order strictness properties* of a function such s-types.

For a pattern θ and an s-type ϕ such that $\epsilon(\theta) = \epsilon(\phi)$, $\mathbf{cs}_{\cong}(\theta, \phi)$ denotes the minimal set of constraints such that, for all instantiation $\mathbf{i}$, $\mathbf{i} \in \mathbf{cs}_{\cong}(\theta, \phi)$ if and only if $\mathbf{i}(\theta) \cong \phi$.

Corollary 21. *Let $\vdash_{\mathrm{T}} M : \rho$ and $\mathcal{W}(M) = \langle \Theta, \theta, \mathcal{E} \rangle$.*

1. *An s-type ϕ can be assigned to M if and only if* $\mathrm{SAT}(\mathcal{E} \cup \mathbf{cs}_{\cong}(\theta, \phi)) \neq \emptyset$.
2. *Let $\rho = \rho_1 \to \cdots \to \rho_n \to \iota$, and $\theta = \theta_1 \to \cdots \to \theta_n \to \alpha^{\iota}$. If ϕ is a first order strictness property and $\epsilon(\phi) = \rho$, then ϕ can be assigned to M if and only if, for some maximal instantiation $\mathbf{i} \in \mathrm{SAT}(\mathcal{E}[\alpha := \perp])$, either $\mathbf{i}(\theta) \cong \perp(\rho)$, i.e., M is an always divergent program, or $\mathbf{i}(\theta) \cong \phi$.*

Example 3. Let $\vdash_{\mathrm{T}}$ twice : ρ be the typing of Example 1. Then $\mathcal{W}(\text{twice}) = \langle \Theta, \theta, \mathcal{E} \rangle$, where $\theta = (\alpha_1{}^{\mathbf{int}} \to \alpha_2{}^{\mathbf{int}}) \to \beta^{\mathbf{int}} \to \alpha_2'{}^{\mathbf{int}}$ and $\mathcal{E}$ is the first set of constraints in Example 2.

Then $\mathcal{E}[\alpha_2' := \perp] \rightsquigarrow \{\mathcal{I}_2\}$, where $\mathcal{I}_2$ is defined in Example 2 and represents the maximum element, $\mathbf{i}_2$, in $\mathrm{SAT}(\mathcal{E}[\alpha_2' := \perp])$. Since $\mathbf{i}_2(\theta) = (\top^{\mathbf{int}} \to \perp^{\mathbf{int}}) \to \top^{\mathbf{int}} \to \perp^{\mathbf{int}}$ we have that twice is strict in its first argument. Moreover, since any instantiation $\mathbf{i}$ such that $\mathbf{i}(\theta) = (\top^{\mathbf{int}} \to \top^{\mathbf{int}}) \to \perp^{\mathbf{int}} \to \perp^{\mathbf{int}}$ is neither $\leq_{\mathrm{a}}$ nor $\geq_{\mathrm{a}}$ than $\mathbf{i}_2$ and $\mathbf{i}_2(\theta) \not\cong \perp(\rho)$, we know that to twice we cannot assign the property $(\top^{\mathbf{int}} \to \top^{\mathbf{int}}) \to \perp^{\mathbf{int}} \to \perp^{\mathbf{int}}$, i.e., twice cannot be proved to be strict in the second argument.

The sets $\mathbf{cs}_{\cong}(\theta, (\perp^{\mathbf{int}} \to \perp^{\mathbf{int}}) \to \perp^{\mathbf{int}} \to \perp^{\mathbf{int}})$ and $\mathbf{cs}_{\cong}(\theta, (\top^{\mathbf{int}} \to \perp^{\mathbf{int}}) \to \top^{\mathbf{int}} \to \perp^{\mathbf{int}})$ are the sets $\mathcal{E}_1$ and $\mathcal{E}_2$ of Example 2. So $\mathcal{E} \cup \mathcal{E}_1 \rightsquigarrow'$ TRUE and $\mathcal{E} \cup \mathcal{E}_2 \rightsquigarrow'$ TRUE.
Finally observe that the set $\mathcal{E}_3$ of Example 2 is $\mathbf{cs}_{\cong}(\theta, (\top^{\mathbf{int}} \to \top^{\mathbf{int}}) \to \perp^{\mathbf{int}} \to \perp^{\mathbf{int}})$. Obviously $\mathcal{E} \cup \mathcal{E}_3 \rightsquigarrow'$ FALSE, since twice cannot be proved to be strict in the second argument. □

As observed in [11, 20], to check that a function has a specified strictness property is more efficient than to look for all its strictness properties. Among other reasons, because some of these properties may be of no interest for the program. However, the first order strictness properties are always interesting, since they specify that some function applications can be done "by value" in every context.
We conclude the section with an example of application of the algorithm involving a fixpoint.

Example 4. The function $F = \mathrm{fix}\, f^{\rho}.M$, where $\rho = \mathbf{int} \to \mathbf{int} \to \mathbf{int} \to \mathbf{int}$ and

$$M = \lambda x^{\mathbf{int}}.\lambda y^{\mathbf{int}}.\lambda z^{\mathbf{int}}.\mathrm{if} \ = z\,0 \ \mathrm{then} \ {+}xy \ \mathrm{else} \ fyx(-z1) \ ,$$

was used in [16] to show the limitation of a type system without conjunction. The same example was considered in [20] and [11] to show the power of the analysis presented there. In our framework we get the same results of [20] and [11], namely: $\emptyset \vdash_{\mathrm{L}} F : \phi_1$, $\emptyset \vdash_{\mathrm{L}} F : \phi_2$, and $\emptyset \vdash_{\mathrm{L}} F : \phi_3$, where $\phi_1 = \perp^{\mathbf{int}} \to \top^{\mathbf{int}} \to \top^{\mathbf{int}} \to \perp^{\mathbf{int}}$, $\phi_2 = \top^{\mathbf{int}} \to \perp^{\mathbf{int}} \to \top^{\mathbf{int}} \to \perp^{\mathbf{int}}$, and $\phi_3 = \top^{\mathbf{int}} \to \top^{\mathbf{int}} \to \perp^{\mathbf{int}} \to \perp^{\mathbf{int}}$. These s-types mean that the function is

strict in each of its 3 arguments.
The derivations $\emptyset \vdash_{\mathbf{L}} F : \phi_1$ and $\emptyset \vdash_{\mathbf{L}} F : \phi_2$ end respectively with the following applications of the rule (Fix):

$$\text{(Fix)}\ \frac{\{f:\phi_1\} \vdash_{\mathbf{L}} M:\phi_2 \quad \{f:\phi_2\} \vdash_{\mathbf{L}} M:\phi_1}{\emptyset \vdash_{\mathbf{L}} \text{fix}\, f^{\rho}.M:\phi_i}$$

for $i \in \{1,2\}$. The derivation $\emptyset \vdash_{\mathbf{L}} F : \phi_3$ ends with the following application of the rule (Fix):

$$\text{(Fix)}\ \frac{\{f:\phi_3\} \vdash_{\mathbf{L}} M:\phi_3}{\emptyset \vdash_{\mathbf{L}} \text{fix}\, f^{\rho}.M:\phi_3}\ .$$

Note that to assign ϕ_3 to F one premise is sufficient, while to assign ϕ_1 or ϕ_2 are needed two premises. So with the rule for fixpoint of Kuo and Mishra, see [16], ϕ_1 and ϕ_2 cannot be assigned to F.

Notice that, for some $\theta = \alpha_{1,0} \to \alpha_{2,0} \to \alpha_{3,0} \to \alpha_{4,0} = \mathbf{fresh}(\rho)$, $\beta_{1,0} = \mathbf{fresh(int)}$, $\beta_{2,0} = \mathbf{fresh(int)}$, $\beta_{3,0} = \mathbf{fresh(int)}$, and $\mathbf{sts}(=) = \langle \gamma_{1,0} \to \gamma_{2,0} \to \gamma_{3,0}, \{\{\gamma_{3,0}\} \Rightarrow \{\{\gamma_{1,0} \leq_{\mathbf{a}} \bot\} \bigvee \{\gamma_{2,0} \leq_{\mathbf{a}} \bot\}\}\}\rangle$, we have

$$\mathcal{W}_0(\{f:\theta,\ x:\beta_{1,0},\ y:\beta_{2,0},\ z:\beta_{3,0}\}, = z\,0) = \langle\, \gamma_{3,0},\ \mathcal{E}_3 \rangle\ ,$$

where

$$\begin{aligned}\mathcal{E}_3 = \{\ \{\gamma_{3,0}\} \Rightarrow \{\ \{\gamma_{3,0}\} \Rightarrow \{\ &\{\gamma_{1,0} \leq_{\mathbf{a}} \bot\} \bigvee \{\gamma_{2,0} \leq_{\mathbf{a}} \bot\},\\ &\beta_{3,0} \leq_{\mathbf{a}} \beta'_{3,0},\\ &\beta'_{3,0} \leq_{\mathbf{a}} \gamma_{1,0} \quad \},\\ \top \leq_{\mathbf{a}} \gamma_{2,0} \quad &\}\}\ .\end{aligned}$$

Then, if $\mathbf{sts}(+) = \langle \beta_{4,0} \to \beta_{5,0} \to \beta_{6,0}, \{\{\beta_{6,0}\} \Rightarrow \{\{\beta_{4,0} \leq_{\mathbf{a}} \bot\} \bigvee \{\beta_{5,0} \leq_{\mathbf{a}} \bot\}\}\}\rangle$, we have

$$\mathcal{W}_0(\{f:\theta,\ x:\beta_{1,0},\ y:\beta_{2,0},\ z:\beta_{3,0}\}, +xy) = \langle\, \beta_{6,0},\ \mathcal{E}_1 \rangle\ ,$$

where

$$\begin{aligned}\mathcal{E}_1 = \{\ \{\beta_{6,0}\} \Rightarrow \{\ \{\beta_{6,0}\} \Rightarrow \{\ &\{\beta_{4,0} \leq_{\mathbf{a}} \bot\} \bigvee \{\beta_{5,0} \leq_{\mathbf{a}} \bot\},\\ &\beta_{1,0} \leq_{\mathbf{a}} \beta'_{1,0},\\ &\beta'_{1,0} \leq_{\mathbf{a}} \beta_{4,0} \quad \},\\ \beta_{2,0} \leq_{\mathbf{a}} \beta'_{2,0},&\\ \beta'_{2,0} \leq_{\mathbf{a}} \beta_{5,0} \quad &\}\}\ .\end{aligned}$$

Moreover, if $\mathbf{sts}(-) = \langle \beta_{7,0} \to \beta_{8,0} \to \beta_{9,0}, \{\{\beta_{9,0}\} \Rightarrow \{\{\beta_{7,0} \leq_{\mathbf{a}} \bot\} \bigvee \{\beta_{8,0} \leq_{\mathbf{a}} \bot\}\}\}\rangle$, we have

$$\mathcal{W}_0(\{f:\theta,\ x:\beta_{1,0},\ y:\beta_{2,0},\ z:\beta_{3,0}\}, fyx(-z1)) = \langle\, \alpha'_{4,0},\ \mathcal{E}_2 \rangle\ ,$$

where

$$\begin{aligned}\mathcal{E}_2 = \{\ \{\alpha'_{4,0}\} \Rightarrow \{\ \{\alpha'_{4,0}\} \Rightarrow \{\ \{\alpha'_{4,0}\} \Rightarrow \{\ \{\alpha'_{4,0}\} \Rightarrow \{\ &\alpha'_{3,0} \leq_{\mathbf{a}} \alpha_{3,0},\\ &\alpha'_{2,0} \leq_{\mathbf{a}} \alpha_{2,0},\\ &\alpha'_{1,0} \leq_{\mathbf{a}} \alpha_{1,0}\ \},\\ \beta_{2,0} \leq_{\mathbf{a}} \beta''_{2,0}&\\ \beta''_{2,0} \leq_{\mathbf{a}} \alpha'_{1,0} \quad &\},\\ \beta_{1,0} \leq_{\mathbf{a}} \beta''_{1,0},&\\ \beta''_{1,0} \leq_{\mathbf{a}} \alpha'_{2,0} \quad &\},\end{aligned}$$

$$\begin{array}{l}\{\beta_{9,0}\} \Rightarrow \{ \quad \{\beta_{9,0}\} \Rightarrow \{ \{\beta_{7,0} \leq_{\mathrm{a}} \bot\} \bigvee \{\beta_{8,0} \leq_{\mathrm{a}} \bot\}, \\ \qquad\qquad\qquad\qquad\qquad \beta_{3,0} \leq_{\mathrm{a}} \beta''_{3,0}, \\ \qquad\qquad\qquad\qquad\qquad \beta''_{3,0} \leq_{\mathrm{a}} \beta_{7,0} \qquad\qquad \}, \\ \qquad\qquad \top \leq_{\mathrm{a}} \beta_{8,0} \qquad\qquad\qquad\qquad\qquad\quad \}, \\ \beta_{9,0} \leq_{\mathrm{a}} \alpha'_{3,0} \qquad\qquad\qquad\qquad\qquad\qquad\qquad\quad \} .\end{array}$$

Lastly we get $\mathcal{W}_0(\{f : \theta\}, M) = \langle \eta, \mathcal{E} \rangle$ where $\eta = \beta_{1,0} \to \beta_{2,0} \to \beta_{3,0} \to \gamma_0$, and $\mathcal{E} = \{ (\mathcal{E}_3 \cup \{\beta'_{3,0} \leq_{\mathrm{a}} \bot\}) \bigvee (\mathcal{E}_1 \cup \mathcal{E}_2 \cup \{\beta_{6,0} \leq_{\mathrm{a}} \gamma_0, \alpha'_{4,0} \leq_{\mathrm{a}} \gamma_0\}) \}$.

So, since $\mathrm{size}(\rho) = 3$, if we apply our type inference algorithm we get (see Fig. 6): $h = 4$ and

$$\mathcal{W}(\emptyset, F) = \langle \emptyset, \eta, \mathrm{CS}(1) \rangle$$

where, for $i \in \{1, \ldots, 4\}$, $\mathcal{E}_i$, θ_i, η_i are obtained from $\mathcal{E}$, θ, η by replacing all the occurrences of the index "0" by the index "i".

Since ϕ_1, ϕ_2 and ϕ_3 are first order strictness properties and $\top^{\mathrm{int}} \to \top^{\mathrm{int}} \to \top^{\mathrm{int}} \to \bot^{\mathrm{int}}$ is not a property of F, we have that $\mathrm{SAT}(\mathrm{CS}(1)[\gamma := \bot])$ has three maximal elements, $\mathbf{i}_1$, $\mathbf{i}_2$, and $\mathbf{i}_3$, corresponding to the first order strictness property of F, i.e., $\mathbf{i}_1(\eta) = \phi_1$, $\mathbf{i}_2(\eta) = \phi_2$, and $\mathbf{i}_3(\eta) = \phi_3$. So, if $\mathcal{I}_1$, $\mathcal{I}_2$, $\mathcal{I}_3$ are the representations of $\mathbf{i}_1$, $\mathbf{i}_2$, $\mathbf{i}_3$, then $\mathrm{CS}(1)[\gamma := \bot] \rightsquigarrow \{\mathcal{I}_1, \mathcal{I}_2, \mathcal{I}_3\}$.

Obviously we also have $\mathrm{CS}(1) \cup \mathbf{cs}_{\cong}(\eta, \phi_1) \rightsquigarrow'$ TRUE, $\mathrm{CS}(1) \cup \mathbf{cs}_{\cong}(\eta, \phi_2) \rightsquigarrow'$ TRUE, and $\mathrm{CS}(1) \cup \mathbf{cs}_{\cong}(\eta, \phi_3) \rightsquigarrow'$ TRUE. □

4 Related Work

We first argue that our system gives the same strictness information as the system introduced by Solberg et al. in [21] (see also [20] Chap. 2). In her PhD thesis Solberg compares that system with the others in the literature and shows that its restriction to strictness is more powerful than the system of Kuo and Mishra, see [16]. Solberg's system infers in addition to strictness also totality properties of programs. As in our system, properties are expressed by annotating types. The annotations allowed are: **b**, $\top$, and **n**. The meaning of the first is to be an undefined computation, the second means absence of knowledge, and the third means to be a terminating computation. Solberg's language of properties is defined by the following grammar: $\phi ::= \mathbf{b}^{\rho} \mid \mathbf{n}^{\rho} \mid \top^{\rho} \mid \phi \to \phi$. (Notice that our s-types $\bot^{\iota}$ and $\top^{\iota}$ coincide with $\mathbf{b}^{\iota}$ and $\top^{\iota}$.) The system infers intersections of such properties. The only rule in which intersection of properties is used in a non trivial way is the rule for assigning a property to a fixpoint. We simulate such intersection allowing multiple premises in the antecedent of the rule (Fix) (see the Example 4 for an application of this rule). As far as combining annotations for strictness and annotation for totality properties, we have proved, in [4], that, for the system of [21], if a term can be assigned ϕ (containing annotations $\top$, **b** and **n**), then it can also be assigned a ϕ' containing only $\top$ and **b**, or $\top$ and **n** such that ϕ' gives the same information of ϕ. So the most efficient way to gather such properties is to address them separately, each one with its inference system.

In [11] Hankin and Le Métayer introduce a type checking algorithm for a system which uses intersections at all levels and fully implements the conjunctive

strictness logic introduced by Jensen, see [14], and (independently) by Benton, see [2]. Given a typed program and a strictness property the algorithm checks whether the program satisfies the property by inspecting directly the code. Even though this is done using a number of optimizations (such as lazy types and the analogous of frontiers for intersection types), in some cases, it is necessary to explore the set of all the properties associated to a type.
Similarly [20] Chap. 3 presents an extension of the system of Chap. 2 having intersections at all levels. The checking algorithm for this system uses the optimization of lazy types. Since it is necessary to have intersection at all the levels to deal with the lazy types, they cannot be used for implementing the system of [20] Chap. 2.
Systems allowing intersections at all levels are more powerful than our system as the following example shows.

Example 5. The function

$$G = (\lambda f^{\mathbf{int}\to\mathbf{int}\to\mathbf{int}}.\lambda x^{\mathbf{int}}.\lambda y^{\mathbf{int}}.\text{if } = x\,0 \text{ then } f1y \text{ else } fy1)+ \ ,$$

is used by Benton in [2] pag. 32 to point out that the type system by Kuo and Mishra [16] gives weaker results than abstract interpretation in the style of [3] even for non-recursive terms. The same example shows also a limitation of our system, in fact we would like to be able to infer that G is strict in its second arguments, i.e., $\emptyset \vdash_{\mathrm{L}} G : \top^{\mathbf{int}} \to \bot^{\mathbf{int}} \to \bot^{\mathbf{int}}$, but this is not derivable in our system, neither in the system in [21] and [20] Chap. 2. Of course this can can be inferred by the system in [11] and [20] Chap. 3, which implement the conjunctive strictness logic of [14] and [2]. □

The algorithm proposed by Hankin and Le Métayer [11] is a *checking* algorithm: it checks whether a program has a particular strictness property by inspecting directly the code.

In this paper we have presented an *inference* algorithm: it returns a principal s-type scheme, representing (by means of a set of constraints) all the s-types that can be assigned to a given program.
The use of constraints in algorithms for static analysis of programs is well known, for instance they are used by Kuo and Mishra in [16]. Our approach of inferring constraints representing s-types has some interesting features. First, generating guarded constraints for codifying uses of the rule ($\to$ E) of Definition 7 (see the treatment of applications in Fig. 6) results in an effect similar to the lazy types optimization used for the checking algorithm in [11]. Moreover, the constraints generated for the rule (Fix) codify all the different alternatives of number of premises needed to analyze a fixpoint. Each alternative is in turn codified by a set of constraints whose solutions represent all the possible uses of the rule with the corresponding number of premises. An algorithm for checking whether the set of constraints is satisfiable can explore the space of alternatives codifying the application of the rule (Fix) from the cheapest to the most expensive stopping as soon as a success is reached. Notice that the number of alternatives is limited, as discussed in Sect. 3.2.

Finally, observe that the extraction of the constraints from a program takes time linear in the size of the program and it is done once. Then the set of constraints can be used to find all the s-types that can be assigned to the program.

Conclusions and Future Work

In this paper we presented a type assignment system and an algorithm for inferring strictness properties of typed functional programs.
It would be easy to extend the algorithm to deal with Curry-style polymorphism, see [13]. So for instance one could think of doing both type and s-type inference at the same time. We can also deal with Milner-style polymorphism, (see [17] and [9]), which is characterized by the presence of a "let $x = N$ in M" construct in which x in M can be assigned different types, all instances of the same type scheme. In this case we have to handle the s-types corresponding to the different types of N in M.

An extension of our system that incorporates rank 2 intersection is in preparation. Using rank 2 intersection makes possible to express a major number of properties of programs (for instance it is possible to give the right s-type to the term G of Example 5). We are also planning to extend the set of data types treated, in particular including lists (as done in [12]). Notice that pairs as treated in [19] can be easily added.

A more general goal of our research (the topic of the first author's PhD thesis) is to present various static analysis of programs: strictness, totality, dead code, binding time, etc. in a common framework. This view is taken in [10], in which the authors identify a basic structure for properties that can be extended to incorporate the various analyses. A general framework for studying program properties is also presented in [7], where, however, the authors deal with an untyped language.

Acknowledgment

During the preparations of the paper we have benefited greatly from discussions with Mario Coppo and Stefano Berardi. We also thank Felice Cardone for reading a preliminary version of this paper, and the anonymous referees and the editors for their suggestions to improve the presentation of the paper.

References

1. H. P. Barendregt, M. Coppo, and M. Dezani-Ciancaglini. A filter lambda model and the completeness of type assignment. *Journal of Symbolic Logic*, 48:931–940, 1983.
2. P. N. Benton. *Strictness Analysis of Lazy Functional Programs*. PhD thesis, University of Cambridge, Pembroke College, 1992.
3. G. L. Burn, C. Hankin, and S. Abramsky. Strictness Analysis for Higher-Order Functions. *Science of Computer Programming*, 7:249–278, 1986.

4. M. Coppo, F. Damiani, and P. Giannini. On Strictness and Totality. Internal report. Dipartimento di Informatica, Universitá di Torino, 1996.
5. M. Coppo, F. Damiani, and P. Giannini. Refinement Types for Program Analysis. In *SAS'96*, LNCS 1145, pages 143–158. Springer, 1996.
6. M. Coppo and M. Dezani-Ciancaglini. An extension of basic functional theory for lambda-calculus. *Notre Dame Journal of Formal Logic*, 21(4):685–693, 1980.
7. M. Coppo and A. Ferrari. Type inference, abstract interpretation and strictness analysis. *Theoretical Computer Science*, 121:113–145, 1993.
8. D. Dussart and F. Henglein and C. Mossin. Polymorphic Recursion and Subtype Qualifications: Polymorphic Binding-Time Analysis in Polynomial Time. In *SAS'95*, LNCS 983, pages 118–135. Springer, 1995.
9. M. J. C. Gordon, R. Milner, and C. P. Wadsworth. *Edinburg LCF*. LNCS 78. Springer, 1979.
10. C. Hankin and D. Le Métayer. A Type-Based Framework for Program Analysis. In *SAS'94*, LNCS 864, pages 380–394. Springer, 1994.
11. C. Hankin and D. Le Métayer. Deriving algorithms for type inference systems: Applications to strictness analysis. In *POPL'94*, pages 202–212. ACM, 1994.
12. C. Hankin and D. Le Métayer. Lazy type inference and program analysis. *Science of Computer Programming*, 25:219–249, 1995.
13. R. Hindley. The principal types schemes for an object in combinatory logic. *Transactions of American Mathematical Society*, 146:29–60, 1969.
14. T. P. Jensen. *Abstract Interpretation in Logical Form*. PhD thesis, University of London, Imperial College, 1992.
15. G. Kahn. Natural semantics. In K. Fuchi and M. Nivat, editors, *Programming Of Future Generation Computer*. Elsevier Sciences B.V. (North-Holland), 1988.
16. T. M. Kuo and P. Mishra. Strictness analysis: a new perspective based on type inference. In *Functional Programming Languages and Computer Architecture*, pages 260–272. ACM, 1989.
17. R. Milner. A Theory of Type Polymorphism in Programming. *Journal of Computer and System Science*, 17:348–375, 1978.
18. J. Palsberg and P. O'Keefe. A Type System Equivalent to Flow Analysis. In *POPL'95*, pages 367–378. ACM, 1995.
19. A. M. Pitts. Operationally-Based Theories of Program Equivalence. Summer School on *Semantics and Logics of Computation*, Cambridge UK, 25-29 Sep 1995.
20. K. L. Solberg. *Annotated Type Systems for Program Analysis*. PhD thesis, Aarhus University, Denmark, 1995.
21. K. L. Solberg, H. R. Nielson, and F. Nielson. Strictness and Totality Analysis. In *SAS'94*, LNCS 864, pages 408–422. Springer, 1994.
22. S. van Bakel. *Intersection Type Disciplines in Lambda Calculus and Applicative Term Rewriting Systems*. PhD thesis, Katholieke Universiteit Nijmegen, 1993.
23. D. A. Wright. *Reduction Types and Intensionality in the Lambda-Calculus*. PhD thesis, University of Tasmania, 1992.

Primitive Recursion for Higher-Order Abstract Syntax

Joëlle Despeyroux[1], Frank Pfenning[2], and Carsten Schürmann[2]

[1] INRIA, F-06902 Sophia-Antipolis Cedex, France
joelle.despeyroux@sophia.inria.fr
[2] Carnegie Mellon University, Pittsburgh PA 15213, USA
{fp|carsten}@cs.cmu.edu

1 Introduction

Higher-order abstract syntax is a central representation technique in many logical frameworks, that is, meta-languages designed for the formalization of deductive systems. The basic idea is to represent variables of the object language by variables in the meta-language. Consequently, object language constructs which bind variables must be represented by meta-language constructs which bind the corresponding variables.

This deceptively simple idea, which goes back to Church [1] and Martin-Löf's system of arities [18], has far-reaching consequences for the methodology of logical frameworks. On one hand, encodings of logical systems using this idea are often extremely concise and elegant, since common concepts and operations such as variable binding, variable renaming, capture-avoiding substitution, or parametric and hypothetical judgments are directly supported by the framework and do not need to be encoded separately in each application. On the other hand, higher-order representations are no longer inductive in the usual sense, which means that standard techniques for reasoning by induction do not apply.

Various attempts have been made to preserve the advantages of higher-order abstract syntax in a setting with strong induction principles [5, 4], but none of these is entirely satisfactory from a practical or theoretical point of view.

In this paper we take a first step towards reconciling higher-order abstract syntax with induction by proposing a system of *primitive recursive functionals* that permits iteration over subjects of functional type. In order to avoid the well-known paradoxes which arise in this setting (see Section 3), we decompose the primitive recursive function space $A \Rightarrow B$ into a modal operator and a parametric function space $(\Box A) \to B$. The inspiration comes from linear logic which arises from a similar decomposition of the intuitionistic function space $A \supset B$ into a modal operator and a linear function space $(!A) \multimap B$.

The resulting system allows, for example, iteration over the structure of expressions from the untyped λ-calculus when represented using higher-order abstract syntax. It is general enough to permit iteration over objects of *any* simple type, constructed over *any* simply typed signature and thereby encompasses Gödel's system T [9]. Moreover, it is conservative over the simply-typed λ-calculus which means that the compositional adequacy of encodings in higher-order abstract syntax is preserved. We view our calculus as an important first step towards a system which allows the methodology of logical frameworks such as LF [10] to be incorporated into systems such as Coq [20] or ALF [12].

The remainder of this paper is organized as follows: Section 2 reviews the idea of higher order abstract syntax and introduces the simply typed λ-calculus ($\lambda^{\rightarrow}$) which we extend to a modal λ-calculus in Section 3. Section 4 then presents the concept of iteration. In Section 5 we sketch the proof of our central result, namely that our extension is conservative over $\lambda^{\rightarrow}$. Finally, Section 6 assesses the results, compares some related work, and outlines future work. A full version of this paper with complete technical developments and detailed proofs is accessible as `http://www.cs.cmu.edu/~carsten/CMU-CS-96-172.ps.gz` [6].

2 Higher-Order Abstract Syntax

Higher-order abstract syntax exploits the full expressive power of a typed λ-calculus for the representation of an object language, where λ-abstraction provides the mechanism to represent binding. In this paper, we restrict ourselves to a simply typed meta-language, although we recognize that an extension allowing dependent types and polymorphism is important future work (see Section 6). Our formulation of the simply-typed meta-language is standard.

$$\begin{array}{lrcl} \text{Pure types:} & B & ::= & a \mid B_1 \to B_2 \\ \text{Objects:} & M & ::= & x \mid c \mid \lambda x{:}A.\, M \mid M_1\, M_2 \\[1ex] \text{Context:} & \Psi & ::= & \cdot \mid \Psi, x : B \\ \text{Signature:} & \Sigma & ::= & \cdot \mid \Sigma, a : \text{type} \mid \Sigma, c : B \end{array}$$

We use a for type constants, c for object constants and x for variables. We also fix a signature Σ for our typing and evaluation judgments so we do not have to carry it around.

Definition 1 Typing judgment. $\Psi \vdash M : B$ is defined by:

$$\frac{\Psi(x) = B}{\Psi \vdash x : B}\,\mathsf{StpVar} \qquad \frac{\Sigma(c) = B}{\Psi \vdash c : B}\,\mathsf{StpConst}$$

$$\frac{\Psi, x : B_1 \vdash M : B_2}{\Psi \vdash \lambda x{:}B_1.\, M : B_1 \to B_2}\,\mathsf{StpLam} \qquad \frac{\Psi \vdash M_1 : B_2 \to B_1 \qquad \Psi \vdash M_2 : B_2}{\Psi \vdash M_1\, M_2 : B_1}\,\mathsf{StpApp}$$

As running examples throughout the paper we use the representation of natural numbers and untyped λ-expressions.

Example 1 Natural numbers.

$$\begin{array}{rcl} & & \text{nat} : \text{type} \\ \ulcorner 0 \urcorner = \mathsf{z} & & \mathsf{z} \;\; : \text{nat} \\ \ulcorner n+1 \urcorner = \mathsf{s}\,\ulcorner n \urcorner & & \mathsf{s} \;\; : \text{nat} \to \text{nat} \end{array}$$

Untyped λ-expressions illustrate the idea of higher-order abstract syntax: object language variables are represented by meta-language variables.

Example 2 Untyped λ-expressions.

$$\text{Expressions: } e ::= x \mid \mathbf{lam}\, x.e \mid e_1 @ e_2$$

$$\begin{array}{rcll} & & & \text{exp} : \text{type} \\ \ulcorner \mathbf{lam}\, x.e \urcorner & = & \text{lam}\ (\lambda x{:}\text{exp}.\ulcorner e \urcorner) & \text{lam} : (\text{exp} \to \text{exp}) \to \text{exp} \\ \ulcorner e_1 @ e_2 \urcorner & = & \text{app}\ \ulcorner e_1 \urcorner \ulcorner e_2 \urcorner & \text{app} : \text{exp} \to (\text{exp} \to \text{exp}) \\ \ulcorner x \urcorner & = & x & \end{array}$$

Not every well-typed object of the meta-language directly represents an expression of the object language. For example, we can see that $\ulcorner e \urcorner$ will never contain a β-redex. Moreover, the argument to lam which has type exp $\to$ exp will always be a λ-abstraction. Thus the image of the translation in this representation methodology is always a β-normal and η-long form. Following [10], we call these forms *canonical* as defined by the following two judgments.

Definition 2 Atomic and canonical forms.

1. $\Psi \vdash V \downarrow B$ (V is atomic of type B in Ψ)
2. $\Psi \vdash V \Uparrow B$ (V is canonical of type B in Ψ)

are defined by:

$$\frac{\Psi(x) = B}{\Psi \vdash x \downarrow B}\,\text{AtVar} \qquad \frac{\Sigma(c) = B}{\Psi \vdash c \downarrow B}\,\text{AtConst} \qquad \frac{\Psi \vdash V_1 \downarrow B_2 \to B_1 \qquad \Psi \vdash V_2 \Uparrow B_2}{\Psi \vdash V_1\ V_2 \downarrow B_1}\,\text{AtApp}$$

$$\frac{\Psi \vdash V \downarrow a}{\Psi \vdash V \Uparrow a}\,\text{CanAt} \qquad \frac{\Psi, x : B_1 \vdash V \Uparrow B_2}{\Psi \vdash \lambda x{:}B_1.\,V \Uparrow B_1 \to B_2}\,\text{CanLam}$$

Canonical forms play the role of "observable values" in a functional language: they are in one-to-one correspondence with the expressions we are trying to represent. For Example 2 (untyped λ-expressions) this is expressed by the following property, which is proved by simple inductions.

Example 3 Compositional adequacy for untyped λ-expressions.

1. Let e be an expression with free variables among $x_1, \ldots, x_n$. Then $x_1 : \text{exp}, \ldots, x_n : \text{exp} \vdash \ulcorner e \urcorner \Uparrow \text{exp}$.
2. Let $x_1 : \text{exp}, \ldots, x_n : \text{exp} \vdash M \Uparrow \text{exp}$. Then $M = \ulcorner e \urcorner$ for an expression e with free variables among $x_1, \ldots, x_n$.
3. $\ulcorner . \urcorner$ is a bijection between expressions and canonical forms where $\ulcorner [e'/x]e \urcorner = [\ulcorner e' \urcorner / x] \ulcorner e \urcorner$.

Since every object in $\lambda^{\to}$ has a unique $\beta\eta$-equivalent canonical form, the meaning of every well-typed object is unambiguously given by its canonical form. Our operational semantics (see Definitions 4 and 7) computes this canonical form and therefore the meaning of every well-typed object. That this property is preserved under an extension of the language by primitive recursion for higher-order abstract syntax may be considered the main technical result of this paper.

3 Modal λ-Calculus

The constructors for objects of type exp from Example 2 are lam : (exp $\rightarrow$ exp) $\rightarrow$ exp and app : exp $\rightarrow$ (exp $\rightarrow$ exp). These cannot be the constructors of an *inductive* type exp, since we have a negative occurrence of exp in the argument type of lam. This is not just a formal observation, but has practical consequences: we cannot formulate a consistent induction principle for expressions in this representation. Furthermore, if we increase the computational power of the meta-language by adding **case** or an iterator, then not every well-typed object of type exp has a canonical form. For example,

$$\cdot \vdash \text{lam}\ (\lambda E{:}\text{exp}.\,\text{case}\ E\ \text{of app}\ E_1\ E_2 \Rightarrow \text{app}\ E_2\ E_1 \mid \text{lam}\ E' \Rightarrow \text{lam}\ E') : \text{exp}$$

but the given object does not represent any untyped λ-expression, nor could it be converted to one. The difficulty with a **case** or iteration construct is that there are many new functions of type exp $\rightarrow$ exp which cannot be converted to a function in $\lambda^{\rightarrow}$. This becomes a problem when such functions are arguments to constructors, since then the extension is no longer conservative even over expressions of base type (as illustrated in the example above).

Thus we must cleanly separate the *parametric function space* exp $\rightarrow$ exp whose elements are convertible to the form $\lambda x : \text{exp}.\,E$ where E is built only from the constructors app, lam, and the variable x, from the *primitive recursive function space* exp $\Rightarrow$ exp which is intended to encompass functions defined through case distinction and iteration. This separation can be achieved by using a modal operator: exp $\rightarrow$ exp will continue to contain only the parametric functions, while exp $\Rightarrow$ exp $=$ ($\Box$exp) $\rightarrow$ exp contains the primitive recursive functions.

Intuitively, we interpret $\Box B$ as the type of *closed* objects of type B. We can iterate or distinguish cases over closed objects, since all constructors are statically known and can be provided for. This is not the case if an object may contain some unknown free variables. The system is non-trivial since we may also abstract over objects of type $\Box A$, but fortunately it is well understood and corresponds (via an extension of the Curry-Howard isomorphism) to the intuitionistic variant of S_4 [3].

In Section 4 we introduce schemas for defining functions by iteration and case distinction which require the subject to be of type $\Box B$. We can recover the ordinary scheme of primitive recursion for type nat if we also add pairs to the language. Pairs (with type $A_1 \times A_2$) are also necessary for the simultaneous definition of mutually recursive functions. Just as the modal type $\Box A$, pairs are lazy and values of these types are not observable—ultimately we are only interested in canonical forms of pure type.

The formulation of the modal λ-calculus below is copied from [3] and goes back to [22]. The language of types includes the pure types from the simply-typed λ-calculus in Section 2.

$$\frac{\Gamma(x) = A}{\Delta;\Gamma \vdash x : A}\,\mathsf{TpVarReg} \qquad \frac{\Delta(x) = A}{\Delta;\Gamma \vdash x : A}\,\mathsf{TpVarMod} \qquad \frac{\Sigma(c) = B}{\Delta;\Gamma \vdash c : B}\,\mathsf{TpConst}$$

$$\frac{\Delta;\Gamma, x : A_1 \vdash M : A_2}{\Delta;\Gamma \vdash \lambda x{:}A_1.\, M : A_1 \to A_2}\,\mathsf{TpLam}$$

$$\frac{\Delta;\Gamma \vdash M_1 : A_2 \to A_1 \qquad \Delta;\Gamma \vdash M_2 : A_2}{\Delta;\Gamma \vdash M_1\ M_2 : A_1}\,\mathsf{TpApp}$$

$$\frac{\Delta;\Gamma \vdash M_1 : A_1 \qquad \Delta;\Gamma \vdash M_2 : A_2}{\Delta;\Gamma \vdash \langle M_1, M_2\rangle : A_1 \times A_2}\,\mathsf{TpPair}$$

$$\frac{\Delta;\Gamma \vdash M : A_1 \times A_2}{\Delta;\Gamma \vdash \mathrm{fst}\ M : A_1}\,\mathsf{TpFst} \qquad \frac{\Delta;\Gamma \vdash M : A_1 \times A_2}{\Delta;\Gamma \vdash \mathrm{snd}\ M : A_2}\,\mathsf{TpSnd}$$

$$\frac{\Delta;\cdot \vdash M : A}{\Delta;\Gamma \vdash \mathrm{box}\ M : \Box A}\,\mathsf{TpBox} \qquad \frac{\Delta;\Gamma \vdash M_1 : \Box A_1 \qquad \Delta, x : A_1;\Gamma \vdash M_2 : A_2}{\Delta;\Gamma \vdash \mathrm{let\ box}\ x = M_1\ \mathrm{in}\ M_2 : A_2}\,\mathsf{TpLet}$$

Fig. 1. Typing judgment $\Delta;\Gamma \vdash M : A$

Types: $A ::= a \mid A_1 \to A_2 \mid \Box A \mid A_1 \times A_2$
Objects: $M ::= c \mid x \mid \lambda x{:}A.\, M \mid M_1\ M_2$
$\mid \mathrm{box}\ M \mid \mathrm{let\ box}\ x = M_1\ \mathrm{in}\ M_2 \mid \langle M_1, M_2\rangle \mid \mathrm{fst}\ M \mid \mathrm{snd}\ M$
Contexts: $\Gamma ::= \cdot \mid \Gamma, x : A$

For the sake of brevity we usually suppress the fixed signature Σ. However, it is important that signatures Σ and contexts denoted by Ψ will continue to contain only pure types, while contexts Γ and Δ may contain arbitrary types. We also continue to use B to range over pure types, while A ranges over arbitrary types. The typing judgment $\Delta;\Gamma \vdash M : A$ uses two contexts: Δ, whose variables range over closed objects, and Γ, whose variables range over arbitrary objects.

Definition 3 Typing judgment. $\Delta;\Gamma \vdash M : A$ is defined in Figure 1.

As examples, we show some basic laws of the (intuitionistic) modal logic S_4.

Example 4 Laws of S_4.

$$\begin{aligned}
\mathrm{funlift} &: \Box(A_1 \to A_2) \to \Box A_1 \to \Box A_2 \\
&= \lambda f{:}\Box(A_1 \to A_2).\, \lambda x{:}\Box A_1. \\
&\qquad \mathrm{let\ box}\ f' = f\ \mathrm{in\ let\ box}\ x' = x\ \mathrm{in\ box}\ (f'\ x')
\end{aligned}$$

$$\begin{aligned}
\mathrm{unbox} &: \Box A \to A \\
&= \lambda x{:}\Box A.\, \mathrm{let\ box}\ x' = x\ \mathrm{in}\ x'
\end{aligned}$$

$$\begin{aligned}
\mathrm{boxbox} &: \Box A \to \Box\Box A \\
&= \lambda x{:}\Box A.\, \mathrm{let\ box}\ x' = x\ \mathrm{in\ box}\ (\mathrm{box}\ x')
\end{aligned}$$

$$\frac{\Psi \vdash M \hookrightarrow V : a}{\Psi \vdash M \Uparrow V : a}\ \mathsf{EcAtomic} \qquad \frac{\Psi, x : B_1 \vdash M\ x \Uparrow V : B_2}{\Psi \vdash M \Uparrow \lambda x{:}B_1.\, V : B_1 \to B_2}\ \mathsf{EcArrow}$$

$$\frac{\Psi(x) = A}{\Psi \vdash x \hookrightarrow x : A}\ \mathsf{EvVar} \qquad \frac{\Sigma(c) = B}{\Psi \vdash c \hookrightarrow c : B}\ \mathsf{EvConst}$$

$$\frac{\cdot; \Psi, x : A_1 \vdash M : A_2}{\Psi \vdash \lambda x{:}A_1.\, M \hookrightarrow \lambda x{:}A_1.\, M : A_1 \to A_2}\ \mathsf{EvLam}$$

$$\frac{\begin{array}{l}\Psi \vdash M_1 \hookrightarrow \lambda x{:}A_2.\, M_1' : A_2 \to A_1\\ \Psi \vdash M_2 \hookrightarrow V_2 : A_2\\ \Psi \vdash [V_2/x](M_1') \hookrightarrow V : A_1\end{array}}{\Psi \vdash M_1\ M_2 \hookrightarrow V : A_1}\ \mathsf{EvApp} \qquad \frac{\begin{array}{l}\Psi \vdash M_1 \hookrightarrow V_1 : B_2 \to B_1\\ \Psi \vdash V_1 \downarrow B_2 \to B_1\\ \Psi \vdash M_2 \Uparrow V_2 : B_2\end{array}}{\Psi \vdash M_1\ M_2 \hookrightarrow V_1\ V_2 : B_1}\ \mathsf{EvAtomic}$$

$$\frac{\cdot; \Psi \vdash M_1 : A_1 \qquad \cdot; \Psi \vdash M_2 : A_2}{\Psi \vdash \langle M_1, M_2 \rangle \hookrightarrow \langle M_1, M_2 \rangle : A_1 \times A_2}\ \mathsf{EvPair}$$

$$\frac{\Psi \vdash M \hookrightarrow \langle M_1, M_2 \rangle : A_1 \times A_2 \qquad \Psi \vdash M_1 \hookrightarrow V : A_1}{\Psi \vdash \mathsf{fst}\ M \hookrightarrow V : A_1}\ \mathsf{EvFst}$$

$$\frac{\Psi \vdash M \hookrightarrow \langle M_1, M_2 \rangle : A_1 \times A_2 \qquad \Psi \vdash M_2 \hookrightarrow V : A_2}{\Psi \vdash \mathsf{snd}\ M \hookrightarrow V : A_2}\ \mathsf{EvSnd}$$

$$\frac{\cdot; \cdot \vdash M : A}{\Psi \vdash \mathsf{box}\ M \hookrightarrow \mathsf{box}\ M : \Box A}\ \mathsf{EvBox}$$

$$\frac{\Psi \vdash M_1 \hookrightarrow \mathsf{box}\ M_1' : \Box A \qquad \Psi \vdash [M_1'/x](M_2) \hookrightarrow V : A_2}{\Psi \vdash \mathsf{let\ box}\ x = M_1\ \mathsf{in}\ M_2 \hookrightarrow V : A_2}\ \mathsf{EvLet}$$

Fig. 2. Evaluation judgments $\Psi \vdash M \hookrightarrow V : A$ and $\Psi \vdash M \Uparrow V : B$

The rules for evaluation must be constructed in such a way that full canonical forms are computed for objects of pure type, that is, we must evaluate under certain λ-abstractions. Objects of type $\Box A$ or $A_1 \times A_2$ on the other hand are not observable and may be computed lazily. We therefore use two mutually recursive judgments for evaluation and conversion to canonical form, written $\Psi \vdash M \hookrightarrow V : A$ and $\Psi \vdash M \Uparrow V : B$, respectively. The latter is restricted to pure types, since only objects of pure type possess canonical forms. Since we evaluate under some λ-abstractions, free variables of pure type declared in Ψ may occur in M and V during evaluation.

Definition 4 Evaluation judgment. $\Psi \vdash M \hookrightarrow V : A$ and $\Psi \vdash M \Uparrow V : B$ are defined in Figure 2.

4 Iteration

The modal operator $\Box$ introduced in Section 3 allows us to restrict iteration and case distinction to subjects of type $\Box B$, where B is a pure type. The technical realization of this idea in its full generality is rather complex. We therefore begin by describing the behavior of functions defined by iteration informally, incrementally developing their formal definition within our system. In the informal presentation we elide the box constructor, but we should convince ourselves that the subject of the iteration or case is indeed assumed to be closed.

Example 5 Addition. The usual type of addition is nat $\rightarrow$ nat $\rightarrow$ nat. This is no longer a valid type for addition, since it must iterate over either its first or second argument and would therefore not be parametric in that argument. Among the possible types for addition, we will be interested particularly in $\Box$nat $\rightarrow$ nat $\rightarrow$ nat and $\Box$nat $\rightarrow$ $\Box$nat $\rightarrow$ $\Box$nat.

$$\begin{aligned} &\text{plus z } n &&= n \\ &\text{plus (s } m) \ n &&= \text{s (plus } m \ n) \end{aligned}$$

Note that this definition cannot be assigned type nat $\rightarrow$ nat $\rightarrow$ nat or $\Box$nat $\rightarrow$ nat $\rightarrow$ $\Box$nat.

In our system we view iteration as replacing constructors of a canonical term by functions of appropriate type, which is also the idea behind *catamorphisms* [8]. In the case of natural numbers, we replace z : nat by a term $M_z : A$ and s : nat $\rightarrow$ nat by a function $M_s : A \rightarrow A$. Thus iteration over natural numbers replaces type nat by A. We use the notation $a \mapsto A$ for a *type replacement* and $c \mapsto M$ for a *term replacement.* Iteration in its simplest form is written as "it $\langle a \mapsto A \rangle$ M $\langle \Omega \rangle$" where M is the subject of the iteration, and Ω is a list containing term replacements for all constructors of type a. The formal typing rules for replacements are given later in this section; first some examples.

Example 6 Addition via iteration. Addition from Example 5 can be formulated in a number of ways with an explicit iteration operator. The simplest one:

$$\begin{aligned} \text{plus}' &: \Box\text{nat} \rightarrow \text{nat} \rightarrow \text{nat} \\ &= \lambda m{:}\Box\text{nat}.\, \lambda n{:}\text{nat}.\, \text{it } \langle \text{nat} \mapsto \text{nat} \rangle\ m\ \langle \text{z} \mapsto n \mid \text{s} \mapsto \text{s} \rangle \end{aligned}$$

Later examples require addition with a result guaranteed to be closed. Its definition is only slightly more complicated.

$$\begin{aligned} \text{plus} &: \Box\text{nat} \rightarrow \Box\text{nat} \rightarrow \Box\text{nat} \\ &= \lambda m{:}\Box\text{nat}.\, \lambda n{:}\Box\text{nat}.\, \text{it } \langle \text{nat} \mapsto \Box\text{nat} \rangle\ m \\ &\qquad \langle \text{z} \mapsto n \\ &\qquad \mid \text{s} \mapsto (\lambda r{:}\Box\text{nat}.\, \text{let box } r' = r \text{ in box (s } r'))\rangle \end{aligned}$$

If the data type is higher-order, iteration over closed objects must traverse terms with free variables. We model this in the informal presentation by introducing new parameters (written as $\nu x{:}B.\,M$) using Odersky's notation [19]. This makes a dynamic extension of the function definition necessary to encompass the new parameters (written as "where $f(x) = M$").

Example 7 Counting variable occurrences. Below is a function which counts the number of occurrences of bound variables in an untyped λ-expression in the representation of Example 2. It can be assigned type $\Box\mathrm{exp} \to \Box\mathrm{nat}$.

$$\begin{array}{ll} \mathrm{cntvar}\ (\mathrm{app}\ e_1\ e_2) & = \mathrm{plus}\ (\mathrm{cntvar}\ e_1)\ (\mathrm{cntvar}\ e_2) \\ \mathrm{cntvar}\ (\mathrm{lam}\ e) & = \nu x{:}\mathrm{exp}.\ \mathrm{cntvar}\ (e\ x)\ \text{where}\ \mathrm{cntvar}\ x = (\mathrm{s}\ \mathrm{z}) \end{array}$$

It may look like the recursive call in the example above is not well-typed since $(e\ x)$ is not closed as required, but contains a free parameter x. Making sense of this apparent contradiction is the principal difficulty in designing an iteration construct for higher-order abstract syntax. As before, we model iteration via replacements. Here, $\mathrm{exp} \mapsto \Box\mathrm{nat}$ and so $\mathrm{lam} \mapsto M_1$ and $\mathrm{app} \mapsto M_2$ where $M_1 : (\Box\mathrm{nat} \to \Box\mathrm{nat}) \to \Box\mathrm{nat}$ and $M_2 : \Box\mathrm{nat} \to (\Box\mathrm{nat} \to \Box\mathrm{nat})$. The types of replacement terms M_1 and M_2 arise from the types of the constructors $\mathrm{lam} : (\mathrm{exp} \to \mathrm{exp}) \to \mathrm{exp}$ and $\mathrm{app} : \mathrm{exp} \to (\mathrm{exp} \to \mathrm{exp})$ by applying the type replacement $\mathrm{exp} \mapsto \Box\mathrm{nat}$. We write

$$\begin{array}{ll} \mathrm{cntvar} & : \Box\mathrm{exp} \to \Box\mathrm{nat} \\ & = \lambda x{:}\Box\mathrm{exp}.\ \mathrm{it}\ \langle \mathrm{exp} \mapsto \Box\mathrm{nat}\rangle\ x \\ & \qquad \langle\ \mathrm{app} \mapsto \mathrm{plus} \\ & \qquad \mid \mathrm{lam} \mapsto \lambda f{:}\Box\mathrm{nat} \to \Box\mathrm{nat}.\ f\ (\mathrm{box}\ (\mathrm{s}\ \mathrm{z}))\rangle \end{array}$$

Informally, the result of $\mathrm{cntvar}\ (\mathrm{lam}\ (\lambda x{:}\mathrm{exp}.\ x))$ can be determined as follows:

$$\begin{array}{l} \mathrm{cntvar}\ (\mathrm{lam}\ (\lambda x{:}\mathrm{exp}.\ x)) \\ = \nu x'{:}\mathrm{exp}.\ \mathrm{cntvar}\ ((\lambda x{:}\mathrm{exp}.\ x)\ x')\ \text{where}\ \mathrm{cntvar}\ x' = (\mathrm{s}\ \mathrm{z}) \\ = \nu x'{:}\mathrm{exp}.\ \mathrm{cntvar}\ x'\ \text{where}\ \mathrm{cntvar}\ x' = (\mathrm{s}\ \mathrm{z}) \\ = \nu x'{:}\mathrm{exp}.\ (\mathrm{s}\ \mathrm{z})\ \text{where}\ \mathrm{cntvar}\ x' = (\mathrm{s}\ \mathrm{z}) \\ = (\mathrm{s}\ \mathrm{z}) \end{array}$$

For the formal operational semantics, see Example 10.

A number of functions can be defined elegantly in this representation. Among them are the conversion from type exp to a representation using de Bruijn indices and one-step parallel reduction. The latter requires mutual iteration and pairs (see [6]).

The following example illustrates two concepts: mutually inductive types and iteration over the form of a (parametric!) function.

Example 8 Substitution in normal forms. Substitution is already directly definable by application, but one may also ask if there is a structural definition in the style of [16]. Normal forms of the untyped λ-calculus are represented by the type nf with an auxiliary definition for atomic forms of type at.

$$\begin{array}{rl} \text{Normal forms:} & N ::= P \mid \mathbf{lam}\ x.N \\ \text{Atomic forms:} & P ::= x \mid P@N \end{array}$$

In this example the represention function $\ulcorner . \urcorner$ acts on normal forms, atomic forms are represented by $\ulcorner\!\ulcorner . \urcorner\!\urcorner$.

$$
\begin{array}{rlll}
 & & \mathrm{nf} & : \mathrm{type} \\
 & & \mathrm{at} & : \mathrm{type} \\
\ulcorner P \urcorner & = \mathrm{atnf}\ \ulcorner\!\ulcorner P \urcorner\!\urcorner & \mathrm{atnf} & : \mathrm{at} \to \mathrm{nf} \\
\ulcorner \mathbf{lam}\ x.N \urcorner & = \mathrm{lm}\ (\lambda x{:}\mathrm{at}.\ulcorner N \urcorner) & \mathrm{lm} & : (\mathrm{at} \to \mathrm{nf}) \to \mathrm{nf} \\
\ulcorner\!\ulcorner P@N \urcorner\!\urcorner & = \mathrm{ap}\ \ulcorner\!\ulcorner P \urcorner\!\urcorner\ \ulcorner N \urcorner & \mathrm{ap} & : \mathrm{at} \to \mathrm{nf} \to \mathrm{at} \\
\ulcorner\!\ulcorner x \urcorner\!\urcorner & = x & &
\end{array}
$$

Substitution of atomic objects for variables is defined by two mutually recursive functions, one with type subnf : $\Box(\mathrm{at} \to \mathrm{nf}) \to \mathrm{at} \to \mathrm{nf}$ and subat : $\Box(\mathrm{at} \to \mathrm{at}) \to \mathrm{at} \to \mathrm{at}$.

$$
\begin{array}{ll}
\mathrm{subnf}\ (\lambda x{:}\mathrm{at}.\,\mathrm{lm}\ (\lambda y{:}\mathrm{at}.\,N\ x\ y))\ Q & = \mathrm{lm}\ (\lambda y{:}\mathrm{at}.\,\mathrm{subnf}\ (\lambda x{:}\mathrm{at}.\,(N\ x\ y))\ Q \\
 & \qquad \text{where subat}\ (\lambda x{:}\mathrm{at}.\,y)\ Q = y) \\
\mathrm{subnf}\ (\lambda x{:}\mathrm{at}.\,\mathrm{atnf}\ (P\ x))\ Q & = \mathrm{atnf}\ (\mathrm{subat}\ (\lambda x{:}\mathrm{at}.\,P\ x)\ Q) \\
\mathrm{subat}\ (\lambda x{:}\mathrm{at}.\,\mathrm{ap}\ (P\ x)\ (N\ x))\ Q & = \mathrm{ap}\ (\mathrm{subat}\ (\lambda x{:}\mathrm{at}.\,P\ x)\ Q) \\
 & \qquad (\mathrm{subnf}\ (\lambda x{:}\mathrm{at}.\,N\ x)\ Q) \\
\mathrm{subat}\ (\lambda x{:}\mathrm{at}.\,x)\ Q & = Q
\end{array}
$$

The last case arises since the parameter x must be considered as a new constructor in the body of the abstraction. The functions above are realized in our calculus by a simultaneous replacement of objects of type nf and at. In other words, the type replacement must account for all mutually recursive types, and the term replacement for all constructors of those types.

$$
\begin{array}{ll}
\mathrm{subnf} & : \Box(\mathrm{at} \to \mathrm{nf}) \to \mathrm{at} \to \mathrm{nf} \\
 & = \lambda N{:}\Box(\mathrm{at} \to \mathrm{nf}).\,\lambda Q{:}\mathrm{at}.\,\mathrm{it}\ \langle \mathrm{nf} \mapsto \mathrm{nf} \mid \mathrm{at} \mapsto \mathrm{at} \rangle\ N \\
 & \qquad \langle\ \mathrm{lm} \mapsto \lambda F{:}\mathrm{at} \to \mathrm{nf}.\,\mathrm{lm}\ (\lambda y{:}\mathrm{at}.\,(F\ y)) \\
 & \qquad \mid \mathrm{atnf} \mapsto \lambda P{:}\mathrm{at}.\,\mathrm{atnf}\ P \\
 & \qquad \mid \mathrm{ap} \mapsto \lambda P{:}\mathrm{at}.\,\lambda N{:}\mathrm{nf}.\,\mathrm{ap}\ P\ N \rangle \\
 & \qquad Q
\end{array}
$$

Via η-contraction we can see that substitution amounts to a structural identity function.

We begin now with the formal discussion and description of the full language. Due to the possibility of mutual recursion among types, the type replacements must be lists (see Example 8).

$$\text{Type replacement: } \omega ::= \cdot \mid (\omega \mid a \mapsto A)$$

Which types must be replaced by an iteration depends on which types are mutually recursive according to the constructors in the signature Σ and possibly the type of the iteration subject itself. If we iterate over a function, the parameter of a function must be treated like a constructor for its type, since it can appear in that role in the body of a function.

Thus, we define the notion of ***type subordination*** which summarizes all dependencies between atomic types by separately considering its ***static*** part $\lhd_\Sigma$ which derives from the dependencies induced by the constructor types from Σ and its ***dynamic*** part $\lhd_B$ which accounts for dependencies induced from the argument types of B. The transitive closure $\blacktriangleleft_{\Sigma;B}$ of static and dynamic type subordination relation defines cleanly all dependencies between types which govern the formation of the subject of iteration. We denote the ***target type*** of a pure type B by $\tau(B)$. All type constants which are mutually dependent with $\tau(B)$, written $\mathcal{I}(\Sigma; B)$, form the domain of the type replacement ω:

$$\mathcal{I}(\Sigma; B) := \{a | \tau(B) \blacktriangleleft_{\Sigma;B} a \text{ and } a \blacktriangleleft_{\Sigma;B} \tau(B)\}$$

In Example 8 of normal and atomic forms we have $\mathcal{I}(\Sigma; \text{at} \rightarrow \text{nf}) = \{\text{at}, \text{nf}\}$. Note that type subordination is built into calculi where inductive types are defined explicitly (such as the Calculus of Inductive Constructions [20]); here it must be recovered from the signature since we impose no ordering constraints except that a type must be declared before it is used. Our choice to recover the type subordination relation from the signature allows us to perform iteration over any functional type, without fixing the possibilities in advance.

Let us now address the question of how the type of an iteration is formed: If the subject of iteration has type B, the iterator object has type $\langle\omega\rangle(B)$, where $\langle\omega\rangle(B)$ is defined inductively by replacing each type constant according to ω, leaving types outside the domain fixed.

A similar replacement is applied at the level of terms: the result of an iteration is an object which resembles the (canonical) subject of the iteration in structure, but object constants are replaced by other objects carrying the intended computational meaning of the different cases. Even though the subject of iteration is closed at the beginning of the replacement process, we need to deal with embedded λ-abstractions due to higher-order abstract syntax. But since such functions are parametric we can simply replace variables x of type B by new variables x' of type $\langle\omega\rangle(B)$.

$$\text{Term replacement: } \Omega ::= \cdot \mid (\Omega \mid c \mapsto M) \mid (\Omega \mid x \mapsto x')$$

Initially the domain of a term replacement is a signature containing all constructors whose target type is in $\mathcal{I}(\Sigma; B)$. We refer to this signature as $\mathcal{S}_{\text{it}}(\Sigma; B)$. The form of iteration follows now quite naturally: We extend the notion of objects by

$$M ::= \ldots \mid \text{it } \langle\omega\rangle\ M\ \langle\Omega\rangle$$

and define the following typing rules for iteration and term replacements.

Definition 5 Typing judgment for iteration. (extending Definition 3)

$$\frac{\Delta; \Gamma \vdash M : \Box B \qquad \Delta; \Gamma \vdash \Omega : \langle\omega\rangle(\mathcal{S}_{\text{it}}(\Sigma; B))}{\Delta; \Gamma \vdash \text{it } \langle\omega\rangle\ M\ \langle\Omega\rangle : \langle\omega\rangle(B)}\ \textsf{TpIt}, \qquad \text{dom}(\omega) = \mathcal{I}(\Sigma; B)$$

$$\frac{}{\Delta; \Gamma \vdash \cdot : \langle\omega\rangle(\cdot)}\ \textsf{TrBase} \qquad \frac{\Delta; \Gamma \vdash \Omega : \langle\omega\rangle(\Sigma) \qquad \Delta; \Gamma \vdash M : \langle\omega\rangle(B')}{\Delta; \Gamma \vdash (\Omega \mid c \mapsto M) : \langle\omega\rangle(\Sigma, c : B')}\ \textsf{TrInd}$$

Example 9. In Example 7 we defined cntvar $=\lambda x{:}\Box\text{exp}.\,\text{it}\ \langle\omega\rangle\ x\ \langle\Omega\rangle$ where

$$\begin{aligned}\omega &= \text{exp} \mapsto \Box\text{nat}\\ \Omega &= \text{app} \mapsto \text{plus}, \text{lam} \mapsto \lambda f{:}\Box\text{nat} \to \Box\text{nat}.\, f\ (\text{box}\ (\text{s z}))\\ \mathcal{S}_{\text{it}}(\Sigma;\text{exp}) &= \text{app} : \text{exp} \to (\text{exp} \to \text{exp}), \text{lam} : (\text{exp} \to \text{exp}) \to \text{exp}\end{aligned}$$

Under the assumption that plus : $\Box$nat $\to$ ($\Box$nat $\to$ $\Box$nat) it is easy to see that

(1) $\cdot; x : \Box\text{exp} \vdash \lambda f{:}\Box\text{nat} \to \Box\text{nat}.\, f\ (\text{box}\ (\text{s z})) : (\Box\text{nat} \to \Box\text{nat}) \to \Box\text{nat}$ by TpLam, etc.

(2) $\cdot; x : \Box\text{exp} \vdash \Omega : \langle\omega\rangle(\mathcal{S}_{\text{it}}(\Sigma;\text{exp}))$ by TrBase, Ass., (1)

(3) $\cdot; x : \Box\text{exp} \vdash x : \Box\text{exp}$ by TpVarReg

(4) $\cdot; x : \Box\text{exp} \vdash \text{it}\ \langle\omega\rangle\ x\ \langle\Omega\rangle : \Box\text{nat}$ by TpIt from (3) (2)

(5) $\cdot; \cdot \vdash \text{cntvar} : \Box\text{exp} \to \Box\text{nat}$ by TpLam from (4)

Applying a term replacement must be restricted to canonical forms in order to preserve types. Fortunately, our type system guarantees that the subject of an iteration can be converted to canonical form. Applying a replacement then transforms a canonical form V of type B into a well-typed object $\langle\omega;\Omega\rangle(V)$ of type $\langle\omega\rangle(B)$. We call this operation *elimination*. It is defined along the structure of V.

Definition 6 Elimination.

$$\langle\omega;\Omega\rangle(c) = \begin{cases} M & \text{if } \Omega(c) = M \\ c & \text{otherwise}\end{cases} \qquad \text{(ElConst)}$$

$$\langle\omega;\Omega\rangle(x) = \Omega(x) \qquad \text{(ElVar)}$$

$$\langle\omega;\Omega\rangle(\lambda x{:}B.\,V) = \lambda u{:}\langle\omega\rangle(B).\,\langle\omega;\Omega \mid x \mapsto u\rangle(V) \qquad \text{(ElLam)}$$

$$\langle\omega;\Omega\rangle(V_1\ V_2) = \langle\omega;\Omega\rangle(V_1)\ \langle\omega;\Omega\rangle(V_2) \qquad \text{(ElApp)}$$

The term resulting from elimination might, of course, contain redices and must itself be evaluated to obtain a final value. Thus we obtain the following evaluation rule for iteration.

Definition 7 Evaluation judgment. (extending Definition 4)

$$\frac{\Psi \vdash M \hookrightarrow \text{box}\ M' : \Box B \qquad \cdot \vdash M' \Uparrow V' : B \qquad \Psi \vdash \langle\omega;\Omega\rangle(V') \hookrightarrow V : \langle\omega\rangle(B)}{\Psi \vdash \text{it}\ \langle\omega\rangle\ M\ \langle\Omega\rangle \hookrightarrow V : \langle\omega\rangle(B)}\ \text{EvIt}$$

Example 10. In Example 7, the evaluation of cntvar (box (lam ($\lambda x{:}\text{exp}.\,x$))) yields box (s z) because

(1) $\cdot \vdash \text{cntvar} \hookrightarrow \text{cntvar} : \Box\text{exp} \to \Box\text{nat}$ by **EvLam**

(2) $\cdot \vdash \text{box}\ (\text{lam}\ (\lambda x{:}\text{exp}.\, x)) \hookrightarrow \text{box}\ (\text{lam}\ (\lambda x{:}\text{exp}.\, x)) : \Box\text{exp}$ by **EvBox**

(3) $\cdot \vdash \text{lam}\ (\lambda x{:}\text{exp}.\, x) \Uparrow \text{lam}\ (\lambda x{:}\text{exp}.\, x) : \text{exp}$ by **EcAtomic**, etc.

(4) $\langle\omega; \Omega\rangle(\text{lam}\ (\lambda x{:}\text{exp}.\, x))$
$= (\lambda f{:}\Box\text{nat} \to \Box\text{nat}.\, f\ (\text{box}\ (\text{s z})))\ (\lambda x'{:}\Box\text{nat}.\, x')$ by elimination

(5) $\cdot \vdash \langle\omega; \Omega\rangle(\text{lam}\ (\lambda x{:}\text{exp}.\, x)) \hookrightarrow \text{box}\ (\text{s z}) : \Box\text{nat}$ by **EvApp**, etc.

(6) $\cdot \vdash \text{it}\ \langle\omega\rangle\ (\text{box}\ (\text{lam}\ (\lambda x{:}\text{exp}.\, x)))\ \langle\Omega\rangle \hookrightarrow \text{box}\ (\text{s z}) : \Box\text{nat}$
by **EvIt** from (2) (3) (5)

(7) $\cdot \vdash \text{cntvar}\ (\text{box}\ (\text{lam}\ (\lambda x{:}\text{exp}.\, x))) \hookrightarrow \text{box}\ (\text{s z}) : \Box\text{nat}$
by **EvApp** from (1) (2) (6)

The reader is invited to convince himself that this operational semantics yields the expected results also on the other examples of this section.

Our calculus also contains a **case** construct whose subject may be of type $\Box B$ for arbitrary pure B. It allows us to distinguish cases based on the intensional structure of the subject. For example, we can test if a given (parametric!) function is the identity or not. The typing rules and operational semantics for **case** are similar, but simpler than those for iteration. We therefore elide it here and refer the interested reader to [6].

5 Meta-Theory

The goal of this subsection is to show that the modal λ-calculus obeys the type preservation property and that it is a conservative extension of the simply typed λ-calculus defined in Section 2. We prove this by Tait's method, often called an argument by *logical relations*. After defining the logical relations we then prove the canonical form theorem for the modal λ-calculus which guarantees that every well-typed object eventually evaluates to a canonical form. The type preservation and the conservative extension property follow directly from this theorem.

We now begin the meta-theoretical discussion with the definition of the logical relation. Due to the lazy character of the modal λ-calculus, the interpretation of a type A is twofold: On the one side we would like it to contain all canonical forms of type A, on the other all objects *evaluating* to a canonical form. This is why we introduce two mutual dependent logical relations: In a context Ψ, $[\![A]\!]$ represents the set of objects evaluating to a value being itself an element of $|A|$. For the definition of the logical relation we require the notion of context extension: $\Psi' \geq \Psi$ holds if every declaration in Ψ also occurs in Ψ'.

Definition 8 Logical relation.

$\Psi \vdash M \in [\![A]\!]$ iff $\cdot; \Psi \vdash M : A$ and $\Psi \vdash M \hookrightarrow V : A$ and $\Psi \vdash V \in |A|$

$\Psi \vdash V \in |A|$ iff

Case: $A = a$ and $\Psi \vdash V \Uparrow a$
Case: $A = A_1 \to A_2$ and either

Case: $V = \lambda x : A_1 . M$ and for all $\Psi' \geq \Psi$: $\Psi' \vdash V' \in |A_1|$ implies $\Psi' \vdash [V'/x](M) \in [\![A_2]\!]$

or

Case: $\Psi \vdash V \downarrow A_1 \to A_2$ and for all $\Psi' \geq \Psi$: $\Psi' \vdash V' \Uparrow A_1$ implies $\Psi' \vdash V\ V' \in |A_2|$

Case: $A = A_1 \times A_2$ and $V = \langle M_1, M_2 \rangle$ and $\Psi \vdash M_1 \in [\![A_1]\!]$ and $\Psi \vdash M_2 \in [\![A_2]\!]$
Case: $A = \Box A'$ and $V = \text{box } M$ and $\cdot \vdash M \in [\![A']\!]$

Since the operational semantics introduced in Section 4 depends on typing information, we must make sure that only well-typed objects are contained in the logical relation. To do so we require that every object $M \in [\![A]\!]$ has type A. As a side effect of this definition the type preservation property is a direct consequence of the canonical form theorem. The proof of the canonical form theorem is split into two parts. In the first part we prove that every element in $[\![A]\!]$ evaluates to a canonical form, in the second part we show that every well-typed object of type A is contained in the logical relation $[\![A]\!]$.

A direct proof of the first property will fail, hence we must generalize its formulation which we can now prove by mutual induction.

Lemma 9 Logical relations and canonical forms.

1. *If* $\Psi \vdash M \in [\![B]\!]$ *then* $\Psi \vdash M \Uparrow V : B$
2. *If* $\Psi \vdash V \downarrow B$ *then* $\Psi \vdash V \in |B|$

The goal of the second part is to show that every well-typed object is in the logical relation, that is, we want to prove that if $\cdot; \Psi \vdash M : A$ then $\Psi \vdash M \in [\![A]\!]$.

It turns out that we cannot prove this property directly by induction over the structure of the typing derivation, either. The reason is that the context Ψ might grow during the derivation and it may also not remain pure. From the definition of the typing relation it also follows quite directly that Δ need not remain empty. Hence we generalize this property by considering a typing derivation $\Delta; \Gamma \vdash M : A$ and a substitution $(\theta; \varrho)$ which maps the variables defined in $\Delta; \Gamma$ into objects satisfying the logical relation to show that $\Psi \vdash [\theta; \varrho](M) \in [\![A]\!]$. The objects in θ — which are only substituting modal variables — might not yet be evaluated due to the lazy character of unobservable objects. On the other hand it is safe to assume that objects in ϱ — which substitute for variables in Γ — are already evaluated since function application follows a call-by-value discipline. To make this more precise we define a logical relation for contexts $\Psi \vdash \theta; \varrho \in [\Delta; \Gamma]$ iff $\vdash \theta \in [\![\Delta]\!]$ and $\Psi \vdash \varrho \in |\Gamma|$ which are defined as follows:

Definition 10 Logical relation for contexts.

$\vdash \theta \in [\![\Delta]\!]$ iff $\Delta = \cdot$ implies $\theta = \cdot$
and $\Delta = \Delta', x : A$ implies $\theta = \theta', M/x$ and $\cdot \vdash M \in [\![A]\!]$ and $\vdash \theta' \in [\![\Delta']\!]$

$\Psi \vdash \varrho \in |\Gamma|$ iff $\Gamma = \cdot$ implies $\varrho = \cdot$
and $\Gamma = \Gamma', x : A$ implies $\varrho = \varrho', V/x$ and $\Psi \vdash V \in |A|$ and $\Psi \vdash \varrho' \in |\Gamma'|$

We can now formulate and prove

Lemma 11 Typing and logical relations.

If $\Delta; \Gamma \vdash M : A$ *and* $\Psi \vdash \theta; \varrho \in [\Delta; \Gamma]$ *then* $\Psi \vdash [\theta; \varrho](M) \in [\![A]\!]$

The proof of this lemma is rather difficult. Due to the restrictions in length we are not presenting any details here, but refer the interested reader to [6]. Now, an easy inductive argument using Lemma 9 shows that the identity substitution $(\cdot, \mathrm{id}_\Psi)$, which maps all variables defined in Ψ to themselves, indeed lies within the logical relation $[\cdot; \Psi]$. The soundness of typing is hence an immediate corollary of Lemma 11.

Theorem 12 Soundness of typing.

If $\cdot; \Psi \vdash M : A$ *then* $\Psi \vdash M \in [\![A]\!]$.

This theorem together with Lemma 9 guarantees that terms of pure type evaluate to a canonical form V.

Theorem 13 Canonical form theorem.

If $\cdot; \Psi \vdash M : B$ *then* $\Psi \vdash M \Uparrow V : B$ *for some* V.

Type preservation now follows easily: We need to show that the evaluation result of a well-typed object possesses the same type A. From Lemma 11 we obtain that M lies within the logical relation $[\![A]\!]$, which guarantees that M evaluates to a term V, also in the logical relation. Further, a simple induction over the structure of an evaluation shows the values are unique and have the right type.

Theorem 14 Type preservation.

If $\cdot; \Psi \vdash M : A$ *and* $\Psi \vdash M \hookrightarrow V : A$ *then* $\cdot; \Psi \vdash V : A$.

On the basis of the canonical form theorem, we can now prove the main result of the paper: Our calculus is a conservative extension of the simply typed λ-calculus $\lambda^{\rightarrow}$ from Section 2. Let M be an object of pure type B, with free variables from a pure context Ψ. M itself need not be pure but rather some term in the modal λ-calculus including iteration and case. We have seen that M has a canonical form V, and an immediate inductive argument shows that V must be a term in the simply typed λ-calculus.

Theorem 15 Conservative extension.

If $\cdot; \Psi \vdash M : B$ *then* $\Psi \vdash M \Uparrow V : B$ *and* $\Psi \vdash V \Uparrow B$.

6 Conclusion and Future Work

We have presented a calculus for primitive recursive functionals over higher-order abstract syntax which guarantees that the adequacy of encodings remains intact. The requisite conservative extension theorem is technically deep and requires a careful system design and analysis of the properties of a modal operator $\Box$ and its interaction with function definition by iteration and cases. To our knowledge, this is the first system in which it is possible to safely program functionally with higher-order abstract syntax representations. It thus complements and refines the logic programming approach to programming with such representations [17, 21].

Our work was inspired by Miller's system [15], which was presented in the context of ML. Due to the presence of unrestricted recursion and the absence of a modal operator, Miller's system is computationally adequate, but has a much weaker meta-theory which would not be sufficient for direct use in a logical framework. The system of Meijer and Hutton [14] and its refinement by Fegaras and Sheard [8] are also related in that they extend primitive recursion to encompass functional objects. However, they treat functional objects extensionally, while our primitives are designed so we can analyze the internal structure of λ-abstractions directly. Fegaras and Sheard also note the problem with adequacy and design more stringent type-checking rules in Section 3.4 of [8] to circumvent this problem. In contrast to our system, their proposal does not appear to have a logical interpretation. Furthermore, they neither claim nor prove type preservation or an appropriate analogue of conservative extension—critical properties which are not obvious in the presence of their internal type tags and **Place** constructor.

Our system is satisfactory from the theoretical point of view and could be the basis for a practical implementation. Such an implementation would allow the definition of functions of arbitrary types, while data constructors are constrained to have pure type. Many natural functions over higher-order representations turn out to be directly definable (e.g., one-step parallel reduction or conversion to de Bruijn indices), others require explicit counters to guarantee termination (e.g., multi-step reduction or full evaluation). On the other hand, it appears that some natural *algorithms* (e.g., a structural equality check which traverses two expressions simultaneously) are not implementable, even though the underlying function is certainly definable (e.g., via a translation to de Bruijn indices). For larger applications, writing programs by iteration becomes tedious and error-prone and a pattern-matching calculus such as employed in ALF [2] or proposed by Jouannaud and Okada [11] seems more practical. Our informal notation in the examples provides some hints what concrete syntax one might envision for an implementation along these lines.

The present paper is a first step towards a system with dependent types in which proofs of meta-logical properties of higher-order encodings can be expressed directly by dependently typed, total functions. The meta-theory of such a system appears to be highly complex, since the modal operators necessitate a *let box* construct which, *prima facie*, requires commutative conversions. Martin Hofmann[3] has proposed a semantical explanation for our iteration operator which has led him to discover an equational formulation of the laws for iteration. This may be the critical insight required for a dependently typed version of our calculus. A similar formulation of these laws is used in [7] for the treatment of recursion. We also plan to reexamine applications in the realm of functional programming [15, 8] and related work on reasoning about higher-order abstract syntax with explicit induction [5, 4] or definitional reflection [13].

Acknowledgments. The work reported here took a long time to come to fruition, largely due to the complex nature of the technical development. During this time we have discussed various aspects of higher-order abstract syntax, iteration, and induction with too many people to acknowledge them individually. Special thanks go to Gérard Huet and Chet Murthy, who provided the original inspiration, and Hao-Chi Wong who helped us understand the nature of modality in this context.

References

1. Alonzo Church. A formulation of the simple theory of types. *Journal of Symbolic Logic*, 5:56–68, 1940.
2. Thierry Coquand, Bengt Nordström, Jan M. Smith, and Björn von Sydow. Type theory and programming. *Bulletin of the European Association for Theoretical Computer Science*, 52:203–228, February 1994.
3. Rowan Davies and Frank Pfenning. A modal analysis of staged computation. In Jr. Guy Steele, editor, *Proceedings of the 23rd Annual Symposium on Principles of Programming Languages*, pages 258–270, St. Petersburg Beach, Florida, January 1996. ACM Press.
4. Joëlle Despeyroux, Amy Felty, and André Hirschowitz. Higher-order abstract syntax in Coq. In M. Dezani-Ciancaglini and G. Plotkin, editors, *Proceedings of the International Conference on Typed Lambda Calculi and Applications*, pages 124–138, Edinburgh, Scotland, April 1995. Springer-Verlag LNCS 902.
5. Joëlle Despeyroux and André Hirschowitz. Higher-order abstract syntax with induction in Coq. In Frank Pfenning, editor, *Proceedings of the 5th International Conference on Logic Programming and Automated Reasoning*, pages 159–173, Kiev, Ukraine, July 1994. Springer-Verlag LNAI 822.
6. Joëlle Despeyroux, Frank Pfenning, and Carsten Schürmann. Primitive recursion for higher-order abstract syntax. Technical Report CMU-CS-96-172, Carnegie Mellon University, September 1996.
7. Thierry Despeyroux and André Hirschowitz. Some theory for abstract syntax and induction. Draft manuscript.
8. Leonidas Fegaras and Tim Sheard. Revisiting catamorphisms over datatypes with embedded functions (or, programs from outer space). In *Proceedings of 23rd Annual Symposium on Principles of Programming Languages*, pages 284–294, St. Petersburg Beach, Florida, January 1996. ACM Press.
9. Kurt Gödel. On an extension of finitary mathematics which has not yet been used. In Solomon Feferman et al., editors, *Kurt Gödel, Collected Works, Volume II*, pages 271–280. Oxford University Press, 1990.

[3] personal communication

10. Robert Harper, Furio Honsell, and Gordon Plotkin. A framework for defining logics. *Journal of the Association for Computing Machinery*, 40(1):143–184, January 1993.
11. Jean-Pierre Jouannaud and Mitsuhiro Okada. A computation model for executable higher-order algebraic specification languages. In Gilles Kahn, editor, *Proceedings of the 6th Annual Symposium on Logic in Computer Science*, pages 350–361, Amsterdam, The Netherlands, July 1991. IEEE Computer Society Press.
12. Lena Magnusson. *The Implementation of ALF—A Proof Editor Based on Martin-Löf's Monomorphic Type Theory with Explicit Substitution.* PhD thesis, Chalmers University of Technology and Göteborg University, January 1995.
13. Raymond McDowell and Dale Miller. A logic for reasoning about logic specifications. Draft manuscript, July 1996.
14. Erik Meijer and Graham Hutton. Bananas in space: Extending fold and unfold to exponential types. In *Proceedings of the 7th Conference on Functional Programming Languages and Computer Architecture*, La Jolla, California, June 1995.
15. Dale Miller. An extension to ML to handle bound variables in data structures: Preliminary report. In *Proceedings of the Logical Frameworks BRA Workshop*, Nice, France, May 1990.
16. Dale Miller. Unification of simply typed lambda-terms as logic programming. In Koichi Furukawa, editor, *Eighth International Logic Programming Conference*, pages 255–269, Paris, France, June 1991. MIT Press.
17. Dale Miller. Abstract syntax and logic programming. In *Proceedings of the First and Second Russian Conferences on Logic Programming*, pages 322–337, Irkutsk and St. Petersburg, Russia, 1992. Springer-Verlag LNAI 592.
18. Bengt Nordström, Kent Petersson, and Jan M. Smith. *Programming in Martin-Löf's Type Theory: An Introduction*, volume 7 of *International Series of Monographs on Computer Science.* Oxford University Press, 1990.
19. Martin Odersky. A functional theory of local names. In *Proceedings of the 21st Annual Symposium on Principles of Programming Languages*, pages 48–59, Portland, Oregon, January 1994. ACM Press.
20. Christine Paulin-Mohring. Inductive definitions in the system Coq: Rules and properties. In M. Bezem and J.F. Groote, editors, *Proceedings of the International Conference on Typed Lambda Calculi and Applications*, pages 328–345, Utrecht, The Netherlands, March 1993. Springer-Verlag LNCS 664.
21. Frank Pfenning. Logic programming in the LF logical framework. In Gérard Huet and Gordon Plotkin, editors, *Logical Frameworks*, pages 149–181. Cambridge University Press, 1991.
22. Frank Pfenning and Hao-Chi Wong. On a modal λ-calculus for S4. In S. Brookes and M. Main, editors, *Proceedings of the Eleventh Conference on Mathematical Foundations of Programming Sematics*, New Orleans, Louisiana, March 1995. To appear in *Electronic Notes in Theoretical Computer Science*, Volume 1, Elsevier.

Eta-Expansions in Dependent Type Theory — The Calculus of Constructions

Neil Ghani
LIENS-DMI, Ecole Normale Superieure
45, Rue D'Ulm, 75230 Paris Cedex 05, France
e-mail:ghani@ens.fr

Abstract. Although the use of expansionary η-rewrite has become increasingly common in recent years, one area where η-contractions have until now remained the only possibility is in the more powerful type theories of the λ-cube. This paper rectifies this situation by applying η-expansions to the Calculus of Constructions — we discuss some of the difficulties posed by the presence of dependent types, prove that every term rewrites to a unique long $\beta\eta$-normal form and deduce the decidability of $\beta\eta$-equality, typeability and type inhabitation as corollaries.

1 Introduction

Extensional equality for the simply typed λ-calculus requires η-conversion, whose interpretation as a rewrite rule has traditionally been as a contraction $\lambda x : T.fx \Rightarrow f$ where $x \not\in \mathtt{FV}(t)$. When combined with the usual β-reduction, the resulting rewrite relation is strongly normalising and confluent, and thus reduction to normal form provides a decision procedure for the associated equational theory.

However η-contractions behave badly when combined with other rewrite rules and the key property of confluence is often lost. For example, if the calculus is extended by a unit type 1 with associated rewrite rule $t \Rightarrow *$ (providing t has type 1), then the divergence $\lambda x{:}1.\, * \Leftarrow \lambda x{:}1.fx \Rightarrow f$ cannot be completed. Another area where η-contractions cannot be used is where the lambda calculus is enriched with base constants and associated rewrite rules which may be confluent on their own, but which fail to be so in the presence of a contractive η-rewrite rule. For example, if we regard 1 as a base type with constants $f : 1 \rightarrow 1$ and $* : 1$ and with rewrite rule $fx \Rightarrow *$, then $\Rightarrow$ is confluent. However, as the above counterexample shows shows, the combination of $\Rightarrow$ with the contractive η-rewrite rule fails to be confluent — see [5] for further details.

Recently several authors [1, 4, 6, 14] have accepted the old proposal [13, 16, 17] that η-conversion be interpreted as an expansion $f \Rightarrow \lambda x.fx$ and the resulting rewrite relation has been shown confluent. Infinite reduction sequences such as

$$\begin{array}{lll} \lambda x.t & \Rightarrow \lambda y.(\lambda x.t)y & \Rightarrow \lambda y.t[y/x] \equiv \lambda x.t \\ tu & \Rightarrow (\lambda x.tx)u & \Rightarrow tu \end{array} \qquad (1)$$

are avoided by imposing syntactic restrictions to limit the possibilities for expansion; namely λ-abstractions cannot be expanded, nor can terms which are applied. This restricted expansion relation is strongly normalising, confluent and generates the same equational theory as the unrestricted expansionary rewrite relation — thus $\beta\eta$-equality can be decided by reduction to normal form in this restricted fragment. In addition, Huet's *long $\beta\eta$-normal forms* [13, 17] are exactly the normal forms of the restricted expansionary rewrite relation and thus η-expansions provide a natural mathematical theory of this important class of terms. Perhaps most pleasingly of all, these properties are maintained if one adds other type constructors, base types and associated rewrite rules [5, 3]. Finally the use of expansions is supported by categorical models of reduction [11] which has led to new expansionary η-rewrite rules for the coproduct and the tensor and !-type constructors of linear logic [12, 11].

This paper extends the initial results in a different direction by generalising the techniques previously developed to the Calculus of Constructions (λC). The major difference between λC and other theories previously studied is the presence of dependent types, leading to several complications

- As a dependent type system, the types of the λC may have terms embedded within them and so λC cannot be separated into firstly a calculus for deriving the types and secondly a calculus for deriving the terms which inhabit these types. Thus all judgement forms (e.g. type inhabitation, equality etc) must be explicitly parameterised by the context in which the judgement occurs.
- Since β and η-redexes may occur in the types of the calculus, terms no longer inhabit unique types and so decidability of $\beta\eta$-equality at the level of types cannot be assumed in proving the decidability of $\beta\eta$-equality on terms. This also means typability and type inhabitation cannot be assumed decidable and hence the reducts of a term, which partially depend on its type, cannot be assumed enumerable.
- In λ-calculi without dependent types all infinite reduction sequences are derived from expansions appearing in the triangle laws given in equation 1. However, in lambda calculi with dependent types there is another source of infinite reduction sequences, namely the domain of the type of the term to be expanded. That is, if we allow expansions

$$\frac{\Gamma \vdash t : \Pi x : A.B}{\Gamma \vdash t \Rightarrow \lambda x : A.tx} \qquad (2)$$

  and define the term $B(x) = (\lambda z : X \rightarrow X.X)(x)$, then there is a typing judgement $X : *, x : X \rightarrow X \vdash x : \Pi z : B(x).X$ and hence an infinite reduction sequence

$$X : *, x : X \rightarrow X \vdash x \Rightarrow \lambda z : B(x).xz \Rightarrow \lambda z : B(\lambda z : B(x).xz).xz \Rightarrow \ldots \qquad (3)$$

  Another problem with the expansions of equation 2 is that β-normal forms are no longer preserved by η-expansion and thus the usual strategy for the calculation of long $\beta\eta$-normal forms no longer works.

The problems associated with the η-expansions of equation 2 lead us to permit a rewrite $\Gamma \vdash t \Rightarrow \eta^A(t)$ iff, in addition to the usual restrictions, the type A is a normal form. Circularity is avoided by axiomatizing the long $\beta\eta$-normal forms of λC and then permitting an expansion only if the domain of the type of the term expanded can be proven to be a such a long $\beta\eta$-normal form.

There have been several attempts to give a satisfactory account of the Calculus of Constructions and other systems in the λ-cube with $\beta\eta$-equality. One of the major problems of doing so is that if one defines $\beta\eta$-equality on pre-terms with the equation $t = \lambda x{:}A.tx$ ($x \not\in \mathtt{FV}(T)$), then the associated equational theory on pre-terms is too coarse, equating semantically distinct entities, e.g.

$$\lambda x{:}A.x =_\beta \lambda x{:}A.(\lambda y{:}B.y)x =_\eta \lambda y{:}B.y$$

where A and B are arbitrary types. However, as η-expansions are best formulated on well-typed terms, we use a typed version of $\beta\eta$-equality and so avoid the inconsistencies inherent in the untyped version. What's more, typed and untyped $\beta\eta$-equality coincide on well-typed terms *of the same type* [9]. Geuvers' paper also proves that typed $\beta\eta$-equality is decidable by orienting the η-equation as a contraction and showing that the resulting rewrite relation, in conjunction with the usual β-redexes, is strongly normalising and confluent. A related paper [7] (which we draw upon) also uses strong normalisation and confluence of β and η-contractions to prove the existence of $\beta\eta$-long normal forms for λC.

After collecting some defintions in section 2, presenting λC in section 3 and defining our restricted rewrite relation in section 4, we prove:

- The methods of [3] are generalised to prove confluence upto an equivalence relation called *type equivalence*. As a corollary we prove that if two long $\beta\eta$-normal forms are $\beta\eta$-equal, then they are the same term.
- We prove that there are no infinite sequences of η-expansions and, since η-expansions preserve β-normal forms, each term normalizes to a unique long $\beta\eta$-normal form.
- Finally, full confluence is proved from which the computability of the long $\beta\eta$-normal form of a term, decidability of $\beta\eta$-equality, typability etc follow as corollaries.

2 Rewriting and Notation

We shall use the following judgement forms:

$\Gamma \vdash t : T$	In context Γ, the pre-term t has type T.
$\Gamma \vdash t =_{\beta\eta} u$	In context Γ, the pre-term t is $\beta\eta$-equal to the pre-term u.
$\Gamma \vdash \mathcal{D}\mathtt{NF}(t)$	In context Γ, the pre-term t is a $\mathcal{D}$-normal form
$\Gamma \vdash t \Rightarrow_{\mathcal{R}} u$	In context Γ, the pre-term t $\mathcal{R}$-rewrites to the pre-term u.
$\Gamma \vdash t \approx u$	In context Γ, the pre-term t is type equivalent to the pre-term u.

We write $\Gamma \vdash J : T$ to indicate that there is a judgement $\Gamma \vdash J$ and that $\Gamma \vdash t : T$ for all the terms mentioned in J, while we write $\Gamma \vdash J_1 \wedge J_2$ if $\Gamma \vdash J_1$ and $\Gamma \vdash J_2$. When the context of a judgement is either clear or unimportant, legibility is increased by omitting the context, e.g. we often speak only of the reducts of a term and omit all reference to the context. We shall use $\eta^A(t)$ as shorthand for $\lambda x : A.tx$ and assume that $x \notin \mathtt{FV}(t)$.

If R is a rewrite relation, then $R^=$ is its ***reflexive closure***, R^+ is its ***transitive closure***, R^* its ***reflexive, transitive closure*** and $=_R$ is its ***reflexive, symmetric, transitive closure***. The relation R^{-1} is its inverse, while $R;S$ denotes the forward relational composition of R with S. We often use $\Rightarrow_R$ as alternate notion for a rewrite relation R. A term t is an *R-normal form* if there is no term t' such that $t \Rightarrow_R t'$, while a relation is ***normalising*** if every term rewrites to a normal form. A relation is ***strongly normalising*** if there are no infinite reduction sequences $t_0 \Rightarrow_R t_1 \Rightarrow_R t_2, \ldots$. A relation $\mathcal{R}$ satisfies the ***diamond property*** iff $\mathcal{R}^{-1}; \mathcal{R} \subseteq \mathcal{R}; \mathcal{R}^{-1}$ while $\mathcal{R}$ is ***confluent*** iff $\mathcal{R}^*$ satisfies the diamond property.

In the rest of this section we briefly review a nice proof of confluence for expansionary $\beta\eta$-rewriting in the simply typed λ-calculus, discuss the problems of applying this proof to λC and use this as motivation to prove an abstract confluence result which will be used later. When considering rewrite relations composed of two fragments R and S (thought of as β-contraction and restricted η-expansion), then confluence of the combined system is often easily established by proving that R and S are separately confluent and that R and S ***commute***

$$Com(R, S) \text{ iff } (R^*)^{-1}; S^* \subseteq R^*; (S^*)^{-1}$$

Unfortunately establishing that R and S commute can be difficult as one must consider not just individual rewrites but also sequences of rewrites. Consequently there is a lot of research into finding sufficient conditions to guarantee that R and S commute and we present one such result from [3, 8]. A relation R is said to ***strongly locally commute*** over S if and only if

$$SLC(R, S) \text{ iff } R^{-1}; S \subseteq S^*; (R^+)^{-1}$$

Lemma 2.1 *If $SLC(R, S)$ and R is strongly normalising, then R and S is commute.*

Applying lemma 2.1 to the simply typed λ-calculus, one may take R to be β-reduction and S to be restricted η-expansion. That β-reduction is strongly normalising and confluent is well known, while η-expansion is confluent as it satisfies the diamond property. Thus one is left to check that β-reduction strongly locally commutes over η-expansion and, although not trivial because of the restrictions on η-expansion, this is no more difficult than checking local confluence. Unfortunately, the following diagram shows that the strong local commutation property no longer holds in λC.

$$\begin{array}{ccc} (\lambda x : B \to B.x)(\lambda y : B.y) & \xrightarrow{\quad\beta\quad} & \lambda y : B.y \\ \Big\downarrow{\scriptstyle \eta} & & \\ (\lambda x : B \to B.\eta^{B_0}(x))(\lambda y : B.y) \underset{\beta}{\rightarrow} \eta^{B_0}(\lambda y : B.y) & \underset{\beta}{\rightarrow} & \lambda y : B_0.y \end{array} \tag{4}$$

All we know is that B_0 and B are $\beta\eta$-equal and hence we cannot obtain completions of the right form. Of course, if these terms were simply typed λ-terms then the above diagram would actually be complete as type uniqueness would force $B_0 = B$.

The solution proposed in this paper is to worry about the types later! Thus we define two terms to be *type equivalent* if they are equal apart from their λ-abstractions which may be over types which are only $\beta\eta$-equal. This defines an equivalence relation and confluence is first proved only modulo type equivalence. There is a considerable body of literature [2, 15] concerning rewriting modulo equational theories and we now present those definitions and results required to generalise lemma 2.1 to rewriting modulo equivalence relations.

In the following R and S are rewrite relations while E is an equivalence relation. Define the relation $R/E = E; R; E$. The relation R is *coherent* with E iff $E; R \subseteq R^*; E$ while R is *strongly coherent* with E iff $E; R \subseteq R; E$ (note the similarity between the notions of *coherence* and *bisimulation* from concurrency theory). A relation R satisfies the diamond property modulo E iff $R^{-1}; R \subseteq R; E; R^{-1}$, while R is confluent modulo E iff R^* satisfies the diamond property modulo E. The following lemmas may be proved by simple diagram chases.

Lemma 2.2 *If R is coherent with E, then R^* is coherent with E and $(R/E)^* \subseteq R^*; E$. If R is strongly normalising and R is strongly coherent with E, then R/E is also strongly normalising.*

Lemma 2.3 *If R satisfies the diamond property modulo E and is coherent with E, then R^* satisfies the diamond property modulo E.*

Lemma 2.4 *Let R be coherent with E. Then R/E is confluent iff R is confluent modulo E.*

Lemma 2.5 *If R is strongly coherent with E, S is coherent with E and $S^{-1}; R \subseteq R^+; E; (S^*)^{-1}$, then R/E strongly locally commutes with S/E.*

3 The Calculus of Constructions — λC

We give a quick presentation of λC [9]. The *sorts* of λC are $\mathcal{S} = \{*, \Box\}$, while $\mathtt{Var}(*)$ and $\mathtt{Var}(\Box)$ are disjoint sets of variables. If x is a variable and $x \in \mathtt{Var}(s)$, this is sometimes indicated by writing x^s. The pre-terms of λC are denoted Λ and are defined by the syntax $T := x \mid s \mid \lambda x : T.T \mid \Pi x : T.T \mid TT$, where x is any variable and s is a sort. The usual notions of *free variable, α-equivalence* and *substitution* are taken for λC and a pre-term is called *neutral* iff it is not a λ-abstraction.

A *pre-context* is a list of pairs of variables and pre-terms, written $x_1 : T_1, \ldots, x_n : T_n$ such that the variables are pairwise distinct. The empty pre-context is denoted $[]$ and the variables declared in a pre-context Γ are denoted $\mathtt{dom}(\Gamma)$. The

typing judgements of λC are of the form $\Gamma \vdash t : T$ and the ***equality judgements*** of λC are of the form $\Gamma \vdash t =_{\beta\eta} u$, where Γ is a pre-context and t, u and T are pre-terms. These judgements are generated by the inference rules of Table 4 presented at the end of this paper. A pre-term t is called a term iff there is a typing judgement $\Gamma \vdash t : T$ and, unless otherwise stated, all future concepts are restricted to well-typed terms. The basic meta-theory of λC may be established using η-expansions but due to lack of space we quote results from [9] which are proven using η-contractions. The first lemma concerns substitution.

Lemma 3.1 *Let $\Gamma \vdash u : A$ and $\Gamma \vdash u =_{\beta\eta} u' : A$*

- *If there is a typing judgement $\Gamma, x : A, \Delta \vdash t : B$, then there is also one $\Gamma, \Delta[u/x] \vdash t[u/x] : B[u/x]$.*
- *If there is an equality judgement $\Gamma, x : A, \Delta \vdash t =_{\beta\eta} t'$, then there is also one $\Gamma, \Delta[u/x] \vdash t[u/x] =_{\beta\eta} t'[u/x]$*
- *If there is a typing judgement $\Gamma, x : A, \Delta \vdash t : B$, then there is also one $\Gamma, \Delta[u/x] \vdash t[u/x] =_{\beta\eta} t[u'/x]$.*

One of the most important lemmas concerning PTS's such as λC is the *Generation lemma.*

Lemma 3.2 *Let $\Gamma \vdash t{:}T$ be a typing judgement. Then*

- *If t is $*$, then $T = \Box$.*
- *If t is a variable x, then there is a declaration $x{:}T' \in \Gamma$ and $\Gamma \vdash T =_{\beta\eta} T'$.*
- *If t is $\lambda x : T'.u$, then there are typing judgements $\Gamma, x : T' \vdash u{:}A$, $\Gamma \vdash \Pi x : T'.A : s$ and $\Gamma \vdash T =_{\beta\eta} \Pi x : T'.A$.*
- *If t is $\Pi x : T'.A$, then there are typing judgements $\Gamma \vdash T' : s_1$, $\Gamma, x : T' \vdash A : s_2$, and $\Gamma \vdash T =_{\beta\eta} s_2$.*
- *If t is FU, then there are typing judgements $\Gamma \vdash F : \Pi x : T'.A$, $\Gamma \vdash U : T'$ and $\Gamma \vdash T =_{\beta\eta} A[U/x]$.*

Lemma 3.3 *If there are typing judgements $\Gamma \vdash t : T$ and $\Gamma \vdash t : T'$, then there is also an equality judgement $\Gamma \vdash T =_{\beta\eta} T'$.*

Another property required of λC is known as *Uniqueness of Products*

Lemma 3.4 *If there is a judgement $\Gamma \vdash \Pi x : A.B =_{\beta\eta} \Pi x : A'.B' : T$, then there are judgements $\Gamma \vdash A =_{\beta\eta} A'$ and $\Gamma, x : A \vdash B =_{\beta\eta} B'$.*

In order to prove some further results about the judgements of λC, we use a notion of *context-equivalence.* Two well formed contexts Γ and Δ are said to be $\beta\eta$-equal, written $\Gamma =_{\beta\eta} \Delta$, iff they can be proved to be so by the following pair of inference rules:

$$\frac{}{[] =_{\beta\eta} []} \qquad \frac{\Gamma_0 =_{\beta\eta} \Gamma_1 \quad \Gamma_0 \vdash A =_{\beta\eta} A' \quad \Gamma_1 \vdash A =_{\beta\eta} A'}{\Gamma_0, x : A =_{\beta\eta} \Gamma_1, x : A'}$$

Lemma 3.5 *Context equivalence is an equivalence relation on well formed contexts. If $\Gamma =_{\beta\eta} \Delta$ and there is a judgement $\Gamma \vdash J$, then there is also one $\Delta \vdash J$.*

Equality judgements are sound for typing, i.e. if one term has a type and is $\beta\eta$-equal to another term, then the second term also has that type.

Lemma 3.6 *If there is a judgement $\Gamma \vdash t_1 =_{\beta\eta} t_2$ such that for $i = 1$ or 2, $\Gamma \vdash t_i : T$ then $\Gamma \vdash t_1 : T \wedge t_2 : T$*

The last lemma we require is a technical one which is needed to provide an induction rank for the construction of long $\beta\eta$-normal forms

Lemma 3.7 *There is a well-founded order $<$ on the terms of λC such that*

- *If u is a subterm of t, then $u < t$*
- *If t is a β-normal form, then there is also a β-normal form T such that there is a judgement $\Gamma \vdash t : T$ and $T < t$.*

Proof The proof follows [7] in using an alternate syntax where a type for a term occurs as a subterm of the term. □

4 A Rewrite Relation

Recall that if t and A are pre-terms, then $\eta^A(t)$ is notational shorthand for the pre-term $\lambda x{:}A.tx$ where $x \not\in \mathtt{FV}(t)$ is a correctly sorted variable. The theory of η-expansions as developed in [] shows that an expansion $\Gamma \vdash t \Rightarrow_{\mathcal{F}} \eta^A(t)$ can only be permitted if t is neither a λ-abstraction nor applied to another term. However in dependent type theories these restrictions on the applicability of expansions are insufficient to prevent unwanted reduction sequences such as equation 3. We therefore impose the additional restriction that the term A is a long $\beta\eta$-normal form by using structural and typing criteria to axiomatise the long $\beta\eta$-normal forms of λC. In Table 1, two judgements $\Gamma \vdash \eta\mathcal{I}\mathtt{NF}(t)$ and $\Gamma \vdash \eta\mathcal{F}\mathtt{NF}(t)$ are defined which characterize those terms which are (internal and full) normal forms of the expansionary η-rewrite rules. These two judgement forms are used to define two others $\Gamma \vdash \mathcal{I}\mathtt{NF}(t)$ and $\Gamma \vdash \mathcal{F}\mathtt{NF}(t)$ which characterise the *internal/full long $\beta\eta$-normal forms* of λC. In Table 1, a pair Γ, t is said to be *not expandable* if either t is a λ-abstraction, or there are no judgements $\Gamma \vdash t : \Pi x : A.B$. Note that as type inhabitation is not assumed decidable, the various normal form judgements of Table 1 cannot be assumed decidable.

The long $\beta\eta$-normal forms of λC as axiomatized in Table 1 are used to define a rewrite relation in two stages. Firstly the purely expansionary η-rewrite relation, denoted $\Rightarrow_{\eta\mathcal{F}}$, is presented in Table 2. As in previous works [11, 10], the condition that terms applied to other terms may not be expanded is enforced by simultaneously defining a further subrelation $\Rightarrow_{\eta\mathcal{I}}$ of $\Rightarrow_{\eta\mathcal{F}}$ which is guaranteed not to include top-level expansions. Thus a subterm which is applied to another subterm may be safely $\Rightarrow_{\eta\mathcal{I}}$-rewritten without the risk of generating the second of reduction loops in equation 1. Note that because we cannot assume the long $\beta\eta$-normal form predicate is decidable, the $\Rightarrow_{\eta\mathcal{F}}$-reducts of a term cannot be assumed enumerable.

Table 1. The Predicates $\mathcal{I}$NF and $\mathcal{F}$NF

$$\frac{t \in \mathcal{S} \cup \mathtt{Var}}{\Gamma \vdash \eta\mathcal{I}\mathtt{NF}(t)} \qquad \frac{\Gamma \vdash \eta\mathcal{I}\mathtt{NF}(t) \quad \Gamma \vdash \eta\mathcal{F}\mathtt{NF}(u)}{\Gamma \vdash \eta\mathcal{I}\mathtt{NF}(tu)}$$

$$\frac{\Gamma \vdash \eta\mathcal{F}\mathtt{NF}(A) \quad \Gamma, x : A \vdash \eta\mathcal{F}\mathtt{NF}(t)}{\Gamma \vdash \eta\mathcal{I}\mathtt{NF}(\lambda x : A.t)} \qquad \frac{\Gamma \vdash \eta\mathcal{F}\mathtt{NF}(A) \quad \Gamma, x : A \vdash \eta\mathcal{F}\mathtt{NF}(B)}{\Gamma \vdash \eta\mathcal{I}\mathtt{NF}(\Pi x : A.B)}$$

$$\frac{\Gamma \vdash \eta\mathcal{I}\mathtt{NF}(t) \quad \Gamma, t \text{ is not expandable}}{\Gamma \vdash \eta\mathcal{F}\mathtt{NF}(t)} \qquad \frac{\Gamma \vdash \eta\mathcal{R}\mathtt{NF}(t) \quad t \text{ is a } \beta\text{-nf} \quad \mathcal{R} \in \{\mathcal{I}, \mathcal{F}\}}{\Gamma \vdash \mathcal{R}\mathtt{NF}(t)}$$

Table 2. Restricted η-expansions

$$\frac{t \text{ neutral} \quad \Gamma \vdash t : \Pi x : A.B \quad \Gamma \vdash \mathcal{F}\mathtt{NF}(\Pi x : A.B)}{\Gamma \vdash t \Rightarrow_{\eta\mathcal{F}} \lambda x : A.tx} \qquad \frac{\Gamma \vdash t \Rightarrow_{\eta\mathcal{I}} t'}{\Gamma \vdash t \Rightarrow_{\eta\mathcal{F}} t'}$$

$$\frac{\Gamma, x : T \vdash t \Rightarrow_{\eta\mathcal{F}} t'}{\Gamma \vdash \lambda x : T.t \Rightarrow_{\eta\mathcal{I}} \lambda x : T.t'} \qquad \frac{\Gamma \vdash T \Rightarrow_{\eta\mathcal{F}} T'}{\Gamma \vdash \lambda x : T.t \Rightarrow_{\eta\mathcal{I}} \lambda x : T'.t}$$

$$\frac{\Gamma \vdash F \Rightarrow_{\eta\mathcal{I}} F'}{\Gamma \vdash FU \Rightarrow_{\eta\mathcal{I}} F'U} \qquad \frac{\Gamma \vdash U \Rightarrow_{\eta\mathcal{F}} U'}{\Gamma \vdash FU \Rightarrow_{\eta\mathcal{I}} FU'}$$

$$\frac{\Gamma, x{:}A \vdash B \Rightarrow_{\eta\mathcal{F}} B'}{\Gamma \vdash \Pi x{:}A.B \Rightarrow_{\eta\mathcal{I}} \Pi x{:}A.B'} \qquad \frac{\Gamma \vdash A \Rightarrow_{\eta\mathcal{F}} A'}{\Gamma \vdash \Pi x{:}A.B \Rightarrow_{\eta\mathcal{I}} \Pi x{:}A'.B}$$

Let $\Rightarrow_\beta$ be the standard β-reduction defined on the set of pre-terms Λ. To keep a standard form of judgement, we introduce a context-parameterised form of β-reduction on typed terms as follows:

$$\frac{t \Rightarrow_\beta t' \quad \Gamma \vdash t : C}{\Gamma \vdash t \Rightarrow_\beta t'} \tag{5}$$

and then define

$$\Gamma \vdash t \Rightarrow_{\mathcal{I}} t' \text{ iff } \Gamma \vdash t \Rightarrow_\beta t' \text{ or } \Gamma \vdash t \Rightarrow_{\eta\mathcal{I}} t' \tag{6}$$

and similarly

$$\Gamma \vdash t \Rightarrow_{\mathcal{F}} t' \text{ iff } \Gamma \vdash t \Rightarrow_\beta t' \text{ or } \Gamma \vdash t \Rightarrow_{\eta\mathcal{F}} t' \tag{7}$$

Lemma 4.1 *Let $\Gamma \vdash t \Rightarrow_{\mathcal{R}} t'$ where $\mathcal{R} \in \{\mathcal{I}, \mathcal{F}\}$. Then*

- *There is a judgement $\Gamma \vdash t =_{\beta\eta} t'$.*
- *If $\Gamma =_{\beta\eta} \Gamma'$, then $\Gamma' \vdash t \Rightarrow_{\mathcal{R}} t'$*
- *If t is a β-normal form and $\Gamma \vdash t \Rightarrow_{\eta\mathcal{F}} t'$, then t' is a β-normal form*

Lemma 4.2 *If $\Gamma \vdash \mathcal{R}\texttt{NF}(t)$ for $\mathcal{R} \in \{\mathcal{I}, \mathcal{F}\}$, then there is no term t' such that $\Gamma \vdash t \Rightarrow_{\mathcal{R}} t'$.*
Proof Induction on the proof that $\Gamma \vdash \mathcal{R}\texttt{NF}(t)$. □

Note however that we cannot yet prove the reverse implication of lemma 4.2 as this would require proving that if there is a judgement $\Gamma \vdash t : \Pi x : A.B$, then there is another judgement $\Gamma \vdash t : \Pi x : A'.B' \wedge \mathcal{F}\texttt{NF}(\Pi x : A'.B')$. This in turn would require proving that every term reduces to a long $\beta\eta$-normal form and this must be postponed until later.

5 Confluence Modulo Type Equivalence

We shall now prove that $\Rightarrow_{\mathcal{F}}$ is confluent modulo a relation called *type equivalence*. As described in the introduction, our proof of confluence is a generalisation of the equivalent theorem for the simply typed λ-calculus as proved in [3]. The main difference is that confluence is proved only upto an equivalence relation stating that terms may differ upto $\beta\eta$-equality in the types abstracted over. In the next section this partial confluence result is used to show successively that $\Rightarrow_{\eta\mathcal{F}}$ is strongly normalising, that $\Rightarrow_{\mathcal{F}}$ is normalising and hence that the restricted expansionary rewrite relation defined in Table 2 is confluent.

Two terms t and t' are said to be *type equivalent* in a context Γ, written $\Gamma \vdash t \approx t'$ iff they can be proved so by the inference rules of Table 3. The inference rule for Π-types requires the types abstracted over to be only $\beta\eta$-equal rather than type equivalent so as to enable the proof of lemma 5.9 to go through. There is no need for some form of weakening for type equivalence as such a rule is admissable.

Table 3. Type-Equivalent Terms

$$\frac{x \in \texttt{dom}(\Gamma)}{\Gamma \vdash x \approx x} \qquad \frac{}{\Gamma \vdash * \approx *}$$

$$\frac{\Gamma \vdash A =_{\beta\eta} A' \quad \Gamma, x : A \vdash B \approx B'}{\Gamma \vdash \lambda x : A.B \approx \lambda x : A'.B'} \qquad \frac{\Gamma \vdash F \approx F' \quad \Gamma \vdash U \approx U'}{\Gamma \vdash FU \approx F'U'}$$

$$\frac{\Gamma \vdash A =_{\beta\eta} A' \quad \Gamma, x : A \vdash B \approx B'}{\Gamma \vdash \Pi x : A.B \approx \Pi x : A'.B'}$$

Lemma 5.1 *The relation $\Gamma \vdash \approx$ is an equivalence relation over well typed terms and if $\Gamma \vdash t \approx t'$, then $\Gamma \vdash t =_{\beta\eta} t'$.*
Proof The only interesting part of the lemma is symmetry which requires an auxilluary lemma stating that if $\Gamma \vdash t \approx t'$ and $\Gamma =_{\beta\eta} \Gamma'$, then $\Gamma' \vdash t \approx t'$. □

Lemma 5.2 *Assume there are typing judgements $\Gamma, x : A, \Delta \vdash t : B$ and $\Gamma \vdash u : A$. If $\Gamma, x : A, \Delta \vdash t \approx t'$ and $\Gamma \vdash u \approx u'$, then $\Gamma, \Delta[u/x] \vdash t[u/x] \approx t'[u'/x]$.*
Proof The proof is by induction on t, using lemma 3.1. □

We would like to apply lemma 2.1 by instantiating R with $\beta/\approx$ and S with $\eta\mathcal{F}/\approx$. However, as lemma 2.5 shows this requires β to be strongly coherent with $\approx$ which is clearly not the case — any β-reduction occurring inside a type abstraction provides a counterexample. What is required is a notion of β-reduction quotiented by type-equivalence, and fortunately this is easily defined:

$$\Gamma \vdash t \Rightarrow_{\beta\approx} t' \text{ iff } \Gamma \vdash t \Rightarrow_{\beta} t' \text{ and } \Gamma \vdash t \not\approx t'$$

Lemma 5.3 *If $\Gamma \vdash (\lambda x : A.t)u$, then there is a reduction $\Gamma \vdash (\lambda x : A.t)u \Rightarrow_{\beta\approx} t[u/x]$*
Proof Assume there is a term $\Gamma \vdash (\lambda x : A.t)u : T$ such that $\Gamma \vdash (\lambda x : A.t)u \approx t[u/x]$ and consider the function E which maps a λC term to an untyped λ-term by erasing all type information in λ and Π-abstractions and which treats sorts and Π's as constants in the untyped calculus. First note that if $\Gamma \vdash t \approx t'$, then $E(t) = E(t')$ and that $E(t[u/x]) = E(t)[E(u)/x]$. Thus we obtain an infinite series of untyped β-reductions

$$E((\lambda x : A.t)u) = (\lambda x.E(t))E(u) \Rightarrow_{\beta} E(t)[E(u)/x] = E(t[u/x]) = E((\lambda x : A.t)u) \Rightarrow_{\beta} \cdots$$

However, the erasure of a typable term is β-SN and hence no such term exists. □

Lemma 5.4 *The relation $\Rightarrow_{\beta\approx}$ is strongly coherent with $\approx$*
Proof Let $\Gamma \vdash t \Rightarrow_{\beta\approx} t'$ and $\Gamma \vdash t \approx u$. The proof is by induction on the term structure of t and we treat only the case where t is of the form $t = FU$. Reductions of proper subterms of t are dealt with by the induction hypothesis, and so the only case left is a reduction $\Gamma \vdash (\lambda x : A.a)b \Rightarrow_{\beta\approx} a[b/x]$. But u must be of the form $(\lambda x : A'.a')b'$ where $\Gamma \vdash b \approx b'$ and $\Gamma, x : A \vdash a \approx a'$. Hence there is a β reduction $\Gamma \vdash (\lambda x : A'.a')b' \Rightarrow_{\beta} a'[b'/x']$ and by lemma 5.3, this is actually a $\Rightarrow_{\beta\approx}$-reduction. Finally, by lemma 5.2, $\Gamma \vdash a[b/x] \approx a'[b'/x]$. □

Lemma 5.5 *The relation $\Rightarrow_{\beta\approx}$ is strongly normalising*
Proof $\Rightarrow_{\beta\approx}$ is strongly normalising because it is a sub relation of $\Rightarrow_{\beta}$. □

Lemma 5.6 *If there is a reduction sequence $\Gamma \vdash t \Rightarrow^{*}_{\beta} t'$, then there is another reduction $\Gamma \vdash t \Rightarrow^{*}_{\beta\approx} t''$ such that $\Gamma \vdash t' \approx t''$.*
Proof The lemma is proved by induction on the length of the reduction sequence $\Gamma \vdash t \Rightarrow^{*}_{\beta} t'$. If the sequence is of length 1, then the lemma holds as either $\Gamma \vdash t \approx t'$, in which case set $t'' = t$, or $\Gamma \vdash t \not\approx t'$, in which case set $t'' = t'$. For

the inductive step:

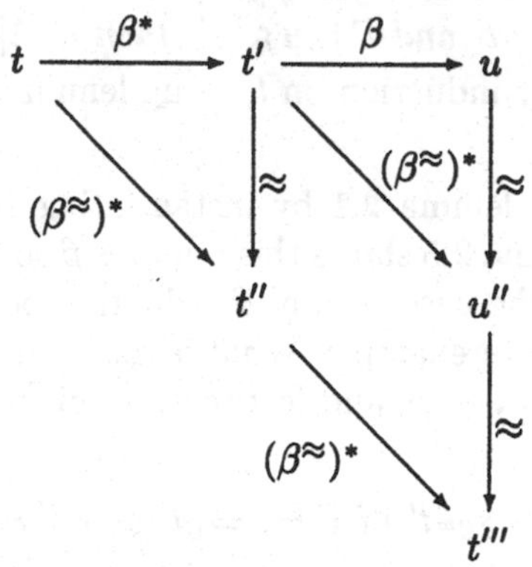

The completions are by induction, the first part of this lemma, and lemma 5.4. □

Lemma 5.7 *$\Rightarrow^*_{\beta\approx}$ satisfies the diamond lemma modulo $\approx$. Thus $\Rightarrow_{\beta\approx}/\approx$ is confluent.*

Proof Let $\Gamma \vdash t \Rightarrow^*_{\beta\approx} u_i$. By the confluence of $\Rightarrow_\beta$, there is a term u such that $\Gamma \vdash u_i \Rightarrow^*_\beta u$. By lemma 5.6 there are terms v_i such that $\Gamma \vdash u_i \Rightarrow^*_{\beta\approx} v_i$ and such that $\Gamma \vdash v_1 \approx u \approx v_2$. Thus $\Gamma \vdash v_1 \approx v_2$. The second part of the lemma follows from lemma 2.4. □

We now prove $\Rightarrow_{\eta\mathcal{F}}/\approx$ is confluent, and as there is no need to prove $\Rightarrow_{\eta\mathcal{F}}/\approx$ is strongly normalising we show that $\Rightarrow_{\eta\mathcal{F}}$ is coherent with type equivalence.

Lemma 5.8 *The relation $\Rightarrow_{\eta\mathcal{F}}$ is coherent with type equivalence.*

Proof Let $\Gamma \vdash t \Rightarrow_{\eta\mathcal{F}} t'$ and $\Gamma \vdash t \approx u$. The proof is by induction on the rewrite $\Gamma \vdash t \Rightarrow_{\eta\mathcal{R}} t'$ and we treat two cases. If $\Gamma \vdash t \Rightarrow_{\eta\mathcal{F}} \eta^A(t)$, then because t is neutral, so is u. By lemma 5.1, if $\Gamma \vdash t \approx u$, then $\Gamma \vdash t =_{\beta\eta} u$ and so by lemma 3.6 t and u have the same types under Γ. Thus there is an expansion $\Gamma \vdash u \Rightarrow_{\eta\mathcal{F}} \eta^A(u)$. If however, t is of the form $\lambda x : A.t_0$, then for $\Rightarrow_{\eta\mathcal{I}}$-reducts induced by reductions of A, set $u' = u$, while for a reduction $\Gamma \vdash \lambda x : A.t_0 \Rightarrow_{\eta\mathcal{I}} \lambda x : A.t_1$, one applies the induction hypothesis. □

Lemma 5.9 *The relations $\Rightarrow_{\eta\mathcal{I}}/\approx$ and $\Rightarrow_{\eta\mathcal{F}}/\approx$ are confluent.*

Proof We prove, by induction on term structure, the stronger property that $\Rightarrow_{\eta\mathcal{I}}$ and $\Rightarrow_{\eta\mathcal{F}}$ satisfy the diamond property modulo type-equivalence. We treat the most interesting cases.

If $\Gamma \vdash t \Rightarrow_{\eta\mathcal{F}} \eta^A(t)$ and $\Gamma \vdash t \Rightarrow_{\eta\mathcal{F}} \eta^{A'}(t)$, then by Uniqueness of Products as stated in lemma 3.4, $\Gamma \vdash A =_{\beta\eta} A'$ and hence the two reducts are type equivalent. The inference rule for the type equivalence of Π-types was chosen to allow this step of the proof to go through — see our previous comments. If $\Gamma \vdash t \Rightarrow_{\eta\mathcal{F}} \eta^A(t)$ and $\Gamma \vdash t \Rightarrow_{\eta\mathcal{I}} t'$, then t' must be neutral and by lemma 3.6, in habit the same types under Γ that t does. Thus there are rewrites $\Gamma \vdash t' \Rightarrow_{\eta\mathcal{F}} \eta^A(t')$ and $\Gamma \vdash \eta^A(t) \Rightarrow_{\eta\mathcal{I}} \eta^A(t')$. □

We now establish the final premise of lemma 2.5.

Lemma 5.10 *If $\Gamma \vdash a \Rightarrow_{\beta\approx} b$ and $\Gamma \vdash a \Rightarrow_{\eta\mathcal{I}} t$. Then there are completions of one of the following forms*

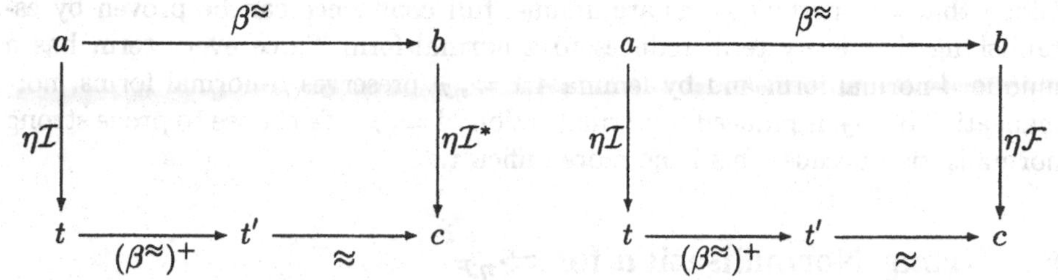

If instead $\Gamma \vdash a \Rightarrow_{\eta\mathcal{F}} t$, then the same diagram holds except $\Gamma \vdash b \Rightarrow^*_{\eta\mathcal{F}} c$ is the only possibility.

Proof Notice that by lemma 5.6 and 5.4, we need only show that there is a reduction sequence $\Gamma \vdash t \Rightarrow^+_\beta t'$ containing at least one $\Rightarrow_{\beta\approx}$-reduction. The proof is by induction on the term a and we consider two cases. If $\Gamma \vdash a \Rightarrow_{\eta\mathcal{F}} \eta^A(a)$, then there is no problem in constructing the required completions unless b is of the form $\lambda x : A_0.a_0$. By lemma's 3.4 and 3.3, $\Gamma \vdash A =_{\beta\eta} A_0$ and so we have completions

$$\Gamma \vdash \eta^A(a) \Rightarrow_{\beta\approx} \eta^A(\lambda x : A_0.a_0) \Rightarrow_\beta \lambda x : A.a_0 \approx \lambda x : A_0.a_0 = b$$

Next consider a $\Rightarrow_{\beta\approx}$-rewrite $\Gamma \vdash (\lambda x : A.t)(u) \Rightarrow_{\beta\approx} t[u/x]$. Clearly any rewrite of the subterm A can be completed correctly, while any $\Rightarrow_{\eta\mathcal{I}}$-rewrite induced by rewrites of either t or u can be completed by the usual analysis, e.g. see the substitutivity lemmas of [11]. □

Lemma 5.11 *The relation $\Rightarrow_{\beta\approx}/\approx \cup \eta\mathcal{F}/\approx$ is confluent. Thus $\Rightarrow^*_\mathcal{F}$ satisfies the diamond lemma modulo type equivalence.*

Proof By lemmas 5.10 and 2.5, the first relation strongly locally commutes over the later. Thus by lemma 2.1 the two relations commute and since we have also established that each relation is confluent, so is their union. The second part of the lemma is easily proved by unraveling the definitions. □

Confluence modulo type equivalence is sufficient to prove the uniqueness of $\Rightarrow_\mathcal{F}$-normal forms.

Lemma 5.12 *Given a judgement $\Gamma \vdash t \approx t'$ where t and t' are $\Rightarrow_\mathcal{I}$-normal forms, then $t = t'$. Similarly, given a judgement $\Gamma \vdash t =_{\beta\eta} t'$ where t and t' are $\Rightarrow_\mathcal{F}$-normal forms, then $t = t'$.*

Proof The two parts of the lemma are proved simultaneously by induction on the terms t and t'. For example if t is of the form $\lambda x : A.u$, then t' must be of the form $\lambda x : A'.u'$ where $\Gamma \vdash A =_{\beta\eta} A'$ and $\Gamma, x : A \vdash u \approx u'$. Since u and u' are

$\Rightarrow_{\mathcal{I}}$-normal forms, by induction $u = u'$. Similarly as A and A' are $\Rightarrow_{\mathcal{F}}$-normal forms, $A = A'$.

For the second part of the lemma, note that by lemma 5.11, t and t' have type equivalent reducts, but as t and t' are $\Rightarrow_{\mathcal{F}}$-normal forms, t and t' must be type equivalent. Since $\Rightarrow_{\mathcal{F}}$-normal forms are also $\Rightarrow_{\mathcal{I}}$-normal forms, by the first part of this lemma t and t' are equal. □

Given that $\Rightarrow_{\mathcal{F}}$-normal forms are unique, full confluence can be proven by establishing that every term reduces to a normal form. Since every term has a unique β-normal form and by lemma 4.1 $\Rightarrow_{\eta\mathcal{F}}$ preserves β-normal forms, normalisation of $\Rightarrow_{\mathcal{F}}$ is reduced to normalisation of $\Rightarrow_{\eta\mathcal{F}}$. We choose to prove strong normalisation because this is no more difficult.

6 Strong Normalisation for $\Rightarrow_{\eta\mathcal{F}}$

In this section we prove that $\Rightarrow_{\eta\mathcal{F}}$ is strongly normalising, and by lemma 4.1 this implies $\Rightarrow_{\mathcal{F}}$ is normalising. The key to proving $\Rightarrow_{\eta\mathcal{F}}$ is strongly normalising lies in calculating the reducts of a variable. Thus given a typing judgement $\Gamma \vdash T : s$ and a variable $z^s \not\in \mathtt{dom}(\Gamma)$, we define a set of terms $\Delta_z(T)$ as follows

$$\begin{aligned} \Delta_z(\Pi x : A.B) &= \{z\} \cup \{\lambda x : A.v[zu/y] \mid u \in \Delta_x(A) \text{ and } v \in \Delta_y(B)\} \\ \Delta_z(T) &= \{z\} \qquad\qquad\qquad\qquad \text{if } T \text{ is not a product} \end{aligned}$$

where the variables x and y are fresh. Note that $\Delta_z(T)$ is always a finite set.

Lemma 6.1 *Let* $\Gamma \vdash \mathcal{F}\mathtt{NF}(T) \wedge T : s$. *Then* $\Gamma, z : T \vdash z \Rightarrow^*_{\eta\mathcal{F}} t$ *iff* $t \in \Delta_z(T)$.

Proof Two containments must be established, both by induction over T. The forward inclusion is proved by showing the set $\Delta_z(T)$ is closed under reduction and there are two sub-cases. Firstly if T is not a product, there can be no judgements of the form $\Gamma \vdash z : \Pi x : A.B \wedge \mathcal{F}\mathtt{NF}(\Pi x : A.B)$ as by lemma 3.3 $\Gamma \vdash T =_{\beta\eta} \Pi x : A.B$ and hence by lemmas 5.12 and 4.2, $T = \Pi x : A.B$. Thus z is a normal form. If however T is of the form $\Pi x : A.B$ and $\Gamma \vdash z \Rightarrow_{\eta\mathcal{F}} \eta^{A'}(z)$, then by lemmas 3.3 and 3.4 $A' = A$ and hence $\eta^{A'}(z) \in \Delta_z(T)$. As A is a long $\beta\eta$-normal form, all reducts of $\lambda x : A.v[zu/y]$ must be caused by reductions of either u or v and hence by the induction hypothesis are contained in $\Delta_z(T)$.

For the other containment, assuming T is of the form $\Pi x : A.B$, there is a reduction sequence

$$\Gamma \vdash z \Rightarrow_{\eta\mathcal{F}} \lambda x : A.zx \Rightarrow^*_{\eta\mathcal{F}} \lambda x : A.zu \Rightarrow^*_{\eta\mathcal{F}} \lambda x : A.v[zu/y]$$

where $u \in \Delta_x(A)$ and $v \in \Delta_y(B)$. The last reduction sequence does not break any of the context sensitive restrictions on expansion as zu is not a λ-abstraction. □

A reduction sequence $\Gamma \vdash t \Rightarrow^*_{\eta\mathcal{F}} t'$ is called an *t-envelope* iff $t = t'$ or there is a judgement $\Gamma \vdash t : T \wedge \mathcal{F}\mathtt{NF}(T)$ and a term $\alpha \in \Delta_z(T)$ such that $t' = \alpha[t/z]$. As $\Delta_z(T)$ is always finite, there are only a finite number of t-envelopes.

Lemma 6.2 *Any reduction sequence $\Gamma \vdash t \Rightarrow^*_{\eta\mathcal{F}} u$ factorises into an internal reduction sequence followed by an envelope, i.e. there is a term t', a reduction sequence $\Gamma \vdash t \Rightarrow^*_{\eta\mathcal{I}} t'$ and an envelope $\Gamma \vdash t' \Rightarrow^*_{\eta\mathcal{F}} u$. In addition, the length of the factorised reduction sequence is the same as that of the original.*

Proof The proof is by induction on the length of the sequence $\Gamma \vdash t \Rightarrow^*_{\eta\mathcal{F}} u$. If this reduction is of length 0, then the lemma clearly holds, while if $\Gamma \vdash t \Rightarrow_{\eta\mathcal{I}} t_0 \Rightarrow^*_{\eta\mathcal{F}}$, then the lemma holds by induction. The only other possibility is a reduction sequence

$$\Gamma \vdash t \Rightarrow_{\eta\mathcal{F}} \lambda x : A.tx \Rightarrow^*_{\eta\mathcal{I}} \lambda x : A.b$$

where there is a reduction sequence $\Gamma, x : A \vdash tx \Rightarrow^*_{\eta\mathcal{F}} b$. By induction this factorises into an internal reduction sequence followed by an envelope

$$\Gamma, x : A \vdash tx \Rightarrow^*_{\eta\mathcal{I}} t'x \Rightarrow^*_{\eta\mathcal{I}} t'u \Rightarrow^*_{\eta\mathcal{F}} b$$

Now the reduction sequence $\Gamma, x : A \vdash t'x \Rightarrow^*_{\eta\mathcal{I}} t'u \Rightarrow^*_{\eta\mathcal{F}} b$ is a $t'x$-envelope and hence the original reduction sequence may be factorised $\Gamma \vdash t \Rightarrow^*_{\eta\mathcal{I}} t' \Rightarrow^*_{\eta\mathcal{F}} \lambda x : A.b$. Examination of the proof shows that the length of the original and factorised reduction sequence is the same. □

Lemma 6.3 *There are no infinite sequences of reductions $\Gamma \vdash t \Rightarrow_{\eta\mathcal{F}} t_1 \Rightarrow_{\eta\mathcal{F}} ...$*

Proof The proof is by induction on the structure of t and, for a given term, we first show that there are no infinite sequences $\Gamma \vdash t \Rightarrow_{\eta\mathcal{I}} t_1 \Rightarrow_{\eta\mathcal{I}} ...$ before obtaining the general result. If t is a variable or sort, then t has no $\Rightarrow_{\eta\mathcal{I}}$-reducts and so is $\Rightarrow_{\eta\mathcal{I}}$-strongly normalising. All $\Rightarrow_{\eta\mathcal{I}}$-reduction sequences of compound terms are induced by $\Rightarrow_{\eta\mathcal{I}}$ or $\Rightarrow_{\eta\mathcal{F}}$-reduction sequences from proper subterms and therefore by the induction hypothesis t is $\Rightarrow_{\eta\mathcal{I}}$-strongly normalising.

Now by lemma 6.2 any reduction sequence $\Gamma \vdash t \Rightarrow^*_{\eta\mathcal{F}} u$ factorises into a reduction sequence of the same length $\Gamma \vdash t \Rightarrow^*_{\eta\mathcal{I}} t' \Rightarrow^*_{\eta\mathcal{F}} u$ where the latter reduction is an envelope. As the are no infinite envelopes and t is $\Rightarrow_{\eta\mathcal{I}}$-strongly normalising, t must be $\Rightarrow_{\eta\mathcal{F}}$ strongly normalising. □

Lemma 6.4 *The relation $\Rightarrow_{\mathcal{F}}$ is normalising.*

Proof If t is a term, consider the term t_0 obtained by contracting all β-redexes in t and then performing as many η-expansions as is permissible. Firstly t_0 is well defined as by lemma 6.3 there are no infinite η-expansion sequences. Secondly t_0 is an $\eta\mathcal{F}$-normal form by definition. Thirdly, by lemma 4.1, our restricted η-expansions preserve β-normal forms and hence t_0 is a β-normal form. Hence t_0 is an $\Rightarrow_{\mathcal{F}}$-normal form. □

7 Confluence and Decidability Results

In this final section we prove the main results of this paper, namely confluence of the expansionary rewrite relation, which together with normalisation imply the decidability of $\beta\eta$-equality. Confluence then allows us to prove the decidability of type inhabitation and typability.

Lemma 7.1 *The relation $\Rightarrow_{\mathcal{F}}$ is confluent*
Proof If $\Gamma \vdash t \Rightarrow_{\mathcal{F}} u_1$ and $\Gamma \vdash t \Rightarrow_{\mathcal{F}} u_2$, then by lemma 6.4, there are terms v_1 and v_2 which are $\Rightarrow_{\mathcal{F}}$-normal forms and are also reducts of u_1 and u_2 respectively. As these terms are $\beta\eta$-equal, lemma 5.12 shows that $v_1 = v_2$. Hence $\Rightarrow_{\mathcal{F}}$ is confluent. □

Not only do $\Rightarrow_{\mathcal{F}}$-normal forms exist, but that they are also computable.

Lemma 7.2 *If there is a judgement $\Gamma \vdash t : T$, then for $\mathcal{R} \in \{\mathcal{I}, \mathcal{F}\}$ there are computable terms $\alpha_{\mathcal{R}}$ such that $\Gamma \vdash t \Rightarrow^*_{\mathcal{R}} \alpha_{\mathcal{R}} \wedge \mathcal{R}\mathtt{NF}(\alpha)$.*
Proof We prove the lemma first for the special case where t is a β-normal form and use induction on the well ordering mentioned in lemma 3.7. Since all the immediate subterms of t are β-normal forms, one may apply the induction hypothesis to these immediate subterms to obtain a term t_0 such that $\Gamma \vdash t \Rightarrow^*_{\mathcal{I}} t_0 \wedge \mathcal{I}\mathtt{NF}(t_0)$. If t is a λ-abstraction, then $\Gamma \vdash \mathcal{F}\mathtt{NF}(t_0)$ and so we are done. If not, by lemma 3.7, there is a β-normal form T_0 such that $T_0 < t$ and $\Gamma \vdash T_0 =_{\beta\eta} T$. Applying the induction hypothesis, we obtain a term T_1 such that $\Gamma \vdash T_0 \Rightarrow^*_{\mathcal{F}} T_1 \wedge \mathcal{F}\mathtt{NF}(T_1)$. Now let α be the biggest term in $\Delta_z(T_1)$ (for some appropriate choice of variable). Then $\Gamma \vdash t \Rightarrow^*_{\mathcal{F}} \alpha[t_0/z] \wedge \mathcal{F}\mathtt{NF}(\alpha[t_0/z])$.

Now given an arbitrary judgement $\Gamma \vdash t : T$, there is also a judgement $\Gamma \vdash \beta(t) : T$ where $\beta(t)$ is the β-normal form of t. By the first part of the lemma we may compute the long $\beta\eta$-normal form of $\beta(t)$ which is of course also the long $\beta\eta$-normal form of t. □

Lemma 7.3 *Given a judgement $\Gamma \vdash t : T$, then t is an $\Rightarrow_{\mathcal{F}}$-normal form iff $\Gamma \vdash \mathcal{F}\mathtt{NF}(t)$*
Proof One half of the lemma is proven in lemma 4.2. The second half of the lemma is proven by induction on t while simultaneously proving that if t is an $\Rightarrow_{\mathcal{I}}$-normal form then $\Gamma \vdash \mathcal{I}\mathtt{NF}(t)$. Note that lemma 7.2 is needed to show that if there is a judgement $\Gamma \vdash t : \Pi x : A.B$, then there is another $\Gamma \vdash t : \Pi x : A'.B' \wedge \mathcal{F}\mathtt{NF}(\Pi x : A'.B')$. □

The computability of the long $\beta\eta$-normal form of a term is the key result required to derive the properties of the decidability of type inhabitation and typability.

Lemma 7.4 *Given a context Γ and pre-terms t and T, then it is decidable whether*

- *There is a pre-term T' such that there is a typing judgement $\Gamma \vdash t : T'$*
- *There is a typing judgement $\Gamma \vdash t : T$*

Proof The two parts of the lemma are proved simultaneously by induction firstly on the structure of t. The only difficult part of the analysis is checking whether certain terms are $\beta\eta$-equal and this may be done by constructing their long $\beta\eta$-normal forms as in lemma 7.2 □

8 Conclusions and Further Work

There are several principle avenues for further research. Firstly, it remains an open question as to whether the relation $\Rightarrow_{\mathcal{F}}$, proved weakly normalising here, is actually strongly normalising as the author suspects. A second conjecture is that, under minimal conditions, the methods developed in this paper can be generalised to the more abstract Pure Type Systems. Finally, we wish to make good our promise that η-expansions can open the way to the smooth combination of dependent type systems with algebraic rewrite systems. The research required here is at an advanced stage and a paper on the subject is forthcoming.

References

1. Y. Akama. On Mints' reduction for ccc-calculus. In *Typed λ-Calculus and Applications*, volume 664 of *Lecture Notes in Computer Science*, pages 1–12. Springer Verlag, 1993.
2. L. Bachmair and N. Dershowitz. Completion for rewriting modulo a congruence. *Theoretical Computer Science*, 67(2-3):173–202, October 1989.
3. R. Di Cosmo. On the power of simple diagrams. To appear in RTA'96.
4. R. Di Cosmo and D. Kesner. Simulating expansions without expansions. *Mathematical Structures in Computer Science*, 4:1–48, 1994.
5. R. Di Cosmo and D. Kesner. Combining algebraic rewriting, extensional λ-calculi and fixpoints. In *TCS*, 1995.
6. D. Dougherty. Some λ-calculi with categorical sums and products. In *Rewriting Techniques and Applications*, volume 690 of *Lecture Notes in Computer Science*, pages 137–151. Springer Verlag, 1993.
7. G. Dowek. On the defintion of the η-long normal form in type systems of the cube. *Informal proceedings of the 1993 Workshop on Types for Proofs and Programs*, 1993.
8. Alfons Geser. *Relative termination.* Dissertation, Fakultät für Mathematik und Informatik, Universität Passau, Germany, 1990. Also available as: Report 91-03, Ulmer Informatik-Berichte, Universität Ulm, 1991.
9. H. Geuvers. The Church-Rosser property for $\beta\eta$-reduction in typed λ-calculi. In *LICS*, pages 453–460. IEEE, 1992.
10. N. Ghani. Eta expansions in f^ω. To appear in proceedings CSL'96, Utrecht.
11. N. Ghani. *Adjoint Rewriting.* PhD thesis, University of Edinburgh, Department of Computer Science, 1995.
12. N. Ghani. $\beta\eta$-equality for coproducts. In *Typed λ-calculus and Applications*, number 902 in Lecture Notes in Computer Science, pages 171–185. Springer Verlag, 1995.
13. G. Huet. *Résolution d'équations dans des langages d'ordre* $1, 2, \ldots, \omega$. Thèse d'Etat, Université de Paris VII, 1976.
14. C. B. Jay and N. Ghani. The virtues of eta-expansion. *Journal of Functional Programming*, Volume 5(2), April 1995, pages 135-154. CUP 1995.
15. Jean-Pierre Jouannaud and Hélène Kirchner. Completion of a set of rules modulo a set of equations. *SIAM Journal on Computing*, 15(4), November 1986.
16. G. E. Mints. Teorija categorii i teoria dokazatelstv.i. *Aktualnye problemy logiki i metodologii nauky*, pages 252–278, 1979.

17. D. Prawitz. Ideas and results in proof theory. In J.E. Fenstad, editor, *Proc. 2nd Scandinavian Logic Symposium*, pages 235–307. North Holland, 1971.

Table 4. Typing and Equality Judgements for λC

Typing Judgements

$$\frac{}{\vdash * : \Box} \qquad \frac{\Gamma \vdash A : s \quad x^s \notin \mathrm{dom}(\Gamma)}{\Gamma, x : A \vdash x : A}$$

$$\frac{\Gamma \vdash A : B \quad \Gamma \vdash C : s \quad x^s \notin \mathrm{dom}(\Gamma)}{\Gamma, x : C \vdash A : B} \qquad \frac{\Gamma \vdash A : B \quad \Gamma \vdash B' : s \quad \Gamma \vdash B =_{\beta\eta} B'}{\Gamma \vdash A : B'}$$

$$\frac{\Gamma \vdash A : s_1 \quad \Gamma, x : A \vdash B : s_2}{\Gamma \vdash \Pi x : A.B : s_2} \qquad \frac{\Gamma \vdash F : \Pi x : A.B \quad \Gamma \vdash U : A}{\Gamma \vdash FU : B[U/x]}$$

$$\frac{\Gamma, x : A \vdash b : B \quad \Gamma \vdash \Pi x : A.B : s}{\Gamma \vdash \lambda x : A.b : \Pi x : A.B}$$

Equality Judgements

$$\frac{\Gamma \vdash t : C}{\Gamma \vdash t =_{\beta\eta} t} \qquad \frac{\Gamma \vdash t =_{\beta\eta} t'}{\Gamma \vdash t' =_{\beta\eta} t}$$

$$\frac{\Gamma \vdash t =_{\beta\eta} t' \quad \Gamma \vdash t' =_{\beta\eta} t''}{\Gamma \vdash t =_{\beta\eta} t''}$$

$$\frac{\Gamma \vdash t : \Pi x : A.B}{\Gamma \vdash t =_{\beta\eta} \eta^A(t)} \qquad \frac{\Gamma \vdash (\lambda x : A.t)u : C}{\Gamma \vdash (\lambda x : A.t)u =_{\beta\eta} t[u/x]}$$

$$\frac{\Gamma, x : A \vdash t =_{\beta\eta} t'}{\Gamma \vdash \lambda x : A.t =_{\beta\eta} \lambda x : A.t'} \qquad \frac{\Gamma \vdash A =_{\beta\eta} A'}{\Gamma \vdash \lambda x : A.t =_{\beta\eta} \lambda x : A'.t}$$

$$\frac{\Gamma \vdash A =_{\beta\eta} A'}{\Gamma \vdash AB =_{\beta\eta} A'B} \qquad \frac{\Gamma \vdash B =_{\beta\eta} B'}{\Gamma \vdash AB =_{\beta\eta} AB'}$$

$$\frac{\Gamma \vdash A =_{\beta\eta} A'}{\Gamma \vdash \Pi x : A.B =_{\beta\eta} \Pi x : A'.B} \qquad \frac{\Gamma, x : A \vdash B =_{\beta\eta} B'}{\Gamma \vdash \Pi x : A.B =_{\beta\eta} \Pi x : A.B'}$$

Proof Nets, Garbage, and Computations*

S. Guerrini[1], S. Martini[2], A. Masini[3]

[1] IRCS, University of Pennsylvania,
3401 Walnut Street, Suite 400A, Philadelphia – USA;
stefanog@saul.cis.upenn.edu.
[2] Dipartimento di Matematica e Informatica, Università di Udine,
Via delle Scienze, 206, I-33100 Udine – Italy; martini@dimi.uniud.it.
[3] Dipartimento di Informatica, Università di Pisa,
Corso Italia, 40, I-56125 Pisa – Italy; masini@di.unipi.it.

ABSTRACT: We study the problem of local and asynchronous computation in the context of multiplicative exponential linear logic (MELL) proof nets. The main novelty is in a complete set of rewriting rules for cut-elimination in presence of weakening (which requires garbage collection). The proposed reduction system is strongly normalizing and confluent.

KEYWORDS: linear logic; typed lambda-calculus; cut-elimination; sharing graphs; proof nets.

1 Introduction

Cut-elimination (or normalization) is the logical description of the computational process of (term) reduction, central to most of the literature on lambda-calculus and related functional languages. From the pioneering description of beta-reduction in terms of normalization of natural deduction proofs (dating back to the sixties, by Curry, Prawitz, and Howard), this logical interpretation has been extended to a variety of functional languages and formal logical systems.

The arrival on the scene of linear logic [Gir87] gave a good momentum to this research area, for the stress on resource conscious computations. Using the key idea that the type constructor for the function space may be understood as a modal operator (explaining the process of duplicating and/or erasing input data) followed by a linear function, it was discovered soon that typed and untyped lambda-calculus may be faithfully embedded into linear logic, thus allowing the use of linear logic computations (in the form of proof net cut-elimination) to perform (or study) lambda-reduction. A crucial step was the discovery [GAL92] that Lamping's graph-reduction algorithm [Lam90] for optimal lambda-reduction (in the sense of Lévy, [Lév78]) could be interpreted as a way of performing proof net cut-elimination in a distributed and local way. The (global) concept of proof net box is replaced with information distributed on the graph (brackets, croissants, and indices). Cut-elimination is performed

* Partially supported by: HCM Project CHRX-CT93-0046 and CNR GNSAGA (Guerrini,Martini); BRA 8130 LOMAPS (Masini).

with a set of completely local graph-rewriting rules, main part of which are those manipulating the information added to the graph to (dynamically) reconstruct the boxes. The (potential) sharing (expressed with new nodes) of common subgraphs is the key to optimal reduction. Cut-elimination in these *sharing graphs* is based on three main ideas. First, in the reduction of a logical cut involving duplication of information, the duplication is not actually performed. It is instead indicated in a lazy way by the introduction of specific new nodes (*fans*). Second, new reduction rules are added to incrementally perform the required duplication. Third, there is a mechanism to recognize when this process of incremental and distributed duplication is over.

Sharing graphs have been revisited from different perspectives: a categorical interpretation (and new notation) [Asp95]; their extension to other logical systems [AL96]; their relations to the geometry of interaction [ADLR94]; a new notation ensuring better properties (in particular, that the normal form of a sharing graph be a proof net) [GMM96].

All these approaches differ in the specific way the bookkeeping information is coded into the sharing graph. However, they agree on their focus on what in [GMM96] we called *restricted* proof nets: weakening is not allowed. There are at least two reasons for this. First, the problems weakening raises during the reduction of an arbitrary proof net do not show up during the reduction of a lambda-term (better: of the proof net corresponding to a lambda-term), even if weakening *is* allowed in the term. Second, the usual syntax for proof nets do not seem to allow for *any* solution to those problems (see Section 2).

We propose in this paper a set of completely local and asynchronous graph rewriting rules for cut-elimination in proof nets for the Multiplicative Exponential Linear Logic (with weakening and contraction but without constants), MELL. The proposed rules are proved strongly normalizing and confluent. Moreover, the normal form of a sharing graph is a proof net. This generalizes to MELL the results of [GMM96].

As in [GMM96], these results rely on a sharp distinction between logic and control. In standard sharing graphs, the nodes used to control the reduction process (fans, brackets, and croissants) have in fact also a static role, to introduce logical formulas. In our approach, instead, new information, in the form of indexes over formulas, are responsible for the static correctness (that is, for *logic*), while the *control* nodes (muxes) are responsible only for the duplication and reindexing during cut-elimination. This logic vs. control separation is rooted in our previous work on indexed systems for linear logic [MM95].

To treat weakening we exploit a well known *permutability* result: In the sequent calculus formulation of MELL, the weakening rule permutes with all the other rules, and hence it can be pushed upwards, to the axioms. Axioms may then be formulated as

$$\vdash p, p^{\perp}, ?\Gamma,$$

dropping an explicit weakening rule. When expressed in a suitable proof net setting, this idea always generates connected proof nets, allowing a local graph-rewriting cut-elimination. The approach may be seen as a specialization of that of Banach [Ban95].

In our setting, the cut-elimination of a box against a weakening may be performed in two (ideal) phases: first, a marking of the box to erase, keeping intact its logical structure; second, the actual erasing of the box, with the reorganization of its (secondary) doors as weakenings.

The structure of the paper is as follows. Section 2 discusses the problems weakening raises, informally introducing the techniques used in the sequel. Section 3 sets the stage with definitions and the relations with more usual formulations of proof nets. Section 4 introduces the rewriting rules. Section 5 states the main properties of the reduction systems. In Section 6 we discuss some relations with optimality and indicate further research lines. Proofs, or even intelligible sketches of them, are well beyond the space limits of this paper.

2 Weakening in Proof Net Reduction

Weakening in linear logic can produce boxes whose contents are disconnected, and, more subtly, such boxes can be generated by the cut-elimination procedure, even starting from proof nets whose boxes are connected. The crucial case (for cut-elimination) is that of a box whose principal door has as premise a weakening link. A more general situation is depicted in Figure 1 (left) (weakening boxes are not shown). The net σ is a correct proof net. The dotted region on the left is built starting from some weakenings and provides the principal door of a box. In a sequent proof, we would first construct the proof σ; then we would proceed with the weakenings; finally we would build π. We may called the dotted region comprising π a *weakening isle*: it is a separate connected component of the net; it is *not* a proof net by itself; the global correctness of the net is thus guaranteed by the presence of the proof net σ.

Cut now the principal door of the box against a contraction, as shown in Figure 1 (right). The reduction of the cut consists of the (global) duplication of the box, and the replacement of the cut with two cuts. But in sharing graphs boxes are not explicit. They may be reconstructed by means of the auxiliary information (brackets, indices, etc.) *through a graph exploration starting from their principal door*. And then we are lost. No matter which rules will be devised to rewrite the cut, if these rules are to be completely local, the σ part of the graph will never be affected by them and, hence, will not be duplicated. The results of any local rewriting of Figure 1 (right), then, cannot be much different from the graph shown in Figure 2. That graph is not the intended reduct of the cut; moreover is not a proof net, and it cannot be made into one by adding exponential boxes. There are two weakening isles, while only one copy of the proof net σ, which cannot validate both.

To solve the connected components problem we change the definition of the axiom link. Beside the dual atoms p and $p^{\perp}$, we attach to an axiom link also a list of weakening formulas. There is no explicit weakening link. In this way a proof net is always connected and there is hope for a local exploration of its boxes. The reduction of the cut of Figure 1 (b) may now be done in the standard

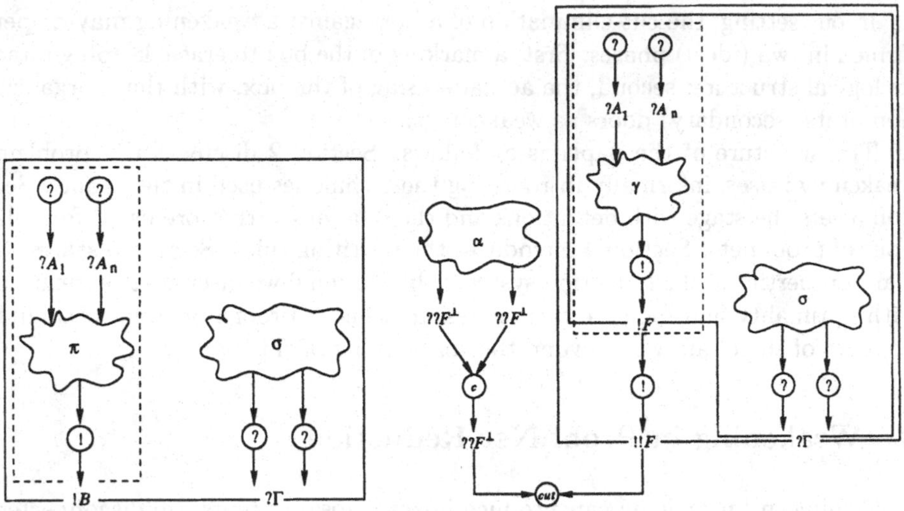

Figure 1. A weakening isle

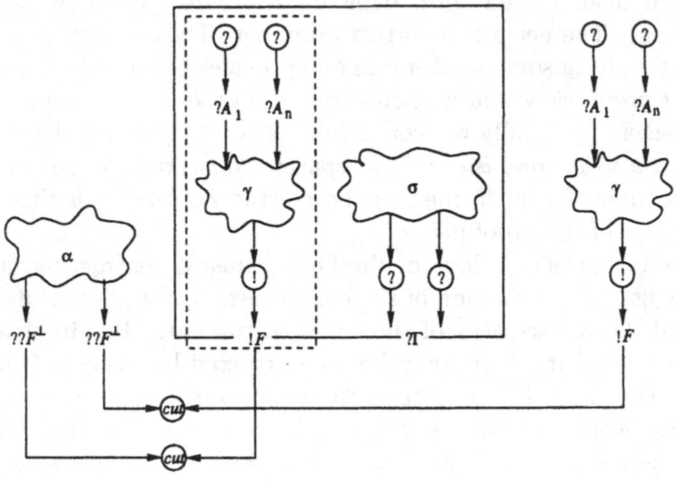

Figure 2. A mistake

sharing graph way: duplicate (and move) the cuts and add a link (a fan, or a *mux* in our terminology) indicating the sharing of the two boxes. The actual duplication will be done incrementally, making the mux travel inside the box. Since any weakening formula is explicitly connected to some axiom, the mux will eventually visit all the box, ensuring the duplication of σ.

This formulation of weakening in proof nets is a variant of the technique introduced by Banach [Ban95]. To prove the so-called sequentialization theorem (that an acyclic and connected proof-structure comes indeed from a sequent linear logic proof and it is thus a proof net) for MELL, Banach introduced the notion of *probe*, an arc pointing "back" from a weakening link to any other link of the net, thus guaranteeing connectedness. In the example of Figure 1 (a), a probe would connect each ? link on top left to one link (anyone is fine) of σ. However, this approach remains too liberal (in the choice of the target link of a probe) for the purpose of a distributed cut-elimination rewriting algorithm. In fact, this freedom is not necessary. Our formulation forces the target of a probe to always be an axiom contained in the same weakening box.

Besides the "weakening isle" case, the other important situation involving weakening is that of a weakening formula (with its probe connecting to an axiom) cut against the principal door of a box, whose reduction is the erasing of the box and the "relocation" of its secondary doors into weakenings. In our approach this will happen in two ideal phases. First (*mark*), weakening and cut are replaced with a mux (connected through the probe to an axiom), which will explore the box, marking the links for deletion, but preserving the logical structure. This mux will stop its marking at the border of the box, like any other mux. Second (*sweep*), starting from the marked axioms, the box will be erased, reducing it to a special "garbage collector" link, which will collect all the secondary doors of the box. At the end, these secondary doors will be transformed into weakenings, with probes toward the axiom connected to the original weakening.

3 Proof Nets and ℓ-nets

3.1 Leveled Structures

We introduce the basic concepts we will use in the paper. The basic notion is that of sharing ℓ-structure, a box-free representation of (shared) proof structures.

Definition 3.1 ($s\ell$-structure). An *$s\ell$-structure* (sharing leveled structure of links) is a finite connected hypergraph whose nodes are labeled with indexed formulas (either a MELL formula or the extra-logical constant $\varnothing$ (*dummy*), decorated with a natural number, the *level* of the formula) and whose hyperedges (also called *links*) are labeled from the set $\{\mathsf{cut}, \mathsf{ax}, \wp, \otimes, !, ?, \bullet\} \cup \{\mathsf{mux}[i] \mid i \geq 0\} \cup \{\mathsf{demux}[i] \mid i \geq 0\} \cup \{\mathsf{gc}[i] \mid i \geq 0\}$; the integer i in (de)muxes and gc's is the *threshold* of the link. Allowed links and nodes are drawn in Figure 3. The source nodes of a link are its *premises*; the target nodes are the *conclusions*. Premises and conclusions are assumed to be distinguishable (i.e., we will have left/right premises, i-th conclusion and so on), with the exception of $\bullet$-links. Those nodes that are not premises of any link are the *net conclusions*; unary (de)muxes are also called *lifts*. Each node is the conclusion of exactly one link and premise of at most one link. For all the links (but gc, mux, demux and weakening) we have the constraint that if a premise/conclusion formula is $\varnothing$ then all the other connected nodes must also be labeled with $\varnothing$. We distinguish the premises (conclusions) of muxes (demuxes) with names (which we call *ports* and are denoted in Figure 3 with $(a_1 \cdots a_k)$); moreover, each port is of kind *identity* or *garbage*.

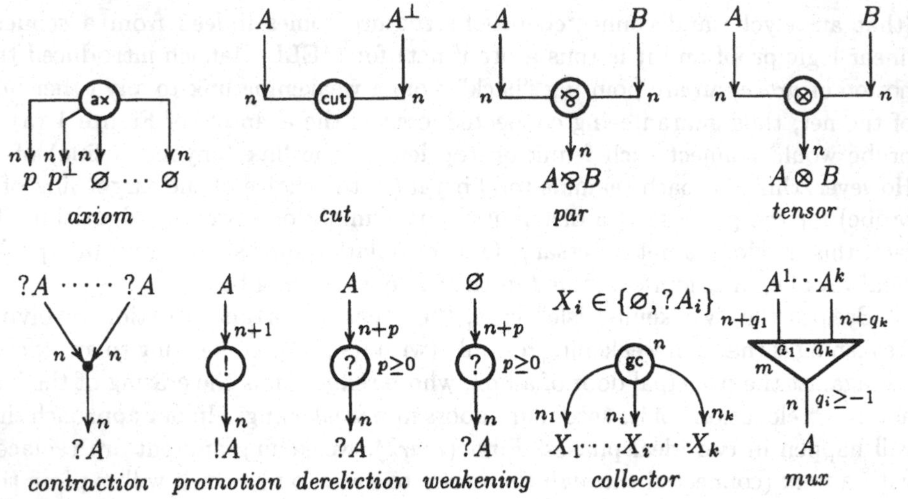

Figure 3. *sℓ*-structure links

In Figure 3 we have represented a generic mux/demux; more precisely in a mux (demux) the nodes $A^1, \ldots, A^k$ are premises (conclusions) and the node A is a conclusion (premise).

With respect to proof nets, *sℓ*-structures have two additional types of link (mux/demux and gc) and a new constant (∅). The ∅ it is used (at first) for introducing weakening formulas. Indeed, the key technical point of our approach is to avoid nullary premise links to introduce weakening formulas. A ∅ premise (always connected to an axiom) distinguishes a ?-link used as a weakening from a ?-link used as ?-introduction. During the cut-elimination process, moreover, ∅ will be used to mark those parts of the proof-structure that have to be discarded (because cut against a weakening).

Muxes (introduced in [Gue96, GMM96]) are responsible for the processes of:

1. reindexing of formulas (that is, the local re-computation of boxes during reduction);
2. local (lazy) duplication;
3. marking of garbage.

The link gc (garbage collector) is designed to collect the garbage—to remove from the net those nodes which have been marked ∅.

Definition 3.2 (proof ℓ-structure). A *proof ℓ-structure* is an *sℓ*-structure without (de)muxes and gc links.

3.2 Decoration

It is well known (see [Gir87], pag. 63) that proof nets may be formulated with *several weakenings in the same weakening box.* With this formulation, it is easy to show (a trivial induction) that by suitable permutations any proof net can be transformed into an equivalent proof net in which each weakening box contains exactly one axiom link as interior. Let $\mathcal{PN}$ be the set of proof nets with such a structure. Since any weakening box contains exactly one axiom link, we may forget the boxes and simply record the weakening formulas with each axiom. Only exponential boxes survive in $\mathcal{PN}$.

We will now show how to associate to each $\mathsf{P} \in \mathcal{PN}$ a (unique!) proof ℓ-structure $\mathscr{D}[\mathsf{P}]$, the *decoration* of P. The proof ℓ-net $\mathscr{D}[\mathsf{P}]$ is obtained by applying the following steps:

1. assign to each node of P a level, corresponding to the number of exponential boxes containing that node;
2. connect each weakening $?A$ to the axiom α belonging to the weakening box of $?A$ by means of a node labeled $\varnothing$;
3. set the level of this $\varnothing$ premise to the number of exponential boxes containing α.

Definition 3.3. A proof ℓ-structure S is a *proof ℓ-net* iff $\mathsf{S} = \mathscr{D}[\mathsf{P}]$ for some $\mathsf{P} \in \mathcal{PN}$.

By using indexes it is possible to "recognize" exponential boxes:

Definition 3.4. Let S be a proof ℓ-structure and let A^k be a premise of a !-link; we call *box* of A^k a sub-hypergraph $\mathsf{bx_S}[A^k]$ of S verifying the following properties:

1. $A^k \in \mathsf{bx_S}[A^k]$ (A^k is the *principal door* of $\mathsf{bx_S}[A^k]$);
2. $\mathsf{bx_S}[A^k]$ is a proof ℓ-net;
3. each net conclusion of $\mathsf{bx_S}[A^k]$ different from the principal door is a premise, in S, of a ?-link with conclusion at level $j < k$ (such ?-premises are the *secondary doors* of the box);
4. for each $B^j \in \mathsf{S}$, if $B^j \in \mathsf{bx_S}[A^k]$, then $j \geq k$.

We denote by BX[S] the set of boxes of S. Because of the definition of ℓ-structure, boxes are connected.

4 Reduction

The new information we added to proof nets allows a clear notion of local and asynchronous computation as a fully distributed execution of the standard cut-elimination process (as defined by [Gir87]). The reduction procedure is described by rules of three kinds: logical reduction (β_l), bookkeeping (π), garbage collection ($\sigma_\varnothing$). The proposed approach extends our previous work [GMM96].

4.1 Logical reduction and bookkeeping

The rules β_ι, drawn in Figure 4, implement a local version of the usual cut-elimination process (which we indicate as β_s). The only difference with this usual process is when an exponential cut is reduced. In this case, no duplication, reindexing, or erasing of boxes is done. Such operations are only indicated by the introduction of suitable muxes.

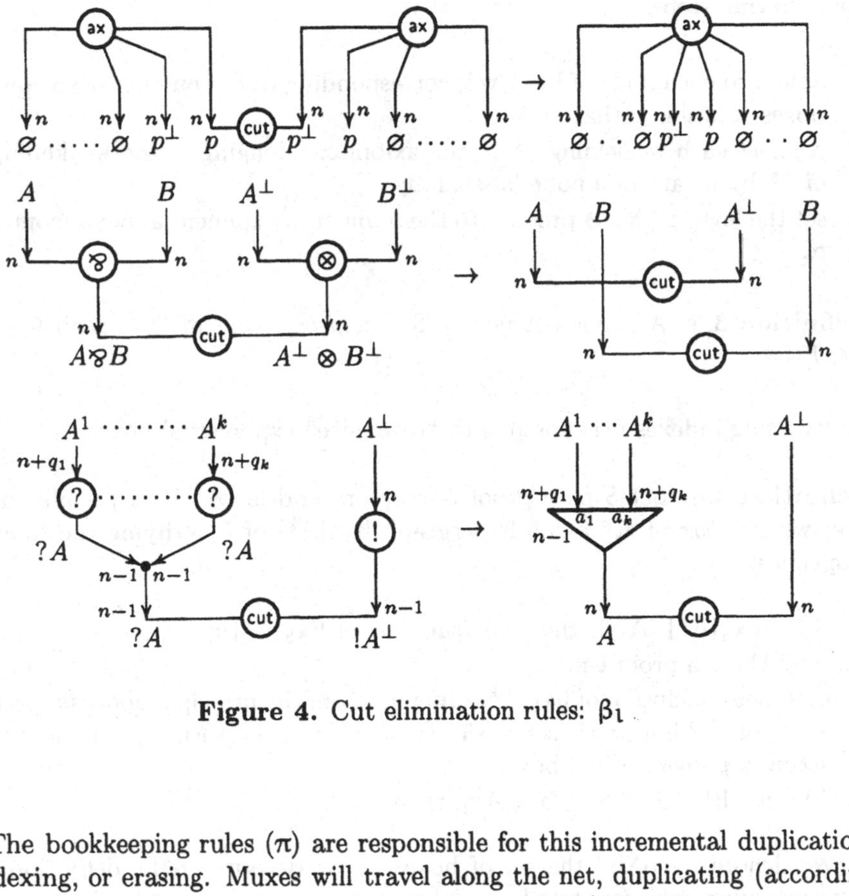

Figure 4. Cut elimination rules: β_ι

The bookkeeping rules (π) are responsible for this incremental duplication, reindexing, or erasing. Muxes will travel along the net, duplicating (according to the mux arity) and reindexing all the links they find (Figure 5 shows a generic reduction, where $\star$ stands for any link, but mux).

When one mux encounters another mux, there are two cases, see Figure 6. In the first one, the two muxes had been generated by the same exponential cut (a fact testified by the same threshold for the two muxes): each one has exhausted its job and they disappear. In the second, the two muxes had been generated by two cuts (and hence they have different thresholds): each mux duplicate the other one.

The propagation of muxes ends when either they annihilate by the rule just seen in Figure 5, or when they reach an auxiliary door of the box on which they

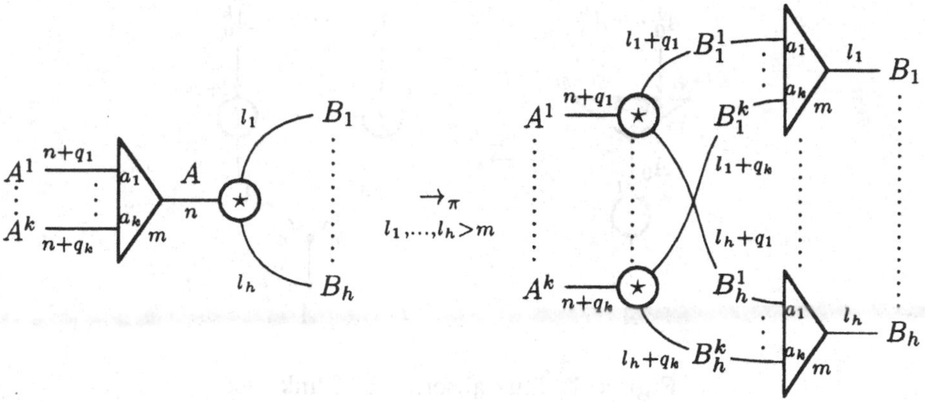

Figure 5. Mux propagation

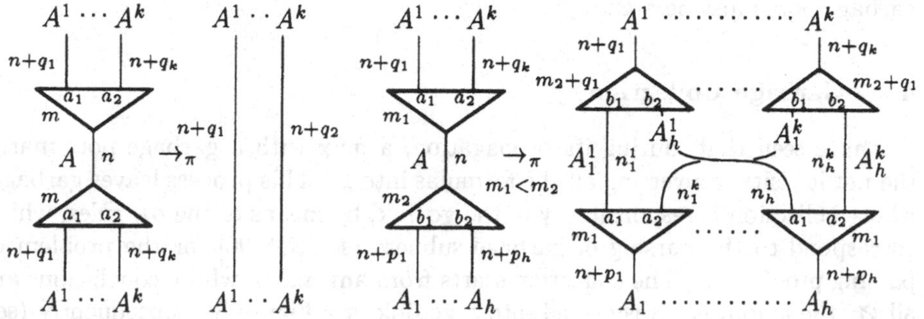

Figure 6. Mux interaction

operate. In this case the mux disappears, absorbed by the corresponding ? links, see Figure 7.

It remains to be discussed how this bookkeeping handles the marking of boxes to be erased (because cut against weakenings). Here come to the stage the *ports* of the muxes (we recall that any mux premise has a name and a kind (*identity* or *garbage*); we call *port* this information). Let us consider the last rule of Figure 4, the creation of a demux.

A port a_i is of kind *garbage* iff the formula A^i is a $\varnothing$ premise of a weakening link (which means that A^i is $\varnothing$ and is the conclusion of an axiom link); otherwise a_i will be an identity port.

Let us consider now Figure 5 and let $A^i = \varnothing$ and let the kind of a_i be *garbage*. We stipulate that in this case, after the reduction, all the $B^i_1, \ldots, B^i_h$ formulas are $\varnothing$. Otherwise (i.e., if a_i is an *identity* port) $B^i_1, \ldots, B^i_h$ are syntactically equal to $B_1, \ldots, B_h$ respectively. The same convention applies to all the other rules involving a mux propagation (Figures 6 and 10).

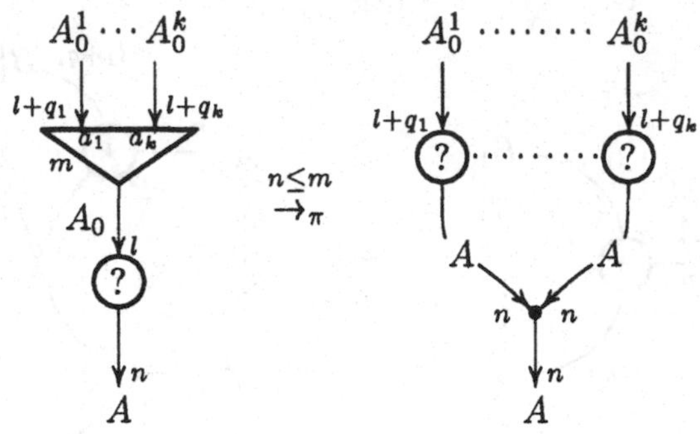

Figure 7. Mux absorption: ? link case

Remark 4.1. A simple rule inspection shows that each node connected to a garbage port must be a $\varnothing$.

4.2 Garbage Collection

We have seen that, during its propagation, a mux with a garbage port marks the net it visits, converting all the formulas into $\varnothing$. This process leaves garbage, whose collection is responsibility of the **gc** link, by means of the $\sigma_\varnothing$ rules, which correspond to the parsing of garbage subnets (see [GM96] for the problem of parsing proof nets). The collection starts from any axiom whose conclusions are all $\varnothing$: the axiom is transformed into a **gc** link, see Figure 8. Subsequently (see Figure 9), the **gc** link keeps "eating" the dummy marked net, collapsing it into a single **gc** link and collecting all the secondary doors of the box to be deleted (see the last rule of Figure 9).

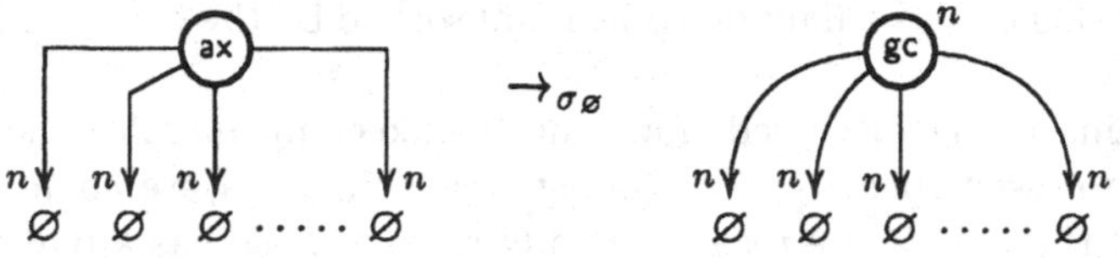

Figure 8. Sweep rules: start collecting

Note that gc does not delete muxes. The interaction between muxes and gc's is regulated by the rule in Figure 10.

The collection ends when the box to be erased is collapsed into a unique gc node. The conclusions of this node are all the secondary doors of the box, plus a single $\varnothing$ conclusion, cut against another $\varnothing$ (coming from the original

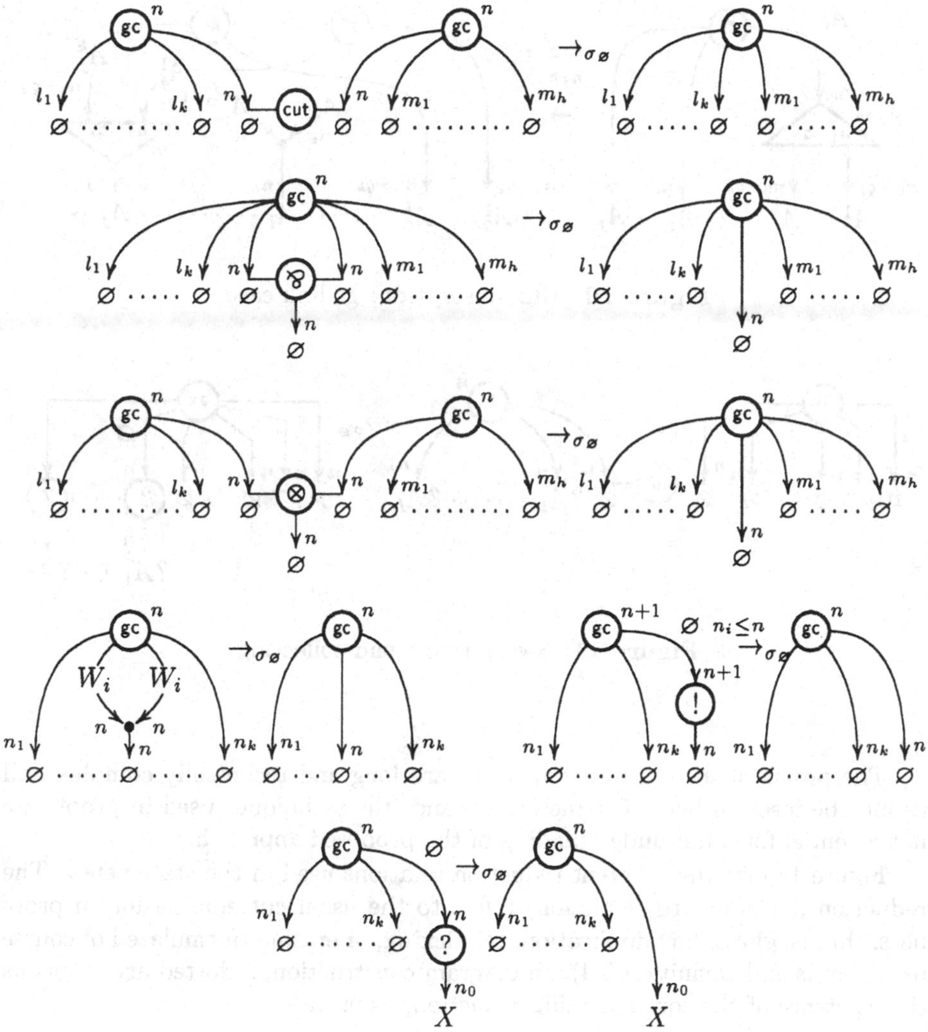

Figure 9. Sweep rules

weakening). This configuration and the corresponding rewriting are depicted in Figure 11. Observe that the gc link disappears and that the secondary doors of the collected box are transformed into weakenings.

5 Results

We may summarize the main result of the paper as: *"cut elimination (with garbage collection) in proof nets may be performed in a completely local and asynchronous way"* .

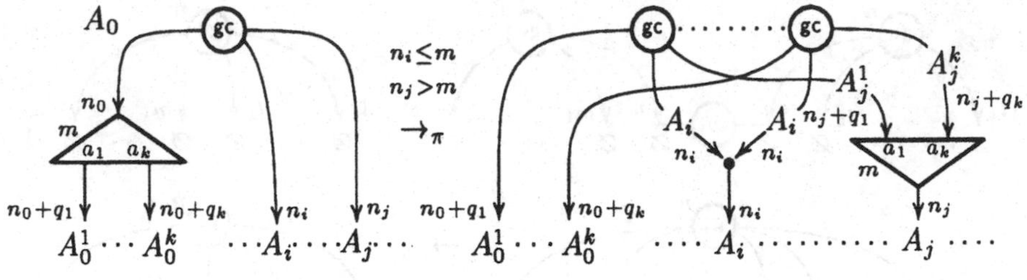

Figure 10. Mux absorption: gc link case

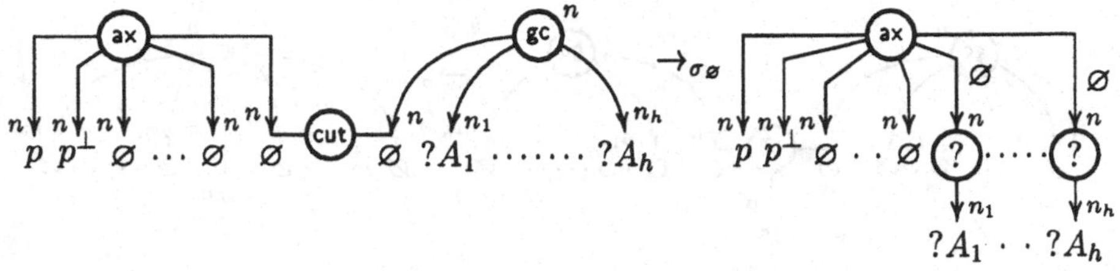

Figure 11. Sweep rules: end collecting

The proofs of theorems stated below are long and technically complex and cannot be inserted here. On the other hand, the techniques used in proofs are not essential for a full understanding of the proposed approach.

Figure 12 lists the different reduction relations used in the statements. The reduction β_s (standard reduction) refers to the usual cut elimination in proof nets, that is, global box duplication, reindexing, or erasing (formulated of course using levels and Definition 3.4). In diagram construction, a dotted arrow means the existence of the corresponding reduction, as usual.

arrow	reduction
$\twoheadrightarrow$	β_s (standard)
$\rightarrow\!\!\triangleright$	β_l (local)
$\Longrightarrow$	$\pi + \sigma_\varnothing$
$\longrightarrow$	$\pi + \sigma_\varnothing + \beta_l$

Figure 12. One-step reductions

The relevant $s\ell$-structures we are interested in are the ones arising along the reduction of a proof ℓ-net.

Definition 5.1 (proof $s\ell$-net). An $s\ell$-structure G is a *proof $s\ell$-net* if there exists a reduction $\mathsf{N} \longrightarrow^{*} \mathsf{G}$, for some proof ℓ-net N.

Theorem 5.2 (existence and uniqueness of readback). *For any proof $s\ell$-net* G, *we have that:*

1. *The $\pi + \sigma_{\varnothing}$ rules are strongly normalizing and confluent on* G.
2. *The normal form of* G *is a proof ℓ-net.*

Definition 5.3 (readback). The *readback* $\mathcal{R}(\mathsf{G})$ of an $s\ell$-net G is the $\pi + \sigma_{\varnothing}$ normal form of G.

Theorem 5.4 (standard strategy). *For any β redex of a proof ℓ-net* N *we have that:*

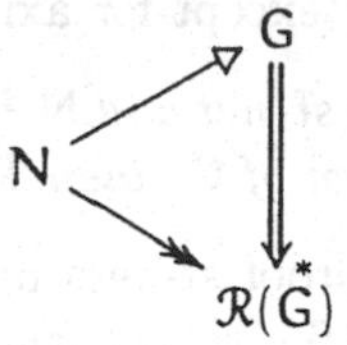

Theorem 5.5 (coherence). *For any proof $s\ell$-net* G *and any proof ℓ-net* N *for which* $\mathsf{N} \longrightarrow^{*} \mathsf{G}$, *we have that:*

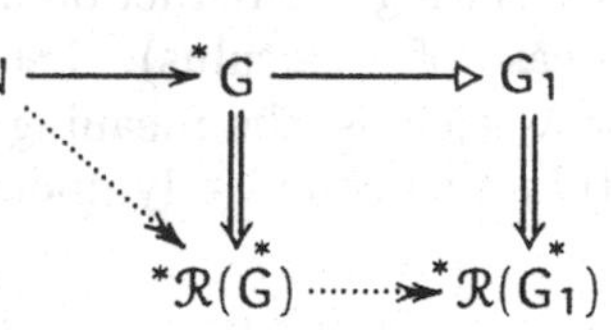

Remark 5.6. The previous result differs from the analogous of [GMM96] in the fact that we cannot ensure that to a β_{ℓ} contraction of G corresponds a non-empty β_{s} reduction of $\mathcal{R}(\mathsf{G})$. In fact, G may not be garbage free, since it may contain part of a box to be erased. Hence, the contraction of a cut in such a part has no correspondence in $\mathcal{R}(\mathsf{G})$. As a consequence, the proof of the strong normalization property becomes more involved.

Theorem 5.7 (strong normalization). *There is no infinite reduction of a proof $s\ell$-net.*

Theorem 5.8 (confluence). *For any proof $s\ell$-net* G *such that* $\mathsf{G} \longrightarrow^{*} \mathsf{G}_1$ *and* $\mathsf{G} \longrightarrow^{*} \mathsf{G}_1$, *there exists an $s\ell$-net* G_0 *such that* $\mathsf{G}_1 \longrightarrow^{*} \mathsf{G}_0$ *and* $\mathsf{G}_2 \longrightarrow^{*} \mathsf{G}_0$.

Corollary 5.9 (unique normal form). *The $\beta_{\ell} + \pi + \sigma_{\varnothing}$ normal form of a proof ℓ-net* N *is unique and coincides with its β_{s} normal form.*

6 Conclusions

6.1 Discussion

We have presented a distributed and local graph-rewriting algorithm for MELL proof nets. The relations of this work with the ideas and techniques developed for the optimal reduction of interaction nets have been discussed in the Introduction. One may wonder why there is no statement on this subject in the paper. The crucial reason is that, in presence of weakening, the very notion of optimal reduction for proof nets is highly an open question.

It is clear that the mux propagation rule, when applied with a duplicating mux facing the premise of a logical node, would duplicate any redex in its scope, thus destroying any optimality. It is easy to split the propagation rule in two rules, one for when the mux faces a premise (call it the *dup* rule) and one for when the mux faces the conclusion (the *odup* rule). Let now π_{opt} be the subset of π including *odup* but not *dup* (except for axioms, cuts, and gc).

Theorem 6.1. *Let* G *be a proof sℓ-net and* N *be its* $\beta+\pi+\sigma$ *normal form. Let* G′ *be a* $\beta+\pi_{opt}+\sigma$ *normal form of* G, *then* $\mathcal{R}(G') = N$.

Therefore, normalization of proof sℓ-nets may be performed in two distinct steps: first "optimal" reduction ($\beta+\pi_{opt}+\sigma$), then read-back reduction. However, this theorem only warrants that no duplication of redexes is done, but nothing is said about the *need* of such redexes (that is, whether these redexes belong to a part of the net which will not be erased). We are missing, in other words, an effective strategy ensuring the reduction of only needed cuts (like the left-most-outermost in the case of λ-calculus). Let us note, incidentally, that even for sharing graphs for λ-calculus, the meaning of an optimal reduction in presence of weakening is still a non completely understood subject.

6.2 Further Work

All the results of this paper hold for full MELL, i.e., MELL with the two constant $\perp, 1$. Such an extension will be developed in a forthcoming full paper. The basic ideas are the following. The constant 1 is introduced by means of a new axiom link, treated as all the other axioms. The $\perp$ constant is treated like weakening formulas, namely: (1) there is a *bottom* link with a $\varnothing$ premise (connected to an axiom) and the $\perp$ constant as conclusion; (2) the concept of box is extended in order to allow $\perp$ as secondary door (such an extension do not increase the expressive power of the logic, [GM96]). All the results stated here extend rather simply to full MELL.

We plan to apply the proposed approach to the case of functional languages (pure and typed λ–calculi) both from a theoretical and practical (implementative) point of view [AG97].

It would be interesting to develop a semantics of the reductions of this papers, along the lines of the geometry of interaction [Gir89], or of the consistent path semantics for sharing graphs of [ADLR94]. For the weakening-free case, see [Gue96].

References

ADLR94. Andrea Asperti, Vincent Danos, Cosimo Laneve, and Laurent Regnier. Paths in the lambda-calculus: three years of communications without understanding. In *Proceedings of 9th Annual Symposium on Logic in Computer Science*, pages 426–436, Paris, France, July 1994. IEEE.

AG97. Andrea Asperti and Stefano Guerrini. *The Optimal Implementation of Functional Programming Languages.* Cambridge University Press, 1997. To appear.

AL96. Andrea Asperti and Cosimo Laneve. Interaction system II: The practice of optimal reductions. *Theoretical Computer Science*, 159(2):191–244, 3 June 1996.

Asp95. Andrea Asperti. Linear logic, comonads and optimal reductions. *Fundamenta infomaticae*, 22:3–22, 1995.

Ban95. Richard Banach. Sequent reconstruction in LLM—A sweepline proof. *Annals of Pure and Applied Logic*, 73:277–295, 1995.

GAL92. Georges Gonthier, Martín Abadi, and Jean-Jacques Lévy. Linear logic without boxes. In *Proceedings of 7th Annual Symposium on Logic in Computer Science*, pages 223–234, Santa Cruz, CA, June 1992. IEEE.

Gir87. Jean-Yves Girard. Linear logic. *Theoretical Computer Science*, 50(1):1–102, 1987.

Gir89. Jean-Yves Girard. Towards a geometry of interaction. Number 92 in Contemporary Mathematics, pages 69–108. AMS, 1989.

GM96. Stefano Guerrini and Andrea Masini. Parsing MELL proof nets. Technical report, Dipartimento di Informatica, Pisa, 1996.

GMM96. Stefano Guerrini, Simone Martini, and Andrea Masini. Coherence for sharing proof-nets. In Harald Ganzinger, editor, *RTA '96*, number 1103 in LNCS, pages 215–229, New Brunswick, NJ, July 1996. Springer-Verlag.

Gue96. Stefano Guerrini. *Theoretical and Practical Issues of Optimal Implementations of Functional Languages.* Phd thesis, Dipartimento di Informatica, Pisa, 1996. TD-3/96.

Lam90. John Lamping. An algorithm for optimal lambda calculus reduction. In *Proceedings of Seventeenth Annual ACM Symposyum on Principles of Programming Languages*, pages 16–30, San Francisco, California, January 1990. ACM.

Lév78. Jean-Jacques Lévy. *Réductions Correctes et Optimales dans le lambda-calcul.* PhD Thesis, Université Paris VII, 1978.

MM95. S. Martini and A. Masini. On the fine structure of the exponential rule. In J.-Y. Girard, Y. Lafont, and L. Regnier, editors, *Advances in Linear Logic*, pages 197–210. Cambridge University Press, 1995. Proceedings of the Workshop on Linear Logic, Ithaca, New York, June 1993.

Recursion from Cyclic Sharing: Traced Monoidal Categories and Models of Cyclic Lambda Calculi

Masahito Hasegawa

LFCS, Department of Computer Science, University of Edinburgh
JCMB, King's Buildings, Edinburgh EH9 3JZ, Scotland
Email: mhas@dcs.ed.ac.uk

Abstract. *Cyclic sharing* (cyclic graph rewriting) has been used as a practical technique for implementing recursive computation efficiently. To capture its semantic nature, we introduce categorical models for *lambda calculi with cyclic sharing* (cyclic lambda graphs), using *notions of computation* by Moggi / Power and Robinson and *traced monoidal categories* by Joyal, Street and Verity. The former is used for representing the notion of sharing, whereas the latter for cyclic data structures. Our new models provide a semantic framework for understanding recursion created from cyclic sharing, which includes traditional models for recursion created from fixed points as special cases. Our cyclic lambda calculus serves as a uniform language for this wider range of models of recursive computation.

1 Introduction

One of the traditional methods of interpreting a recursive program in a semantic domain is to use the least fixed-point of continuous functions. However, in the real implementations of programming languages, we often use some kind of shared cyclic structure for expressing recursive environments efficiently. For instance, the following is a call-by-name operational semantics of the recursive call, in which free x may appear in M and N. We write $E \vdash M \Downarrow V$ for saying that evaluating a program M under an environment E results a value V; in a call-by-name strategy an environment assigns a free variable to a pair consisting of an environment and a program.

$$\frac{E' \vdash N \Downarrow V \quad \text{where } E' = E \cup \{x \mapsto (E', M)\}}{E \vdash \mathsf{letrec}\ x = M\ \mathsf{in}\ N \Downarrow V}$$

That is, evaluating a recursive program letrec $x = M$ in N under an environment E amounts to evaluating the subprogram N under a cyclic environment E' which references itself. One may see that it is reasonable and efficient to implement the recursive (self-referential) environment E' as a cyclic data structure as below.

or equivalently

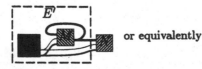

Also it is known that if we implement a programming language using the technique of sharing, the use of the fixed point combinator causes some unexpected duplication of resources; it is more efficient to get recursion by cycles than by the fixed point combinator in such a setting. This fact suggests that there is a gap between the traditional approach based on fixed points and cyclically created recursion.

The aim of this paper is to introduce semantic models for understanding recursive computation created by such a cyclic data structure, especially cyclic lambda graphs as studied in [AK94]. Our task is to deal with the notion of values/non-values (which provides the notion of sharing) and the notion of cycles at the semantic level. This is done by combining Moggi's *notions of computation* [Mog88] and the notion of *traced monoidal categories* recently introduced by Joyal, Street and Verity [JSV96]. The former has been used for explaining computation and values systematically, which we apply for interpreting the notion of sharing. The latter has originally been invented for analyzing cyclic structures arising from mathematics and physics, notably knot theory (e.g. [RT90]); it is then natural to use this concept for modeling cyclic graph structure. We claim that our new models are natural objects for the study of recursive computation because they unify several aspects on recursion in just one semantic framework. The leading examples are

- the *graphical (syntactical) interpretation* of recursive programs by cyclic data structures motivated as above,
- the *domain-theoretic interpretation* in which the meaning of a recursive program $\mathsf{letrec}\ x = F(x)\ \mathsf{in}\ x$ is given by the least fixed point $\bigcup_n F^n(\perp)$, and
- the *non-deterministic interpretation* where the program $\mathsf{letrec}\ x = F(x)\ \mathsf{in}\ x$ is interpreted by $\{x \mid x = F(x)\}$, the set of all possible solutions of the equation $x = F(x)$.

Each of them has its own strong tradition in computer science. However, to our knowledge, this is the first attempt to give a uniform account on these well-known, but less-related, interpretations. Moreover, our cyclic lambda calculus serves as a uniform language for them.

Construction of this paper

We recall the definition of traced monoidal categories in Section 2. In Section 3 we observe that traces and fixed point operators are closely related in two practically interesting situations - in cartesian categories, and in a special form of monoidal adjunction known as *notions of computation*.The motivating examples above are shown to be instances of our setting. Armed with these semantic observations, in Section 4 we give the models for two simply typed lambda calculi with cyclic sharing - one with unrestricted substitution, and the other with restricted substitution on values. The two settings studied in the previous section correspond to the models of these calculi respectively; the soundness and completeness results are stated. As an application, we analyze fixed point operators definable in our calculi (Section 5).

Related work

On fixed point operators. Axiomatizations of feedback operators similar to Theorem 3.1 have been given by Bloom and Ésik in [BÉ93] where they study the dual situation (categories with coproducts). Also the same authors have considered a similar axiomatization of fixed point operators in cartesian closed categories [BÉ96]. Ignoring the difference of presentations, their "Conway cartesian categories" exactly correspond to traced cartesian categories (see the remark after Theorem 3.1). Their "Conway cartesian closed categories" are then traced cartesian closed categories with an additional condition called "abstraction identity".

On cyclic lambda calculi. Our source of cyclic lambda calculi is the version presented in [AK94]. The use of the letrec-syntax for representing cyclic sharing is not new; our presentation is inspired by a graph rewriting system in [AA95] and the call-by-need $\lambda_{\mathrm{letrec}}$-calculus in [AF96]. In this paper we concentrate on the equational characterization of the calculi; the connection between rewriting-theoretic aspects and our work remains as an important future issue. We think the relation to operational semantics should be established in this direction, especially in the connection with the call-by-need strategy [Lau93, AF96]. Also we note that our approach is applicable not only to cyclic lambda calculi but also to general cyclic graph rewriting systems.

On action calculi. The syntax and models in this paper have arisen from the study of Milner's *action calculi* [Mil96], a proposed framework for interactive computation. The use of notions of computation as models of *higher-order action calculi* [Mil94a] is developed in [GH96], whereas the relation between traced categories and *reflexive action calculi* [Mil94b] is studied by Mifsud [Mif96] and the author - axioms for reflexion are proved to be equivalent to those of trace. In fact, our cyclic lambda calculus can be seen as a fragment of a higher-order reflexive action calculus. A further study of action calculi in this paper's direction will appear in the author's forthcoming thesis (also see Example 3.10).

On Geometry of Interaction. It has been pointed out that several models of *Geometry of Interaction* [Gir89] can be regarded as traced monoidal categories (see Abramsky's survey [Abr96]). We expect that there are potential applications of our results in this direction.

2 Traced Monoidal Categories

The notion of trace we give here for symmetric monoidal categories is adopted from the original definition of traces for balanced monoidal categories [JS93] in [JSV96].

For ease of presentation, in this section we write as if our monoidal categories are strict (i.e. monoidal products are strictly associative and coherence isomorphisms are identities).

Definition 2.1. (Traced symmetric monoidal categories [JSV96])
A symmetric monoidal category $(\mathcal{T}, \otimes, I, c)$ (where c is the symmetry; $c_{X,Y} : X \otimes Y \longrightarrow Y \otimes X$) is said to be *traced* if it is equipped with a natural family of functions, called a *trace*,

$$Tr^X_{A,B} : \mathcal{T}(A \otimes X, B \otimes X) \longrightarrow \mathcal{T}(A, B)$$

subject to the following three conditions.

- **Vanishing:**

$$Tr^I_{A,B}(f) = f : A \to B$$

  where $f : A \longrightarrow B$, and

$$Tr^{X \otimes Y}_{A,B}(f) = Tr^X_{A,B}(Tr^Y_{A \otimes X, B \otimes X}(f)) : A \longrightarrow B$$

  where $f : A \otimes X \otimes Y \longrightarrow B \otimes X \otimes Y$
- **Superposing:**

$$Tr^X_{C \otimes A, C \otimes B}(id_C \otimes f) = id_C \otimes Tr^X_{A,B}(f) : C \otimes A \longrightarrow C \otimes B$$

  where $f : A \otimes X \longrightarrow B \otimes X$
- **Yanking:**

$$Tr^X_{X,X}(c_{X,X}) = id_X : X \longrightarrow X \qquad ■$$

We present the graphical version of these axioms to help with the intuition of traced categories as categories with cycles (or feedback, reflexion). Such graphical languages for various monoidal categories have been developed in [JS91].

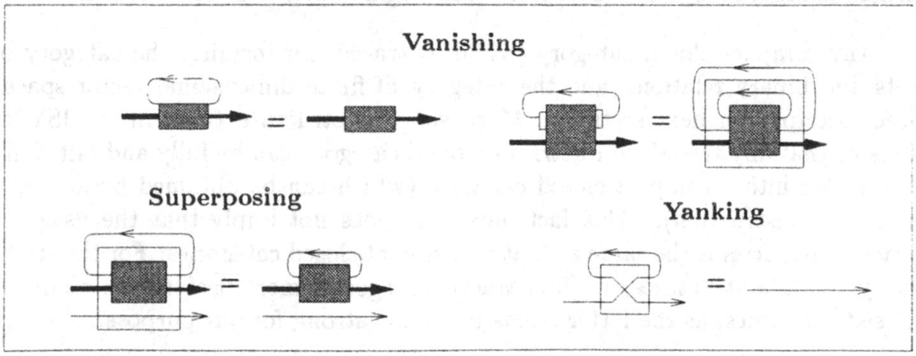

Note that naturality of a trace can be axiomatized as follows.

- Naturality in A (**Left Tightening**)

$$Tr^X_{A,B}((g \otimes id_X); f) = g; Tr^X_{A',B}(f) : A \longrightarrow B$$

  where $f : A' \otimes X \longrightarrow B \otimes X$, $g : A \longrightarrow A'$

- Naturality in B **(Right Tightening)**

$$Tr^X_{A,B}(f;(g\otimes id_X)) = Tr^X_{A,B'}(f);g : A \longrightarrow B$$

where $f : A\otimes X \longrightarrow B'\otimes X$, $g : B' \longrightarrow B$

- Naturality in X **(Sliding)**

$$Tr^X_{A,B}(f;(id_B\otimes g)) = Tr^{X'}_{A,B}((id_A\otimes g);f) : A \longrightarrow B$$

where $f : A\otimes X \longrightarrow B\otimes X'$, $g : X' \longrightarrow X$

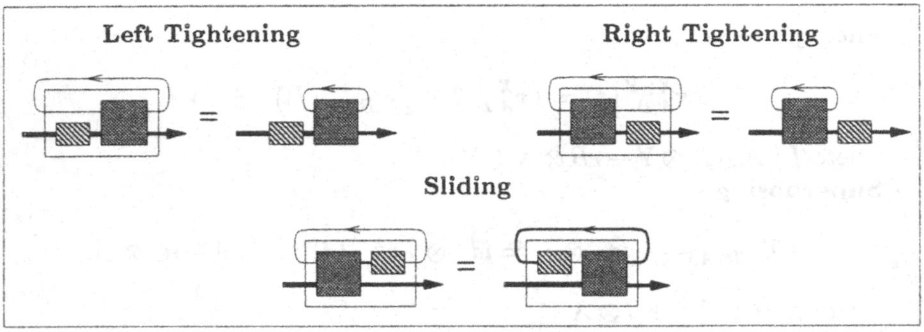

Remark 2.2. The axiom **Superposing** is slightly simplified from the original version in [JSV96]

$$Tr^X_{A\otimes C,B\otimes D}((id_A\otimes c_{C,X});(f\otimes g);(id_B\otimes c_{X,D})) = Tr^X_{A,B}(f)\otimes g$$

where $f : A\otimes X \longrightarrow B\otimes X$, $g : C \longrightarrow D$. Assuming axioms **Left & Right Tightenings**, ours is derivable from this original one, and vice versa. ■

Any compact closed category [KL80] is traced, for instance the category of sets and binary relations, and the category of finite dimensional vector spaces (see examples in next section). Moreover, the structure theorem in [JSV96] tells us that any traced symmetric monoidal category can be fully and faithfully embedded into a compact closed category (which can be obtained by a simple fraction construction). This fact, however, does not imply that the usage of traced categories is the same as that of compact closed categories. For the study of cyclic data structures, we find traced categories more useful than compact closed categories, as the latter seems to be too strong for our purpose.

3 Recursion from Traces

In this section we observe that traced categories can support recursive computation under certain circumstances. These results form the basis of our semantic analysis of "recursion created by cyclic structures" where we regard traced categories as the models of cyclic sharing.

3.1 Fixed Point Operators in Traced Cartesian Categories

Compact closed categories whose monoidal product is cartesian are trivial. This is not the case for traced categories. In fact, in [JSV96] it is shown that the category of sets and binary relations with its biproduct as the monoidal product is traced. Actually we find traced cartesian categories interesting in the context of semantics for recursive computation:

Theorem 3.1. *A cartesian category $\mathcal{C}$ is traced if and only if it has a family of functions*

$$(-)^{\dagger_{A,X}} : \mathcal{C}(A \times X, X) \longrightarrow \mathcal{C}(A, X)$$

(in below, parameters A, X may be omitted) such that

1. *$(-)^\dagger$ is a parametrized fixed point operator; for $f : A \times X \longrightarrow X$, $f^\dagger : A \longrightarrow X$ satisfies $f^\dagger = \langle id_A, f^\dagger \rangle; f$.*
2. *$(-)^\dagger$ is natural in A; for $f : A \times X \longrightarrow X$ and $g : B \longrightarrow A$, $((g \times id_X); f)^\dagger = g; f^\dagger : B \longrightarrow X$.*
3. *$(-)^\dagger$ is natural in X; for $f : A \times X \longrightarrow Y$ and $g : Y \longrightarrow X$, $(f; g)^\dagger = ((id_A \times g); f)^\dagger; g : A \longrightarrow X$.*
4. *$(-)^\dagger$ satisfies Bekič's lemma; for $f : A \times X \times Y \longrightarrow X$ and $g : A \times X \times Y \longrightarrow Y$, $\langle f, g \rangle^\dagger = \langle id_A, (\langle id_{A \times X}, g^\dagger \rangle; f)^\dagger \rangle; \langle \pi'_{A,X}, g^\dagger \rangle : A \longrightarrow X \times Y$.*

Sketch of the proof: From a trace operator Tr, we define a fixed point operator $(-)^\dagger$ by

$$f^\dagger \;=\; Tr^X(\langle f, f \rangle) \;:\; A \longrightarrow X$$

for $f : A \times X \longrightarrow X$. Conversely, from a fixed point operator $(-)^\dagger$ we define a trace Tr by

$$Tr^X(f) \;=\; \langle id_A, (f; \pi'_{B,X})^\dagger \rangle; f; \pi_{B,X} \;:\; A \longrightarrow B$$

(equivalently $((id_A \times \pi'_{B,X}); f)^\dagger; \pi_{B,X}$) for $f : A \times X \longrightarrow B \times X$. We note that these constructions are mutually inverse. □

This theorem was proved by Martin Hyland and the author independently. There are several equivalent formulations of this result. For instance, in the presence of other conditions, we can restrict 3 to the case that g is a symmetry (c.f. Lemma 1.1. of [JSV96]). For another – practically useful – example, Hyland has shown that axioms 1∼4 are equivalent to 2 and

- *(parametrized) dinaturality:* $(\langle \pi_{A,X}, f \rangle; g)^\dagger = \langle id_A, (\langle \pi_{A,Y}, g \rangle; f)^\dagger \rangle; g : A \longrightarrow X$ for $f : A \times X \longrightarrow Y$ and $g : A \times Y \longrightarrow X$
- *diagonal property:* $(f^\dagger)^\dagger = ((id_A \times \langle id_X, id_X \rangle); f)^\dagger$ for $f : A \times X \times X \longrightarrow X$.

This axiomatization is the same as that of "Conway cartesian categories" in [BÉ96]. Further variations are: 2,4 with dinaturality; and 1,2,4 with the symmetric form of 4.

Perhaps the simplest example is the opposite of the category of sets and partial functions with coproduct as the monoidal product; the trace is given by the *feedback* operator which maps a partial function $f : X \to A + X$ to $f^\dagger : X \to A$, determined by iterating f until we get an answer in A if exists. Such a setting is studied in [BÉ93].

An immediate consequence of Theorem 3.1 is the close relationship between traces and the least fixed point operators in traditional domain theory.

Example 3.2. (the least fixed point operator on domains)
Consider the cartesian closed category **Dom** of Scott domains and continuous functions. The least fixed point operator satisfies conditions 1~4, thus determines a trace operator given by $Tr^X(f) = \lambda a^A.\pi(f(a, \bigsqcup^n (\lambda x^X.\pi'(f(a,x)))^n(\perp_X)))$ $: A \to B$ for $f : A \times X \to B \times X$. Since the least fixed point operator is the unique dinatural fixed point operator on **Dom**, the trace above is the unique one on **Dom**. ■

The same is true for several cartesian closed categories arising from domain theory. In fact, a systematic account is possible. Simpson [Sim93] has shown that, under a mild condition, in cartesian closed full subcategories of algebraic cpos, the least fixed point operator is characterized as the unique dinatural fixed point operator. On the other hand, it is easy to see that the least fixed point operators satisfy the conditions in Theorem 3.1. Therefore, in many such categories, a trace uniquely exists and is determined by the least fixed point operator. However, we note that there are at least two traces in the category of continuous lattices, an instance which does not satisfy Simpson's condition; this category has two fixed point operators which satisfy our conditions – the least one and the greatest continuous one.

Further justification of our axiomatization of fixed point operator comes from recent work on *axiomatic domain theory* which provides a more abstract and systematic treatment of domains and covers a wider range of models of domain theory than the traditional order-theoretic approach. For this, we assume some working knowledge of this topic as found in [Sim92]. Readers who are not familar with this topic may skip over to next subsection.

Example 3.3. (axiomatic domain theory)
Consider a cartesian closed category $\mathcal{C}$ (category of "predomains") equipped with a commutative monad L (the "lift") such that the Kleisli category $\mathcal{C}_L$ (category of predomains and "partial maps") is *algebraically compact* [Fre91]. This setting provides a canonical fixed point operator (derived from the *fixpoint object* [CP92]) on the category of "domains" (obtained as the co-Kleisli category of the induced comonad on the Eilenberg-Moore category $\mathcal{C}^L$) which satisfies our axioms – Bekič's lemma is proved from the algebraic compactness of $\mathcal{C}_L$ [Mog95] (this idea is due to Plotkin). Thus the requirement for solving recursive domain equations (algebraic compactness) implies that the resulting category of domains is traced. ■

Regarding these facts, we believe that traces provide a good characterization of fixed point operators in traditional denotational semantics.

We conclude this subsection by observing an attractive fact which suggests how natural our trace-fixpoint correspondence is (this is rather a digression in this paper, however). Our correspondence preserves a fundamental concept on fixed point operators called *uniformity*, also known as Plotkin's condition. This is important because fixed point operators are often canonically and uniquely characterized by this property.

Proposition 3.4. *In a traced cartesian category, the following two conditions are equivalent for any* $h : X \longrightarrow Y$.

- *(Uniformity of the trace operator) For any* f *and* g,

$$\text{if}\quad \begin{array}{ccc} A\times X & \xrightarrow{\;f\;} & B\times X \\ {\scriptstyle A\times h}\downarrow & & \downarrow{\scriptstyle B\times h} \\ A\times Y & \xrightarrow[\;g\;]{} & B\times Y \end{array} \quad \text{commutes then } Tr^X(f) = Tr^Y(g).$$

- *(Uniformity of the fixed point operator) For any* f *and* g,

$$\text{if}\quad \begin{array}{ccc} A\times X & \xrightarrow{\;f\;} & X \\ {\scriptstyle A\times h}\downarrow & & \downarrow{\scriptstyle h} \\ A\times Y & \xrightarrow[\;g\;]{} & Y \end{array} \quad \text{commutes then } f^{\dagger}; h = g^{\dagger}. \qquad \square$$

In the case of domain-theoretic categories, the second condition is equivalent to saying that h is a strict map ($\perp$-preserving map). This fact suggests the possibility of studying the notion of strict maps and uniformity of fixed points in more general settings as in the following subsection. In particular, the first condition makes sense in any traced monoidal category.

3.2 Trace and Notions of Computation

Our observation so far says that to have an abstract trace is to have a fixed point operator in the traditional sense, provided the monoidal product is cartesian. However, regarding our motivation to model cyclic sharing, this setting is somewhat restrictive – in a cartesian category (regarded as an algebraic theory) arbitrary substitution is justified, thus there is no non-trivial notion of sharing.

To overcome this, we consider a mild generalization. Now our traced category may not be cartesian, but it is assumed to have a sub-cartesian category such that the inclusion functor preserves symmetric monoidal structure and has a right adjoint (examples will be given below). Intuitively, the sub-cartesian category is the category of "values" which can be substituted freely whereas the symmetric monoidal category part is the category of all cyclic structures which cannot be copied in general because they may contain shared resources. In this weaker setting, there still exists a fixed point operator.

Let $F : \mathcal{C} \longrightarrow \mathcal{T}$ be a faithful, identity-on-objects strict symmetric monoidal functor from a cartesian category $\mathcal{C}$ to a traced symmetric monoidal category $\mathcal{T}$, with a right adjoint. Thus we identify the objects in $\mathcal{C}$ and in $\mathcal{T}$, and F is identity as a function on objects. However, for readability, we write $A \times B$ and $A \otimes B$ for cartesian product in $\mathcal{C}$ and tensor product in $\mathcal{T}$ respectively though they are identical as F is strict symmetric monoidal. We assume a similar convention for the terminal object 1 and the unit object I.

Theorem 3.5. *Given $F : \mathcal{C} \longrightarrow \mathcal{T}$ as above, there is a family of functions*

$$(-)^{\dagger_{A,X}} : \mathcal{T}(A \otimes X, X) \longrightarrow \mathcal{T}(A, X)$$

such that

1. *$(-)^{\dagger}$ is a parametrized fixed point operator in the sense that, for $f : A \otimes X \longrightarrow X$ in $\mathcal{T}$, $f^{\dagger} : A \longrightarrow X$ satisfies $f^{\dagger} = \Delta_A; (id_A \otimes f^{\dagger}); f$ where $\Delta_A = F(\langle id_A, id_A \rangle) : A \longrightarrow A \otimes A$.*
2. *$(-)^{\dagger}$ is natural in A in $\mathcal{C}$; for $f : A \otimes X \longrightarrow X$ in $\mathcal{T}$ and $g : B \longrightarrow A$ in $\mathcal{C}$, $((F(g) \otimes id_X); f)^{\dagger} = F(g); f^{\dagger} : B \longrightarrow X$.*
3. *$(-)^{\dagger}$ is natural in X in $\mathcal{T}$; for $f : A \otimes X \longrightarrow Y$ in $\mathcal{T}$ and $g : Y \longrightarrow X$ in $\mathcal{T}$, $(f; g)^{\dagger} = ((id_A \otimes g); f)^{\dagger}; g : A \longrightarrow X$.*

Sketch of the proof: Let us write $U : \mathcal{T} \longrightarrow \mathcal{C}$ for the right adjoint of F, and $\epsilon_X : UX \longrightarrow X$ (in $\mathcal{T}$) for the counit. By definition, we have a natural isomorphism $(-)^* : \mathcal{T}(A, B) \xrightarrow{\sim} \mathcal{C}(A, UB)$. We also define $\theta_{A,X} : A \times UX \longrightarrow U(A \otimes X)$ in $\mathcal{C}$ by $\theta_{A,X} = (id_A \otimes \epsilon_X)^*$. Now we define $(-)^{\dagger}$ by

$$f^{\dagger} = Tr^{UX}(F(\theta_{A,X}; Uf); \Delta_{UX}); \epsilon_X : A \longrightarrow X \quad \text{in } \mathcal{T}$$

for $f : A \otimes X \longrightarrow X$ in $\mathcal{T}$. □

Observe that an easier construction (c.f. Theorem 3.1) $Tr^X(f; \Delta_X) : A \longrightarrow X$ from $f : A \otimes X \longrightarrow X$ in $\mathcal{T}$ does not work as a fixed point operator – the construction in Theorem 3.5 uses the adjunction in a crucial manner.

It is in general impossible to recover a trace operator from a fixed point operator which satisfies the conditions of Theorem 3.5; for instance, if $\mathcal{T}$ has a zero object 0 such that $0 \otimes A \simeq 0$ (e.g. **Rel** below), the zero map satisfies these conditions. It is an interesting question to ask if we can strengthen the conditions so that we can recover a trace operator.

A careful inspection of our construction reveals that we need the trace operator just on objects of the form UX (equivalently $F(UX)$ as F is identity-on-objects); actually it is sufficient if the full subcategory of $\mathcal{T}$ whose objects are of the form of $UX_1 \otimes \ldots \otimes UX_n$ is traced. Thus such a fixed point operator exists even in a weaker setting. It would be interesting to see if this fixed point operator determines this sub-trace structure. It would be more interesting to see if there is a good connection between such a fixed point operator and fixed point operators in models of intuitionistic linear logic as studied in [Bra95].

An observation corresponding to Proposition 3.4 is as follows: for any h in $\mathcal{T}$, if $F(U(h))$ satisfies the uniformity condition for the trace operator then h satisfies the uniformity condition for the fixed point operator.

Note that our setting is equivalent to saying that we have a cartesian category $\mathcal{C}$ with a monad $U \circ F$ on it, which satisfies the mono-requirement and has a commutative tensorial strength θ, such that the Kleisli category $\mathcal{T}$ is traced. In other words, we are dealing with some *notions of computation* in the sense of Moggi [Mog88] with extra structure (trace). Our definition is inspired by a recent reformulation of notions of computation by Power and Robinson [PR96].

Definition 3.6. A *traced computational model* is a faithful, identity-on-object strict symmetric monoidal functor $F : \mathcal{C} \longrightarrow \mathcal{T}$ where $\mathcal{C}$ is a cartesian category and $\mathcal{T}$ a traced symmetric monoidal category, such that the functor $F(-) \otimes X : \mathcal{C} \longrightarrow \mathcal{T}$ has a right adjoint $X \Rightarrow (-) : \mathcal{T} \longrightarrow \mathcal{C}$: thus $\mathcal{T}(F(-) \otimes X, -) \simeq \mathcal{C}(-, X \Rightarrow -)$. ■

$X \Rightarrow Y$ is the so-called *Kleisli exponent*; if $\mathcal{C}$ is cartesian closed, $X \Rightarrow Y$ is obtained as $(UY)^X$. As a traced computational model satisfies the assumption in Theorem 3.5 (a right adjoint of F is given by $I \Rightarrow (-)$), there is a fixed point operator in its traced category. The right adjoint $X \Rightarrow (-)$ can be used to interpret higher-order (higher-type) computation. Thus traced computational models have enough structure to interpret higher-order recursive computation; later we see how they can be used as the models of a simply typed lambda calculus with cyclic sharing.

To help with the intuition, we shall give a selection of traced computational models below. Most of them have already been mentioned in Section 1.

Example 3.7. (traced cartesian closed categories)
A traced cartesian closed category is a traced computational model in which the cartesian category part and the traced category part are identical. Examples include many domain-theoretic categories such as Example 3.2. ■

Example 3.8. (non-deterministic model)
The inclusion from the category **Set** of sets and functions to the category **Rel** of sets and binary relations (with the direct product of sets as the symmetric monoidal product) forms a traced computational model: $\mathbf{Rel}(A \otimes X, B) \simeq \mathbf{Set}(A, \mathbf{Rel}(X, B))$. The trace operator on **Rel**, induced by the compact closed structure of **Rel**, is given as follows: for a relation $R : A \otimes X \longrightarrow B \otimes X$, we define a relation $Tr^X(R) : A \longrightarrow B$ by $(a, b) \in Tr^X(R)$ iff $((a, x), (b, x)) \in R$ for an $x \in X$ (here a relation from A to B is given as a subobject of $A \times B$). The parametrized fixed point operator $(-)^\dagger$ on **Rel** is given by

$$R^\dagger = \{(a, x) \mid \exists S \subseteq X \; S = \{y \mid \exists x \in S \; ((a, x), y) \in R\} \;\&\; x \in S\} \; : \; A \longrightarrow X$$

for $R : A \otimes X \longrightarrow X$ (and $R^\dagger$ is not the zero map!). Note that we can use an elementary topos instead of **Set**, which may provide a computationally more sophisticated model. ■

Example 3.9. (finite dimensional vector spaces over a finite field)
Let F_2 be the field with just two elements (thus its characteristic is 2), and $\mathbf{Vect}_{F_2}^{\mathrm{fin}}$ be the category of finite dimensional vector spaces (with chosen bases) over F_2. There is a strict symmetric monoidal functor from the category of finite sets to $\mathbf{Vect}_{F_2}^{\mathrm{fin}}$ which maps a set S to a vector space with the basis S, and this functor has a right adjoint (the underlying functor). Since $\mathbf{Vect}_{F_2}^{\mathrm{fin}}$ is traced (in the very classical sense), this is an instance of traced computational models. Note that this example is similar to the previous one – compare the matrix representation of binary relations and that of linear maps. ■

Example 3.10. (higher-order reflexive action calculi)
Recent work [GH96, Mif96] on action calculi [Mil96] shows that the higher-order reflexive extension of an action calculus [Mil94a, Mil94b] forms a traced computational model. In this calculus the fixed point operator $(-)^\dagger$ is given by

$$t^\dagger \;=\; \uparrow_{\epsilon\Rightarrow n}((x^{\epsilon\Rightarrow n})\ulcorner(\mathbf{id}_m \otimes \langle x\rangle \cdot \mathbf{ap}_{\epsilon,n}) \cdot t\urcorner \cdot \mathbf{copy}_{\epsilon\Rightarrow n}) \cdot \mathbf{ap}_{\epsilon,n} \;:\; m \longrightarrow n$$

for $t : m \otimes n \longrightarrow n$. Mifsud gives essentially the same operator $\mathbf{iter}_f(t)$ in his thesis [Mif96]. Using this, we can present recursion operators in various process calculi, typically the replication operator. ■

4 Semantics of Lambda Calculi with Cyclic Sharing

We introduce two simply typed lambda calculi enriched with the notion of cyclic sharing, the *simply typed* $\lambda_{\mathrm{letrec}}$*-calculus* and $\lambda^{v}_{\mathrm{letrec}}$*-calculus* in which cyclically shared resources are represented in terms of the letrec syntax. It is shown that traced cartesian closed categories and traced computational models are sound and complete models of these calculi respectively.

4.1 The Syntax and Axioms

As the semantic observation we have seen suggests, the simply typed $\lambda^{v}_{\mathrm{letrec}}$-calculus is designed as a modification of Moggi's computational lambda calculus [Mog88]; we replace the let-syntax by the letrec-syntax which allows cyclic bindings.

In this section, we fix a set of *base types*.

Types

$$\sigma, \tau \ldots \;::=\; b \mid \sigma \Rightarrow \tau \quad \text{(where } b \text{ is a base type)}$$

Syntactic Domains

Variables	$x, y, z \ldots$	
Raw Terms	$M, N \ldots$	$::= x \mid \lambda x.M \mid MN \mid \mathsf{letrec}\ D\ \mathsf{in}\ N$
Values	$V, W \ldots$	$::= x \mid \lambda x.M$
Declarations	$D \ldots$	$::= x = M \mid x = M, D$

In a declaration, binding variables are assumed to be disjoint.

Typing

$$\frac{}{\Gamma, x:\sigma \vdash x:\sigma}\ \text{Variable} \qquad \frac{\Gamma, x:\sigma, y:\sigma', \Gamma' \vdash M:\tau}{\Gamma, y:\sigma', x:\sigma, \Gamma' \vdash M:\tau}\ \text{Exchange}$$

$$\frac{\Gamma, x:\sigma \vdash M:\tau}{\Gamma \vdash \lambda x.M : \sigma \Rightarrow \tau}\ \text{Abstraction} \qquad \frac{\Gamma \vdash M:\sigma \Rightarrow \tau \quad \Gamma \vdash N:\sigma}{\Gamma \vdash MN:\tau}\ \text{Application}$$

$$\frac{\Gamma, x_1:\sigma_1, \ldots, x_n:\sigma_n \vdash M_i:\sigma_i\ (i=1,\ldots,n) \quad \Gamma, x_1:\sigma_1, \ldots, x_n:\sigma_n \vdash N:\tau}{\Gamma \vdash \mathsf{letrec}\ x_1 = M_1, \ldots, x_n = M_n\ \mathsf{in}\ N : \tau}\ \mathsf{letrec}$$

Axioms

Identity	$\mathsf{letrec}\ x = M\ \mathsf{in}\ x$	$= M \quad (x \notin FV(M))$
Associativity	$\mathsf{letrec}\ y = (\mathsf{letrec}\ D_1\ \mathsf{in}\ M), D_2\ \mathsf{in}\ N$	$= \mathsf{letrec}\ D_1, y = M, D_2\ \mathsf{in}\ N$
	$\mathsf{letrec}\ D_1\ \mathsf{in}\ \mathsf{letrec}\ D_2\ \mathsf{in}\ M$	$= \mathsf{letrec}\ D_1, D_2\ \mathsf{in}\ M$
Permutation	$\mathsf{letrec}\ D_1, D_2\ \mathsf{in}\ N$	$= \mathsf{letrec}\ D_2, D_1\ \mathsf{in}\ N$
Commutativity	$(\mathsf{letrec}\ D\ \mathsf{in}\ M)N$	$= \mathsf{letrec}\ D\ \mathsf{in}\ MN$
	$M(\mathsf{letrec}\ D\ \mathsf{in}\ N)$	$= \mathsf{letrec}\ D\ \mathsf{in}\ MN$
β	$(\lambda x.M)N$	$= \mathsf{letrec}\ x = N\ \mathsf{in}\ M$
σ_v	$\mathsf{letrec}\ x = V, D[x]\ \mathsf{in}\ M$	$= \mathsf{letrec}\ x = V, D[V]\ \mathsf{in}\ M$
	$\mathsf{letrec}\ x = V, D\ \mathsf{in}\ M[x]$	$= \mathsf{letrec}\ x = V, D\ \mathsf{in}\ M[V]$
	$\mathsf{letrec}\ x = V\ \mathsf{in}\ M$	$= M \quad (x \notin FV(V) \cup FV(M))$
η_0	$\lambda x.yx$	$= y$

Both sides of equations must have the same type under the same typing context; we will work just on well-typed terms. We assume the usual conventions on variables.

We remark that axioms Identity, Associativity, Permutation and Commutativity ensure that two $\lambda^v_{\mathsf{letrec}}$-terms are identified if they correspond to the same cyclic directed graph; thus they are a sort of structural congruence, rather than representing actual computation. β creates a sharing from a function application. σ_v describes the substitution of values (the first two for the dereference, the last one for the garbage collection). In $M[x]$ and $D[x]$, $[x]$ denotes a free occurrence of x. From β, σ_v and η_0, we have the "call-by-value" $\beta\eta$-equations:

Lemma 4.1. *In $\lambda^v_{\mathsf{letrec}}$-calculus, the following are derivable.*

$$\begin{array}{lll} \beta_v & (\lambda x.M)V = M\{V/x\} & \\ \eta_v & (\lambda x.Vx) \ = V & (x \notin FV(V)) \quad \square \end{array}$$

We think it is misleading to relate this calculus to the call-by-value operational semantics; restricting substitutions on values does not mean that this calculus is for call-by-value. Rather, our equational theory is fairly close to the *call-by-need* calculus proposed in [AF96], which corresponds to a version of lazy implementations of the call-by-name operational semantics. We expect that this connection is the right direction to relate our calculus to an operational semantics.

Also we define a "strengthened" version in which arbitrary substitution and η-reduction are allowed (thus any term is a value):

$$
\begin{array}{lll}
\sigma & \mathsf{letrec}\ x = N, D[x]\ \mathsf{in}\ M = \mathsf{letrec}\ x = N, D[N]\ \mathsf{in}\, M & \\
& \mathsf{letrec}\ x = N, D\ \mathsf{in}\ M[x] = \mathsf{letrec}\ x = N, D\ \mathsf{in}\, M[N] & \\
& \mathsf{letrec}\ x = N\ \mathsf{in}\ M \qquad\quad = M \quad (x \notin FV(M)) & \\
\eta & \lambda x.Mx \qquad\qquad\qquad\quad\; = M \quad (x \notin FV(M)) &
\end{array}
$$

We shall call this version the *simply typed* $\lambda_{\mathsf{letrec}}$*-calculus* – this corresponds to the calculus in [AK94] ignoring the typing and the extensionality (η-axiom).

4.2 Interpretation into Traced Computational Models

We just present the case of the $\lambda^v_{\mathsf{letrec}}$-calculus; the case of the $\lambda_{\mathsf{letrec}}$-calculus is obtained just by replacing a traced computational model by a traced cartesian closed category.

Let us fix a traced computational model $F : \mathcal{C} \longrightarrow \mathcal{T}$, and choose an object $[\![b]\!]$ for each base type b. The interpretation of arrow types is then defined by $[\![\sigma \Rightarrow \tau]\!] = [\![\sigma]\!] \Rightarrow [\![\tau]\!]$. We interpret a $\lambda^v_{\mathsf{letrec}}$-term (with its typing environment) $x_1 : \sigma_1, \ldots, x_n : \sigma_n \vdash M : \tau$ to an arrow $[\![x_1 : \sigma_1, \ldots, x_n : \sigma_n \vdash M : \tau]\!] : [\![\sigma_1]\!] \otimes \ldots \otimes [\![\sigma_n]\!] \longrightarrow [\![\tau]\!]$ in $\mathcal{T}$ as follows.

$$
\begin{array}{ll}
[\![x_1 : \sigma_1, \ldots, x_n : \sigma_n \vdash x_i : \sigma_i]\!] & = F(\pi_i) \text{ where } \pi_i \text{ is the } i\text{-th projection} \\
[\![\Gamma \vdash \lambda x.M : \sigma \Rightarrow \tau]\!] & = F(\mathbf{cur}([\![\Gamma, x : \sigma \vdash M : \tau]\!])) \\
[\![\Gamma \vdash M^{\sigma \Rightarrow \tau} N^{\sigma} : \tau]\!] & = \Delta; ([\![\Gamma \vdash M : \sigma \Rightarrow \tau]\!] \otimes [\![\Gamma \vdash N : \tau]\!]); \mathbf{ap} \\
\multicolumn{2}{l}{[\![\Gamma \vdash \mathsf{letrec}\ x_1 = M_1^{\sigma_1}, .., x_k = M_k^{\sigma_k}\ \mathsf{in}\ N : \tau]\!] =} \\
\multicolumn{2}{l}{\quad \Delta; (id \otimes Tr^{[\![\sigma_1]\!] \otimes .. \otimes [\![\sigma_k]\!]}(\Delta_k; ([\![\Gamma' \vdash M_1 : \sigma_1]\!] \otimes \ldots \otimes [\![\Gamma' \vdash M_k : \sigma_k]\!]); \Delta)); [\![\Gamma' \vdash N : \tau]\!]}
\end{array}
$$

where $\mathbf{ap}_{A,B} : (A \Rightarrow B) \otimes A \longrightarrow B$ is the counit of the adjoint $F(-) \otimes A \dashv A \Rightarrow (-)$, and $\mathbf{cur} : \mathcal{T}(FA \otimes B, C) \longrightarrow \mathcal{C}(A, B \Rightarrow C)$ the associated natural bijection. In the last case, Γ' is $\Gamma, x_1 : \sigma_1, \ldots, x_k : \sigma_k$ and Δ_k is the k-times copy ($\Delta_{k\,A} = F(\langle \underbrace{id, \ldots, id}_{k \text{ times}} \rangle) : A \longrightarrow \underbrace{A \otimes \ldots \otimes A}_{k \text{ times}}$). Note that values are first interpreted in $\mathcal{C}$ (following Moggi's account, $\mathcal{C}$ is the category of values) and then lifted to $\mathcal{T}$ via F.

A straightforward calculation shows that traced computational models are sound for the $\lambda^v_{\mathsf{letrec}}$-calculus (and the same for traced cartesian closed categories and the $\lambda_{\mathsf{letrec}}$-calculus):

Theorem 4.2. *(Soundness)*

- *For any traced computational model with chosen object* $[\![b]\!]$ *for each base type* b*, this interpretation is sound; if* $\Gamma \vdash M : \sigma$*,* $\Gamma \vdash N : \sigma$ *and* $M = N$ *in the* $\lambda^v_{\mathsf{letrec}}$*-calculus then* $[\![\Gamma \vdash M : \sigma]\!] = [\![\Gamma \vdash N : \sigma]\!]$.
- *For any traced cartesian closed category with chosen object* $[\![b]\!]$ *for each base type* b*, this interpretation is sound; if* $\Gamma \vdash M : \sigma$*,* $\Gamma \vdash N : \sigma$ *and* $M = N$ *in the* $\lambda_{\mathsf{letrec}}$*-calculus then* $[\![\Gamma \vdash M : \sigma]\!] = [\![\Gamma \vdash N : \sigma]\!]$. □

Example 4.3. (domain-theoretic model)
As we already noted, **Dom** is a traced cartesian closed category (hence also a traced computational model). The interpretation of a $\lambda_{\mathsf{letrec}}$-term $\vdash \mathsf{letrec}\ x = M\ \mathsf{in}\ x : \sigma$ in **Dom** is just the least fixed point $\bigcup_n F^n(\bot)$ where $F : [\![\sigma]\!] \longrightarrow [\![\sigma]\!]$ is the interpretation of $x : \sigma \vdash M : \sigma$. ■

Example 4.4. (non-deterministic model)
In **Rel** (Example 3.8), a $\lambda^{v}_{\mathsf{letrec}}$-term is interpreted as the set of "all possible solutions of the recursive equation". The interpretation of $\vdash \mathsf{letrec}\ x = M\ \mathsf{in}\ x : \sigma$ is just the set $\{x \in [\![\sigma]\!] \mid (x, x) \in [\![x : \sigma \vdash M : \sigma]\!]\}$ (a subobject of $[\![\sigma]\!] = 1 \times [\![\sigma]\!]$). For instance,

$$\begin{array}{lll} [\![\vdash \mathsf{letrec}\ x = x\ \mathsf{in}\ x : \sigma]\!] & = [\![\sigma]\!] & : 1 \longrightarrow [\![\sigma]\!] \\ [\![\vdash \mathsf{letrec}\ x = x^2\ \mathsf{in}\ x : \mathsf{nat}]\!] & = \{0, 1\} & : 1 \longrightarrow \mathbf{N} \\ [\![\vdash \mathsf{letrec}\ x = x + 1\ \mathsf{in}\ x : \mathsf{nat}]\!] & = \emptyset & : 1 \longrightarrow \mathbf{N} \end{array}$$

(for the latter two cases we enrich the calculus with natural numbers). Note that this model is sound for the $\lambda^{v}_{\mathsf{letrec}}$-calculus, but not for the $\lambda_{\mathsf{letrec}}$-calculus – since we cannot copy non-deterministic computation, this model is "resource-sensitive". ■

Moreover, we can construct a term model (enriched with the unit and product types) to which the $\lambda^{v}_{\mathsf{letrec}}$-calculus (or $\lambda_{\mathsf{letrec}}$-calculus) is faithfully interpreted. Actually it is possible to show that the $\lambda^{v}_{\mathsf{letrec}}$-calculus is faithfully embedded into the higher-order reflexive action calculus (Example 3.10) which is an instance of traced computational models. Thus we also have completeness:

Theorem 4.5. *(Completeness)*

- *If* $[\![\Gamma \vdash M : \sigma]\!] = [\![\Gamma \vdash N : \sigma]\!]$ *for every traced computational model, then* $M = N$ *in the* $\lambda^{v}_{\mathsf{letrec}}$*-calculus.*
- *If* $[\![\Gamma \vdash M : \sigma]\!] = [\![\Gamma \vdash N : \sigma]\!]$ *for every traced cartesian closed category, then* $M = N$ *in the* $\lambda_{\mathsf{letrec}}$*-calculus.* □

Remark 4.6. To represent the parametrized fixed point operator given in Theorem 3.5 we have to extend the $\lambda^{v}_{\mathsf{letrec}}$-calculus with a *unit type* unit which has a unique value $*$:

$$\frac{}{\Gamma \vdash * : \mathsf{unit}}\ \mathrm{Unit} \qquad V = * \quad (V : \mathsf{unit})$$

The interpretation of the unit type in a traced computational model is just the terminal object (unit object). The type constructor $\mathsf{unit} \Rightarrow (-)$ then plays the role of the right adjoint of the inclusion from the category of values to the category of terms. We define the parametrized fixed point operator by

$$\frac{\Gamma, x : \sigma \vdash M : \sigma}{\Gamma \vdash \mu x^{\sigma}.M \equiv \mathsf{letrec}\ f^{\mathsf{unit} \Rightarrow \sigma} = \lambda y^{\mathsf{unit}}.((\lambda x^{\sigma}.M)(f*))\ \mathsf{in}\ f* \ : \sigma}$$

which satisfies $\mu x.M = (\lambda x.M)(\mu x.M)$, but may not satisfy $\mu x.M = M\{\mu x.M/x\}$ in the $\lambda^{v}_{\mathsf{letrec}}$-calculus because $\mu x.M$ may not be a value in general. The operator Y_3 in the next section is essentially same as this fixed point operator, except for avoiding to use unit. ■

We could give the untyped version and its semantic models – by a reflexive object in a traced computational model (or a traced cartesian closed category). Regarding the results in Section 3, we can establish the connection between the dinatural diagonal fixed point operator in a model of the untyped $\lambda_{\mathrm{letrec}}$-calculus and the trace operator of the cartesian closed category. It would be interesting to compare recursion created by untypedness and recursion created by trace (cyclic sharing) in such models.

5 Analyzing Fixed Points

In the $\lambda^{v}_{\mathrm{letrec}}$-calculus, several (weak) fixed point operators are definable – this is not surprising, because there are several known encodings of fixed point operators in terms of cyclic sharing. However, it is difficult to see that they are not identified by our equational theory – syntactic reasoning for cyclic graph structures is not an easy task, as the non-confluency result in [AK94] suggests. On the other hand, in many traditional models for recursive computation, all of them have the same denotational meaning mainly because we cannot distinguish values from non-values in such models.

One purpose of developing the traced computational models is to give a clear semantic account for these several recursive computations created from cyclic sharing. Though this topic has not yet been fully developed, we shall give some elementary analysis using the $\lambda^{v}_{\mathrm{letrec}}$-calculus and a traced computational model (**Rel**).

We define $\lambda^{v}_{\mathrm{letrec}}$-terms $\Gamma \vdash Y_i(M) : \sigma$ $(i = 1, 2, 3)$ for given term $\Gamma \vdash M : \sigma \Rightarrow \sigma$ as follows.

$$\begin{aligned}
Y_1 &= \mathsf{letrec}\ \mathsf{fix}^{(\sigma\Rightarrow\sigma)\Rightarrow\sigma} = \lambda f^{\sigma\Rightarrow\sigma}.f(\mathsf{fix}\ f)\ \mathsf{in}\ \mathsf{fix}\\
Y_2 &= \lambda f^{\sigma\Rightarrow\sigma}.\mathsf{letrec}\ x^{\sigma} = fx\ \mathsf{in}\ x\\
Y_3(M) &= \mathsf{letrec}\ g^{\tau\Rightarrow\sigma} = \lambda y^{\tau}.M(gy)\ \mathsf{in}\ gN\\
&\quad (N \text{ is a closed term of type } \tau, \text{ e.g. } \mathsf{letrec}\ x = x\ \mathsf{in}\ x\ : \tau)
\end{aligned}$$

Each of them can be used as a fixed point operator, but their behaviours are not the same. For instance, it is known that Y_2 is more efficient than others, under the call-by-need evaluation strategy [Lau93]. Y_1 satisfies the fixed point equation $YV = V(YV)$ for any value $V : \sigma \Rightarrow \sigma$.

$$\begin{aligned}
Y_1 M &= \mathsf{letrec}\ \mathsf{fix} = \lambda f.f(\mathsf{fix}\ f)\ \mathsf{in}\ \mathsf{fix} M && \text{Commutativity}\\
&= \mathsf{letrec}\ \mathsf{fix} = \lambda f.f(\mathsf{fix}\ f)\ \mathsf{in}\ (\lambda f.f(\mathsf{fix}\ f))M && \sigma_v\\
&= \mathsf{letrec}\ \mathsf{fix} = \lambda f.f(\mathsf{fix}\ f)\ \mathsf{in}\ \mathsf{letrec}\ f' = M\ \mathsf{in}\ f'(\mathsf{fix}\ f') && \beta\\
&= \mathsf{letrec}\ f' = M\ \mathsf{in}\ \mathsf{letrec}\ \mathsf{fix} = \lambda f.f(\mathsf{fix}\ f)\ \mathsf{in}\ f'(\mathsf{fix}\ f') && \text{Associativity, Permutation}\\
&= \mathsf{letrec}\ f' = M\ \mathsf{in}\ f'((\mathsf{letrec}\ \mathsf{fix} = \lambda f.f(\mathsf{fix}\ f)\ \mathsf{in}\ \mathsf{fix})f') && \text{Commutativity}\\
&= \mathsf{letrec}\ f' = M\ \mathsf{in}\ f'(Y_1 f')\\
&(= M(Y_1 M) \quad \text{if } M \text{ is a value})
\end{aligned}$$

Y_2 satisfies $Y_2 M = M(Y_2 M)$ only when Mx is equal to a value (hence M is supposed to be a higher-order value). If $M = \lambda y.V$ for some value V,

$$\begin{array}{rll} Y_2 M & = \mathsf{letrec}\ x = (\lambda y.V)x\ \mathsf{in}\ x & \beta_v \\ & = \mathsf{letrec}\ x = V\{x/y\}\ \mathsf{in}\ x & \beta_v \\ & = \mathsf{letrec}\ x = V\{x/y\}\ \mathsf{in}\ V\{x/y\} & \sigma_v \\ & = \mathsf{letrec}\ x = (\lambda y.V)x\ \mathsf{in}\ (\lambda y.V)x & \beta_v \\ & = (\lambda y.V)(\mathsf{letrec}\ x = (\lambda y.V)x\ \mathsf{in}\ x) & \text{Commutativity} \\ & = M(Y_2 M) & \end{array}$$

Y_3 satisfies $Y_3(M) = M(Y_3(M))$ for any term $M : \sigma \Rightarrow \sigma$ (thus is a "true" fixed point operator).

$$\begin{array}{rll} Y_3(M) & = \mathsf{letrec}\ g = \lambda y.M(gy)\ \mathsf{in}\ (\lambda y.M(gy))N & \sigma_v \\ & = \mathsf{letrec}\ g = \lambda y.M(gy)\ \mathsf{in}\ \mathsf{letrec}\ y' = N\ \mathsf{in}\ M(gy') & \beta \\ & = \mathsf{letrec}\ g = \lambda y.M(gy)\ \mathsf{in}\ M(g(\mathsf{letrec}\ y' = N\ \mathsf{in}\ y')) & \text{Commutativity} \\ & = M(\mathsf{letrec}\ g = \lambda y.M(gy)\ \mathsf{in}\ g(\mathsf{letrec}\ y' = N\ \mathsf{in}\ y')) & \text{Commutativity} \\ & = M(\mathsf{letrec}\ g = \lambda y.M(gy)\ \mathsf{in}\ gN) & \text{Identity} \\ & = M(Y_3(M)) & \end{array}$$

The interpretation of these operators in a traced computational model is as follows.

$$\begin{array}{ll} [\![\vdash Y_1]\!] & = Tr^{(A\Rightarrow A)\Rightarrow A}(F(\mathbf{cur}((id \otimes \Delta); (\mathbf{ap} \otimes id); c; \mathbf{ap})); \Delta) \\ [\![\vdash Y_2]\!] & = F(\mathbf{cur}(Tr^A(\mathbf{ap}; \Delta))) \\ [\![\Gamma \vdash Y_3(M)]\!] & = (Tr^{B\Rightarrow A}(F(\mathbf{cur}(([\![\Gamma \vdash M : \sigma \Rightarrow \sigma]\!] \otimes \mathbf{ap}); \mathbf{ap})); \Delta) \times [\![\vdash N : \tau]\!]); \mathbf{ap} \end{array}$$

where $A = [\![\sigma]\!]$ and $B = [\![\tau]\!]$. They have the different interpretations in **Rel**, hence are not identified in the $\lambda^v_{\mathsf{letrec}}$-calculus. Assume that $S = [\![\vdash M : \sigma \Rightarrow \sigma]\!] \subseteq \mathbf{Rel}(A, A)$. Then

$$[\![\vdash Y_1(M) : \sigma]\!] = \bigcup_{f\in S} \bigcup_{(A';f)=A'\subseteq A} A' \qquad [\![\vdash Y_2(M) : \sigma]\!] = \bigcup_{f\in S} \{x \mid (x,x) \in f\}$$

whereas

$$[\![\vdash Y_3(M) : \sigma]\!] = \bigcup_{(A';\bigcup S)=A'\subseteq A} A'$$

(In the definition of Y_3, we take $N : \tau$ as $\mathsf{letrec}\ x = x\ \mathsf{in}\ x : \tau$.)

6 Conclusion

We have presented new semantic models for interpreting cyclic sharing in terms of traced monoidal categories and notions of computation, and shown the connections with cyclic lambda calculi and with traditional semantics for recursive computation. We have also demonstrated that our framework covers a wider range of models of recursion than the traditional approach. We summarize this situation, together with examples in this paper, in the diagram below.

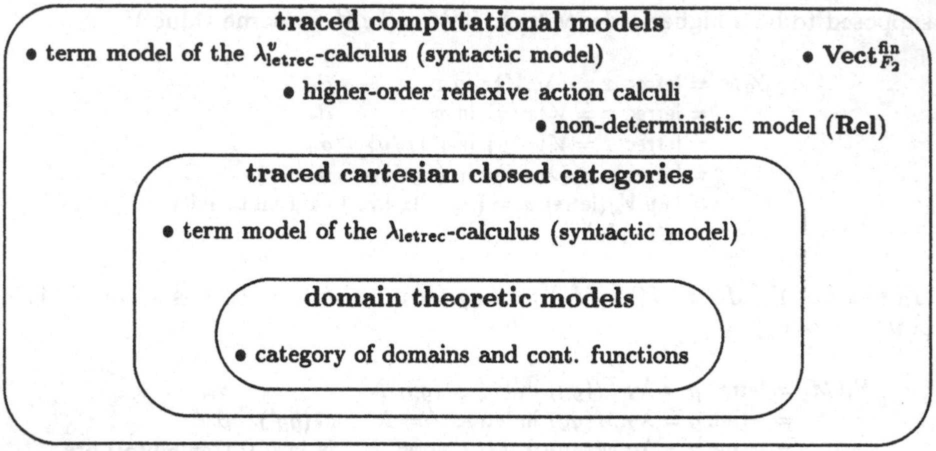

Acknowledgements

I am deeply grateful to Martin Hyland and John Power for helpful discussions, suggestions and encouragement. I also thank Philippa Gardner, Alex Mifsud, Marcelo Fiore, Alex Simpson and Gordon Plotkin for their comments and encouragement.

References

Abr96. S. Abramsky, Retracing some paths in process algebra. In *Proc. 7th Int. Conf. Concurrency Theory (CONCUR'96)*, Springer LNCS 1119, pages 1-17, 1996.

AA95. Z. Ariola and Arvind, Properties of a first-order functional language with sharing. *Theoretical Computer Science* 146, pages 69-108, 1995.

AF96. Z. Ariola and M. Felleisen, A call-by-need lambda calculus. Technical report CIS-TR-96-97, 1996. To appear in *Journal of Functional Programming.*

AK94. Z. Ariola and J. Klop, Cyclic lambda graph rewriting. In *Proc. 9th Symposium on Logic in Computer Sciece (LICS'94)*, pages 416-425, 1994.

BÉ93. S. L. Bloom and Z. Ésik, *Iteration Theories.* EATCS Monographs on Theoretical Computer Science, Springer-Verlag, 1993.

BÉ96. S. L. Bloom and Z. Ésik, Fixed point operators on ccc's. Part I. *Theoretical Computer Science* 155, pages 1-38, 1996.

Bra95. T. Braüner, The Girard translation extended with recursion. In *Proc. Computer Science Logic 1994 (CSL'94)*, Springer LNCS 933, pages 31-45, 1995.

CP92. R. L. Crole and A. M. Pitts, New foundations for fixpoint computations: Fix hyperdoctrines and the fix logic. *Information and Computation* 98, pages 171-210, 1992.

Fre91. P. Freyd, Algebraically complete categories. In *Proc. 1990 Como Category Theory Conference*, Springer LNM 1144, pages 95-104, 1991.

GH96. P. Gardner and M. Hasegawa, On higher-order action calculi and notions of computation. Draft, LFCS, University of Edinburgh, 1996.

Gir89. J. -Y. Girard, Geometry of interaction I: interpretation of system F. In *Logic Colloquium '88*, pages 221-260, North-Holland, 1989.

JS91. A. Joyal and R. Street, The geometry of tensor calculus I. *Advances in Mathematics* 88, pages 55-113, 1991.

JS93. A. Joyal and R. Street, Braided tensor categories. *Advances in Mathematics* 102, pages 20-78, 1993.

JSV96. A. Joyal, R. Street and D. Verity, Traced monoidal categories. *Mathematical Proceedings of the Cambridge Philosophical Society* 119(3), pages 447-468, 1996.

KL80. M. Kelly and M. L. Laplaza, Coherence for compact closed categories. *Journal of Pure and Applied Algebra* 19, pages 193-213, 1980.

Lau93. J. Launchbury, A natural semantics for lazy evaluation. In *Proc. 21st ACM Symp. Principles of Programming Languages (POPL'93)*, pages 144-154, 1993.

Mif96. A. Mifsud, *Control structures*. PhD thesis, LFCS, University of Edinburgh, 1996.

Mil94a. R. Milner, Higher-order action calculi. In *Proc. Computer Science Logic 1992 (CSL'92)*, Springer LNCS 832, pages 238-260, 1994.

Mil94b. R. Milner, Action calculi V: reflexive molecular forms (with Appendix by O. Jensen). Third draft, July 1994.

Mil96. R. Milner, Calculi for interaction. *Acta Informatica* 33(8), pages 707-737, 1996.

Mog88. E. Moggi, Computational lambda-calculus and monads. Technical report ECS-LFCS-88-66, LFCS, University of Edinburgh, 1988.

Mog95. E. Moggi, Metalanguages and applications. Draft, 1995.

PR96. A. J. Power and E. P. Robinson, Premonoidal categories and notions of computation. 1996. To appear in *Mathematical Structures in Computer Science*.

RT90. N. Yu. Reshetikhin and V. G. Turaev, Ribbon graphs and their invariants derived from quantum groups. *Communications in Mathematical Phisics* 127, pages 1-26, 1990.

Sim92. A. Simpson, Recursive types in Kleisli categories. Manuscript, LFCS, University of Edinburgh, 1992.

Sim93. A. Simpson, A characterisation of the least-fixed-point operator by dinaturality. *Theoretical Computer Science* 118, pages 301-314, 1993.

Games and Weak-Head Reduction for Classical PCF

Hugo Herbelin *

LITP, University Paris 7, 2 place Jussieu, 75252 Paris Cedex 05, France
INRIA-Rocquencourt, B.P. 105, 78153 Le Chesnay Cedex, France
Hugo.Herbelin@inria.fr

Abstract. We present a game model for classical PCF, a finite version of PCF extended by a `catch/throw` mechanism. This model is build from E-dialogues, a kind of two-players game defined by Lorenzen. In the E-dialogues for classical PCF, the strategies of the first player are isomorphic to the Böhm trees of the language.
We define an interaction in E-dialogues and show that it models the weak-head reduction in classical PCF. The interaction is a variant of Coquand's debate and the weak-head reduction is a variant of the reduction in Krivine's Abstract Machine.
We then extend E-dialogues to a kind of games similar to Hyland-Ong's games. Interaction in these games also models weak-head reduction. In the intuitionistic case (i.e. without the `catch/throw` mechanism), the extended E-dialogues are Hyland-Ong's games where the innocence condition on strategies is now a rule.
Our model for classical PCF is different from Ong's model of Parigot's lambda-mu-calculus. His model works by adding new moves to the intuitionistic case while ours works by relaxing the game rules.

Introduction

We investigate the links between Lorenzen's and Coquand's game-theoretic approach of provability, Hyland-Ong's game-theoretic approach of λ-calculus, and weak-head reduction as implemented by Krivine's Abstract Machine.

To exemplify these links, we choose as framework the finite Böhm trees of a variant of PCF extended by a catch/throw mechanism. This variant of PCF is classical in the sense that its typing system includes implicational classical logic.

We refer to Felscher [8] for the works of Lorenzen and his school. According to Felscher, the goal of Lorenzen was to give a game-based foundation of intuitionistic logic. Several kinds of two-players games parametrized by formulas were defined. Of these games, we note one in particular, called E-dialogues in Felscher [8]. As noticed by Lorenzen & Schwemmer [16], the strategies for the first player in E-dialogues have a structure of proofs in a certain cut-free sequent

* This research was partly supported by ESPRIT Basic Research Action "Types for Proofs and Programs" and by Programme de Recherche Coordonnées "Mécanisation du raisonnement".

calculus. The propositional fragment of this calculus is described in [10]. It is a variant of Gentzen's calculus LJ.

More generally, we call E-dialogue the kind of game which game-theoretically expresses the terms or proofs typed by a system having the subformula property. In particular, Coquand's games [3] (inspired from Gentzen [2]) for infinitary propositional logic are E-dialogues. Similarly, the typing system of Böhm trees for PCF has the subformula property and we can define E-dialogues for PCF.

We focus in section 1 on Böhm trees for classical PCF. Before defining E-dialogues for classical PCF, we define in section 2 a generic notion of two-players games parametrized by types. It's only by restricting the rules of the generic games that we get E-dialogues (see section 3). An isomorphism between Böhm trees for classical PCF and strategies for the first player in E-dialogues can now be stated. However, the rules of the E-dialogues are not the same for both players. So we define in section 4 spread E-dialogues where the players have dual roles. In spread E-dialogues both the strategies of the first player and of the second player are in one-to-one correspondence with Böhm trees.

In section 5, we recover E-dialogues and spread E-dialogues for intuitionistic PCF by another extra restriction, called *last asked first answered* condition. Intuitionistic spread E-dialogues are Hyland-Ong's dialogues where the innocence condition on strategies is now a rule of the game. The observation that the *last asked first answered* condition distinguishes between classical and intuitionistic type systems comes from Lorenz [13].

It is possible to define an interaction between strategies in an E-dialogue (in a way similar to Coquand's interaction [3]), but also in a spread E-dialogue (in this case it is just a "ping-pong"-like process). On the other side, we can evaluate the application of a Böhm tree to another by using a variant of Krivine's Abstract Machine. This is the weak-head reduction. We show that both interactions model the weak-head reduction.

The correspondence between weak-head reduction and interaction in Hyland-Ong's style of games is also proven in Danos *et al* [6]. The framework was the simply-typed pure λ-calculus and the starting point was the analogy between the justification pointers in Hyland-Ong's games and the pointers introduced in Danos & Regnier [7] to implement Krivine's Abstract Machine.

A game-theoretic model of simply-typed pure $\lambda\mu$-calculus (see Parigot [18]) has been given by Ong [17]. It derives from the intuitionistic game by adding new moves. In contrast, our model for classical PCF results from a liberalization of the rules of the intuitionistic game.

Interaction between strategies can be formalized through abstract machines too. The relations between these machines and Krivine's Abstract Machine are shown in [5].

1 Classical Simply-Typed Böhm Trees

We consider a language of Böhm trees for a simply-typed λ-calculus including constants, a case operator and a `catch`/`throw` mechanism (with static binding).

We adopt for the **catch** and **throw** operators a syntax (and later a behaviour) reminiscent of Parigot's $\lambda\mu$-calculus [18]. The operator $\texttt{catch}_\alpha$ is written $\mu\alpha$ and the operator $\texttt{throw}_\alpha$ is written $[\alpha]$. Classical Böhm trees are defined by the following grammar:

$$t ::= \textbf{case } x(u, ..., u) \textbf{ of } (c \to t, ..., c \to t) \mid [\alpha]c$$
$$u ::= \lambda x, ..., x.\mu\alpha.t$$

The letters x and α range over two distinct domains of names and c over the constants in the base types. The x's are called **λ-variables**. They are supposed to be distinct in the expression $\lambda x, ..., x.\mu\alpha.$. The α's are called **μ-variables** (they roughly correspond to entry points of (local) functions). In the **case** construct, the c's are supposed to be distinct and the t's are called **continuations**. We will often use a vector notation as in $\lambda\vec{x}.\mu\alpha.t$ or **case** $x(\vec{u})$ **of** $(\vec{c} \to \vec{t})$.

The objects defined by the entries t and u are respectively called **classical evaluable Böhm trees** and **classical functional Böhm trees.**

Our Böhm trees are simply-typed. Types are built on a family $\mathcal{V}_C$ of base types, each one being inhabited by a finite number of elements. The following grammar defines types

$$A ::= A, ..., A \to C$$

where the sequence $A, ..., A$ may be empty and where C ranges over $\mathcal{V}_C$.

In a type $A = B_1, ..., B_p \to C$ the base type C is called **conclusion** of A and each B_i is called **premise** of A.

The typing system has two kinds of sequents. The sequents $(\Gamma \vdash \Delta)$ type evaluable Böhm trees and the sequents $(\Gamma \vdash \Delta; A)$ type functional Böhm trees. The Γ's are sequences of types annotated by λ-variables. The Δ's are sequences of base types annotated by μ-variables. The typing rules are:

$$\frac{t : (\Gamma, A_1^{x_1}, ..., A_n^{x_n} \vdash \Delta, C^\alpha)}{\lambda x_1, ..., x_n.\mu\alpha.t : (\Gamma \vdash \Delta; A_1, ..., A_n \to C)}\ \textit{Abs} \qquad \frac{c \text{ is in } C}{[\alpha]c : (\Gamma \vdash \Delta, C^\alpha)}\ \textit{Cst}$$

$$\frac{u_1 : (\Gamma \vdash \Delta; A_1) \ \ \ldots \ \ u_n : (\Gamma \vdash \Delta; A_n) \qquad t_1 : (\Gamma \vdash \Delta) \ \ \ldots \ \ t_p : (\Gamma \vdash \Delta)}{\textbf{case } x(u_1, ..., u_n) \textbf{ of } (c_1 \to t_1, ..., c_p \to t_p) : (\Gamma \vdash \Delta)}\ \textit{App}$$

with $(A_1, ..., A_n \to C)^x$ in Γ and $C = \{c_1, ..., c_p\}$.

If $u : (\vdash; A)$, we say that u is a **closed Böhm tree of type** A.

Remarks: 1) The extension to infinite set of constants and/or to non well-founded Böhm trees poses no difficulties. The extension to Böhm trees with undefined nodes is also direct.

2) We justify our langage of Böhm trees as follows. Let

$$\begin{aligned} t ::= & \; x \mid (t\ t) \mid \lambda x.t \\ & \mid c \mid \mathbf{case}\ t\ \mathbf{of}\ (c \to t, ..., c \to t) \\ & \mid \mathtt{catch}_\alpha t \mid \mathtt{throw}_\alpha t \end{aligned}$$

be a grammar for a finite version of PCF with a catch/throw mechanism. In fact, we intend $\mathtt{catch}_\alpha t$ to represent the construction $\mu\alpha[\alpha]t$ of $\lambda\mu$-calculus and $\mathtt{throw}_\alpha t$ the construction $\mu\delta[\alpha]t$ with δ not in t. This is justified by the following derived rule of $\lambda\mu$-calculus (δ and α not in t)

$$E(\mu\alpha[\alpha]E'(\mu\delta[\alpha]t)) \to E(t)$$

where E and E' are applicative contexts and $\mu\delta[\alpha]t$ is in evaluation position in E'.

Consider the theory of λ-calculus with constants and case operators

$$\begin{aligned} (\lambda x.t\ u) &= t[x{:=}u] \\ t &= \lambda x.(t\ x) && x \text{ fresh for } t \\ t &= \mathbf{case}\ t\ \mathbf{of}\ (\vec{c} \to \vec{c}) \\ \mathbf{case}\ c_i\ \mathbf{of}\ (\vec{c} \to \vec{t}) &= t_i \\ (\mathbf{case}\ t\ \mathbf{of}\ (\vec{c} \to \vec{u})\ v) &= \mathbf{case}\ t\ \mathbf{of}\ (\vec{c} \to \overrightarrow{(u\ v)}) \\ \mathbf{case}\ (\mathbf{case}\ t\ \mathbf{of}\ (\vec{c} \to \vec{t}))\ \mathbf{of}\ (\vec{c'} \to \vec{t'}) &= \mathbf{case}\ t\ \mathbf{of}\ (\vec{c} \to \overrightarrow{\mathbf{case}\ t\ \mathbf{of}\ (\vec{c'} \to \vec{t'})}) \end{aligned}$$

enriched by equations coming from the theory of $\lambda\mu$-calculus

$$\begin{aligned} t &= \mathtt{catch}_\alpha t && \alpha \text{ fresh for } t \\ (\mathtt{catch}_\alpha t\ u) &= \mathtt{catch}_\alpha(t[\mathtt{throw}_\alpha v{:=}\mathtt{throw}_\alpha(v\ u)]\ u) \\ (\mathtt{throw}_\alpha t\ u) &= \mathtt{throw}_\alpha t \\ \mathbf{case}\ (\mathtt{catch}_\alpha t)\ \mathbf{of}\ (\vec{c} \to \vec{u}) &= \mathtt{catch}_\alpha(\mathbf{case}\ t'\ \mathbf{of}\ (\vec{c} \to \vec{u})) \\ & \quad \text{where } t' = t[\mathtt{throw}_\alpha v{:=}\mathtt{throw}_\alpha(\mathbf{case}\ v\ \mathbf{of}\ (\vec{c} \to \vec{u}))] \\ \mathbf{case}\ (\mathtt{throw}_\alpha t)\ \mathbf{of}\ (\vec{c} \to \vec{u}) &= \mathtt{throw}_\alpha t \\ \mathtt{throw}_\alpha(\mathbf{case}\ t\ \mathbf{of}\ (\vec{c} \to \vec{t})) &= \mathbf{case}\ t\ \mathbf{of}\ (\vec{c} \to \overrightarrow{\mathtt{throw}_\alpha t}) \\ \mathtt{throw}_\alpha(\mathtt{throw}_\beta t) &= \mathtt{throw}_\alpha t \\ \mathtt{throw}_\alpha(\mathtt{catch}_\beta t) &= \mathtt{throw}_\alpha(t[\beta{:=}\alpha]) \\ \mathtt{catch}_\alpha(\mathtt{catch}_\beta t) &= \mathtt{catch}_\alpha(t[\beta{:=}\alpha]) \\ \mathtt{catch}_\alpha c &= \mathtt{catch}_\alpha \mathtt{throw}_\alpha c \end{aligned}$$

where the various substitutions are defined as in Parigot [18] (replacing $\mathtt{throw}_\alpha$ by $[\alpha]$).

Orient the rules from left to right to get a rewriting system. We can show that any typed term reduced to a typed Böhm tree. Up to the 2nd, 3rd and 7th rules, typed Böhm trees are normal. Thus, assuming the confluence of the rewriting system, the typed Böhm trees describe the equivalence classes of typed terms. This justifies the terminology.

3) According to the theory of $\lambda\mu$-calculus in Parigot [18], any term of the form $\mu\alpha[\beta]t$ is equivalent to $\mu\alpha[\alpha]\mu\delta[\beta]t$ with δ fresh for t. This justifies to abbreviate our $\mathtt{catch}_\alpha t$ (i.e. $\mu\alpha[\alpha]t$) by $\mu\alpha.t$ and our $\mathtt{throw}_\alpha t$ (i.e. $\mu\delta[\alpha]t$ with δ not in t) by $[\alpha]t$.

A precise study of the relations between $\lambda\mu$-calculus and λ-calculus extended by **catch** and **throw** can be found in Crolard [4].

Numeric Böhm trees are a special case of Böhm trees. In numeric Böhm trees, the λ-variables range over pairs of natural numbers and μ-variables on natural numbers. We will use numeric Böhm trees in section 3.1 to prove the correspondence with strategies. The numbers are imposed by the following annotated typing system:

$$\frac{t : (\Gamma, A_1^{\binom{p}{1}}, ..., A_n^{\binom{p}{n}} \overset{p}{\vdash} \Delta, C^\alpha)}{\lambda\binom{p}{1}, ..., \binom{p}{n}.\mu\alpha.t : (\Gamma \overset{p}{\vdash} \Delta; A_1, ..., A_n \to C)}\ Abs \qquad \frac{c \text{ is in } C}{[p]c : (\Gamma \overset{p}{\vdash} \Delta, C^p)}\ Cst$$

$$\frac{u_1 : (\Gamma \overset{p+1}{\vdash} \Delta; A_1) \ \ \dots \ \ u_n : (\Gamma \overset{p+1}{\vdash} \Delta; A_n) \quad t_1 : (\Gamma \overset{p+1}{\vdash} \Delta) \ \ \dots \ \ t_p : (\Gamma \overset{p+1}{\vdash} \Delta)}{\texttt{case } \binom{q}{j}(\mathrm{u}_1, ..., \mathrm{u}_n) \texttt{ of } (c_1 \to t_1, ..., c_p \to t_p) : (\Gamma \overset{p}{\vdash} \Delta)}\ App$$

with $(A_1, ..., A_n \to C)^{\binom{q}{j}}$ in Γ and $C = \{c_1, ..., c_p\}$.

1.1 Weak-Head Reduction

We now define a computation on classical Böhm trees: the weak-head computation of a functional Böhm tree applied to arguments which are themselves Böhm trees. For this purpose, we define a variant of Krivine's Abstract Machine. The machine is described in a syntax reminiscent of $\lambda\sigma$-calculi (see Abadi *et al* [1]), in the style of Leroy [12] or Hardin *et al* [9].

$$\begin{aligned} s &::= t[e] \\ e &::= w; ...; w \\ w &::= (\vec{x} \leftarrow \vec{u}; [\alpha]\vec{c} \to \vec{t})[e] \end{aligned}$$

A **state** of the machine (entry s of the grammar) consists of an evaluable Böhm tree in an environment. An **environment** e is a sequence of windows. A **window** w contains bindings of two kinds. First the bindings of variables to arguments. Second the bindings of a μ-variable and of constants to continuations. Both terms and continuations of a window have meaning in an environment local to them.

There are two kinds of rules. The first rule applies when the computation needs to know the value of a variable x_{ij}. If x_{ij} is bound to $u_{ij} = \lambda\vec{y}.\mu\beta.t'$ in e then its arguments $\vec{u}$ are bound to the formal parameters $\vec{y}$ and the current continuation $\vec{t}$ (what to do when t' returns a constant) is bound to the entry point β of u_{ij}.

$$\texttt{case } x_{ij}(\vec{u}) \texttt{ of } (\vec{c} \to \vec{t})[e] \quad \overset{wh}{\to}_1 \quad t'[(\vec{y} \leftarrow \vec{u}; [\beta]\vec{c} \to \vec{t})[e]; e_i]$$

with $e = w_1; \ldots; w_r$ and $w_i = (\vec{x_i} \leftarrow \vec{u_i}; [\alpha_i]\vec{c_i} \rightarrow \vec{t_i})[e_i]$ and $u_{ij} = \lambda\vec{y}.\mu\beta.t'$.

The second rule applies when a constant is return to some entry point.

$$[\alpha_i]c_j\ [e] \quad \overset{wh}{\rightarrow}_1 \quad t_{ij}[e_i]$$

with $e = w_1; \ldots; w_r$ and $w_i = (\vec{x_i} \leftarrow \vec{u_i}; [\alpha_i]\vec{c_i} \rightarrow \vec{t_i})[e_i]$.

The weak-head reduction works by repeatedly applying the two rules. Let $u_0 = \lambda\vec{x_0}.\mu\alpha_0.t_0$ be a Böhm tree of type $\vec{A} \rightarrow C$ and $\vec{v_0}$ a family of Böhm trees of respective types $A_1, \ldots, A_n$. The **weak-head reduction of u_0 applied to $\vec{v_0}$ is the following sequence:**

$$r_0 = t_0[(\vec{x_0} \leftarrow \vec{v_0}; [\alpha_0]\epsilon)[\,]\ \overset{wh}{\rightarrow}_1 r_1 \overset{wh}{\rightarrow}_1 \ldots \overset{wh}{\rightarrow}_1 r_n \overset{wh}{\rightarrow}_1 \ldots$$

The symbol ϵ denotes the empty sequence of continuations bindings. When u_0 and $\vec{v_0}$ are closed (both for λ-variables and μ-variables), only the constants returned on α_0 are not bound in the environment. This property is preserved from step to step. Thus, if it stops, the sequence stops in a state of the form $[\alpha_0]c_i\ [e]$. Since no continuation is bound to $[\alpha_0]c_i$, no reduction rule applies.

Remark: Our machine arises as a stack-free form of Krivine's Abstract Machine. The typing ensures that the arity of arguments always matches the arity of formal parameters. This is what allows to avoid a stack. The way to handle case operators without stack comes from [5]. More generally, we refer to [5] for a comparison of our machine with Danos-Regnier Pointer Abstract Machine (see [7, 6]) and for an extension to pure λ-calculus. A short proof of correctness of the machine w.r.t. Coquand's debate (as done in section 3.2) also appears in [5].

2 Games

Games interpret types. If A is a type, we write G_A the game which interprets it. It is a game between two players called *Player* and *Opponent*. Moves consist in attacks of subtypes of A or in answers to attacks. Plays are alternating sequences of numbered Player's and Opponent's moves starting from an initial attack of A by Opponent.

Assume A is $B_1, \ldots, B_n \rightarrow C_0$. The **initial attack** of A is written $[_*$ and numbered 0. This attack means both a question on the conclusion C_0 of A and the assertion of the premises B_i of A. All subsequent **attacks** are written $[_i^p$. The exponent p is called **justification.** It is a reference to a previous attack of the other player, say the attack of $B'_1, \ldots, B'_n \rightarrow C'$. This attack was asserting B'_1, ...,B'_n and i is the index of one of the B'_1, ..., B'_n. Here again, the attack means both a question on the conclusion of B'_i and the assertion of each premise of B'_i.

An attack is waiting for an answer. An **answer** is written $]_c^p$. The exponent p is also a **justification.** It corresponds to the number of the attack to which it answers. This attack was questioning a base type, say C'' and c is a constant in C''.

Formally, the **game** G_A is a set of legal positions on A. A **legal position** on A is a finite or infinite sequence $d_0d_1d_2...$ of moves. For $n \geq 1$, it may be convenient to write d_n as $m_n^{p_n}$. The number p_n is the justification and m_n is either $[_i$ or $]_c$. To be a legal position, the sequence has to satisfy the following properties:

- Initial attack of Opponent
  We have $d_0 = [_*$.
- Moves are justified by previous moves of the other player
  For all $n > 0$, p_n is less than n and of distinct parity.
- Correctness of attacks
  For $n \geq 1$, if $d_n = [_i^p$, the move d_p is an attack of some type $B_1, ..., B_m \to C$ and $1 \leq i \leq m$. We say that d_n is an attack of B_i.
- Correctness of answers
  For $n \geq 1$, if $d_n =]_c^p$, the move d_p is an attack of some type $B_1, ..., B_m \to C$ and c is in C.

The moves d_n, with n odd, are called **Player's moves** or **P-moves** and the ones with n even are called **Opponent's moves** or **O-moves**.

2.1 Strategies

Unformally, a strategy for a player is a function mapping legal positions (at which the player is to move) to a move of this player. In a legal position, only the moves of the other player are useful to determine what to move. This leads to the following definition.

Assume A is $B_1, ..., B_{n_0} \to C$. A **P-strategy** (resp **O-strategy**) ϕ for G_A is a function which maps finite sequences of O-moves (resp P-moves) to P-moves (resp O-moves). The domain **Dom**(ϕ) of ϕ is structured as a tree. It satisfies the following clauses:

- For a *P*-strategy: the sequence reduced to the single O-move $[_*$ is in **Dom**(ϕ).
- For an *O*-strategy: the one-P-move sequences $]_c^0$ (with c in C^0) and $[_i^0$ (with $1 \leq i \leq n_0$) are in **Dom**(ϕ).
- If the sequence $d_0d_1...d_n$ is in **Dom**(ϕ) then,
  1- $d_0\phi(d_0)d_1...d_n\phi(d_0...d_n)$ forms a legal position,
  2- $d_0d_1...d_{n+1}$ is in **Dom**(ϕ) if and only if $d_0\phi(d_0)d_1...d_n\phi(d_0...d_n)d_{n+1}$ forms a legal position.

P-strategies are strategies for Player while O-strategies are strategies for Opponent.

Alternatively, a P-strategy can be seen as a (possibly infinite) tree where branches are labelled by O-moves and nodes by P-moves. Moreover, it makes sense to restrict a strategy to what it determines after some point in a play. This leads to the definition of substrategy in tree form beyond a legal position.

A **P-substrategy in tree form beyond** π is inductively defined by:

Let d be a P-move such that πd is legal. If, for any O-move d' such that $\pi dd'$ is legal, $\phi_{d'}$ is a P-substrategy beyond $\pi dd'$, then the tree $(d, (\phi_{d'})_{d'})$ is a P-substrategy beyond π.

The move d labels the root node of the tree and the $\phi_{d'}$ are the branches. When no O-move is allowed after d, the family $(\phi_{d'})_{d'}$ is empty and the P-substrategy is restricted to a leaf. A **P-strategy in tree form** is a P-substrategy in tree form beyond the legal position restricted to the initial move $[_*$.

Proposition 1. *There is a one-to-one correspondence between strategies (as defined by the first definition) and strategies in tree-form. This correspondence preserves the tree structure of the domains of the strategies.*

The notion of P-substrategies in tree form beyond a legal position is a technical notion used to prove the correspondence with Böhm trees.

3 The Model of E-Dialogues

Coquand [3] interprets proofs of the "Calculus of Novikoff" as strategies in a two-players game. The calculus is a (cut-free) sequent calculus (as LJ and LK of Gentzen) for infinitary logic. The game is as in section 2 except for Opponent: the O-moves must be justified by the preceding P-move. Such a game is similar to Lorenzen's classical E-dialogues in Felscher [8].

Here, we define E-dialogues for classical Böhm trees.

3.1 E-Dialogues

The E-dialogue G^E_A interpreting the type A is defined by its set of legal E-positions. A **legal E-position** on A is a legal position in G_A which satisfies the following extra condition:

- O-moves are justified by the preceding P-move:

  For all even $n \neq 0$, we have $p_n = n - 1$.

The justifications of O-moves are trivial in E-dialogues. We do not write them in the sequel.

An **E-P-strategy for** A is a P-strategy for the E-dialogue G^E_A.

Proposition 2. *E-P-strategies for A and closed functional Böhm trees of type A are isomorphic.*

Proof. We show rather the correspondence between E-P-strategies in tree form and numeric Böhm trees. Let π be a legal E-position of odd length. Let Γ be the set of types asserted by Opponent (and thus attackable by Player) in π. Let Δ be the types of the questions asked by Opponent. We tag each type in Γ by a pair consisting of the move number when the type was asserted and of the premise index (as in the definition of numeric Böhm trees). Similarly, we tag each base type in Δ by the move number when the question was asked. We show that E-P-substrategies in tree form beyond π are isomorphic to evaluable Böhm trees typed by the sequent $(\Gamma \vdash \Delta)$.

For well-founded strategies and Böhm trees, the definition of the correspondence is by recursion on the tree structure:

- A P-attack $[_i^p$ corresponds to an occurrence of the *App* rule with head-variable $(_i^p)$. Branches indexed by an O-move correspond to subderivations of the *App* rule.
  - An O-attack $[_i$ corresponds to the ith left premise of the rule *App* (together with the subsequent *Abs* rule).
  - An O-answer $]_{c_i}$ corresponds to the ith right premise of the rule *App*.
- A P-answer $]_c^p$ corresponds to an occurrence of the *Cst* rule with constant c and type C^p.

Finally, the initial move $[_*$ is in correspondence with the top *Abs* rule of functional Böhm trees. Then, it is direct to show that the correspondence is an isomorphism respecting the tree structure.

Hereafter, we write ϕ_u for the E-P-strategy associated to the Böhm tree u.

3.2 Coquand's Debate

Let ϕ be an E-P-strategy for $A_1, ..., A_n \rightarrow C$ and $\vec{\psi}$ be a family of E-P-strategies for $A_1, ..., A_n$ respectively. Since neither ϕ and $\vec{\psi}$ are forced to play moves justified by the preceding opponent's move, it is not possible to directly let them interact.

Coquand [3] proposed a way to let ϕ and $\vec{\psi}$ interact in such a way that the interaction computes a result for C. The idea of Coquand is as follows: at each step of the debate (which is a sequence of moves), there is a canonical way to extract a subsequence which is a legal E-position. From this legal E-position, the E-strategies can be applied. Following Hyland-Ong's terminology [11], we call *view* the extracted E-position. However, in contrast with [11], when Player (resp Opponent) is to move, we keep in the view only the moves of Opponent (resp Player). This is sufficient to apply the E-strategies.

Roughly, the view (typically for P) of a legal position is obtained by forgetting the moves which occur between an O-move and the P-move which justified it.

Formally, the view $\mathcal{V}(\pi)$ of a legal sequence $\pi = d_0 d_1 ... d_n$ is recursively defined as follows:

- If $d_n = m^p$ with $p \neq 0$ then $\mathcal{V}(d_0...d_n) = \mathcal{V}(d_0...d_{p-1})m$

- If $d_n = m^0$ then $\mathcal{V}(d_0...d_n) = m$
- $\mathcal{V}(d_0) = d_0$

If the last move of π is a P-move, the view is an **O-view**. Otherwise, it is a **P-view**.

Each move in the view of π is a move in π with the justification dropped. The **view renumbering sequence** v_π, defined as follows, tells the number that the moves of the view have in π.

- If $d_n = m^p$ with $p \neq 0$ then $v_{d_0...d_n} = v_{d_0...d_{p-1}} n$
- If $d_n = m^0$ then $v_{d_0...d_n} = n$
- $v_{d_0} = 0$

If $v_\pi = v_0 v_1 ... v_{n'}$ is the view renumbering sequence of π. If $d = m^p$ is a move with $p \leq n'$, we note $v_\pi(d)$ for the move m^{v_p}.

The **debate** $d_E(\phi, \vec{\psi})$ between ϕ and $\vec{\psi}$ forms a legal position. Both ϕ and the ψ_i in $\vec{\psi}$ are E-P-strategies but in the resulting play, who plays according to ϕ is Player and who plays according to the ψ_i is Opponent. The first move is an initial attack of Opponent questioning C. Then, both players play in turn:

$$\begin{aligned} d_0 &= [_* \\ d_{2n+1} &= v_{d_0...d_{2n}}(\phi(\mathcal{V}(d_0...d_{2n}))) \\ d_{2n+2} &= v_{d_0...d_{2n+1}}(\vec{\psi}(\mathcal{V}(d_0...d_{2n+1}))) \\ d_E(\phi, \psi) &= d_0 d_1 ... d_n ... \end{aligned}$$

where $\vec{\psi}(m_0 m_1 ... m_q)$ is $\psi_i([_* m_1 ... m_q)$ when m_0 is $[_i$.

Thus, the view transposes the current state of the debate into a subsequence of O-moves in an E-dialogue. From this, the player who is to move can apply its strategy. The view renumbering sequences serve to transpose back the justification in the whole position.

3.3 Debate Models Weak-Head Reduction

Let $u_0 = \lambda \vec{x_0}.\mu\alpha_0.t_0$ be a Böhm tree of type $A_1, ..., A_n \to C$ and $\vec{v_0}$ a family of Böhm trees of respective types $A_1, ..., A_n$. The debate $d_E(\phi_{u_0}, \vec{\phi_{v_0}})$ follows step by step the weak-head reduction of u_0 applied to $\vec{v_0}$.

To express this correspondence, we consider the transposition $d_{\text{wh}}(u_0, \vec{v_0})$ of the weak-head reduction into a legal position. We annotate instances of the first reduction rule by attacks and instances of the second by answers. A superscript on windows is necessary too. At the end, we add a dummy window ()[ ] in the second reduction rule (this simplifies the numbering of windows in the next proof).

$$\textbf{case } x_{ij}(\vec{u}) \textbf{ of } (\vec{c} \to \vec{t})[e] \quad \xrightarrow{wh(n)=[_j^p}{}_1 \quad t'[(\vec{y} \leftarrow \vec{u}; [\beta]\vec{c} \to \vec{t})^n[e]; e_i]$$

with $e = w_1; ...; w_r$ and $w_i = (\vec{x_i} \leftarrow \vec{u_i}; [\alpha_i]\vec{c_i} \rightarrow \vec{t_i})^p[e_i]$ and $u_{ij} = \lambda\vec{y}.\mu\beta.t'$.

$$[\alpha_i]c_j[e] \quad \overset{wh(n)=]^p_{c_i}}{\longrightarrow}_1 \quad t_{ij}[()[\,]; e_i]$$

with $e = w_1; ...; w_r$ and $w_i = (\vec{x_i} \leftarrow \vec{u_i}; [\alpha_i]\vec{c_i} \rightarrow \vec{t_i})^p[e_i]$.

Starting from $r_0 = t_0[(\vec{x_0} \leftarrow \vec{v_0}; [\alpha_0]\epsilon)^0[\,]]$, we get the sequence

$$r_0 \overset{wh(1)=d_1}{\longrightarrow}_1 r_1 \overset{wh(2)=d_2}{\longrightarrow}_1 \cdots \overset{wh(n)=d_n}{\longrightarrow}_1 r_n \overset{wh(n+1)=d_{n+1}}{\longrightarrow}_1 \cdots$$

We write $d_{\mathrm{wh}}(u_0, v_0)$ for the sequence $[_*, d_1, ..., d_n, ...$

We now state the correspondence.

Theorem 3. *If u_0 is a closed Böhm tree of type $A_1, ..., A_n \rightarrow C$ and $\vec{v_0}$ a family of closed Böhm trees of respective types $A_1, ..., A_n$ then we have*

$$d_{\mathrm{wh}}(u_0, \vec{v_0}) = d_E(\phi_{u_0}, \phi_{\vec{v_0}})$$

We need first some definitions.

We define **occurrences** of evaluable Böhm trees and **hat occurrences** of functional Böhm trees in a functional Böhm tree u. Böhm subtrees and P-views are in correspondence. To enforce this link, occurrences are taken to be P-views, i.e. sequences of O-moves with the justification dropped.

- $\lambda\vec{x}.\mu\alpha.t$ has hat occurrence $\widehat{[_*}$ in itself
- t has occurrence $[_*$ in $\lambda\vec{x}.\mu\alpha.t$
- If $\mathtt{case}\ x_p(\lambda\vec{y_1}.\mu\alpha_1.t_1, ..., \lambda\vec{y_n}.\mu\alpha_n.t_n)\ \mathtt{of}\ (c_1 \rightarrow t'_1, ..., c_q \rightarrow t'_q)$ has occurrence π in u then $\lambda\vec{y_i}.\mu\alpha_i.t_i$ has hat occurrence $\widehat{\pi[_i}$ in u, t_i has occurrence $\pi[_i$ in u and t'_i has occurrence $\pi]_{c_i}$ in u.

If u is a Böhm tree and π an occurrence of t in u, we define $u_{|\pi}$ as t. If $\vec{v}$ is a family of Böhm trees and $\pi = [_* m_1 ... m_n$ is an occurrence of t in v_i, we let $\vec{v}_{|[_i m_1 ... m_n} = (v_i)_{|\pi}$. Similarly for hat occurrences.

We can now give the proof.

Proof. Let $u_0 = \lambda\vec{x_0}.\mu\alpha_0.t_0$. Let

$$r_0 \overset{wh(1)=d_1}{\longrightarrow}_1 r_1 \overset{wh(2)=d_2}{\longrightarrow}_1 \cdots \overset{wh(n)=d_n}{\longrightarrow}_1 r_n \overset{wh(n+1)=d_{n+1}}{\longrightarrow}_1 \cdots$$

be the weak-head reduction originating from $r_0 = t_0[(\vec{x_0} \leftarrow \vec{v_0}; [\alpha_0]\epsilon)^0[\,]]$. Let π be the debate $d_E(\phi_{u_0}, \phi_{\vec{v_0}})$. We note $\pi_{|n}$ for the restriction of the debate to the moves 0 to n.

We show by induction on n that $r_n = t_n[e_n]$ where

- if n is even, $t_n = (u_0)_{|\mathcal{V}(\pi_{|n})}$
  if n is odd, $t_n = (v_0)_{|\mathcal{V}(\pi_{|n})}$

- e_n is $w_{q_r};...;w_{q_1}$ where $r = |\mathcal{V}(\pi_{|n})|$ and $q_{i+1} = v_{\pi_{|n}}(i)$
- w_0 is $(\vec{x_0} \leftarrow \vec{v_0}; [\alpha_0]\epsilon)^0[\,]$
  if q is odd (resp q is even) and $(\vec{v_0})_{|\mathcal{V}(\pi_{|q})}$ (resp $(u_0)_{|\mathcal{V}(\pi_{|q})}$) is a constant c then $w_{q+1} = ()[\,]$ otherwise $w_{q+1} = (\vec{x_q} \leftarrow \vec{u_q}; [\alpha_q]\vec{c_q} \rightarrow \vec{t'_q})^q[e_q]$ where
  - if q is even, $u_{qj} = (u_0)_{|\widehat{\mathcal{V}(\pi_{|q})}[_j}$ and $t'_{qj} = (u_0)_{|\mathcal{V}(\pi_{|q})]_{c_j}}$ $(q \neq 0)$
  - if q is odd, $u_{qj} = (\vec{v_0})_{|\widehat{\mathcal{V}(\pi_{|q})}[_j}$ and $t'_{qj} = (\vec{v_0})_{|\mathcal{V}(\pi_{|q})]_{c_j}}$

Clearly, r_0 satisfies this property.

Now if $r_n = t_n[e_n]$, what happens for r_{n+1}? We suppose n odd (the case $n \neq 0$ even is similar). We have to consider the two possible reduction steps:

- $\left| \begin{array}{l} t_n = \texttt{case}\ x(\vec{u'})\ \texttt{of}\ (\vec{c''} \rightarrow \vec{t''}) \text{ with } x = x_{q_i j} \text{ in } e_n \text{ and } u_{q_i j} = \lambda \vec{x'}.\mu\alpha'.t' \\ r_{n+1} = t'[(\vec{x'} \leftarrow \vec{u'}; [\alpha']\vec{c''} \rightarrow \vec{t''})^{n+1}[e_n]; e_{q_i}] \end{array} \right.$
  In this case, we have $wh(n+1) = [^{q_i}_j$ and $\mathcal{V}(\pi_{|n+1}) = \mathcal{V}(\pi_{|q_i})\ [_j$.
  We show first that $t' = t_{n+1}$. We have $u_{q_i j} = (\vec{v_0})_{|\widehat{\mathcal{V}(\pi_{|q})}[_j}$. By definition of the view, we get $u_{q_i j} = (\vec{v_0})_{|\widehat{\mathcal{V}(\pi_{|n+1})}}$ and $t' = (\vec{v_0})_{|\mathcal{V}(\pi_{|n+1})}$.
  Then, we show $[(\vec{x'} \leftarrow \vec{u'}; [\alpha']\vec{c''} \rightarrow \vec{t''})^{n+1}[e_n]; e_{q_i}] = [e_{n+1}] = [w_{q_r};...;w_{q_1}]$ with $r = |\mathcal{V}(\pi_{|n+1})|$ and $q_{i+1} = v_{\pi_{|n+1}}(i)$. By definition of the view, we actually have $r = |\mathcal{V}(\pi_{|q_i})| + 1 = |\mathcal{V}(\pi_{|n+1})|$. By definition of the view and of e_{q_i}, the r first windows actually are the windows w_{q_1} to $w_{q_{r-1}}$ with $q_{k+1} = v_{\pi_{|n+1}}(k)$. But also $v_{\pi_{|n+1}}(r) = n+1$. Then, u'_j $(= (u_0)_{|\widehat{\mathcal{V}(\pi_{|q})}[_j}$ by definition of hat occurrences) is actually $u_{(n+1)j}$ and t''_j $(= (u_0)_{|\mathcal{V}(\pi_{|q})]_{c_j}}$ by definition of occurrences) is actually $t'_{(n+1)j}$.
- $\left| \begin{array}{l} t_n = [\alpha]c_j \text{ with } \alpha = \alpha_{q_i} \text{ in } e_n \\ r_{n+1} = t'_{q_i j}[()[\,]; e_{q_i}] \end{array} \right.$
  Then we have $wh(n+1) =]^{q_i}_{c_j}$ and $\mathcal{V}(\pi_{|n+1}) = \mathcal{V}(\pi_{|q_i})\ ]_{c_j}$.
  We show first that $t'_{q_i j} = t_{n+1}$. We have $t'_{q_i j} = (\vec{v_0})_{|\mathcal{V}(\pi_{|q_i})]_{c_j}}$. By definition of the view, this means $t'_{q_i j} = (\vec{v_0})_{|\mathcal{V}(\pi_{|n+1})}$ as wanted.
  We show then that $[()[\,]; e_{q_i}] = [e_{n+1}] = [w_{q_r};...;w_{q_1}]$ with $r = |\mathcal{V}(\pi_{|n+1})|$ and $q_{i+1} = v_{\pi_{|n+1}}(i)$. As above, this comes by induction hypothesis for windows in e_{q_i}. Moreover, $t_n = (u_0)_{|\mathcal{V}(\pi_{|n})}$ is a constant and w_{n+1} is actually dummy.

This ends the proof

4 The Model of Spread E-Dialogues

The E-dialogues are highly asymmetrical between Player and Opponent. A liberalization of the rules leads to the spread E-dialogues. These dialogues can be seen as a variant for classical PCF of Hyland-Ong's dialogues (see section 5).

4.1 Spread E-Dialogues

We now allow Opponent to play moves justified by a P-move which is not the last move of Player. However, to keep a kind of dialogue which constitutes a model, we internalize the determinism w.r.t. the view in the rules of the game. This is the *same view same move* condition.

The **spread E-dialogue** G_A^S associated to A is defined by its set of legal spread E-positions. A **legal spread E-position** on A is a legal position $\pi = d_0d_1...d_q$ in G_A which satisfies the following extra condition:

- *same view same move* condition
  For all $n, n' \leq q$, if $\mathcal{V}(\pi_{|n}) = \mathcal{V}(\pi_{|n'})$ then there is a move m^p such that both $d_{n+1} = v_{\pi_{|n}}(m^p)$ and $d_{n'+1} = v_{\pi_{|n'}}(m^p)$.

A **spread E-P-strategy for** A is a P-strategy for the spread E-dialogue G_A^S. Similarly for an E-O-strategy. A **spread E-strategy** is either a spread E-P-strategy or a spread E-O-strategy.

4.2 E-P-Strategies and Spread E-Strategies

Since the moves in spread E-strategies are determined by the views (which are sequences of O-moves in E-dialogues), we can expect a bijection between E-P-strategies and spread E-strategies (whatever the strategy is for Player or for Opponent).

In the rest of the section, we consider P-views of sequences of O-moves and O-views of sequences of P-moves. This makes sense since only the knowledge of the moves of other player are relevant in the definition of the view.

Let ϕ be an E-P-strategy for A. Let $\mathbf{Dom}(\phi^*)$ be the set of sequences π of O-moves such that $\mathcal{V}(\pi)$ is in $\mathbf{Dom}(\phi)$. Let ϕ^* be the extension of ϕ on $\mathbf{Dom}(\phi^*)$ defined by $\phi^*(\pi) = v_\pi(\phi(\mathcal{V}(\pi)))$.

Proposition 4. *If ϕ is an E-P-strategy for A then ϕ^* is a spread E-P-strategy for A.*

Similarly, let $\vec{\psi}$ be a family of E-P-strategies for $A_1, ..., A_n$ respectively. Let $\mathbf{Dom}(\vec{\phi}^+)$ be the set of sequences π of P-moves (replacing the first move $[_i^0$ by $[_*$) such that $\mathcal{V}(\pi)$ is in $\mathbf{Dom}(\phi_i)$. Let $\vec{\phi}^+$ be defined on $\mathbf{Dom}(\vec{\phi}^+)$ by $\phi^+(\pi) = v_\pi(\phi_i(\mathcal{V}(\pi')))$ when π begins with $[_i^0$ and π' is obtained from π by replacing $[_i^0$ by $[_*$.

Proposition 5. *If $\vec{\phi}$ is a family of E-P-strategies for $A_1, ..., A_n$ then $\vec{\phi}^+$ is a spread E-O-strategy for $A_1, ..., A_n \rightarrow C$.*

Conversely, a spread E-strategy can be restricted into a E-P-strategy by keeping in the domain only the sequences which come from a legal E-position. Let ϕ be a spread E-P-strategy for A. We define $\mathbf{Dom}(\phi)^-$ as the set of sequences

$[_* m_1 ... m_q$ of O-moves such that $[_* m_1^1 ... m_q^{2q-1}$ is in $\mathbf{Dom}(\phi)$. Similarly, let ψ be a spread E-O-strategy for $A_1, ..., A_n \rightarrow C$. For each i, we define $\mathbf{Dom}(\phi)_i^-$ as the set of sequences $[_* m_1 ... m_q$ of O-moves such that $[_i^0 m_1^2 ... m_q^{2q}$ is a sequence of P-moves in $\mathbf{Dom}(\phi)$.

Proposition 6. *1. If ϕ is a spread E-P-strategy for A then the restriction of ϕ on $\mathbf{Dom}(\phi)^-$ is an E-P-strategy for A.*

2. If ψ is a spread E-O-strategy for $A_1, ..., A_n \rightarrow C$ then the restriction of ψ on $\mathbf{Dom}(\phi)_i^-$ is an E-P-strategy for A_i

3. If ϕ is an E-P-strategy for A then the restriction of ϕ^* on $\mathbf{Dom}(\phi^*)^-$ is ϕ.

4. If ψ is a family of E-P-strategies for $A_1, ..., A_n$ respectively then the restriction of $\vec{\psi}^+$ on $\mathbf{Dom}(\phi^*)_i^-$ is ψ_i.

Corollary 7. *The followings are in bijection:*

- *closed Böhm trees of type A*
- *E-P-strategies for A*
- *spread E-P-strategies for A.*

Also, the followings are in bijection:

- *families of closed Böhm trees of respective types $A_1, ..., A_n$*
- *families of E-P-strategies for $A_1, ..., A_n$*
- *families of spread E-P-strategies for $A_1, ..., A_n$*
- *spread E-O-strategies for $A_1, ..., A_n \rightarrow C$*

4.3 Interaction Between Spread E-Strategies

Let ϕ be a spread E-P-strategy for $A_1, ..., A_n \rightarrow C$ and $\vec{\psi}$ a family of spread E-P-strategies for $A_1, ..., A_n$. By corollary 7, $\vec{\psi}$ can be seen as a spread E-O-strategy Ψ for $A_1, ..., A_n \rightarrow C$. Therefore, it is direct to define an interaction between ϕ and $\vec{\psi}$ by letting ϕ and Ψ play the one against the other.

We define the debate $d_S(\phi, \vec{\psi})$ as follows:

$$\begin{aligned} d_0 &= [_* \\ d_{2n+1} &= \phi(d_0 d_2 ... d_{2n}) \\ d_{2n+2} &= \Psi(d_1 d_3 ... d_{2n+1}) \end{aligned}$$

The interaction in a spread E-dialogue follows step by step the debate in an E-dialogue.

Proposition 8. *If ϕ is an E-P-strategy for $A_1, ..., A_n \rightarrow C$ and $\vec{\psi}$ a family of E-P-strategies for $A_1, ..., A_n$ respectively, then we have*

$$d_E(\phi, \vec{\psi}) = d_S(\phi^*, \vec{\psi}^+)$$

Proof. Directly since $\phi^*(\pi) = v_\pi(\phi(\mathcal{V}(\pi)))$ and similarly for $\vec{\psi}$.

Corollary 9. *If u_0 is a Böhm tree of type $A_1, ..., A_n \rightarrow C$ and $\vec{v_0}$ is a family of Böhm trees of types $A_1, ..., A_n$ respectively then we have*

$$d_{\mathrm{wh}}(u_0, \vec{v_0}) = d_E(\phi_{u_0}, \vec{\phi_{v_0}}) = d_S(\phi^*_{u_0}, \vec{\phi_{v_0}}^+)$$

5 Intuitionistic PCF

5.1 Syntax and Typing of Intuitionistic Böhm Trees

Intuitionistic Böhm trees for PCF are defined by the following restricted syntax:

$$\begin{aligned} t &::= \mathtt{case}\ x(u, ..., u)\ \mathtt{of}\ (c \to t, ..., c \to t) \mid c \\ u &::= \lambda x, ..., x.t \end{aligned}$$

These Böhm trees correspond, up to the undefined Ω, to Hyland-Ong's Finite Canonical Forms of PCF [11].

The typing rules for intuitionistic Böhm trees are:

$$\frac{t : (\Gamma, A_1^{x_1}, ..., A_n^{x_n} \vdash C)}{\lambda x_1, ..., x_n.t : (\Gamma \vdash A_1, ..., A_n \to C)}\ \mathit{Abs} \qquad\qquad \frac{c \text{ is in } C}{\vdash c : (\Gamma \vdash C)}\ \mathit{Cst}$$

$$\frac{u_1 : (\Gamma \vdash A_1) \quad \dots \quad u_n : (\Gamma \vdash A_n) \qquad t_1 : (\Gamma \vdash C) \quad \dots \quad t_p : (\Gamma \vdash C)}{\mathtt{case}\ x(u_1, ..., u_n)\ \mathtt{of}\ (c_1 \to t_1, ..., c_p \to t_p) : (\Gamma \vdash C)}\ \mathit{App}$$

with $(A_1, ..., A_n \to C')^x$ in Γ and $C' = \{c_1, ..., c_p\}$

5.2 Weak-Head Reduction

Environments in the intuitionistic case do not name the continuations:

$$\begin{aligned} s &::= t[e] \\ e &::= w; ...; w \\ w &::= (\vec{x} \leftarrow \vec{u}; \vec{c} \to \vec{t})[e] \end{aligned}$$

The (annotated) rules of reduction differ slightly from the ones of classical case. The first rules becomes

$$\mathtt{case}\ x_{ij}(\vec{u})\ \mathtt{of}\ (\vec{c} \to \vec{t})[e] \quad \overset{wh(n)=[_j^p}{\longrightarrow}{}_1 \quad t'[(\vec{y} \leftarrow \vec{u}; \vec{c} \to \vec{t})^n[e]; e_i]$$

with $e = w_1; ...; w_r$ and $w_i = (\vec{x_i} \leftarrow \vec{u_i}; \vec{c_i} \to \vec{t_i})^p[e_i]$ and $u_{ij} = \lambda\vec{y}.t'$. The second becomes

$$c_j[e] \quad \overset{wh(n)=]_{c_i}^p}{\longrightarrow}{}_1 \quad t_{ij}[()^n[\,]; e_i]$$

with $e = w_1; ...; w_r$ and $w_i = (\vec{x_i} \leftarrow \vec{u_i}; \vec{c_i} \to \vec{t_i})^p[e_i]$ is the first non dummy window (starting from w_1). Thus, intuitionistic PCF returns constants to the last pushed continuation while classical PCF allows to bypass an arbitrary number of continuations.

5.3 Intuitionistic Games

Intuitionistic E-dialogues are E-dialogues where legal positions should satisfy another extra rule.

- *Last asked first answered* condition.
  Let $d_n =]_c^{p_n}$ an answer, and n' such that $p_n < n' < n$. If $d_{n'}$ is an attack then there is n'' such that $n' < n'' < n$ and $d_{n''} =]_{c''}^{n'}$ for some c''. If $d_{n'} =]_{c'}^{p_{n'}}$ is an answer then $p_{n'} \neq p_n$.

Similarly, we get intuitionistic spread E-dialogues by adding the above extra condition to the rules of (classical) spread E-dialogues.

5.4 Debate

The definitions of the debates d_E and d_S are the same for intuitionistic and classical games. Moreover, the propositions 2 and 3 and the corollaries 7 and 9 still hold in the intuitionistic case.

Intuitionistic spread E-dialogues can be understood as Hyland-Ong's games where the innocence condition on strategies is now a rule of the game. As a consequence, all spread E-O-strategies are innocent in our games and therefore in one-to-one correspondence with families of intuitionistic Böhm trees. On the other side, definability for Hyland-Ong's games states that even against a non-innocent opponent (for instance a non-deterministic player), we still have the bijection between P-strategies and Böhm trees.

Acknowledgement

This work has benefited from the discussions with P.-L. Curien, V. Danos and L. Regnier. I specially thank P.-L. Curien for his fruitful feedback on the paper.

References

1. M. Abadi, L. Cardelli, P.-L. Curien and J.-J. Lévy, *Explicit Substitutions*, Journal of Functional Programming 1, pp 375-416, 1991.
2. P. Bernays, *On the Original Gentzen Consistency Proof for Number Theory*, in Intuitionism and Proof Theory, Kino, Myhill & Vesley eds, pp 409-417. Original proof printed in the *The Collected Papers of Gerhard Gentzen* by M. E. Szabo, North Holland, 1969.
3. T. Coquand, *A Semantics of Evidence for Classical Arithmetic, revised version*, Journal of Symbolic Logic, Vol 60, 1995, pp 325-337. First version in: Proceedings of the CLICS workshop, Århus, 1992.
4. T. Crolard, *Extension de l'isomorphisme de Curry-Howard au traitement des exceptions (application d'une étude de la dualité en logique intuitionniste)*, thèse de doctorat, Université Paris 7, 1996.

5. P.-L. Curien and H. Herbelin, *Computing with Abstract Böhm Trees*, submitted, 1996.
6. V. Danos, H. Herbelin, L. Regnier, *Game Semantics and Abstract Machines*, Proceedings, 11th Annual IEEE Symposium on Logic in Computer Science (LICS'96), pp 394-405, 1996.
7. V. Danos, L. Regnier, *Deus Ex Machina*, unpublished paper, 1990.
8. W. Felscher, *Dialogues as a Foundation of Intuitionistic Logic*, Handbook of Philosophical Logic, Vol 3, pp 341-372, 1986.
9. T. Hardin, L. Maranget and B. Pagano, *Functional Back-Ends within the Lambda-Sigma Calculus*, Proceedings, International Conference on Functional Programming (ICFP'96), ACM Press, pp 25-33, 1996.
10. H. Herbelin, *Séquents qu'on calcule: de l'interprétation du calcul des séquents comme calcul de λ-termes et comme calcul de stratégies gagnantes*, thèse de doctorat, Université Paris 7, 1995.
11. M. Hyland, C.-H. L. Ong, *On Full Abstraction for PCF*, submitted, currently available at ftp://ftp.comlab.ox.ac.uk/pub/Documents/techpapers/Luke.Ong/.
12. X. Leroy, *The ZINC Experiment*, Technical Report, number 117, INRIA, 1990.
13. K. Lorenz, *Arithmetik und Logik als Spiele*, Dissertation, Universität Kiel, 1961. Partially reprinted in [15].
14. P. Lorenzen, *Logik und Agon*, in Atti Congr. Internat. di Filosofia, Vol 4, Sansoni, Firenze, pp 187-194, 1960. Reprinted in [15].
15. P. Lorenzen, K. Lorenz, *Dialogische Logik*, Wissentschaftliche Buchgesellschaft, Darmstadt, 1978.
16. P. Lorenzen, K. Schwemmer, *Konstriktive Logik, Ethik und Wissenschafttheorie*, Bibliograph. Institut Mannheim, 1973.
17. C.-H. L. Ong, *A Semantic View of Classical Proofs*, Proceedings, 11th Annual IEEE Symposium on Logic in Computer Science (LICS'96), pp 230-241, 1996.
18. M. Parigot, *λμ-Calculus: an Algorithmic Interpretation of Classical Natural Deduction*, Springer Lecture Notes in Computer Sciences 624, pp 190-201.

A Type Theoretical View of Böhm-Trees *

Toshihiko Kurata

Department of Mathematical and Computing Sciences
Tokyo Institute of Technology
Ookayama, Meguro-ku, Tokyo 152, Japan
kurata@is.titech.ac.jp

Abstract. Two variations of the intersection type assignment system are studied in connection with Böhm-trees. One is the intersection type assignment system with a non-standard subtype relation, by means of which we characterize whereabouts of $\perp$ in Böhm-trees. The other is a refinement of the intersection type assignment system whose restricted typability is shown to coincide with finiteness of Böhm-trees.

1 Introduction

The intersection type assignment system $\lambda\wedge$ is introduced in [4] as an extension of Curry's simple type assignment system. The set of types of this system is defined by adding the constructor $\wedge$ for intersection and the type-constant ω to the formation of simple types. This extension induces the invariance of types under β-conversion (especially β-expansion), which is one of the reasons why the intersection type assignment system enjoys nice theoretical properties. For example, the invariance property is indispensable to the following characterization of solvability and normalizability.

Proposition (Coppo et al. [4, Theorems 4 and 5]).
(i) *M is solvable if and only if there exist a type $A_1 \to \cdots \to A_n \to a$ and a basis Γ such that $\Gamma \vdash_{\lambda\wedge} M : A_1 \to \cdots \to A_n \to a$ and $a \not\equiv \omega$.*
(ii) *M is normalizable if and only if there exist a type A and a basis Γ such that $\Gamma \vdash_{\lambda\wedge} M : A$ and ω does not appear in Γ and A positively.*

Our purpose in this paper is, as in this proposition, to give type-theoretical characterizations of some fundamental properties of λ-terms. We shall study two properties concerning Böhm-trees through variations of the system $\lambda\wedge$. First we introduce a non-standard subtype relation to $\lambda\wedge$, by which we shall characterize the property: whether or not $\perp$ appears in the n-th level or above of the Böhm-tree of M. Second we shall refine the system $\lambda\wedge$, and show that a restricted typability of the system so obtained coincides with the finiteness of Böhm-trees.

This paper is organized as follows. In Sect. 2, we shall recall the definition of the intersection type assignment systems $\lambda\wedge_\nabla$ with subtype relations $\leq_\nabla$, and

* This research was supported in part by JSPS Research Fellowships for Young Scientists.

their relation to filter λ-models. We shall also note the isomorphism between filter domains and Scott's inverse limit spaces, which is due to [5, 9]. In Sect. 3, we shall extend the notion of logical relation for simple types [11] to one for intersection types. This notion will be a generalization of the computability predicates used, for example, in the proof of the approximation theorem in [5] and in the proof of the strong normalizability of λ-terms typable in the intersection type assignment system without ω in [1]. From a model-theoretical viewpoint, we can consider logical relations as a variation of the simple semantics of intersection types. We shall prove the completeness of the system $\lambda\wedge_\nabla$ under the interpretation based on logical relations. In Sect. 4, we shall investigate the characterization of whereabouts of $\perp$ in Böhm-trees. Through the correspondence, mentioned in Sect. 2, of the systems $\lambda\wedge_\nabla$ to filter λ-models and inverse limit spaces, our result immediately implies model-theoretical characterizations of the syntactical property. In Sect. 5, we shall study a characterization of the finiteness of Böhm-trees in terms of typability in another variation of $\lambda\wedge$. Roughly speaking, this system is defined so as to preserve invariance of types under β-conversion even though it has only a weak (ω)-axiom.

2 Preliminaries

We first recall the *intersection type assignment systems with subtype relation.* The set $\mathrm{T}_\wedge$ of *types* is inductively defined by the following abstract grammar;

$$\mathrm{T}_\wedge ::= a \mid \omega \mid \mathrm{T}_\wedge \to \mathrm{T}_\wedge \mid \mathrm{T}_\wedge \wedge \mathrm{T}_\wedge$$

where a ranges over a set of type-variables. We write Atom for the set consisting of type-variables and the type-constant ω, and call its elements *atomic-types.* We use letters $A, B, C, \ldots$ as meta-variables standing for types in $\mathrm{T}_\wedge$. As usual, we omit parentheses in types under the assumption that the intersection associates with types more strongly than the arrow and that the arrow associates to the right. Let ∇ be a set of axioms and rules for binary relations over $\mathrm{T}_\wedge$. Then the *subtype relation* $\leq_\nabla$ based on ∇ is defined as the least binary relation satisfying axioms and rules in ∇ and those listed below:[2]

$(a1)\ A \leq A \wedge A,$	$(a5)\ (A \to B) \wedge (A \to C) \leq A \to B \wedge C,$
$(a2)\ A_1 \wedge A_2 \leq A_i \quad (i = 1, 2),$	$(r1)\ A \leq B\ \&\ B \leq C \Longrightarrow A \leq C,$
$(a3)\ A \leq \omega,$	$(r2)\ A \leq B\ \&\ C \leq D \Longrightarrow A \wedge C \leq B \wedge D,$
$(a4)\ \omega \leq \omega \to \omega,$	$(r3)\ A \leq B\ \&\ C \leq D \Longrightarrow B \to C \leq A \to D\ .$

The binary relation $\leq_\nabla$ is a pre-order for any ∇, and hence we can naturally introduce an equivalence relation $\sim_\nabla$ as the intersection of $\leq_\nabla$ and its inverse. Since this equivalence relation satisfies the associative law and the commutative law with respect to $\wedge$, we identify intersections of types $A_1, A_2, \ldots, A_n$ in any

[2] In the case where ∇ is empty, we call $\leq_\emptyset$ the *standard subtype relation.* The system $\lambda\wedge_\emptyset$ is extensively studied by H. P. Barendregt et al. [3].

association and order, and write it as $\bigwedge_{i=1}^{n} A_i$. For a basis[3] Γ, a (type-free) λ-term M and $A \in \mathrm{T}_\wedge$, the *judgment* $\Gamma \vdash M : A$ in the intersection type assignment system with subtype relation $\leq_\nabla$ is generated by axioms and inference rules below:

$$(\text{var})\ \Gamma \vdash x : A \quad \text{if } x : A \in \Gamma \qquad (\omega)\quad \Gamma \vdash M : \omega$$

$$(\to \mathrm{I})\ \frac{\Gamma, x : A \vdash M : B}{\Gamma \vdash \lambda x.M : A \to B} \qquad (\to \mathrm{E})\ \frac{\Gamma \vdash M : A \to B \quad \Gamma \vdash N : A}{\Gamma \vdash MN : B}$$

$$(\wedge \mathrm{I})\ \frac{\Gamma \vdash M : A \quad \Gamma \vdash M : B}{\Gamma \vdash M : A \wedge B} \qquad (\leq_\nabla)\ \frac{\Gamma \vdash M : A}{\Gamma \vdash M : B}\ \text{if } A \leq_\nabla B$$

We denote this system by $\lambda\wedge_\nabla$, and use the symbol $\vdash_\nabla$ to indicate the subtype relation $\leq_\nabla$ underlying the system.

One of the advantages of the preceding systems is that we can construct λ-models[4] based on some of these systems. We briefly recall the construction. For given ∇, we write F_∇ for the set of *filters*[5] over $\mathrm{T}_\wedge$ with respect to the subtype relation $\leq_\nabla$. The application $\cdot$ and the mapping $[\![]\!]^{\mathcal{F}_\nabla} : \Lambda \times \mathrm{Env}_{\mathcal{F}_\nabla} \longrightarrow F_\nabla$[6] are defined by

$$\alpha \cdot \beta = \{B \mid A \to B \in \alpha \text{ for some } A \in \beta\},$$
$$[\![M]\!]_\xi^{\mathcal{F}_\nabla} = \{A \mid \Gamma \vdash_\nabla M : A \text{ for some } \Gamma \subseteq \mathrm{B}_\xi\}$$

where $\mathrm{Env}_{\mathcal{F}_\nabla}$ is the set of mappings from Var to F_∇[7] and $\mathrm{B}_\xi =_{\mathrm{def}} \{x : A \mid A \in \xi(x)\}$. For some ∇, the structure $\mathcal{F}_\nabla = \langle F_\nabla, \cdot, [\![]\!]^{\mathcal{F}_\nabla}\rangle$ turns out to be a λ-model, which we call a *filter λ-model*. However it is not the case that all ∇ yield λ-models. In this regard, the following equivalence is well-known.

Proposition 1 (Coppo et al. [5, Corollary 2.7]).
(i) *The structure $\mathcal{F}_\nabla$ is a λ-model if and only if types are invariant under β-conversion in the system $\lambda\wedge_\nabla$; that is, if $\Gamma \vdash_\nabla M : A$ and $M =_\beta N$ then $\Gamma \vdash_\nabla N : A$.*
(ii) *The structure $\mathcal{F}_\nabla$ is an extensional λ-model if and only if types are invariant under $\beta\eta$-conversion in the system $\lambda\wedge_\nabla$.*

Moreover, the filter λ-models so obtained have the following neat correspondence with Scott's inverse limit spaces generated from ω-algebraic complete lattices. (As for the construction of inverse limit spaces, see [2, Sect. 18.2].)

---

[3] A finite set $\Gamma = \{x_1 : A_1, \ldots, x_n : A_n\}$ is called a basis if x_i's are distinct variables and A_i's are types, and we define $\mathrm{pred}(\Gamma) = \{A_1, \ldots, A_n\}$. The set of all bases is denoted by Basis.

[4] In this paper, by a λ-model, we mean a *syntactical λ-model* in [2, Sect. 5.3].

[5] A filter is a non-empty subset of $\mathrm{T}_\wedge$ closed under $\wedge$ and upper-closed with respect to $\leq_\nabla$. We use Greek letters $\alpha, \beta, \gamma, \ldots$ as meta-variables for filters.

[6] In this paper, we fix a set Var of term-variables, and write Λ for the set of λ-terms generated therefrom.

[7] As usual, these mappings are called *environments* to $\mathcal{D}$. We use Greek letter ξ for environments.

Proposition 2 (Coppo et al. [5, Theorem 2.9]). *Let* $(D_0, \sqsubseteq)$ *be an* ω*-algebraic complete lattice with a bijection* $\tau : \text{Atom} \longrightarrow \mathsf{K}(D_0)$[8] *satisfying* $\tau(\omega) = \bot$, *and* (φ_0, ψ_0) *a projection of* $[D_0 \to D_0]$ *onto* D_0. *Then the inverse limit space* $(D_\infty, \cdot)$ *constructed from* D_0 *and* (φ_0, ψ_0) *is isomorphic, both as ordered sets and as applicative structures, to* $(F_\nabla, \cdot)$ *where*

$$\nabla = \{a \le b \mid \tau(b) \sqsubseteq \tau(a)\} \cup \{a \sim \textstyle\bigwedge_{i=0}^{n}(o_i \to p_i) \mid \varphi_0(\tau(a)) = \bigsqcup_{i=0}^{n}(\tau(o_i)\searrow \tau(p_i))\}\ .^{9}$$

Proof. Let us define the mapping $h : \mathrm{T}_\wedge \longrightarrow \mathsf{K}(D_\infty)$, as follows:

$$h(a) = \Phi_{0,\infty}(\tau(a)),^{10} \quad h(A \to B) = G(h(A)\searrow h(B)),^{11}$$
$$h(\omega) = \bot_{D_\infty}, \quad h(A \wedge B) = h(A) \sqcup h(B)\ .$$

Then the isomorphism $g : F_\nabla \longrightarrow D_\infty$ is given by $g(\alpha) = \bigsqcup\{h(A) \mid A \in \alpha\}$ for each $\alpha \in F_\nabla$. □

Corollary 3. *For* D_∞, ∇ *and* g *above,*
(i) *The mapping* g *is a* λ*-model isomorphism between* $\mathcal{F}_\nabla$ *and* D_∞; *that is,* $[\![M]\!]^{D_\infty}_{g\circ\xi} = g([\![M]\!]^{\mathcal{F}_\nabla}_{\xi})$ *for each* $M \in \Lambda$ *and* $\xi \in \mathrm{Env}_{\mathcal{F}_\nabla}$.
(ii) *The structure* $\mathcal{F}_\nabla$ *is an extensional* λ*-model.*

3 Logical Relations for Intersection Types

The notion of logical relation can be traced back, to the author's knowledge, to [15], in which W. W. Tait formalized Gödel's reckonability of functionals of finite type. Since then the notion has successfully been applied to various typed λ-calculi, yielding important syntactical and semantical results. For example, J-Y. Girard [6, 7] proved strong normalizability of the second order typed λ-calculus, and G. Plotkin [12] extracted λ-definable elements from the set-theoretical models of simply typed λ-calculus. In this section, we introduce the notion of logical relation for intersection types by extending the idea of [11, 14]. This is also a generalization of the computability predicates in [1, 5].

We first introduce logical relations over λ-models, and later we shall extend it to more general structures. Let $\mathcal{D} = \langle D, \cdot, [\![]\!]^{\mathcal{D}}\rangle$ be a λ-model. Then a mapping

[8] For a cpo D, we say $d \in D$ is *compact* if, for each directed subset U of D, $d \sqsubseteq \bigsqcup U$ implies the existence of $e \in U$ such that $d \sqsubseteq e$. We write $\mathsf{K}(D)$ for the set of compact elements of D.

[9] For $d \in \mathsf{K}(D)$ and $e \in \mathsf{K}(E)$, we write $(d\searrow e)$ for *the step function* defined by d and e; that is, for each $d' \in D$, if $d \sqsubseteq d'$ then $(d\searrow e)(d') = e$ else $(d\searrow e)(d') = \bot_E$. In the description of ∇, a and b range over type-variables, and o_i and p_i over atomic-types (i.e., type-variables and ω).

[10] The mapping $\Phi_{0,\infty} : D_0 \longrightarrow D_\infty$ is the embedding defined in [2, Definition 18.2.5].

[11] The mapping $G : [D_\infty \to D_\infty] \longrightarrow D_\infty$ is the isomorphism defined in [2, Theorem 18.2.16].

R : Basis $\times$ $\mathrm{T}_\wedge \longrightarrow \wp D$[12] is said to be a (*unary*[13]) *logical relation* over $\mathcal{D}$ if it satisfies

$$\begin{array}{ll}\text{(LR1)} & R^\Gamma(\omega) = D, \\ \text{(LR2)} & R^\Gamma(A \to B) = \{d \in D \mid d \cdot e \in R^{\Gamma \uplus \Delta}(B) \\ & \qquad\qquad \text{for each basis } \Delta \text{ and } e \in R^\Delta(A)\}, \\ \text{(LR3)} & R^\Gamma(A \wedge B) = R^\Gamma(A) \cap R^\Gamma(B)\end{array}$$

for each basis Γ. Here, for each basis Γ and Δ, we define

$$\begin{array}{rl}\Gamma \uplus \Delta = & \{x : A \mid x : A \in \Gamma \,\&\, (x : B \in \Delta \Longrightarrow A \equiv B)\} \cup \\ & \{x : A \mid x : A \in \Delta \,\&\, (x : B \in \Gamma \Longrightarrow A \equiv B)\} \cup \\ & \{x : A \wedge B \mid x : A \in \Gamma \,\&\, x : B \in \Delta \,\&\, A \not\equiv B\} .\end{array}$$

Note that a logical relation is uniquely determined from its values at type-variables. We write $\mathrm{LR}_\mathcal{D}$ for the set of logical relations over $\mathcal{D}$.

Logical relations so defined may be considered as a variation of simple semantics of types. Note however that the value of a logical relation for a type A is determined only when a basis Γ is supplied, and it is equal to the set of interpretations of λ-terms which have the type A under the basis Γ.

In what follows, we show that the system $\lambda\wedge_\nabla$ is complete with respect to the interpretation based on logical relations, provided types are invariant under β-conversion in $\lambda\wedge_\nabla$. We begin by defining two notions of monotonicity for logical relations. A logical relation R is said to be *monotone with respect to subtype relation* $\leq_\nabla$ if $A \leq_\nabla B$ implies $R^\Gamma(A) \subseteq R^\Gamma(B)$ for each Γ. Next, defining a binary relation $\Subset$ between bases by

$$\Gamma \Subset \Delta \Longleftrightarrow \Delta = \Gamma \uplus \Sigma \text{ for some } \Sigma,$$

we say R is *monotone with respect to* $\Subset$ if $\Gamma \Subset \Delta$ implies $R^\Gamma(A) \subseteq R^\Delta(A)$ for each $A \in \mathrm{T}_\wedge$. Then we define the following notion of *validity*, which is in accordance with [5];

$$\begin{array}{rl}\Delta \models_\nabla M : A \Longleftrightarrow & (\mathcal{D}, \xi, R^\Gamma \models \Delta \Longrightarrow [\![M]\!]^\mathcal{D}_\xi \in R^\Gamma(A)) \\ & \text{for each } \lambda\text{-model } \mathcal{D}, \xi \in \mathrm{Env}_\mathcal{D}, \text{basis } \Gamma \text{ and} \\ & R \in \mathrm{LR}_\mathcal{D} \text{ which is monotone with respect to } \leq_\nabla \text{ and } \Subset ,\end{array}$$

where $\mathcal{D}, \xi, R^\Gamma \models \Delta$ means that $\xi(x) \in R^\Gamma(A)$ for each $x : A \in \Delta$.

Theorem 4 (Soundness). *If* $\Gamma \vdash_\nabla M : A$ *then* $\Gamma \models_\nabla M : A$.

Proof. By induction on the length of the derivation. □

Now we prove completeness. For this, we take the (*open*) *term model* $\mathcal{T}_\beta = \langle \Lambda/{=_\beta}, \cdot, [\![]\!]^{\mathcal{T}_\beta} \rangle$. When types are invariant under β-conversion in the system $\lambda\wedge_\nabla$, we define the logical relation $R_\nabla \in \mathrm{LR}_{\mathcal{T}_\beta}$ by

[12] $\wp D$ stands for the power-set of D.

[13] Though we consider only unary logical relations in this paper, our argument can easily be generalized to logical relations of arbitrary arity (cf. [11, 14]).

$$R^{\Gamma}_{\nabla}(a) = \{[M] \mid \Gamma \vdash_{\nabla} M : a\}^{14}$$

for each basis Γ and type-variable a.

Lemma 5. *Suppose types are invariant under β-conversion in the system $\lambda\wedge_{\nabla}$.*
(i) *For each A and Γ, $R^{\Gamma}_{\nabla}(A) = \{[M] \mid \Gamma \vdash_{\nabla} M : A\}$.*
(ii) *R_{∇} is monotone with respect to the subtype relation $\leq_{\nabla}$ and the relation $\Subset$.*

Proof. (i) By induction on the structure of A, we prove that the equality holds for each Γ.
Consider the case where A is an arrow type, say $B \to C$. To show that the left-hand side of the equality includes the right-hand side, suppose $\Gamma \vdash_{\nabla} M : B \to C$. Then, by induction hypothesis, for each Δ and $[N] \in R^{\Delta}_{\nabla}(B)$, we have $\Delta \vdash_{\nabla} N : B$. Hence we have $\Gamma \uplus \Delta \vdash_{\nabla} MN : C$, which implies $[M]\cdot[N] = [MN] \in R^{\Gamma\uplus\Delta}_{\nabla}(C)$ again by induction hypothesis. Thus we get $[M] \in R^{\Gamma}_{\nabla}(B \to C)$.
To see the reverse direction, suppose $[M] \in R^{\Gamma}_{\mathcal{T}_\beta,\nabla}(B \to C)$. For a fresh term-variable z, because of the judgment $z : B \vdash_{\nabla} z : B$, we have $[z] \in R^{z:B}_{\nabla}(B)$ by induction hypothesis. Hence it follows from the definition of logical relations that $[Mz] = [M] \cdot [z] \in R^{\Gamma,z:B}_{\nabla}(C)$, which implies $\Gamma, z : B \vdash_{\nabla} Mz : C$ again by induction hypothesis. Then, by generation lemma [5, Lemma 1.7], we can find a type B' such that $\Gamma \vdash_{\nabla} M : B' \to C$ and $B \leq_{\nabla} B'$. Hence $\Gamma \vdash_{\nabla} M : B \to C$.
The other cases are easier and their proofs are omitted.
(ii) is immediate from (i). □

Theorem 6 (Completeness). *Suppose types are invariant under β-conversion in the system $\lambda\wedge_{\nabla}$. Then $\Gamma \models_{\nabla} M : A$ if and only if $\Gamma \vdash_{\nabla} M : A$.*

Proof. If part is by Theorem 4. To show only-if part, define $\iota \in \mathrm{Env}_{\mathcal{T}_\beta}$ by $\iota(x) = [x]$ for each term-variable x. Then we have $\mathcal{T}_\beta, \iota, R^{\Gamma}_{\nabla} \models \Gamma$. This is because, for each $x : B \in \Gamma$, the judgment $\Gamma \vdash_{\nabla} x : B$ and Lemma 5 (i) imply $\iota(x) = [x] \in R^{\Gamma}_{\nabla}(B)$. Hence, from the assumption, it follows that $[M] = [\![M]\!]^{\mathcal{T}_\beta}_{\iota} \in R^{\Gamma}_{\nabla}(A)$, which together with Lemma 5 (i) implies $\Gamma \vdash_{\nabla} M : A$. □

The proof above is simpler than that of the completeness for simple semantics in [8, 5] in the sense that our proof does not need the notion of extension of basis.

Next we generalize the notion of logical relation and prove that $\lambda\wedge_{\nabla}$ is still sound for the semantics defined thereby. This soundness, when we restrict our attention to simple types, is called the basic lemma in [11].

We say a triple $\mathcal{D} = \langle D, \cdot, [\![]\!]^{\mathcal{D}} \rangle$ is a *weak semantic structure* if D is a set with the binary operator $\cdot$ and the mapping $[\![]\!]^{\mathcal{D}} : \Lambda \times \mathrm{Env}_{\mathcal{D}} \longrightarrow D$ satisfies

1. $[\![x]\!]^{\mathcal{D}}_{\xi} = \xi(x)$,
2. $[\![MN]\!]^{\mathcal{D}}_{\xi} = [\![M]\!]^{\mathcal{D}}_{\xi} \cdot [\![N]\!]^{\mathcal{D}}_{\xi}$.

Clearly this notion of structure is a generalization of λ-model. For a weak semantic structure $\mathcal{D} = \langle D, \cdot, [\![]\!]^{\mathcal{D}} \rangle$, a mapping $R :$ Basis $\times\ \mathrm{T}_{\wedge} \longrightarrow \wp D$, which satisfies (LR1) through (LR3), is said to be a *logical relation* over $\mathcal{D}$ provided that

[14] We write $[M]$ for the equivalence class under $=_\beta$ containing M.

$$\exists\Sigma\ (\Sigma, x : A \vdash_\nabla M : B \ \&\ \mathcal{D}, \xi, R^\Gamma \models \Sigma^{15}\ )\ \&$$
$$\forall\Delta\,\forall d \in R^\Delta(A)\ [\![M]\!]^{\mathcal{D}}_{\xi(x:d)} \in R^{\Gamma\uplus\Delta}(B)$$
$$\Longrightarrow \forall\Delta\,\forall d \in R^\Delta(A)\ [\![\lambda x.M]\!]^{\mathcal{D}}_\xi \cdot d \in R^{\Gamma\uplus\Delta}(B)\ .$$

The condition above is called the ***admissibility*** of R. Note that, in case where $\mathcal{D}$ is a λ-model, the admissibility condition is always satisfied, because the equation $[\![\lambda x.M]\!]^{\mathcal{D}}_\xi \cdot d = [\![M]\!]^{\mathcal{D}}_{\xi(x:d)}$ holds for the valuation mapping.

Lemma 7 (Soundness of $\lambda\wedge_\nabla$ over weak semantic structures). *If $\Gamma \vdash_\nabla M : A$ then $\mathcal{D}, \xi, R^\Delta \models \Gamma$ implies $[\![M]\!]^{\mathcal{D}}_\xi \in R^\Delta(A)$, for any weak semantic structure $\mathcal{D}$, $\xi \in \mathrm{Env}_{\mathcal{D}}$, basis Δ and logical relation R over $\mathcal{D}$ which is monotone with respect to $\leq_\nabla$ and $\Subset$.*

Proof. By induction on the length of the derivation. The condition of admissibility is necessary for the case of ($\to$I)-rule. □

As an example of weak semantic structure, the *applicative structure* $\mathcal{L} = \langle\Lambda, \cdot, [\![]\!]^{\mathcal{L}}\rangle$ *of λ-terms* is important, where application $\cdot$ and the mapping $[\![]\!]^{\mathcal{L}}$ are defined by

$$M \cdot N \equiv MN,$$
$$[\![M]\!]^{\mathcal{L}}_\xi \equiv M[x_1 := \xi(x_1), \ldots, x_n := \xi(x_n)]$$

where $\mathrm{FV}(M) = \{x_1, \ldots, x_n\}$. We shall abbreviate the instance of simultaneous substitution in the latter identity to $M\theta_\xi$.

Logical relations over the weak semantic structure $\mathcal{L}$ have been used to show syntactical properties, such as the strong normalizability and the Church-Rosser property, of typed λ-terms. They will also play a central role in our argument in later sections. In this regard, we note the next lemma which ensures that the closure under head β-expansion is sufficient for logical relations to be admissible.

Lemma 8. *Suppose $R : \mathrm{Basis} \times \mathrm{T}_\wedge \longrightarrow \wp\Lambda$ satisfies* (LR1) *through* (LR3) *and the condition*

$$M[x := N]\bar{P} \in R^{\Gamma\uplus\Delta}(a) \text{ for each } \Delta \text{ and } N \in R^\Delta(A)$$
$$\Longrightarrow (\lambda x.M)N\bar{P} \in R^{\Gamma\uplus\Delta}(a) \text{ for each } \Delta \text{ and } N \in R^\Delta(A)\ .^{16}$$

Then R is admissible; hence, R is a logical relation over $\mathcal{L}$.

Proof. We prove the following stronger statement by induction on the structure of type B.

$$[\![M]\!]^{\mathcal{L}}_{\xi(x:N)}\bar{P} \in R^{\Gamma\uplus\Delta}(B) \text{ for each } \Delta \text{ and } N \in R^\Delta(A)$$
$$\Longrightarrow [\![\lambda x.M]\!]^{\mathcal{L}}_\xi N\bar{P} \in R^{\Gamma\uplus\Delta}(B) \text{ for each } \Delta \text{ and } N \in R^\Delta(A)\ .$$

---

[15] I.e., $\xi(y) \in R^\Gamma(C)$ for each $y : C \in \Sigma$.

[16] For $m \geq 0$, we often abbreviate $\lambda x_1 \cdots x_m.M$ and $MN_1 \cdots N_m$ to $\lambda\bar{x}.M$ and $M\bar{N}$, respectively.

For the space limitation, we verify the statement only in the case where B is a type-variable, say a. Assume the left-hand side, and take a fresh variable z. Then we have $(M[x := z]\theta_{\xi(z:z)})[z := N]\bar{P} = [\![M]\!]^{\mathcal{L}}_{\xi(x:N)}\bar{P} \in R^{\Gamma \uplus \Delta}(a)$ for each Δ and $N \in R^{\Delta}(A)$. From the assumption displayed above in the lemma, it follows that $(\lambda z.(M[x := z]\theta_{\xi(z:z)}))N\bar{P} \in R^{\Gamma \uplus \Delta}(a)$ for each Δ and $N \in R^{\Delta}(A)$. This completes the proof of this case. □

4 Whereabouts of ⊥ in Böhm-Trees

As mentioned in Sect. 2 Propositions 1 and 2, there are close connections among the systems $\lambda\wedge_{\nabla}$, filter λ-models and inverse limit spaces. These connections enable us to study filter λ-models and inverse limit spaces from a type-theoretical point of view. Indeed, M. Coppo et al. [5, Sect. 4] succeeded in this approach to construct a filter λ-model characterizing the set of normalizable terms, and moreover a non-standard inverse limit space whose local structure is not Hilbert-Post complete. Along the same line, in this section we will investigate a filter λ-model and an inverse limit space characterizing the following sets

$$\mathrm{SOL}_n = \{M \mid \bot \text{ does not appear in } n\text{-th level or above in } \mathrm{BT}(M)\}$$

where $n \in \mathbb{N}(=_{\mathrm{def}} \{0, 1, \ldots\})$. For example, normalizable terms and fixed point combinators are all contained in $\bigcap_{n=0}^{\infty} \mathrm{SOL}_n$, and $\lambda x.x(\lambda x.x\Omega) \in \mathrm{SOL}_1 \setminus \mathrm{SOL}_2$ where $\Omega \equiv (\lambda x.xx)(\lambda x.xx)$. Note that SOL_0 is identical with the set of solvable terms.

We begin with the definition of an intersection type assignment system characterizing the sets SOL_n. Let us fix the set of type-variables and the set of axioms for subtype relations by analogy with those in [5, Definition 3.1]. Type-variables consist of three series $a_0, a_1, \ldots, b_0, b_1, \ldots, c_1, c_2, \ldots$, and the set $\star$ of axioms is defined by

$$\begin{array}{ll} (\star 1)\ b_n \leq a_n, & (\star 4)\ b_0 \sim \omega \to b_0, \\ (\star 2)\ a_0 \sim b_0 \to a_0, & (\star 5)\ b_{n+1} \sim a_n \to b_{n+1}, \\ (\star 3)\ a_{n+1} \sim c_{n+1} \to a_{n+1}, & (\star 6)\ c_{n+1} \sim (\omega \to b_0) \wedge \bigwedge_{i=0}^{n}(a_i \to b_{i+1}), \end{array}$$

where $n \in \mathbb{N}$. Informally speaking, one may consider the type-variables a_n and b_n corresponding to the set SOL_n and the set SOL_n^+, respectively, where

$$\begin{aligned} \mathrm{SOL}_0^+ &= \{M \mid MN_1 \cdots N_m \in \mathrm{SOL}_0 \text{ for each } m \in \mathbb{N} \text{ and all } N_1, \ldots, N_m \in \Lambda\}, \\ \mathrm{SOL}_{n+1}^+ &= \{M \mid MN_1 \cdots N_m \in \mathrm{SOL}_{n+1} \\ &\qquad \text{for each } m \in \mathbb{N} \text{ and all } N_1, \ldots, N_m \in \mathrm{SOL}_n\}\ . \end{aligned}$$

Indeed, in this respect one may consider that the axiom $(\star 1)$ corresponds to the inclusion $\mathrm{SOL}_n^+ \subseteq \mathrm{SOL}_n$, and the axiom $(\star 4)$ to the equivalence

$$M \in \mathrm{SOL}_0^+ \iff MN \in \mathrm{SOL}_0^+ \text{ for each } N \in \Lambda\ .$$

Some reader may wonder if type-variables c_{n+1}'s are redundant in the sense that $\bigwedge_{i=0}^{n+1} b_i$ may be substituted for c_{n+1}. Indeed, without c_{n+1}'s we obtain a filter λ-model which satisfies our goal. However introducing c_{n+1}'s enables us to construct an inverse limit space corresponding to the system $\lambda\wedge_{\star}$, as follows. Let us define $(D_0^{\star}, \sqsubseteq)$ as illustrated in Fig. 1.

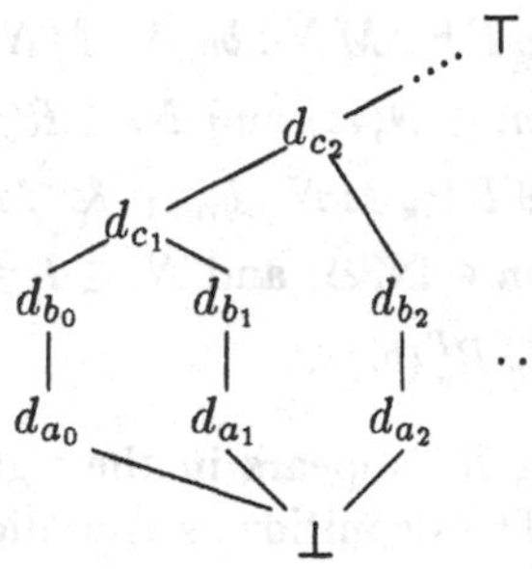

Fig. 1. Hasse-diagram of $(D_0^\star, \sqsubseteq)$

This is indeed an ω-algebraic complete lattice, in which each element other than $\top$ is compact. The projection $(\varphi_0^\star, \psi_0^\star)$ of $[D_0^\star \to D_0^\star]$ on $D_0^\star$ is defined by

$$\begin{array}{ll}
\varphi_0^\star(\bot) = \bot \searrow \bot, & \varphi_0^\star(d_{b_{n+1}}) = d_{a_n} \searrow d_{b_{n+1}}, \\
\varphi_0^\star(d_{a_0}) = d_{b_0} \searrow d_{a_0}, & \varphi_0^\star(d_{c_{n+1}}) = (\bot \searrow d_{b_0}) \sqcup \bigsqcup_{i=0}^{n} (d_{a_i} \searrow d_{b_{i+1}}), \\
\varphi_0^\star(d_{a_{n+1}}) = d_{c_{n+1}} \searrow d_{a_{n+1}}, & \varphi_0^\star(\top) = (\bot \searrow d_{b_0}) \sqcup \bigsqcup_{i=0}^{\infty} (d_{a_i} \searrow d_{b_{i+1}}), \\
\varphi_0^\star(d_{b_0}) = \bot \searrow d_{b_0}, &
\end{array}$$

$$\psi_0^\star(f) = \sqcup\{d \in D_0^\star \mid \varphi_0^\star(d) \sqsubseteq f\} \qquad \text{for each } f \in [D_0^\star \to D_0^\star] \ .$$

We write $D_\infty^\star$ for the inverse limit space constructed from $D_0^\star$ and $(\varphi_0^\star, \psi_0^\star)$. Then it follows from Corollary 3 that $D_\infty^\star$ is isomorphic to the structure $\mathcal{F}_\star$ as a λ-model. Furthermore Proposition 1 and Corollary 3 ensure the following.

Theorem 9.
(i) *The structure $\mathcal{F}_\star$ is an extensional λ-model.*
(ii) *If $\Gamma \vdash_\star M : A$ and $M =_{\beta\eta} N$, then $\Gamma \vdash_\star N : A$.*

In what follows, for each $n \in \mathbb{N}$, we shall characterize the set SOL_n in terms of typability in the system $\lambda\wedge_\star$, which will be accomplished in Theorem 10. Then this will immediately induce a characterization of SOL_n by means of the λ-models $\mathcal{F}_\star$ and $D_\infty^\star$ through the propositions in Sect. 2.

First let us define logical relation $R_\star$ over the weak semantic structure $\mathcal{L}$ in Sect. 3, by analogy with the computability predicate in [5, Definition A.1]. The mapping $R_\star$: Basis $\times$ $\mathrm{T}_\wedge \longrightarrow \wp\Lambda$ is recursively defined by the following and (LR1) through (LR3):

$$\begin{array}{rl}
M \in R_\star^\Gamma(a_0) \Longleftrightarrow & \biguplus_{i=1}^{m} \Delta_i \uplus \Gamma \vdash_\star M\bar{N} : a_0 \ \& \ M\bar{N} \in \mathrm{SOL}_0 \\
& \text{for each } m \in \mathbb{N}, \Delta_i \text{ and } N_i \in R_\star^{\Delta_i}(b_0) \ (i = 1, \ldots, m), \\
M \in R_\star^\Gamma(a_{n+1}) \Longleftrightarrow & \biguplus_{i=1}^{m} \Delta_i \uplus \Gamma \vdash_\star M\bar{N} : a_{n+1} \ \& \ M\bar{N} \in \mathrm{SOL}_{n+1} \\
& \text{for each } m \in \mathbb{N}, \Delta_i \text{ and } N_i \in R_\star^{\Delta_i}(c_{n+1}) \ (i = 1, \ldots, m),
\end{array}$$

$$M \in R_\star^\Gamma(b_0) \iff \biguplus_{i=1}^m \Delta_i \uplus \Gamma \vdash_\star M\bar{N} : b_0 \;\;\&\;\; M\bar{N} \in \mathrm{SOL}_0$$
$$\text{for each } m \in \mathbb{N}, \Delta_i \text{ and } N_i \in R_\star^{\Delta_i}(\omega)\ (i = 1, \ldots, m),$$
$$M \in R_\star^\Gamma(b_{n+1}) \iff \biguplus_{i=1}^m \Delta_i \uplus \Gamma \vdash_\star M\bar{N} : b_{n+1} \;\;\&\;\; M\bar{N} \in \mathrm{SOL}_{n+1}$$
$$\text{for each } m \in \mathbb{N}, \Delta_i \text{ and } N_i \in R_\star^{\Delta_i}(a_n)\ (i = 1, \ldots, m),$$
$$M \in R_\star^\Gamma(c_{n+1}) \iff M \in \bigcap_{i=0}^{n+1} R_\star^\Gamma(b_i)\ .$$

Note that, though the mapping $R_\star$ appears in the right-hand side of the definition, there is no circularity. The definition is complicated mainly in order to guarantee monotonicity with respect to $\leq_\star$.

Lemma 10.
(i) *$R_\star$ is admissible; hence $R_\star$ is a logical relation.*
(ii) *$R_\star$ is monotone with respect to the relation* $\Subset$.
(iii) *$R_\star$ is monotone with respect to the subtype relation $\leq_\star$.*

Proof. (i) From Lemma 8, it is sufficient just to show that, for each type-variable a, $M[x := N]\bar{P} \in R_\star^\Gamma(a)$ implies $(\lambda x.M)N\bar{P} \in R_\star^\Gamma(a)$. This follows from the definition of $R_\star$ and Theorem 9 (ii).
(ii) By induction on the structure of types.
(iii) By induction on the generation of the subtype relation $\leq_\star$. (cf. [5, Lemma A.4].) □

The following is the main theorem of this section, in which for simplicity we write c_0 for b_0. (Thus $c_0, c_1, c_2, \ldots$ stands for the sequence $b_0, c_1, c_2, \ldots$.)

Theorem 11. *For each λ-term M and $n \in \mathbb{N}$, the following are equivalent:*
(i) $M \in \mathrm{SOL}_n$.
(ii) *For some basis $\Gamma \subseteq \{x : c_n \mid x \in \mathrm{Var}\}$, $\Gamma \vdash_\star M : a_n$.*

Proof. (i) $\Longrightarrow$ (ii): By induction on n.
Basis: Suppose $M \in \mathrm{SOL}_0$ and $\lambda x_1 \cdots x_p.yN_1 \cdots N_q$ is the head-normal form of M. Let $\Gamma = \{x_i : b_0 \mid 1 \leq i \leq p\} \cup \{y : b_0\}$. Then we can deduce $\Gamma \vdash_\star y : \omega \to \cdots \to \omega \to b_0$ by ($\leq_\star$)-rule and ($\star$4). On the other hand, we have $\Gamma \vdash_\star N_i : \omega$ $(i = 1, \ldots, q)$ by (ω)-axiom. These judgments imply $\Gamma \vdash_\star y\bar{N} : a_0$, from which we obtain $\Delta \vdash_\star \lambda\bar{x}.y\bar{N} : b_0 \to \cdots \to b_0 \to a_0$ where $\Delta = \Gamma \backslash \{x_i : b_0 \mid 1 \leq i \leq p\}$. Thus we get $\Delta \vdash_\star M : a_0$ because of the equivalence $a_0 \sim_\star b_0 \to \cdots \to b_0 \to a_0$ and Theorem 9 (ii).
Induction step: Suppose $M \in \mathrm{SOL}_{n+1}$ and $\lambda x_1 \cdots x_p.yN_1 \cdots N_q$ is the head-normal form of M. Then $N_1, \ldots, N_q \in \mathrm{SOL}_n$ follows from the definition of SOL_{n+1}. Therefore, by induction hypothesis, we can find basis $\Delta_1, \ldots, \Delta_q \subseteq \{x : c_n \mid x \in \mathrm{Var}\}$ such that $\Delta_j \vdash_\star N_j : a_n$ $(j = 1, \ldots, q)$. Now, let

$$\Gamma = \{x : c_{n+1} \mid x \in \bigcup_{j=1}^q \mathrm{subj}(\Delta_j)\} \cup \{x_i : c_{n+1} \mid 1 \leq i \leq p\} \cup \{y : c_{n+1}\}\ .$$

Then we have $\Gamma \vdash_\star y : a_n \to \cdots \to a_n \to b_{n+1}$ by ($\leq_\star$)-rule and axioms ($\star$5) and ($\star$6). On the other hand, we have $\Gamma \vdash_\star N_j : a_n$ $(j = 1, \ldots, q)$ since $\Delta_1, \ldots, \Delta_q \Subset \Gamma$. Thus we have $\Gamma \vdash_\star y\bar{N} : a_{n+1}$. Hence we obtain $\Delta \vdash_\star \lambda\bar{x}.y\bar{N}$:

$c_{n+1} \to \cdots \to c_{n+1} \to a_{n+1}$ where $\Delta = \Gamma \setminus \{x_i : c_{n+1} \mid 1 \leq i \leq p\}$. As in the base case, it means that M satisfies (ii).
(ii) $\Longrightarrow$ (i): For each $x : c_n \in \Gamma$, we have $\Gamma \vdash_\star x : b_i$ $(i = 0, \ldots, n)$. So it follows from the definition of $\leq_\star$ and $R_\star$ that $x \in \bigcap_{i=0}^{n} R_\star^\Gamma(b_i) = R_\star^\Gamma(c_n)$. This means $\mathcal{L}, \iota, R_\star^\Gamma \models \Gamma$ where $\iota \in \mathrm{Env}_\mathcal{L}$ is the identity mapping. Thus $M \equiv [\![M]\!]_\iota^\mathcal{L} \in R_\star^\Gamma(a_n)$ follows from Lemmas 7 and 10, which implies $M \in \mathrm{SOL}_n$. □

Corollary 12. *For each $n \in \mathbb{N}$, The following are equivalent:*
(i) $M \in \mathrm{SOL}_n$.
(ii) $a_n \in [\![M]\!]_\xi^{\mathcal{F}_\star}$ *where* $\xi(x) = \{A \mid c_n \leq_\nabla A\}$ *for each* $x \in \mathrm{Var}$.
(iii) $\Phi_{0,\infty}(d_{a_n}) \sqsubseteq [\![M]\!]_\xi^{D_\infty^\star}$ *where* $\xi(x) = \Phi_{0,\infty}(d_{c_n})$ *for each* $x \in \mathrm{Var}$.

5 Finiteness of Böhm-Trees

As mentioned in earlier sections, logical relations have been useful to show syntactical properties of various typed λ-calculi. By observing proofs using logical relations for simple types, J. C. Mitchell [11, Sect. 3.4.3] extracted three conditions (TC1) through (TC3) below for a set Θ ($\subseteq \Lambda$) of λ-terms, and clarified their essential role in the proofs;

$$\begin{array}{ll}(\mathrm{TC1}) & M_1, \ldots, M_m \in \Theta \Longrightarrow xM_1 \cdots M_m \in \Theta, \\ (\mathrm{TC2}) & Mz \in \Theta \ \&\ z \notin \mathrm{FV}(M) \Longrightarrow M \in \Theta, \\ (\mathrm{TC3}) & N, M[x := N]\bar{P} \in \Theta \Longrightarrow (\lambda x.M)N\bar{P} \in \Theta\ .\end{array}$$

A set which satisfies (TC1) through (TC3) is said to be *type-closed.*

Proposition 13 (Mitchell [11, Theorem 3.4.7]). *If M is typable under $\lambda\!\to$ (that is, $\Gamma \vdash_{\lambda\to} M : A$ for some Γ and A), then $M \in \Theta$ for each type-closed set Θ.*

The preceding theorem captures the essence of various applications of logical relations for simple types. For example, the strong normalizability of all terms typable under $\lambda\!\to$ is immediately implied from this theorem, because the set SN_β[17] is shown to be type-closed.

Now recall the intersection type assignment system $\lambda\wedge$ without subtype relation defined in [4]. We write $\lambda\wedge^{-\omega}$ for the system obtained from $\lambda\wedge$ by eliminating the type constant ω. One of the remarkable properties of the system $\lambda\wedge^{-\omega}$ is that its typability coincides with strong normalizability. This has been studied by several researchers (see [1], [10] and [13] for example). In their proof, the notion of logical relations is applied to show that typability implies strong normalizability, and all the properties of strong normalizability required in the proof are also (TC1) through (TC3). So this implies the following equivalence.

Proposition 14. *The following are mutually equivalent:*
(1) *M is typable under the system $\lambda\wedge^{-\omega}$.*
(2) *$M \in \Theta$ for each type-closed set Θ.*
(3) $M \in \mathrm{SN}_\beta$.

[17] We write SN_β for the set of strongly normalizable terms with respect to β-reduction.

Corollary 15. SN_β *is the intersection of all type-closed sets.*

This section is devoted to characterizing, as in the proposition above, finiteness of Böhm-trees in terms of an intersection type assignment system. We write $\mathrm{BT_{fin}}$ for the set of terms with finite Böhm-tree; that is,

$$\mathrm{BT_{fin}} = \{M \mid M \twoheadrightarrow_\beta N \;\&\; N \in \mathrm{NF}_\Omega \text{ for some } N\}$$

where NF_Ω is the smallest set satisfying

1. $M \in \mathrm{NF}_\Omega$ if M is unsolvable,
2. $M_1, \ldots, M_n \in \mathrm{NF}_\Omega \Longrightarrow \lambda\bar{x}.yM_1 \cdots M_n \in \mathrm{NF}_\Omega$.

As a first attempt at a system whose typability characterizes $\mathrm{BT_{fin}}$, one may think of the intersection type assignment system obtained from the system $\lambda\wedge$ in [4] by restricting the (ω)-axiom to

$$(\Omega) \quad \Gamma \vdash M : \omega \qquad \text{if } M \text{ is unsolvable} .$$

The typability of the system so obtained lies between $\lambda\wedge$ and $\lambda\wedge^{-\omega}$, and in fact it is shown that all terms typable under this system belong to $\mathrm{BT_{fin}}$. However the reverse inclusion does not hold. This is because, in this system, not only types but typability is no longer preserved under β-expansion. For example, consider the β-reduction $(\lambda y.x(yZ))Z \to_\beta x\Omega$ where $Z \equiv \lambda x.xx$. In the restricted system, although $x : \omega \to a \vdash x\Omega : a$, the term $(\lambda y.x(yZ))Z$ is not typable. This sharply contrasts with the fact that, by assigning ω to solvable terms, we can deduce $x : \omega \to a \vdash_{\lambda\wedge} (\lambda y.x(yZ))Z : a$, as follows;

$$\dfrac{\dfrac{\dfrac{\dfrac{}{\Gamma \vdash x : \omega \to a}\,(\mathrm{var}) \quad \dfrac{}{\Gamma \vdash yZ : \omega}\,(\omega)}{\Gamma \vdash x(yZ) : a}\,(\to\mathrm{E})}{x : \omega \to a \vdash \lambda y.x(yZ) : \omega \to a}\,(\to\mathrm{I}) \quad \dfrac{}{x : \omega \to a \vdash Z : \omega}\,(\omega)}{x : \omega \to a \vdash (\lambda y.x(yZ))Z : a}\,(\to\mathrm{E})$$

where $\Gamma = \{x : \omega \to a, y : \omega\}$.

Taking the above-mentioned difficulty into consideration, we shall now define the system whose typability precisely captures the finiteness of Böhm-trees. We add a special term-variable $*$ not in Var, called *edge-variable*, to the formation of λ-terms. Then we say M is a $\lambda*$-term if M has at least one free-occurrence of the edge-variable.[18] We use the Greek letter π for finite sequences of $\lambda*$-terms and ϵ for the empty sequence. For each $\xi \in \mathrm{Env}_{\mathcal{L}}$ (namely, a mapping ξ from Var to Λ), we define simultaneous substitution θ_ξ for sequences of $\lambda*$-terms by

1. $\epsilon\,\theta_\xi = \epsilon$,
2. $(\pi, N)\theta_\xi = \pi\theta_\xi, N[x_1 := \xi(x_1), \ldots, x_n := \xi(x_n)]$,

where $\{x_1, \ldots, x_n\} = \mathrm{FV}(N) \setminus \{*\}$. Note that no λ-term is substituted for $*$ by the substitution θ_ξ. With these preparations, we now define the set T_Ω of types and their free (term-)variables by simultaneous recursion:

[18] Note that $\lambda*$-terms are not in Λ.

1. For any type-variable a, $a \in \mathrm{T}_\Omega$ and $\mathrm{FV}(a) = \emptyset$,
2. For any finite sequence π of $\lambda *$-terms, $\omega_\pi \in \mathrm{T}_\Omega$ and $\mathrm{FV}(\omega_\pi) = \bigcup\{\mathrm{FV}(M) \mid M \text{ is a component of } \pi\}$,
3. If $A, B \in \mathrm{T}_\Omega$ then $A \wedge B \in \mathrm{T}_\Omega$ and $\mathrm{FV}(A \wedge B) = \mathrm{FV}(A) \cup \mathrm{FV}(B)$,
4. If $A, B \in \mathrm{T}_\Omega$ and $x \in \mathrm{Var} \setminus \mathrm{FV}(A)$ then $\prod x : A.B$, $\bigwedge x : A.B \in \mathrm{T}_\Omega$ and $\mathrm{FV}(\prod x : A.B) = \mathrm{FV}(\bigwedge x : A.B) = \mathrm{FV}(A) \cup (\mathrm{FV}(B) \setminus \{x\})$.

Our intention for the type ω_π is the set $\{M \in \Lambda \mid \pi \circledast M \text{ is unsolvable}\}$ where

$$\pi \circledast M \equiv \begin{cases} M & \text{if } \pi = \epsilon, \\ N_1[* := N_2[* := \cdots N_n[* := M] \cdots]] & \text{if } \pi = N_1, \ldots, N_n \ (n \geq 1) . \end{cases}$$

The type $\prod x : A.B$ (and $\bigwedge x : A.B$) may be thought of the cartesian product (and the intersection, resp.) of all types of the form $B[x := N]$ where N ranges over terms of type A. We shall consider types modulo α-conversion for $\prod$- and $\bigwedge$-binding, modulo associative law and commutative law with respect to binary $\wedge$, and also modulo the following equivalence

$$\omega_\pi \equiv \omega_{\pi'} \quad \text{where } \pi \circledast M \text{ is unsolvable if and only if so is } \pi' \circledast M \text{ for each } \lambda\text{-term } M \ .$$

We write ω for ω_ϵ and $A \to B$ for $\prod x : A.B$ such that $x \notin \mathrm{FV}(B)$. Thus the set $\mathrm{T}_\wedge$ of types of $\lambda\wedge$ is naturally embedded into T_Ω. Type-variables and the types of the form ω_π are called atomic-types of T_Ω. For each $\xi \in \mathrm{Env}_\mathcal{L}$, we extend the substitution θ_ξ to types by

1. $a\theta_\xi = a$,
2. $\omega_\pi \theta_\xi = \omega_{\pi\theta_\xi}$,
3. $(A \wedge B)\theta_\xi = A\theta_\xi \wedge B\theta_\xi$,
4. $(\prod x : A.B)\theta_\xi = \prod x : A\theta_\xi . B\theta_{\xi(x:x)}$,
5. $(\bigwedge x : A.B)\theta_\xi = \bigwedge x : A\theta_\xi . B\theta_{\xi(x:x)}$.

We write $A[x := N]$ for $A\theta_{\iota(x:N)}$ where $\iota \in \mathrm{Env}_\mathcal{L}$ is the identity mapping. Basis Γ is defined as in Sect. 2, except that each statement $x : A \in \Gamma$ must satisfy $x \notin \mathrm{FV}(A)$. Typing judgments are generated by the axiom (var) and rule ($\wedge$I) in $\lambda\wedge_\nabla$, the axiom (Ω) mentioned above, and the rules listed below.

$$(\textstyle\prod \mathrm{I}) \ \frac{\Gamma, x : A \vdash M : B}{\Gamma \vdash \lambda x.M : (\prod x : A.B)} \quad \text{if } x \notin \textstyle\bigcup_{C \in \mathrm{pred}(\Gamma)} \mathrm{FV}(C)$$

$$(\textstyle\prod \mathrm{E}) \ \frac{\Gamma \vdash M : (\prod x : A.B) \quad \Gamma \vdash N : A}{\Gamma \vdash MN : B[x := N]} \qquad (\wedge \mathrm{E}) \ \frac{\Gamma \vdash M : A_1 \wedge A_2}{\Gamma \vdash M : A_i} \ (i = 1, 2)$$

$$(*\mathrm{I}) \ \frac{\Gamma \vdash M[x := N] : \omega_\pi}{\Gamma \vdash N : \omega_{\pi, M[x:=*]}} \ \text{if } x \in \mathrm{FV}(M) \qquad (*\mathrm{E}) \ \frac{\Gamma \vdash x : \omega_{\pi, M[x:=*]}}{\Gamma \vdash M : \omega_\pi}$$

$$(\textstyle\bigwedge \mathrm{I}) \ \frac{\Gamma, x : A \vdash M : B}{\Gamma \vdash M : (\bigwedge x : A.B)} \quad \text{if } x \notin \textstyle\bigcup_{C \in \mathrm{pred}(\Gamma)} \mathrm{FV}(C) \cup \mathrm{FV}(M)$$

$$(\textstyle\bigwedge \mathrm{E}) \ \frac{\Gamma \vdash M : (\bigwedge x : A.B) \quad \Gamma \vdash N : A}{\Gamma \vdash M : B[x := N]}$$

We denote this system by $\lambda\Omega$ and write $\Gamma \vdash_\Omega M : A$ for judgments in $\lambda\Omega$. With these complicated devices, we now have $x : \omega \to a \vdash_\Omega (\lambda y.x(yZ))Z : a$, as follows;

$$\dfrac{\dfrac{\dfrac{\dfrac{}{\Gamma \vdash x : \omega \to a}\,(\mathrm{var}) \qquad \dfrac{\dfrac{}{\Gamma \vdash y : \omega_{*Z}}\,(\mathrm{var})}{\Gamma \vdash yZ : \omega}\,(*\mathrm{E})}{\Gamma \vdash x(yZ) : a}\,(\textstyle\prod \mathrm{E})}{x : \omega \to a \vdash \lambda y.x(yZ) : \omega_{*Z} \to a}\,(\textstyle\prod \mathrm{I}) \qquad \dfrac{\dfrac{}{x : \omega \to a \vdash \Omega : \omega}\,(\Omega)}{x : \omega \to a \vdash Z : \omega_{*Z}}\,(*\mathrm{I})}{x : \omega \to a \vdash (\lambda y.x(yZ))Z : a}\,(\textstyle\prod \mathrm{E})$$

where $\Gamma = \{x : \omega \to a, y : \omega_{*Z}\}$. A term M is said to be $\mathrm{T}_\wedge$-*typable* in the system $\lambda\Omega$ if there exist a basis Γ and a type $A \in \mathrm{T}_\wedge$ such that $\mathrm{pred}(\Gamma) \subseteq \mathrm{T}_\wedge$ and $\Gamma \vdash_\Omega M : A$. In the rest of this section, we will prove that this restricted typability is equivalent to the finiteness of Böhm-trees. (Note that, unrestricted typability of $\lambda\Omega$ is trivial in the sense that all terms have type $\omega_{\Omega*}$.)

We first show that finiteness of Böhm-trees implies $\mathrm{T}_\wedge$-typability, which will be given in Theorem 19. To show it, we restrict our attention to a specific form of derivation. A derivation δ in the system $\lambda\Omega$ is said to be ***substitution-free*** if, for each application of ($\prod$E)-rule in δ, the predicate $\prod x : A.B$ of its major premise is $A \to B$ (i.e., $x \notin \mathrm{FV}(B)$). We use the symbol $\vdash_{\underline{\Omega}}$ to denote substitution-free derivations in $\lambda\Omega$. Proofs of the following two lemmas are easy and omitted.

Lemma 16. *If* $M \in \mathrm{NF}_\Omega$ *then there exists* Γ *and* $A \in \mathrm{T}_\wedge$ *such that* $\mathrm{pred}(\Gamma) \subseteq \mathrm{T}_\wedge$ *and* $\Gamma \vdash_{\underline{\Omega}} M : A$.

Lemma 17.
(i) *If* $\Gamma \vdash_{\underline{\Omega}} M : A$ *and* $\Gamma \Subset \Delta$ *then* $\Delta \vdash_{\underline{\Omega}} M : A$.[19]
(ii) *If* $\Gamma, x : A, y : B \vdash_{\underline{\Omega}} M : C$ *then* $\Gamma, x : A, y : (\bigwedge x : A.B) \vdash_{\underline{\Omega}} M : C$.

Lemma 18.
(i) *If* $\Gamma \vdash_{\underline{\Omega}} M[x := N] : A$ *then* $\Gamma \vdash_{\underline{\Omega}} (\lambda x.M)N : A$.
(ii) *If* $\Gamma \vdash_{\underline{\Omega}} M : A$ *and* $N \to_\beta M$ *then* $\Gamma \vdash_{\underline{\Omega}} N : A$.

Proof. (i) We claim that the derivation δ of $\Gamma \vdash_{\underline{\Omega}} M[x := N] : A$ implies $\Gamma, z : B \vdash_{\underline{\Omega}} M[x := z] : A$ and $\Gamma \vdash_{\underline{\Omega}} N : B$ for a type $B \in \mathrm{T}_\Omega$ and a fresh-variable z. Once this claim is verified, we have $\Gamma \vdash_{\underline{\Omega}} \lambda x.M : B \to A$, and this completes the proof by ($\prod$E)-rule. We show the claim above by induction on the length of δ. We distinguish cases according to whether or not $x \in \mathrm{FV}(M)$.
Case 1: Suppose $x \notin \mathrm{FV}(M)$. Then, applying Lemma 17 (i) to δ, we obtain $\Gamma, z : \omega_{\Omega*} \vdash_{\underline{\Omega}} M[x := z] : A$. On the other hand, we can deduce $\Gamma \vdash_{\underline{\Omega}} N : \omega_{\Omega*}$ from $\Gamma \vdash_{\underline{\Omega}} \Omega N : \omega$ by means of (*I)-rule.
Case 2: Suppose $x \in \mathrm{FV}(M)$. Then we distinguish cases again according to the last rule applied in δ.
Subcase 2.1: Suppose the last step of δ is by (ω), (*I) or (*E). Then $A \equiv \omega_\pi$ for some π, and we have $\Gamma \vdash_{\underline{\Omega}} N : \omega_{\pi, M[x:=*]}$ from δ by (*I)-rule. On the other hand,

[19] For bases in the system $\lambda\Omega$, we define the relation $\Subset$ as before.

we can deduce $\Gamma, z : \omega_{\pi,M[x:=*]} \vdash_{\underline{\Omega}} M[x := z] : \omega_\pi$ from $\Gamma, z : \omega_{\pi,M[x:=*]} \vdash_{\underline{\Omega}} z : \omega_{\pi,M[x:=z][z:=*]}$ by means of ($*$E)-rule.
Subcase 2.2: Suppose the last step of δ is by ($\prod$I), yielding $\Gamma \vdash_{\underline{\Omega}} (\lambda y.P)[x := N] : (\prod y : B.C)$ from $\Gamma, y : B \vdash_{\underline{\Omega}} P[x := N] : C$. Note that $y \notin \mathrm{FV}(N)$, and $y \not\equiv x$ since $x \in \mathrm{FV}(\lambda y.P)$. By induction hypothesis, we have

$$\Gamma, y : B, z : D \vdash_{\underline{\Omega}} P[x := z] : C \text{ and} \quad (1)$$
$$\Gamma, y : B \vdash_{\underline{\Omega}} N : D \quad (2)$$

for some D. From (1), we obtain $\Gamma, y : B, z : (\bigwedge y : B.D) \vdash_{\underline{\Omega}} P[x := z] : C$ by Lemma 17 (ii), which implies $\Gamma, z : (\bigwedge y : B.D) \vdash_{\underline{\Omega}} (\lambda y.P)[x := z] : (\prod y : B.C)$ by ($\prod$I)-rule. On the other hand, from (2) and ($\bigwedge$I)-rule, we get $\Gamma \vdash_{\underline{\Omega}} N : (\bigwedge y : B.D)$.
The other cases are similar and their proofs are omitted.
(ii) By induction on the length of the derivation. If M is the contractum of N then (i) guarantees the lemma. Otherwise we distinguish cases according to the last rule in the derivation. We show the statement only in the case of ($*$E)-rule. Suppose $\Gamma \vdash_{\underline{\Omega}} M : \omega_\pi$ is a direct consequence of $\Gamma \vdash_{\underline{\Omega}} x : \omega_{\pi,M[x:=*]}$ and $N \rightarrow_\beta M$. Then $\omega_{\pi,M[x:=*]} \equiv \omega_{\pi,N[x:=*]}$ since, for any $P \in \Lambda$, $\pi \circledast M[x := P]$ is unsolvable if and only if so is $\pi \circledast N[x := P]$. Thus we obtain $\Gamma \vdash_{\underline{\Omega}} N : \omega_\pi$ from $\Gamma \vdash_{\underline{\Omega}} x : \omega_{\pi,N[x:=*]}$. □

Theorem 19. *If $M \in \mathrm{BT_{fin}}$ then M is $\mathrm{T}_\wedge$-typable under the system $\lambda\Omega$.*

Proof. Suppose $M \in \mathrm{BT_{fin}}$. Then there exists N such that $M \twoheadrightarrow_\beta N$ and $N \in \mathrm{NF}_\Omega$. Combining Lemmas 16 and 18 (ii), we can find Γ and $A \in \mathrm{T}_\wedge$ such that $\mathrm{pred}(\Gamma) \subseteq \mathrm{T}_\wedge$ and $\Gamma \vdash_{\underline{\Omega}} M : A$. □

Next we show the direction from $\mathrm{T}_\wedge$-typability to finiteness of Böhm-trees. This is accomplished by extending the notion of logical relation over the structure $\mathcal{L}$ in Sect. 3, and by investigating its relation to a stronger condition for type-closed sets. We say a set Θ ($\subseteq \Lambda$) is *Ω-type-closed* if it satisfies (TC1) and (TC2) before and the following (TC4) and (TC5):

$$\begin{array}{ll} \text{(TC4)} & M[x := N]\bar{P} \in \Theta \Longrightarrow (\lambda x.M)N\bar{P} \in \Theta, \\ \text{(TC5)} & M \in \Theta \text{ if } M \text{ is unsolvable} \ . \end{array}$$

For example, $\mathrm{BT_{fin}}$ is Ω-type-closed. For each Ω-type-closed set Θ, we define the mapping $R_\Theta : \mathrm{T}_\Omega \longrightarrow \wp\Lambda$ by

$$\begin{aligned} &R_\Theta(a) = \Theta, \\ &R_\Theta(\omega_\pi) = \begin{cases} \Theta & \text{if } \pi = \epsilon, \\ \{M \mid \pi \circledast M \text{ is unsolvable}\} & \text{otherwise,} \end{cases} \\ &R_\Theta(A \wedge B) = R_\Theta(A) \cap R_\Theta(B), \\ &R_\Theta(\textstyle\prod x : A.B) = \{M \mid MN \in R_\Theta(B[x := N]) \text{ for each } N \in R_\Theta(A)\}, \\ &R_\Theta(\textstyle\bigwedge x : A.B) = \{M \mid M \in R_\Theta(B[x := N]) \text{ for each } N \in R_\Theta(A)\} \ . \end{aligned}$$

Lemma 20.
(i) *For each* $A \in \mathrm{T}_\Omega$, *if* $[\![M]\!]^{\mathcal{L}}_{\xi(x:N)} \in R_\Theta(A)$ *then* $[\![\lambda x.M]\!]^{\mathcal{L}}_\xi N \in R_\Theta(A)$.
(ii) *If* $A \in \mathrm{T}_\wedge$ *and* $M_1, \ldots, M_n \in \Theta$ *then* $xM_1 \cdots M_n \in R_\Theta(A)$.
(iii) *If* $A \in \mathrm{T}_\wedge$ *and* $M \in R_\Theta(A)$ *then* $M \in \Theta$.

Proof. For each atomic-type $o \in \mathrm{T}_\Omega$, it is clear from (TC4) that $M[x := N]\bar{P} \in R_\Theta(o)$ implies $(\lambda x.M)N\bar{P} \in R_\Theta(o)$. Then we can obtain (i) as Lemma 8. The parts (ii) and (iii) can be verified by simultaneous induction on the structure of A. □

Lemma 21. *Suppose* $\Gamma \vdash_\Omega M : A$ *and* $\xi \in \mathrm{Env}_{\mathcal{L}}$. *If* $\xi(x) \in R_\Theta(B\theta_\xi)$ *for each* $x : B \in \Gamma$ *then* $[\![M]\!]^{\mathcal{L}}_\xi \in R_\Theta(A\theta_\xi)$.

Proof. By induction on the length of the derivation.
Case 1: Suppose the last step of the derivation is by ($\prod$I), yielding $\Gamma \vdash_\Omega \lambda x.N : (\prod x : B.C)$ from $\Gamma, x : B \vdash_\Omega N : C$. Suppose $\xi(y) \in R_\Theta(D\theta_\xi)$ for each $y : D \in \Gamma$ and $N \in R_\Theta(B\theta_\xi)$. Then we have $\xi(x : N)(y) \in R_\Theta(D\theta_\xi) = R_\Theta(D\theta_{\xi(x:N)})$ for each $y : D \in \Gamma$ because of the side-condition of ($\prod$I)-rule. We also have $\xi(x : N)(x) \in R_\Theta(B\theta_\xi) = R_\Theta(B\theta_{\xi(x:N)})$. Hence $[\![N]\!]^{\mathcal{L}}_{\xi(x:N)} \in R_\Theta(C\theta_{\xi(x:N)})$ follows from induction hypothesis. This implies $[\![\lambda x.N]\!]^{\mathcal{L}}_\xi N \in R_\Theta(C\theta_{\xi(x:N)}) = R_\Theta(C\theta_{\xi(x:x)}[x := N])$ because of Lemma 20 (i) and because we may assume $x \notin \bigcup_{y \in \mathrm{FV}(C)\setminus\{x\}} \mathrm{FV}(\xi(y))$ without loss of generality. Therefore $[\![\lambda x.N]\!]^{\mathcal{L}}_\xi \in R_\Theta((\prod x : B.C)\theta_\xi)$.
Case 2: Suppose the last step of the derivation is by ($*$I), yielding $\Gamma \vdash_\Omega Q : \omega_{\pi, P[x:=*]}$ from $\Gamma \vdash_\Omega P[x := Q] : \omega_\pi$. Suppose $\xi(y) \in R_\Theta(B\theta_\xi)$ for each $y : B \in \Gamma$. Then, by induction hypothesis, we have $P[x := z]\theta_{\xi(z:z)}[z := *][* := Q\theta_\xi] \equiv [\![P[x := Q]]\!]^{\mathcal{L}}_\xi \in R_\Theta(\omega_\pi\theta_\xi)$ for a fresh-variable z. This implies $[\![Q]\!]^{\mathcal{L}}_\xi \in R_\Theta(\omega_{\pi\theta_\xi, P[x:=z]\theta_{\xi(z:z)}[z:=*]}) = R_\Theta(\omega_{\pi, P[x:=*]}\theta_\xi)$.
The other cases are simpler and their proofs are omitted. □

Theorem 22. *If* M *is* $\mathrm{T}_\wedge$*-typable under the system* $\lambda\Omega$, *then* $M \in \Theta$ *for each* Ω*-type-closed set* Θ.

Proof. Suppose $\Gamma \vdash_\Omega M : A$ where $A \in \mathrm{T}_\wedge$ $\mathrm{pred}(\Gamma) \subseteq \mathrm{T}_\wedge$. Then Lemma 20 (ii) implies $\iota(x) \equiv x \in R_\Theta(B) = R_\Theta(B\theta_\iota)$ for the identity environment $\iota \in \mathrm{Env}_{\mathcal{L}}$ and each $x : B \in \Gamma$. Hence it follows from Lemma 21 that $M \equiv [\![M]\!]^{\mathcal{L}}_\iota \in R_\Theta(A\theta_\iota) = R_\Theta(A)$. Therefore $M \in \Theta$ by Lemma 20 (iii). □

Corollary 23. *The following are mutually equivalent*:
(i) M *is* $\mathrm{T}_\wedge$*-typable under the system* $\lambda\Omega$.
(ii) $M \in \Theta$ *for each* Ω*-type-closed set* Θ.
(iii) $M \in \mathrm{BT}_{\mathrm{fin}}$.

Acknowledgements

I am grateful to Prof. Masako Takahashi and Prof. J. Roger Hindley for their valuable advice towards this paper. I also thank anonymous referees of the conference, who gave me a lot of helpful comments.

References

1. S. van Bakel, Complete restrictions of the intersection type descipline, *Theoretical Computer Science* **102** (1992), 135-163.
2. H. P. Barendregt, *The Lambda Calculus: Its Syntax and Semantics*, revised edition, North-Holland, Amsterdam, 1984.
3. H. P. Barendregt, M. Coppo and M. Dezani-Ciancaglini, A filter lambda model and the completeness of type assignment, *Journal of Symbolic Logic* **48** (1983), 931–940.
4. M. Coppo, M. Dezani-Ciancaglini and B. Venneri, Functional characters of solvable terms, *Zeitschrift für Mathematische Logik und Grundlagen der Mathmatik* **27** (1981), 45–58.
5. M. Coppo, M. Dezani-Ciancaglini and M. Zacchi, Type theories, normal forms and D_∞ lambda-models, *Information and Computation* **72** (1987), 85–116.
6. J. Y. Girard, *Interprétation fonctionnelle et élimination des coupures dans l'arithmétique d'ordre supérieur*, Thèse de doctorat d'état, Université Paris VII, 1972.
7. J. Y. Girard, P. Taylor and Y. Lafont, *Proofs and Types*, Cambridge University Press, 1989.
8. J. R. Hindley, The completeness theorem for typing λ-terms, *Theoretical Computer Science* **22** (1983), 1–17.
9. F. Honsell and S. Ronchi Della Rocca, An approximation theorem for topological lambda models and the topological incompleteness of lambda calculus, *Journal of Computer and System Sciences* **45** (1992), 49–75.
10. J. L. Krivine, *Lambda-Calculus, Types and Models*, Ellis Horwood, 1993.
11. J. C. Mitchell, Type systems for programming languages, *Handbook of Theoretical Computer Science Volume B: Formal Models and Semantics*, The MIT Press / Elsevier, 1990.
12. G. D. Plotkin, λ-definability in the full type hierarchy, in: *To H. B. Curry: Essays on Combinatory Logic, Lambda Calculus and Formalism*, ed. J. R. Hindley and J. P. Seldin, Academic Press, New York, 363–373.
13. G. Pottinger, A type assignment for the strongly normalizable λ-terms, in: *To H. B. Curry: Essays on Combinatory Logic, Lambda Calculus and Formalism*, ed. J. R. Hindley and J. P. Seldin, Academic Press, New York, 363–373.
14. R. Statman, Logical relations and the typed lambda calculus, *Information and Control* **65** (1985), 85–97.
15. W. W. Tait, Intensional interpretation of functionals of finite type, *Journal of Symbolic Logic* **32** (1967), 198–212.

Semantic Techniques for Deriving Coinductive Characterizations of Observational Equivalences for λ-calculi*

Marina Lenisa

Dipartimento di Matematica e Informatica, Università di Udine, Italy.
lenisa@dimi.uniud.it

Abstract. Coinductive (applicative) characterizations of various observational congruences which arise in the semantics of λ-calculus, for various reduction strategies, are discussed. Two semantic techniques for establishing the coincidence of the applicative and the contextual equivalences are analyzed. The first is based on *intersection types*, the second is based on a *mixed induction-coinduction principle*.

Introduction

This paper is part of a general project on finding *elementary proof principles* for reasoning *rigorously* on possibly *infinite* objects [8, 13]. Often, in dealing with infinite or circular objects, pure structural induction can be applied only after cumbersome encodings. Thus, alternative reasoning principles are called for, such as *coinduction* and *mixed induction-coinduction principles* (see [16, 8]). In this paper, we focus on the behaviour of λ-terms when they are evaluated according to a given *reduction strategy*. In particular, we introduce two general semantic techniques for deriving the coincidence of the *applicative* equivalence with the *observational* (operational, contextual) equivalence for various reduction strategies. This amounts to a *coinductive* characterization of the observational equivalence and hence it gives rise immediately to a *coinduction principle*; the particular nature of one of the techniques gives rise also to a *mixed induction-coinduction principle*. This paper is a companion to [14], where syntactic techniques are discussed.

A reduction strategy is a procedure for determining, for each λ-term, a specific β-redex in it, to contract. Let $\Lambda(C)$ ($\Lambda^0(C)$) denote the set of (closed) λ-terms, where C is a set of basic constants. When $C = \emptyset$, we write Λ (Λ^0). A (possibly non-deterministic) strategy can be formalized as a relation $\rightarrow_\sigma \subseteq \Lambda(C) \times \Lambda(C)$ ($\Lambda^0(C) \times \Lambda^0(C)$) such that, if $(M, N) \in \rightarrow_\sigma$ (also written infix as $M \rightarrow_\sigma N$), then N is a possible result of applying $\rightarrow_\sigma$ to M. A particular set of terms which do not belong to the domain of $\rightarrow_\sigma$ are called *σ-values*, denoted by Val_σ. Given $\rightarrow_\sigma$, we can define the *evaluation relation* $\Downarrow_\sigma \subseteq \Lambda(C) \times \Lambda(C)$

* Work supported by CNR, EC HCM Project No. CHRX-CT92.0046 Lambda-Calcul Typé, and MURST 40% grant.

$(\Lambda^0(C) \times \Lambda^0(C))$, such that $M \Downarrow_\sigma N$ holds if and only if there exists a reduction path from M to the σ-value N. If there exists N such that $M \Downarrow_\sigma N$, then $\to_\sigma$ ***halts successfully*** (terminates) on M, otherwise it ***diverges*** on M ($M \not\Downarrow_\sigma$).

Each reduction strategy induces an ***operational semantics***, in that we can imagine a machine which evaluates terms by implementing the given strategy. The ***observational equivalence*** arises if we consider programs as ***black boxes*** and only observe their "halting properties".

Definition 1 (σ-observational Equivalence). Let $\to_\sigma$ be a reduction strategy and let $M, N \in \Lambda^0(C)$. The ***observational equivalence*** $\approx_\sigma$ is defined by $M \approx_\sigma N$ iff $\forall C[\,].(C[M], C[N] \in \Lambda^0(C) \Rightarrow (C[M] \Downarrow_\sigma \Leftrightarrow C[N] \Downarrow_\sigma))$.

Showing σ-observational equivalences by induction on computation steps is difficult. Powerful proof-principles, allowing to factorize this task, are precious. Coinduction and mixed induction-coinduction principles for establishing $\approx_\sigma$ follow from the fact that $\approx_\sigma = \approx_\sigma^{app}$, where $\approx_\sigma^{app}$ denotes the ***applicative equivalence*** induced by $\to_\sigma$ (see Section 1).

In this paper, we introduce two ***general*** methods for showing $\approx_\sigma = \approx_\sigma^{app}$:

1. ***domain logic*** method based on the ***intersection types*** presentation of a suitable computationally adequate CPO-λ-model;
2. ***logical relations*** method based on a mixed induction-coinduction principle.

Both methods are semantical, in the sense that they are based on a suitable ***computationally adequate*** CPO-λ-model. The "domain logic method" is the generalization of the method originally used by Abramsky and Ong in [1] for the special case of the lazy reduction strategy. The "logical relations method" is the generalization of the method originally used by Pitts in [16, 17, 18] for the special cases of by-name and by-value lazy reduction strategies. Pitts' technique is based on the ***minimal invariance property*** of the model, as is the case for initial models. In this paper we show how to extend it to many more strategies and models. Both methods presented in this paper can be used to show that $\approx_\sigma = \approx_\sigma^{app}$ for all the reduction strategies in [8], thus solving some open questions raised there. The strategies in [8] are:

- $\to_l \subseteq \Lambda^0 \times \Lambda^0$, the ***lazy call-by-name*** strategy, which reduces the leftmost β-redex not appearing within a λ-abstraction (see [1]);
- $\to_v \subseteq \Lambda^0 \times \Lambda^0$, the ***lazy call-by-value*** strategy ([19]), which reduces the ***leftmost*** β-redex, not within an abstraction, whose argument is an abstraction;
- $\to_o \subseteq \Lambda^0(\{\Omega\}) \times \Lambda^0(\{\Omega\})$, a non-deterministic strategy, which rewrites λ-terms containing occurrences of the constant Ω by reducing any β-redex. The congruence $\approx_o$ cannot be realized by a continuously complete CPO model ([10]);
- $\to_h \subseteq \Lambda \times \Lambda$, the ***head call-by-name*** strategy, which reduces the leftmost β-redex, if the term is not in head normal form;
- $\to_n \subseteq \Lambda \times \Lambda$, the leftmost***$\beta$-normalizing*** strategy; $\approx_n$ is modeled in [4];
- $\to_p \subseteq \Lambda \times \Lambda$, Barendregt's ***perpetual*** strategy, which reduces the leftmost β-redex not in the operator of a β-redex, which is either an I-redex, or a K-redex whose argument is a normal form; $\to_p$ terminates exactly on ***strongly normalizing*** terms; the model in [7] for the λN^0-calculus is fully abstract w.r.t. $\approx_p$ ([9]).

In this paper, we outline the two methods in full generality, but, for lack of space, we give details only for $\rightarrow_v$ and $\rightarrow_h$. These are representative of the difficulties one has to deal with in the generalization of [1] and [16, 17, 18]. The latter is a prototype of a strategy which has to take into account also *open* terms and does not have an "appropriate" initial model.

In the paper we use λ-calculus concepts and notation as defined in [2, 8]. The paper is organized as follows. In Section 1 we introduce the problem of characterizing coinductively contextual equivalences via applicative equivalences. In Section 2 we define $\rightarrow_v$ and $\rightarrow_h$. In Section 3 we give a description via intersection types of the computationally adequate models for $\approx_h$ and $\approx_v$. In Section 4 we present in general the domain logic method and the logical relations method, and we carry out the details for the case of $\approx_v$ and $\approx_h$, respectively. Final remarks appear in Section 5.

The author is grateful to Mariangiola Dezani, Furio Honsell, and Andrew Pitts for useful discussions.

1 Coinductive Characterizations via Applicative Equivalences

Given a reduction strategy $\rightarrow_\sigma$, the *σ-applicative equivalence*, $\approx_\sigma^{app}$, differs from the observational one, in that the behaviour of the programs is tested only on applicative (closed) contexts.

Definition 2. Let $\approx_\sigma^{app} \subseteq \Lambda^0(C) \times \Lambda^0(C)$ be the *applicative equivalence*:
$M \approx_\sigma^{app} N \iff \forall P_1, \ldots, P_n \in \Lambda^0(C).\ (MP_1 \ldots P_n \Downarrow_\sigma \Leftrightarrow NP_1 \ldots P_n \Downarrow_\sigma)$.

The equivalence $\approx_\sigma^{app}$ has a coinductive characterization:

Lemma 3. *The applicative equivalence $\approx_\sigma^{app}$ can be viewed as the greatest fixed point of the monotone operator $\Psi_\sigma : \mathcal{P}(\Lambda^0(C) \times \Lambda^0(C)) \rightarrow \mathcal{P}(\Lambda^0(C) \times \Lambda^0(C))$*

$$\Psi_\sigma(R) = \{(M,N) \mid (M \not\Downarrow_\sigma \ \wedge N \not\Downarrow_\sigma \ \wedge \forall P \in \Lambda^0(C).\ (MP, NP) \in R) \vee (M \Downarrow_\sigma \ \wedge N \Downarrow_\sigma \ \wedge \forall P \in \Lambda^0(C).\ (MP, NP) \in R)\}.$$

An immediate consequence is the validity of the *coinduction principle*:

$$\frac{(M,N) \in R \quad R \text{ is a } \Psi_\sigma\text{-bisimulation}}{M \approx_\sigma^{app} N},$$

where a *Ψ_σ-bisimulation* is a relation $R \subseteq \Lambda^0(C) \times \Lambda^0(C)$ s.t. $R \subseteq \Psi_\sigma(R)$.

Remark. i) Another important relation on $\Lambda^0(C) \times \Lambda^0(C)$ is the *applicative preequivalence* $\lesssim_\sigma^{app}$ defined as follows:
$M \lesssim_\sigma^{app} N \iff \forall P_1, \ldots, P_n \in \Lambda^0(C).\ (MP_1 \ldots P_n \Downarrow_\sigma \Rightarrow NP_1 \ldots P_n \Downarrow_\sigma)$.
Then $\approx_\sigma^{app} = \lesssim_\sigma^{app} \cap (\lesssim_\sigma^{app})^{-1}$. Using a suitable monotone operator $\Psi_\sigma^{\lesssim}$, similar to that of Lemma 3, one can give a coinductive characterization also to $\lesssim_\sigma^{app}$.
ii) When $\Downarrow_\sigma$ is axiomatized on the whole Λ, we can define the *applicative equivalence extended to open contexts* $\approx_\sigma^{app,ext}$: let $M, N \in \Lambda^0(C)$,
$M \approx_\sigma^{app,ext} N \iff \forall P_1, \ldots, P_n \in \Lambda(C).\ (MP_1 \ldots P_n \Downarrow_\sigma \Leftrightarrow NP_1 \ldots P_n \Downarrow_\sigma)$.
$\approx_\sigma^{app,ext}$ will be useful in the sequel for reasoning on $\approx_\sigma^{app}$, since the two equivalences coincide in many interesting cases.

If $\approx_\sigma = \approx_\sigma^{app}$, then the coinduction principle above can be used to establish directly the observational equivalence. Hence the natural question arises: for which strategies σ's do the two equivalences coincide? Notice that there are σ's such that $\approx_\sigma \not\supseteq \approx_\sigma^{app}$:

Examples. 1) Let $Val_\tau = \{\lambda x.xM_1M_2 \mid M_1 =_\beta M_2\}$, where $=_\beta$ is β-conversion. Let $\to_\tau \subseteq \Lambda \times \Lambda$ be the reduction strategy which reduces the leftmost β-redex, if it exists, and if the term is not in Val_τ.
Then one can easily show that:
i) $M \approx_\tau N$ if and only if $M =_\beta N$, but
ii) let P, Q be two β-different fixed point of K ($K \equiv \lambda xy.x$), i.e. $\forall M.\ KPM =_\beta P$, $\forall M.\ KQM =_\beta Q$, but $P \neq_\beta Q$, then $P \approx_\tau^{app} Q$, but $P \not\approx_\tau Q$.
2) A closed term $\lambda x.M$ is said to be an "eraser" if $x \notin FV(M)$. The strategy $\to_e \subseteq \Lambda^0 \times \Lambda^0$ rewrites closed λ-terms not in normal form which are not erasers by reducing any β-redex. $Val_e = \{M \in \Lambda^0 \mid M$ is an eraser$\}$.
One can easily show that $\approx_e^{app} \neq \approx_e$. In fact, let e.g. $M \equiv \lambda x.(\Delta\Delta(xO))$, and $N \equiv \lambda x.(\Delta\Delta x)$, where $\Delta \equiv \lambda x.xx$ and $O \equiv \lambda x.I$. Then, it is easy to check that $M \approx_e^{app} N$, but if we take $C[\,] \equiv \lambda y.([\,]\lambda z.zy)$, then $C[M] \Downarrow_e$, while $C[N] \not\Downarrow_e$.
However, $\approx_e$ can be characterized coinductively since it coincides with $\approx_o^{app}$.

However, for many interesting strategies in the literature, one can show that $\approx_\sigma = \approx_\sigma^{app}$, see e.g. [1, 5, 8, 11, 17, 15]. The techniques introduced in this paper are rather general and both can be used for establishing the coincidence for all the strategies listed in the Introduction.

2 Two Interesting Strategies

In this section we briefly discuss the reduction strategies $\to_v$ and $\to_h$, which we shall focus on in Section 4. In particular we axiomatize the corresponding evaluation relations.

$\to_v$ strategy. The *lazy call-by-value* strategy $\to_v \subseteq \Lambda^0 \times \Lambda^0$ reduces the *leftmost* β-redex, not appearing within a λ-abstraction, whose argument is a λ-abstraction. $Val_v = \{\lambda x.M \mid M \in \Lambda\} \cap \Lambda^0$. The evaluation $\Downarrow_v$ is the least binary relation over $\Lambda^0 \times Val_v$ satisfying the following rules:

$$\frac{}{\lambda x.M \Downarrow_v \lambda x.M} \qquad \frac{M \Downarrow_v \lambda x.P \quad N \Downarrow_v Q \quad P[Q/x] \Downarrow_v U}{MN \Downarrow_v U}$$

The notion of β-reduction which is correct w.r.t. $\approx_v$[2] is the $\to_{\beta_v} \subseteq \Lambda \times \Lambda$ [19], i.e. $(\lambda x.M)N \to_{\beta_v} M[N/x]$, if N is a variable or an abstraction.
The initial solution in $CPO_\perp$, D^v, of the domain equation $D \simeq [D \to_\perp D]_\perp$, where $[D \to_\perp D]_\perp$ denotes the lifted space of strict Scott-continuous functions, gives a computationally adequate model for $\approx_v$. This model is studied in [5].

[2] The β-reduction $\to_{\beta_r}$ is correct w.r.t. $\approx_\sigma$ if, for all $M, N \in \Lambda^0(C)$, $M =_{\beta_r} N \Rightarrow M \approx_\sigma N$.

$\to_h$ **strategy.** The *head call-by-name* strategy $\to_h \subseteq \Lambda \times \Lambda$ reduces the *leftmost* β-redex, if the term is not in head normal form. Val_h is the set of λ-terms in head normal form. The evaluation $\Downarrow_h$ is the least binary relation over $\Lambda \times Val_h$ satisfying the following rules, for $n \geq 0$:

$$\frac{}{xM_1 \ldots M_n \Downarrow_h xM_1 \ldots M_n} \qquad \frac{M \Downarrow_h N}{\lambda x.M \Downarrow_h \lambda x.N} \qquad \frac{M[N/x]M_1 \ldots M_n \Downarrow_h P}{(\lambda x.M)NM_1 \ldots M_n \Downarrow_h P}$$

Classical β-reduction is correct w.r.t. $\approx_h$. The canonical D_∞ model of Scott is fully abstract w.r.t. $\approx_h$ (see [2]). We will refer to the model D_∞ as D^h.

3 Computationally Adequate Models

In this section we phrase in terms of intersection types ([3]) the computational adequacy of an algebraic lattice λ-model D^σ. We assume that the finitary logical presentation of D^σ is given by the intersection types theory $\mathcal{T}_\sigma = (T_\sigma, \leq_\sigma)$.

Definition 4. Let $\mathcal{T}_\sigma = (T_\sigma, \leq_\sigma)$ be the intersection type theory corresponding to D^σ. Then $\quad (T_\sigma \ni)\ \phi ::= \alpha \mid \phi \wedge \phi \mid \phi \to \phi,$
where $\alpha \in TB_\sigma$, and TB_σ is a non-empty set of base types.

The relation $\leq_\sigma$ on types is the least relation containing the set IB_σ of inequalities involving the base types in TB_σ, and closed under the rules:

$$\phi \leq_\sigma \phi \qquad \phi_1 \wedge \phi_2 \leq_\sigma \phi_1 \qquad \phi_1 \wedge \phi_2 \leq_\sigma \phi_2$$

$$\frac{\phi_1 \leq_\sigma \phi_2 \quad \phi_2 \leq_\sigma \phi_3}{\phi_1 \leq_\sigma \phi_3} \qquad \frac{\phi \leq_\sigma \phi_1 \quad \phi \leq_\sigma \phi_2}{\phi \leq_\sigma \phi_1 \wedge \phi_2} \qquad \frac{\phi_2 \leq_\sigma \phi_1 \quad \psi_1 \leq_\sigma \psi_2}{\phi_1 \to \psi_1 \leq_\sigma \phi_2 \to \psi_2}$$

$$\phi \to \phi_1 \wedge \phi_2 \leq_\sigma (\phi \to \phi_1) \wedge (\phi \to \phi_2) \qquad (\phi \to \phi_1) \wedge (\phi \to \phi_2) \leq_\sigma \phi \to \phi_1 \wedge \phi_2$$

We put $\phi_1 =_\sigma \phi_2$ if and only if $\phi_1 \leq_\sigma \phi_2 \ \wedge\ \phi_2 \leq_\sigma \phi_1$. A *type environment* is a partial function $\Gamma : Vars \longrightarrow T_\sigma$. Let Env_Γ denote the set of type environments. The type assignment system S_σ includes the following rules and, possibly, a nonempty set SB_σ of axioms involving base types, and base constants of the language: $\qquad \Gamma[\phi/x] \vdash_\sigma x : \phi \quad var$

$$\frac{\Gamma \vdash_\sigma M : \phi \quad \Gamma \vdash_\sigma M : \psi}{\Gamma \vdash_\sigma M : \phi \wedge \psi} \wedge I \qquad \frac{\Gamma \leq_\sigma \Delta \quad \Delta \vdash_\sigma M : \phi \quad \phi \leq_\sigma \psi}{\Gamma \vdash_\sigma M : \psi} \leq$$

$$\frac{\Gamma[\phi/x] \vdash_\sigma M : \psi}{\Gamma \vdash_\sigma \lambda x.M : \phi \to \psi} \to I \qquad \frac{\Gamma \vdash_\sigma M : \phi \to \psi \quad \Gamma \vdash_\sigma N : \phi}{\Gamma \vdash_\sigma MN : \psi} \to E$$

The canonical interpretation $[\![\ ]\!]^{D^\sigma}$ in the filter model D^σ can be characterized as follows: $[\![M]\!]_\rho^{D^\sigma} = \{\phi \mid \exists \Gamma \sqsubseteq_{D^\sigma} \rho.\ \Gamma \vdash_\sigma M : \phi\}$, where $\rho : Var \to VD^\sigma$, for a suitable $VD^\sigma \subseteq D^\sigma$. $\Gamma \sqsubseteq_{D^\sigma} \rho$ means that $\forall x \in Var.\ \Gamma(x) \sqsubseteq_{D^\sigma} \rho(x)$, and $\sqsubseteq_{D^\sigma}$ denotes the order relation on D^σ.

The computational adequacy of the model D^σ (i.e. for $M, N \in \Lambda^0(C)$, $[\![M]\!]^{D^\sigma} \sqsubseteq_{D^\sigma} [\![N]\!]^{D^\sigma} \Rightarrow M \leq_\sigma N$) can now take the strong form:

Theorem 5 (Computational Adequacy). *There exists a filter of types $T_\sigma^{conv} \subseteq T_\sigma$, and a subset Env_Γ^{conv} of T_σ^{conv}-type environments such that*

$$M \Downarrow_\sigma \iff \exists \Gamma \in Env_\Gamma^{conv}.\ \exists \phi \in T_\sigma^{conv}.\ (\Gamma \vdash_\sigma M : \phi).$$

Intersection Type Description of the Model D^v. We recall the intersection type description of the model D^v of [5].

Definition 6. Let $\mathcal{T}_v = (T_v, \leq_v)$ be the intersection type theory:

$$(T_v \ni)\ \phi ::= \ \nu \mid \phi \wedge \phi \mid \phi \rightarrow \phi.$$

The axiomatization of $\leq_v$ consists of the standard rules of Definition 4, and of the following rule involving the base type ν: $\quad \phi \leq_v \nu$.

The type system S_v consists of the rules of Definition 4, and moreover:

$$\Gamma \vdash_v \lambda x.M : \nu \quad \nu\,.$$

Theorem 7 (Comp. Ad. [5]). $M \Downarrow_v \iff \exists \phi \in T_v.\ \vdash_v M : \phi\,.$

Intersection Type Description of the Model D^h. An intersection type description of the model D^h is:

Definition 8. Let $\mathcal{T}_h = (T_h, \leq_h)$ be the intersection type theory:

$$(T_h \ni)\ \phi ::= \ \omega \mid 1 \mid \phi \wedge \phi \mid \phi \rightarrow \phi.$$

The axiomatization of $\leq_h$ consists of the standard rules of Definition 4, and moreover of the following rules involving the base types ω and 1:
$\phi \leq_h \omega \quad \omega \leq_h \omega \rightarrow \omega \quad 1 =_h \omega \rightarrow 1$.

The type system S_h consists of the rules of Definition 4 and moreover:

$$\Gamma \vdash_h M : \omega \quad \omega\,.$$

Theorem 9 (Comp. Ad.). $M \Downarrow_h \iff \exists \Gamma\ \exists \phi \neq_h \omega.\ \Gamma \vdash_h M : \phi\,.$

4 Two Techniques for Showing $\approx_\sigma^{app} = \approx_\sigma$

In this section we analyze two general techniques for establishing $\approx_\sigma^{app} = \approx_\sigma$. The inclusion $\approx_\sigma \subseteq \approx_\sigma^{app}$ is immediate; the other inclusion is in general difficult to prove. The techniques we present are:
i) *Domain logic* method based on the *intersection types* presentation of a *computationally adequate* model;
ii) *Logical relations* method based on an induction-coinduction principle.

Both techniques are based on the existence of a suitable computationally adequate *CPO*-model. A special case of the former was described in the case of $\rightarrow_l$ by Abramsky and Ong in [1]. A version of the latter, applicable only to computationally adequate *CPO*-models which satisfy a "minimal invariance property", appears in [16, 17, 18]. We generalize and streamline both techniques.

Both methods are used to show that $\approx_\sigma^{app}$ is a congruence w.r.t. application. In fact, in order to prove that $\approx_\sigma^{app} \subseteq \approx_\sigma$, it is sufficient to show that (Theorem 13):
1. $\approx_\sigma^{app}$ is a congruence w.r.t. application, i.e. for all $M, N, P, Q \in \Lambda^0(C)$,
$$M \approx_\sigma^{app} N \wedge P \approx_\sigma^{app} Q \implies MP \approx_\sigma^{app} NQ\ ;$$
2. $\approx_\sigma^{app}$ is a congruence w.r.t. λ-abstraction, i.e., $\forall M, N \in \Lambda(C)$ such that $FV(M, N) \subseteq \{x_1, \ldots, x_n\}$, $\forall P_1, \ldots, P_n \in \Lambda^0(C)$. $M[P_i/x_i] \approx_\sigma^{app} N[P_i/x_i] \implies$
$$\lambda x_1 \ldots . x_n.M \approx_\sigma^{app} \lambda x_1 \ldots . x_n.N\ .$$
(In case the strategy is by-value, i.e. for $\sigma = v, p$, $P_1, \ldots, P_n$ are chosen to be convergent terms.)

The congruence of $\approx_\sigma^{app}$ w.r.t. application is in general much more difficult to show than that w.r.t. λ-abstraction. The latter follows immediately from the extensionality of $\approx_\sigma^{app}$ (Theorem 10). The proof of Theorem 10 is problematic only for $\sigma = n$, where one needs to exploit extensively the separability technique of [12]. For lack of space, we omit this proof.

Theorem 10 (Extensionality of $\approx_\sigma^{app}$). *i) Let $\sigma = l, v$. Let $M, N \in \Lambda^0$ be such that $M \Downarrow_\sigma$, $N \Downarrow_\sigma$. If, for all $P \in \Lambda^0$, $MP \approx_\sigma^{app} NP$, then $M \approx_\sigma^{app} N$.*
ii) Let $\sigma = o, h, p, n$. Let $M, N \in \Lambda^0(C)$. If, for all $P \in \Lambda^0(C)$, $MP \approx_\sigma^{app} NP$, then $M \approx_\sigma^{app} N$.

Theorem 11. *$\approx_\sigma^{app}$ is a congruence w.r.t. λ-abstraction, for $\sigma \in \{l, v, o, h, n, p\}$.*

Proof. We show that, for $M, N \in \Lambda$ s.t. $FV(M, N) \subseteq \{x\}$,
$\forall P \in \Lambda^0$*(possibly convergent)*. $M[P/x] \approx_\sigma^{app} N[P/x] \Rightarrow \lambda x.M \approx_\sigma^{app} \lambda x.N$.
For $\sigma = l, v$ this is immediate. For $\sigma = o, h, n$ the proof follows from the Extensionality Theorem, using the fact that $(\lambda x.M)P \approx_\sigma^{app} M[P/x]$, which in turn follows from the correctness of the β-reduction w.r.t. $\approx_\sigma^{app}$. For $\sigma = p$, the proof follows from the fact that, for all $M \in \Lambda$, $(\exists P \in \Lambda^0.\ M[P/x] \Downarrow_p) \iff M \Downarrow_p$. The implication ($\Rightarrow$) in this latter fact follows since $\rightarrow_p$ is perpetual. The other implication is proved by computation induction, choosing as P a suitable permutator. □

Corollary 12. *$\approx_\sigma^{app,ext} = \approx_\sigma^{app}$, for $\sigma = h, n, p$.*

Theorem 13. *Suppose that $\approx_\sigma^{app}$ is a congruence w.r.t. λ-abstraction and application. Then $\approx_\sigma^{app} \subseteq \approx_\sigma$.*

Proof. We prove by induction on the context $C[\,]$ that:
$M \approx_\sigma^{app} N \implies \forall C[\,].(C[M], C[N] \in \Lambda(C) \wedge FV(C[M], C[N]) \subseteq \{x_1, \ldots, x_n\} \Rightarrow$
$\forall P_1 \ldots P_n \in \Lambda^0(C).\ C[M][P_i/x_i] \approx_\sigma^{app} C[N][P_i/x_i])$. □

4.1 Domain Logic Method

This is a semantical method for showing $\approx_\sigma^{app} \subseteq \approx_\sigma$, based on the logical description via intersection types of a computationally adequate CPO-model. This method can be also used to derive the computational adequacy of D^σ, for

$\sigma = l, v, o, p$. However, for $\sigma = h, n$, the computational adequacy must be assumed at the outset. It generalizes that in [1] in two main respects:
i) we relax the condition on type interpretation on applicative structures, by requiring only that the type interpretation be *adequate*. In fact, in order to show the crucial Theorem 27, a uniformly defined notion of type interpretation on the whole class of applicative structures (like in [1]) is not necessary. An *adequate* type interpretation on the applicative structure $\mathcal{A}$ is enough;
ii) we use, in place of λ-model, the weaker notion of σ-combinatory algebra.

In order to state the crucial result of this method, i.e. Theorem 27, we need a number of results concerning the notions of ***applicative σ-structure with convergence*** (Definition 14) and of ***combinatory σ-algebra*** (Definition 23).

Definition 14 (Applicative σ-structure with Convergence). An *applicative σ-structure with convergence* is a structure $\mathcal{A} = (A, \bullet_{\mathcal{A}}, \Downarrow_{\mathcal{A}})$ endowed with a notion of applicative equivalence $\approx^{app}_{\mathcal{A}} \subseteq A \times A$ such that
• $\{a \in A \mid a \Downarrow_{\mathcal{A}}\} \neq \emptyset, A$;
• if the application between λ-terms is right (left) strict w.r.t. $\Downarrow_\sigma$, then also $\bullet_{\mathcal{A}}$ is right (left) strict w.r.t. $\Downarrow_{\mathcal{A}}$;
• $a \approx^{app}_{\mathcal{A}} b \iff \forall k \geq 0\ \forall c_1, \ldots, c_k \in A\ (\forall i.\ c_i \Downarrow_{\mathcal{A}} \Rightarrow (ac_1 \ldots c_k \Downarrow_{\mathcal{A}} \Leftrightarrow bc_1 \ldots c_k \Downarrow_{\mathcal{A}}))$, if the application between λ-terms is right strict w.r.t. $\Downarrow_\sigma$,
$a \approx^{app}_{\mathcal{A}} b \iff \forall k \geq 0\ \forall c_1, \ldots, c_k \in A\ (ac_1 \ldots c_k \Downarrow_{\mathcal{A}} \Leftrightarrow bc_1 \ldots c_k \Downarrow_{\mathcal{A}})$, otherwise.

The model D^σ can be viewed as an applicative σ-structure whose notion of convergence is: $a \Downarrow_{D^\sigma} \iff \exists \phi \in T'_\sigma.\ a \sqsupseteq_{D^\sigma} \phi$, for any filter $T'_\sigma \subseteq T_\sigma$. Throughout this section we shall focus on the filter T^{conv}_σ of Theorem 5. This is not strictly necessary until Theorem 27, below. Notice that until then no mention is made of the fact that D^σ is computationally adequate w.r.t. precisely that T^{conv}_σ; and even then this assumption is not necessary in all cases.

We can define various notions of interpretations for the types in $\mathcal{T}_\sigma$ on the class of applicative σ-structures with convergence. In the following definition, we isolate the class of type interpretations we are interested in:

Definition 15. Let $\mathcal{A} = (A, \bullet_{\mathcal{A}}, \Downarrow_{\mathcal{A}})$ be an applicative σ-structure with convergence. An interpretation of the type theory $\mathcal{T}_\sigma$, $[\![\ ]\!]^{t\mathcal{A}} : T_\sigma \to \mathcal{P}(A)$, is ***adequate*** on $\mathcal{A}$, if $[\![\ ]\!]^{t\mathcal{A}}$ satisfies the following properties:

– $[\![\phi \wedge \psi]\!]^{t\mathcal{A}} = [\![\phi]\!]^{t\mathcal{A}} \cap [\![\psi]\!]^{t\mathcal{A}}$

$$[\![\phi \to \psi]\!]^{t\mathcal{A}} =$$

$$= \begin{cases} \{a \in A \mid a \Downarrow_{\mathcal{A}}\ \wedge \forall c \in [\![\phi]\!]^{t\mathcal{A}}.\ (c \Downarrow_{\mathcal{A}} \Rightarrow\ a \bullet_{\mathcal{A}} c \in [\![\psi]\!]^{t\mathcal{A}})\} & \text{if (1)} \\ \{a \in A \mid a \Downarrow_{\mathcal{A}}\ \wedge \forall c \in [\![\phi]\!]^{t\mathcal{A}}.\ a \bullet_{\mathcal{A}} c \in [\![\psi]\!]^{t\mathcal{A}}\} & \text{if (2)} \\ \{a \in A \mid \forall c \in [\![\phi]\!]^{t\mathcal{A}}.\ (c \Downarrow_{\mathcal{A}} \Rightarrow\ a \bullet_{\mathcal{A}} c \in [\![\psi]\!]^{t\mathcal{A}})\} & \text{if (3)} \\ \{a \in A \mid \forall c \in [\![\phi]\!]^{t\mathcal{A}}.\ a \bullet_{\mathcal{A}} c \in [\![\psi]\!]^{t\mathcal{A}}\} & \text{otherwise} \end{cases}$$

where: (1) $\bullet_{\mathcal{A}}$ is left and right strict w.r.t. $\Downarrow_{\mathcal{A}}$;
(2) $\bullet_{\mathcal{A}}$ is left strict w.r.t. $\Downarrow_{\mathcal{A}}$; (3) $\bullet_{\mathcal{A}}$ is right strict w.r.t. $\Downarrow_{\mathcal{A}}$.

– $\phi \leq_\sigma \psi \Rightarrow [\![\phi]\!]^{tA} \subseteq [\![\psi]\!]^{tA}$
– $a \Downarrow_{\mathcal{A}} \iff \exists \phi \in T^{conv}_\sigma.\ a \in [\![\phi]\!]^{tA}$.

Proposition 16. *The type interpretation $[\![\ ]\!]^{tD^\sigma}$ on D^σ, defined as, $a \in [\![\phi]\!]^{tD^\sigma}$ if and only if $a \sqsupseteq_{D^\sigma} \phi$, is adequate.*

An intersection type interpretation on an applicative structure naturally induces a *logical equivalence*:

Definition 17. Let $\mathcal{A} = (A, \bullet_{\mathcal{A}}, \Downarrow_{\mathcal{A}}, [\![\ ]\!]^{tA})$ be an applicative σ-structure with type interpretation, the *logical equivalence* $\approx^{\mathcal{L}}_{\mathcal{A}} \subseteq A \times A$ is defined as follows

$$a \approx^{\mathcal{L}}_{\mathcal{A}} b \iff \forall \phi.\ a \in [\![\phi]\!]^{tA} \Leftrightarrow b \in [\![\phi]\!]^{tA}.$$

Definition 18. Let $\mathcal{A} = (A, \bullet_{\mathcal{A}}, \Downarrow_{\mathcal{A}})$ be an applicative σ-structure. An applicative σ-substructure with convergence of $\mathcal{A}$ is an applicative σ-structure $\mathcal{A}' = (A', \bullet_{\mathcal{A}'}, \Downarrow_{\mathcal{A}'})$, such that $A' \subseteq A$, $\bullet_{\mathcal{A}'} = \bullet_{\mathcal{A}|A' \times A'}$, and $\Downarrow_{\mathcal{A}'} = \Downarrow_{\mathcal{A}|A'}$.

Theorem 19. *Let $\mathcal{A} = (A, \bullet_{\mathcal{A}}, \Downarrow_{\mathcal{A}}, [\![\ ]\!]^{tA})$ be an applicative σ-structure with adequate type interpretation. Let $\mathcal{A}' = (A', \bullet_{\mathcal{A}'}, \Downarrow_{\mathcal{A}'})$ be an applicative σ-substructure of $\mathcal{A}$ such that $\approx^{app}_{\mathcal{A}'} = \approx^{app}_{\mathcal{A}}{}_{|A' \times A'}$. If, for all $\phi \in TB_\sigma$, and for all $a, b \in A$, $a \approx^{app}_{\mathcal{A}} b \Rightarrow (a \in [\![\phi]\!]^{tA} \Leftrightarrow b \in [\![\phi]\!]^{tA})$, then $\approx^{app}_{\mathcal{A}'} \subseteq \approx^{\mathcal{L}}_{\mathcal{A}'}$.*

Approximable σ-applicative structures are important structures, because, under the hypothesis of Theorem 19 on the applicative substructure, the applicative equivalence on them coincides with the logical equivalence:

Definition 20. Let $\mathcal{A} = (A, \bullet_{\mathcal{A}}, \Downarrow_{\mathcal{A}}, [\![\ ]\!]^{tA})$ be an applicative σ-structure with type interpretation, and $\mathcal{A}' = (A', \bullet_{\mathcal{A}'}, \Downarrow_{\mathcal{A}'})$ be an applicative σ-substructure. $\mathcal{A}'$ is *approximable* if, for all $a, b_1, \ldots, b_n \in A'$, if $ab_1 \ldots b_n \Downarrow_{\mathcal{A}'}$, then

$$\begin{cases} \exists \phi_1, \ldots, \phi_n \in T_\sigma\ \exists \phi \in T^{conv}_\sigma.\ (a \in [\![\phi_1 \to \ldots \phi_n \to \phi]\!]^{tA} \wedge & \\ \forall i.\ b_i \in [\![\phi_i]\!]^{tA}) & \text{if } (1) \\ \exists \phi_1, \ldots, \phi_n \in T_\sigma\ \exists \phi \in T^{conv}_\sigma.\ (a \in [\![\phi_1 \to \ldots \phi_n \to \phi]\!]^{tA} \wedge & \\ \forall i.\ (b_i \Downarrow_{\mathcal{A}} \wedge\ b_i \in [\![\phi_i]\!]^{tA})) & \text{if } (2) \end{cases}$$

where: (1) $\Downarrow_{\mathcal{A}}$ is not right strict w.r.t. $\bullet_{\mathcal{A}}$; (2) $\Downarrow_{\mathcal{A}}$ is right strict w.r.t. $\bullet_{\mathcal{A}}$.

D^σ with the type interpretation $[\![\ ]\!]^{tD^\sigma}$ of Proposition 16 is approximable.

Theorem 21. *Let $\mathcal{A} = (A, \bullet_{\mathcal{A}}, \Downarrow_{\mathcal{A}}, [\![\ ]\!]^{tA})$ be an applicative σ-structure with adequate type interpretation. Let $\mathcal{A}' = (A', \bullet_{\mathcal{A}'}, \Downarrow_{\mathcal{A}'})$ be an approximable applicative σ-substructure of $\mathcal{A}$:*
i) Then we have $\approx^{app}_{\mathcal{A}'} \supseteq \approx^{\mathcal{L}}_{\mathcal{A}'}$.
ii) Moreover, if $\mathcal{A}$ and $\mathcal{A}'$ are such that $\approx^{app}_{\mathcal{A}'} = \approx^{app}_{\mathcal{A}}{}_{|A' \times A'}$ and, $\forall \phi \in TB_\sigma, \forall a, b \in A$, $a \approx^{app}_{\mathcal{A}} b \Rightarrow (a \in [\![\phi]\!]^{tA} \Leftrightarrow b \in [\![\phi]\!]^{tA})$, then $\approx^{app}_{\mathcal{A}'} = \approx^{\mathcal{L}}_{\mathcal{A}'}$.

Theorem 22. *Let $\mathcal{A} = (A, \bullet_{\mathcal{A}}, \Downarrow_{\mathcal{A}}, [\![\ ]\!]^{t\mathcal{A}})$ be an applicative σ-structure with adequate type interpretation. Let $\mathcal{A}' = (A', \bullet_{\mathcal{A}'}, \Downarrow_{\mathcal{A}'})$ be an approximable applicative σ-substructure of $\mathcal{A}$. If $\mathcal{A}$ and $\mathcal{A}'$ are such that $\approx^{app}_{\mathcal{A}'} = \approx^{app}_{\mathcal{A}}{}_{|A' \times A'}$, and for all $\phi \in TB_\sigma$, for all $a, b \in A$, $a \approx^{app}_{\mathcal{A}} b \Rightarrow (a \in [\![\phi]\!]^{\mathcal{A}} \Leftrightarrow b \in [\![\phi]\!]^{\mathcal{A}})$, then $\approx^{app}_{\mathcal{A}'}$ is a congruence w.r.t. $\bullet_{\mathcal{A}'}$.*

Proof. It is sufficient to show that, for all $k \geq 0$, for all $d_1, \ldots, d_k \in A'$, $abd_1 \ldots d_k \Downarrow_{\mathcal{A}'} \Rightarrow acd_1 \ldots d_k \Downarrow_{\mathcal{A}'}$.
From $abd_1 \ldots d_k \Downarrow_{\mathcal{A}'}$, by approximability of $\mathcal{A}'$, there exist $\phi_0, \ldots, \phi_k \in T_\sigma$, $\phi \in T^{conv}_\sigma$ s.t. $a \in [\![\phi_0 \to \ldots \phi_k \to \phi]\!]^{t\mathcal{A}} \wedge b \in [\![\phi_0]\!]^{t\mathcal{A}}$ and $\forall i = 1, \ldots, k.\ b_i \in [\![\phi_i]\!]^{t\mathcal{A}}$.
By Theorem 21, $c \in [\![\phi_0]\!]^{t\mathcal{A}}$, hence $acd_1 \ldots d_k \in [\![\phi]\!]^{t\mathcal{A}}$, and $acd_1 \ldots d_k \Downarrow_{\mathcal{A}'}$. □

The notion of *combinatory σ-algebra* generalizes the standard notion of combinatory λ-algebra (see e.g. [2]).

Definition 23 (Combinatory σ-algebra). A combinatory σ-algebra $\mathcal{A}$ is a structure $(A, \bullet_{\mathcal{A}}, \Downarrow_{\mathcal{A}}, [\![\ ]\!]^{\mathcal{A}})$ such that
$(A, \bullet_{\mathcal{A}}, \Downarrow_{\mathcal{A}})$ is an applicative σ-structure with convergence;
$[\![\ ]\!]^{\mathcal{A}} : \Lambda(C) \times Env \to A$ is an interpretation function such that, for

$$Env \ni \rho : Var \to \begin{cases} \{a \in A \mid a \Downarrow_{\mathcal{A}}\} & \text{if } \bullet_{\mathcal{A}} \text{ is right strict} \\ A & \text{otherwise} \end{cases}$$

- $[\![x]\!]^{\mathcal{A}}_\rho = \rho(x)$
- $[\![MN]\!]^{\mathcal{A}}_\rho = [\![M]\!]^{\mathcal{A}}_\rho \bullet_{\mathcal{A}} [\![N]\!]^{\mathcal{A}}_\rho$
- $[\![\lambda x.M]\!]^{\mathcal{A}}_\rho \bullet_{\mathcal{A}} a = [\![M]\!]^{\mathcal{A}}_{\rho[a/x]}$, for all $a \in \{a \in A \mid a \Downarrow_{\mathcal{A}}\}$, if $\bullet_{\mathcal{A}}$ is right strict, for all $a \in A$, otherwise;
- $\forall M \in \Lambda^0(C).\ (M \Downarrow_\sigma \Rightarrow [\![M]\!]^{\mathcal{A}} \Downarrow_{\mathcal{A}})$.

A combinatory σ-algebra $\mathcal{A}$ is *adequate* w.r.t. D^σ, if
$\forall M \in \Lambda^0(C).\ ([\![M]\!]^{\mathcal{A}} \Downarrow_{\mathcal{A}} \Rightarrow [\![M]\!]^{D^\sigma} \Downarrow_{D^\sigma})$.

The kind of combinatory σ-algebras we are interested in are those which arise from applicative structures with adequate type interpretations.

Theorem 24 (Soundness). *Let $\mathcal{A}$ be a combinatory σ-algebra with adequate $\mathcal{T}_\sigma$-interpretation. If, for all axioms ax in SB_σ, and $\forall M \in \Lambda(C)$, Γ, ρ,
$\Gamma \vdash_\sigma M : \phi\ \ ax \implies \forall \rho\ (\forall x.\rho(x) \in [\![\Gamma(x)]\!]^{t\mathcal{A}} \Rightarrow [\![M]\!]^{\mathcal{A}}_\rho \in [\![\phi]\!]^{t\mathcal{A}})$.
Then, for all $\phi \in T_\sigma$, and for all $M \in \Lambda(C)$, Γ,*

$$\Gamma \vdash_\sigma M : \phi \implies \forall \rho\ (\forall x.\rho(x) \in [\![\Gamma(x)]\!]^{t\mathcal{A}} \Rightarrow [\![M]\!]^{\mathcal{A}}_\rho \in [\![\phi]\!]^{t\mathcal{A}})\ .$$

Proof. The proof is by induction on the length of the derivation $\Gamma \vdash_\sigma M : \phi$. We consider the case $\bullet_{\mathcal{A}}$ not right strict (the other case is similar).
Base Case (axioms in SB_σ): immediate.
Inductive Step: If the last rule in the derivation of $\Gamma \vdash_\sigma M : \phi$ is $\wedge I$, the thesis follows immediately from the induction hypothesis. If the last rule is $\leq$, then the

thesis follows from the induction hypothesis and Definition 15. If the last rule is *var*, then the thesis follows from the definition of the interpretation of variables in a combinatory σ-algebra.
If the last rule is $\rightarrow I$, i.e. $\dfrac{\Gamma[\phi/x] \vdash_\sigma M : \psi}{\Gamma \vdash_\sigma \lambda x.M : \phi \rightarrow \psi}$, then, by induction hypothesis, $\forall \rho\ (\forall y.\rho(y) \in [\![\Gamma[\phi/x](y)]\!]^{t\mathcal{A}} \Rightarrow [\![M]\!]^{\mathcal{A}}_{\rho} \in [\![\psi]\!]^{t\mathcal{A}})$. We have to show that $\forall \rho\ (\forall y.\rho(y) \in [\![\Gamma(y)]\!]^{t\mathcal{A}} \Rightarrow [\![\lambda x.M]\!]^{\mathcal{A}}_{\rho} \in [\![\phi \rightarrow \psi]\!]^{t\mathcal{A}})$, i.e., if ρ is such that $\forall y.\ \rho(y) \in [\![\Gamma(y)]\!]^{t\mathcal{A}}$, then $\forall a \in [\![\phi]\!]^{t\mathcal{A}}.\ [\![\lambda x.M]\!]^{\mathcal{A}}_{\rho} \bullet_{\mathcal{A}} a \in [\![\psi]\!]^{t\mathcal{A}}$. But, from the definition of term interpretation in a combinatory σ-algebra, we have $[\![\lambda x.M]\!]^{\mathcal{A}}_{\rho} \bullet_{\mathcal{A}} a = [\![M]\!]^{\mathcal{A}}_{\rho[a/x]}$, and moreover $\forall y.\ \rho[a/x](y) \in [\![\Gamma[\phi/x](y)]\!]^{t\mathcal{A}}$. Hence, using the induction hypothesis, we get the thesis.
Finally, if the last rule is $\rightarrow E$, i.e. $\dfrac{\Gamma \vdash_\sigma M : \phi \rightarrow \psi \quad \Gamma \vdash_\sigma N : \phi}{\Gamma \vdash_\sigma MN : \psi}$, then, by induction hypothesis, $\forall \rho\ (\forall y.\rho(y) \in [\![\Gamma(y)]\!]^{t\mathcal{A}} \Rightarrow ([\![M]\!]^{\mathcal{A}}_{\rho} \in [\![\phi \rightarrow \psi]\!]^{t\mathcal{A}} \wedge [\![N]\!]^{\mathcal{A}}_{\rho} \in [\![\phi]\!]^{t\mathcal{A}}))$. Hence $[\![M]\!]^{\mathcal{A}}_{\rho} \bullet_{\mathcal{A}} [\![N]\!]^{\mathcal{A}}_{\rho} \in [\![\psi]\!]^{t\mathcal{A}}$. □

The model D^σ is a combinatory σ-algebra with adequate $\mathcal{T}_\sigma$-interpretation. Moreover, D^σ is sound and complete w.r.t. the type assignment system $\mathcal{T}_\sigma$:

Theorem 25 (Soundness and Completeness). *For all* $M \in \Lambda(C), \Gamma$,
$$\Gamma \vdash_\sigma M : \phi \iff \forall \rho\ (\forall x.\ \rho(x) \in [\![\Gamma(x)]\!]^{tD^\sigma} \Rightarrow [\![M]\!]^{D^\sigma}_{\rho} \in [\![\phi]\!]^{tD^\sigma})\ .$$

Proof. The only non trivial fact is that D^σ with term interpretation $[\![\ ]\!]^{D^\sigma}$ is a combinatory σ-algebra. This is shown following the argument for D^h in [3]. □

Remark. The type assignment system $\mathcal{T}_\sigma$ is sound and complete w.r.t. the class of combinatory σ-algebras with adequate type interpretation, which satisfy the hypotheses of Theorem 24. This is the generalization of the *Completeness* of [3].

Definition 26. Let $\mathcal{A}$ be a combinatory σ-algebra. D^σ is *computationally adequate w.r.t.* $\mathcal{A}$ if, for all $M \in \Lambda^0(C)$, $[\![M]\!]^{\mathcal{A}} \Downarrow_{\mathcal{A}} \iff [\![M]\!]^{D^\sigma} \Downarrow_{D^\sigma}$.

Item ii) of the following theorem generalizes Proposition 7.2.4 of [1].

Theorem 27. *Let* $\mathcal{A} = (A, \bullet_{\mathcal{A}}, \Downarrow_{\mathcal{A}}, [\![\ ]\!]^{\mathcal{A}})$ *be a combinatory* σ*-algebra with an adequate* $\mathcal{T}_\sigma$*-interpretation* $[\![\ ]\!]^{t\mathcal{A}}$ *such that (1) the Soundness Theorem holds for* $\mathcal{A}$*, and (2)* $\mathcal{A}$ *is adequate w.r.t. the filter model* D^σ*. Then*
i) D^σ *is computationally adequate w.r.t.* $\mathcal{A}$*;*
ii) let $\mathcal{A}^0 = (A^0, \bullet_{\mathcal{A}^0}, \Downarrow_{\mathcal{A}^0})$ *be the applicative* σ*-substructure of* $\mathcal{A}$*, where* $A^0 \subseteq A$ *denotes the interpretation domain of closed* λ*-terms. Then* $\mathcal{A}^0$ *is approximable.*

Proof. i) Use Theorem 25.
ii) We carry out the proof in the case $\bullet_{\mathcal{A}}$ is not right strict (the other case is similar). We show that, for all $M, N_1, \ldots, N_k \in \Lambda^0(C)$,
$[\![MN_1 \ldots N_k]\!]^{\mathcal{A}} \Downarrow_{\mathcal{A}^0} \Longrightarrow \exists \phi_1, \ldots, \phi_k \in \mathcal{T}_\sigma, \exists \phi \in \mathcal{T}^{conv}_\sigma$.
$([\![M]\!]^{\mathcal{A}} \in [\![\phi_1 \rightarrow \ldots \phi_k \rightarrow \phi]\!]^{t\mathcal{A}} \wedge \forall i.\ [\![N_i]\!]^{\mathcal{A}} \in [\![\phi_i]\!]^{t\mathcal{A}})$. From $[\![MN_1 \ldots N_k]\!]^{\mathcal{A}} \Downarrow_{\mathcal{A}^0}$,

since $\mathcal{A}$ is adequate w.r.t. D^σ, it follows that $[\![MN_1 \ldots N_k]\!]^{D^\sigma} \Downarrow_{D^\sigma}$. By approximability of D^σ, $\exists \phi_1, \ldots, \phi_k \in T_\sigma, \exists \phi \in T_\sigma^{conv}$ such that
$[\![M]\!]^{D^\sigma} \in [\phi_1 \to \ldots \phi_k \to \phi]^{tD^\sigma} \wedge \forall i = 1, \ldots, k.\ [\![N_i]\!]^{D^\sigma} \in [\phi_i]^{tD^\sigma}$.
From the completeness of D^σ, $\exists \phi_1, \ldots, \phi_k \in T_\sigma, \exists \phi \in T_\sigma^{conv}$ such that
$\vdash_\sigma M : \phi \to \ldots \phi_k \to \phi \wedge \forall i = 1, \ldots, k.\ \vdash_\sigma N_i : \phi_i$.
By soundness of $\mathcal{A}$, $[\![M]\!]^{\mathcal{A}} \in [\phi_1 \to \ldots \phi_k \to \phi]^{t\mathcal{A}} \wedge \forall i = 1, \ldots, k.\ [\![N_i]\!]^{\mathcal{A}} \in [\phi_i]^{t\mathcal{A}}$. □

Finally, after having presented all these technical results, we are in the position of explaining how to use these results in order to show that the applicative equivalence $\approx_\sigma^{app}$ is a congruence. For the reader's convenience, we outline the:

General pattern of the proof that $\approx_\sigma^{app}$ is a congruence:

1. Let $\to_{\beta_\sigma}$ be a notion of β-reduction which is correct w.r.t. $\approx_\sigma$. Naturally endow $\Lambda(C)/=_{\beta_\sigma}$ (or $\Lambda^0(C)/=_{\beta_\sigma}$) with a structure of combinatory σ-algebra with convergence, i.e. $\mathcal{A}_\sigma = (\Lambda(C)/=_{\beta_\sigma}, \bullet_{=_{\beta_\sigma}}, \Downarrow_\sigma, [\![\ ]\!]^{\mathcal{A}_\sigma})$, where

- by abuse of notation, $\Downarrow_\sigma$ denotes convergence of $=_{\beta_\sigma}$-classes;
- $[\![\ ]\!]^{\mathcal{A}_\sigma} : \Lambda(C) \times Env \to \Lambda(C)/=_{\beta_\sigma}$;
- $Env \ni \rho : Var \to \Lambda(C)/=_{\beta_\sigma}$, if $\bullet_{=_{\beta_\sigma}}$ is not right strict w.r.t. $\Downarrow_\sigma$
  $Env \ni \rho : Var \to Val_\sigma/=_{\beta_\sigma}$, otherwise;
- $[\![M]\!]_\rho^{\mathcal{A}_\sigma} = [\rho(M)]_{=_{\beta_\sigma}}$, where
  $\rho(M) = M[\rho(x_1)/x_1, \ldots, \rho(x_n)/x_n]$, with $FV(M) = \{x_1, \ldots, x_n\}$.

2. Endow $\mathcal{A}_\sigma$ with an adequate interpretation of the type theory $\mathcal{T}_\sigma$.
3. Use Theorem 27ii) for showing that the applicative σ-substructure of $\mathcal{A}_\sigma$, $\mathcal{A}_\sigma^0 = (\Lambda^0(C)/=_{\beta_\sigma}, \bullet_{=_{\beta_\sigma}}, \Downarrow_\sigma)$, is approximable. (Use Theorem 27i to derive the computational adequacy of D^σ).
4. Use Theorem 22 to show that the σ-structure $\mathcal{A}_\sigma^0$ is such that the applicative equivalence $\approx_{\mathcal{A}_\sigma^0}^{app}$ is a congruence w.r.t. $\bullet_{=_{\beta_\sigma}}$.
5. Use correctness of $\to_{\beta_\sigma}$ reduction w.r.t. $\approx_\sigma$ to deduce that $\approx_\sigma^{app}$ is a congruence w.r.t. λ-term application.

Domain Logic Method: Case $\sigma = v$. In this section we apply the general pattern of the "domain logic method" to the case of the $\to_v$-strategy. Our presentation somewhat simplifies that of [1]. We do not require Λ^0 to be endowed with a structure of lazy call-by-value model, but simply with a v-combinatory algebra structure. Since $\Downarrow_v$ is axiomatized on closed λ-terms, we consider the following syntactical combinatory v-algebra $\mathcal{A}_v^0$ on closed λ-terms:

Definition 28. Let $\mathcal{A}_v^0 = (\Lambda^0/=_{\beta_v}, \bullet_{=_{\beta_v}}, \Downarrow_v, [\![\ ]\!]^{\mathcal{A}_v^0})$ be the combinatory v-algebra defined as follows:
• $[\![\ ]\!]^{\mathcal{A}_v^0} : \Lambda \times Env \to \Lambda^0/=_{\beta_v}$, where $Env \ni \rho : Var \to Val_v/=_{\beta_v}$
• $[\![M]\!]_\rho^{\mathcal{A}_v^0} = [\rho(M)]_{=_{\beta_v}}$, where
$\rho(M) = M[\rho(x_1)/x_1, \ldots, \rho(x_n)/x_n]$, with $FV(M) = \{x_1, \ldots, x_n\}$.

Now we define a notion of type interpretation on applicative v-structures with convergence, which is proved to be adequate on the whole class of applicative v-structures with convergence. Thus the type assignment system $\mathcal{T}_v$ is sound and complete w.r.t. a large class of structures (see the Remark after Theorem 25).

Definition 29. Let $\mathcal{A} = (A, \bullet_{\mathcal{A}}, \Downarrow_{\mathcal{A}})$ be an applicative v-structure with convergence. We define the type interpretation as follows:
$[\![\nu]\!]^{t\mathcal{A}} = \{a \in A \mid a \Downarrow_{\mathcal{A}}\}$
$[\![\phi \wedge \psi]\!]^{t\mathcal{A}} = [\![\phi]\!]^{t\mathcal{A}} \cap [\![\psi]\!]^{t\mathcal{A}}$
$[\![\phi \to \psi]\!]^{t\mathcal{A}} = \{a \in A \mid a \Downarrow_{\mathcal{A}} \ \wedge \forall c \in [\![\phi]\!]^{t\mathcal{A}}.\ (c \Downarrow_{\mathcal{A}} \Rightarrow\ a \bullet_{\mathcal{A}} c \in [\![\psi]\!]^{t\mathcal{A}})\}$.

Theorem 30. *The notion of type interpretation of Definition 29 is adequate on the whole class of applicative v-structures with convergence. Moreover, it coincides with the canonical type interpretation on the model D^v.*

Applying Theorem 27, we get:

Theorem 31. *i) D^v is computationally adequate.*
ii) $\mathcal{A}_v^0$ is approximable.

Theorem 32. *$\approx_v^{app}$ is a congruence w.r.t. application.*

Proof. From Theorem 22 it follows that $\approx_{\mathcal{A}_v^0}^{app}$ is a congruence w.r.t. application. Hence, using the correctness of $\to_{\beta_v}$ w.r.t. $\approx_v^{app}$, we get the thesis. □

The Domain Logic Method for $\sigma = h, n, p, o$. The application of this method to $\sigma = h, n, p, o$ differs from the one above in two main respects:
• We have to take into account also open terms for $\sigma = h, n, p$, and define accordingly the combinatory σ-algebra $\mathcal{A}_\sigma$, with adequate type interpretation, on the whole $\Lambda/=_{\beta_\sigma}$. Moreover, for $\sigma = h, n$, in order to prove the adequacy of the type interpretation, we need to assume that D^σ is computationally adequate.
• Because of the "non-initial" nature of the model D^σ, for $\sigma = h, n, p, o$, and hence the presence of equations over base types in the corresponding intersection type description, the natural type interpretation on $\mathcal{A}_\sigma$ does not coincide with the canonical type interpretation on D^σ. Hence we have to consider two different type interpretations, the first adequate on $\mathcal{A}_\sigma$ and such that the Soundness Theorem holds, the second adequate on D^σ and realizing both soundness and completeness (Theorem 25).

4.2 Logical Relations Method

This is a semantical method, which uses a computationally adequate *CPO*-model, obtained as inverse limit. The core of this method consists in defining a suitable relation $\lhd_\sigma \subseteq D^\sigma \times \Lambda^0(C)$. This definition usually requires a deep analysis of the inverse limit model. Then, one proceeds to show

1. $\lhd_\sigma$ is a congruence w.r.t. application;
2. $M \leq_\sigma^{app} N \iff [\![M]\!]^{D^\sigma} \lhd_\sigma N$.

This method extends that originally introduced by Pitts in [16, 17, 18], so that it can be applied to all the strategies mentioned in the Introduction. A related technique appears in [5] for $\to_v$. The technique in [5] (which applies also to $\to_o$), requires to know already that the model is computationally adequate, and also the λ-definability of the projections of the inverse limit. This technique is based on the construction of a relation $\sqsubseteq_\sigma \subseteq \Lambda^0(C) \times \Lambda^0(C)$, which plays the role of the relation $\lhd_\sigma$. The relation $\sqsubseteq_\sigma$ is built inductively in terms of relations $\sqsubseteq_\sigma^n \subseteq (\Lambda^0(C))_n \times (\Lambda^0(C))_n$, where $(\Lambda^0(C))_n$ denotes the set of n-th projections of closed λ-terms. An interesting by-product of this technique is a *purely syntactical* mixed induction-coinduction principle for establishing $\approx_\sigma^{app}$ ([8]).

The technique presented in [16, 17, 18] does not require the λ-definability of projections, but it requires the initiality of D^σ, i.e. the validity of the following "minimal invariance property" à la Plotkin: D^σ is an invariant object of a suitable functor F_σ on the category $CPO_\perp$ via the isomorphism $i_\sigma : D^\sigma \to F_\sigma(D^\sigma)$, and, moreover, id_{D^σ} is the least fixed point of the continuous function $\delta_\sigma : [D^\sigma \to D^\sigma] \to [D^\sigma \to D^\sigma]$ defined as $\delta_\sigma(e) = i_\sigma^{-1} \circ F_\sigma(e) \circ i_\sigma$. Categorically, this amounts to say that D^σ is the initial algebra for a suitable functor on the algebraically compact category $CPO_\perp$ ([6]). This last property guarantees the derivability of an induction-coinduction principle characterizing the relation $\lhd_\sigma \subseteq D^\sigma \times \Lambda^0(C)$. This mixed principle is the essential ingredient for proving that $M \leq_\sigma^{app} N \iff [\![M]\!]^{D^\sigma} \lhd_\sigma N$. Unlike in [5], where $\sqsubseteq_\sigma$ is explicitly characterized inductively in terms of relations $\sqsubseteq_\sigma^n$, here the relation $\lhd_\sigma$ is implicitly defined as the *unique* fixed point of an appropriate mixed variance operator. Hence the relation $\lhd_\sigma$ is a congruence w.r.t. application by definition. The computational adequacy of D^σ can be obtained as a by-product of the method in [16, 17, 18], while it is required for the case of [5]. Because of the requirement that D^σ is initial, Pitts' technique can be applied directly only to $\to_l$ and $\to_v$.

The logical relations method which we present here can be viewed as a generalization and a strengthening of both the two techniques above. We follow the pattern of [16, 17, 18], aiming at the derivation of a suitable induction-coinduction principle for the relation $\lhd_\sigma$, but without assuming the strong minimal invariance property. This is achieved by giving an explicit inductive characterization of $\lhd_\sigma$ in terms of relations $\lhd_\sigma^n \subseteq D_n^\sigma \times \Lambda^0(C)$, in the line of [5]. This technique can be applied successfully to all the strategies $\to_\sigma$ mentioned in the Introduction. It is not always the case, however, that the computational adequacy of the model need not to be assumed at the outset (e.g. for $\sigma = n$).

General pattern of the proof that $\approx_\sigma^{app}$ is a congruence using the logical relations method:

1. Define relations $\trianglelefteq_\sigma^n \subseteq D_n^\sigma \times \Lambda^0(C)$, $n \in \mathbf{N}$, where C contains constants representing the points ϕ of D_0^σ. A constant $c_\phi \in C$ representing ϕ is necessary when no term in Λ^0 is typable with ϕ. The relations $\trianglelefteq_\sigma^n$ are defined inductively, in terms of $\Downarrow_\sigma$. Let $\trianglelefteq_\sigma = \bigcup_{n \in \mathbf{N}} \trianglelefteq_\sigma^n$. Moreover, let $\lhd_\sigma^n = \{(d, P) \in$

$D^\sigma \times \Lambda^0(C) \mid d_n \trianglelefteq^n_\sigma P\}$, and $\lhd_\sigma = \bigcap_{n\in\mathbf{N}} \lhd^n_\sigma$.

Definition 33. A relation $R \subseteq D^\sigma \times \Lambda^0(C)$ is called *limit-closed*, if, whenever for all $n \in \mathbf{N}$ $(d_n, P) \in R$, then also $(d, P) \in R$.

Given a relation $R \subseteq D^\sigma \times \Lambda^0(C)$, $\overline{R}$ denotes the least limit-closed relation including R. The following lemma should then be provable:

Lemma 34. *$\lhd_\sigma$ is the least limit-closed relation including $\trianglelefteq_\sigma$.*

2. Define a monotone "mixed variance" operator $T_\sigma : \mathcal{R}_\sigma \to \mathcal{R}_\sigma$, where $\mathcal{R}_\sigma$ is a complete lattice of pairs of relations such that
i) $\mathcal{R}_\sigma \subseteq \{((R^-, R^+), (\supseteq, \subseteq)) \mid R^-, R^+ \subseteq \bigcup_{n\in\mathbf{N}}(D^\sigma_n \times \Lambda^0(C)) \ \wedge \ R^-, R^+ \supseteq \trianglelefteq^0_\sigma\}$,
ii) for all $(R^-, R^+) \in \mathcal{R}_\sigma$,
$(\trianglelefteq^0_\sigma, \trianglelefteq^0_\sigma) \subseteq T_\sigma(R^-, R^+)$
$(d, P) \in \pi_1(T_\sigma(R^-, R^+)) \implies \forall (d', P') \in R^+.\ (dd', PP') \in R^-$
$(d, P) \in \pi_2(T_\sigma(R^-, R^+)) \implies \forall (d', P') \in R^-.\ (dd', PP') \in R^+$.
3. Prove that $(\trianglelefteq_\sigma, \trianglelefteq_\sigma)$ is the least fixed point of T_σ on $\mathcal{R}_\sigma$, i.e.
$\trianglelefteq_\sigma = \bigcup_{R^- \subseteq \pi_1(T_\sigma(R^-, R^+))\ R^+ \supseteq \pi_2(T_\sigma(R^-, R^+))} R^-$ and
$\trianglelefteq_\sigma = \bigcap_{R^- \subseteq \pi_1(T_\sigma(R^-, R^+))\ R^+ \supseteq \pi_2(T_\sigma(R^-, R^+))} R^+$.
The proof will be carried out by computing the least fixed point of T_σ explicitly.
4. Since $\lhd_\sigma = \overline{\trianglelefteq_\sigma}$, deduce the following induction-coinduction principle:

$$\frac{R^- \subseteq \pi_1(T_\sigma(R^-, R^+)) \quad R^+ \supseteq \pi_2(T_\sigma(R^-, R^+))}{R^- \subseteq \lhd_\sigma \subseteq \overline{R^+}}.$$

5. Prove that $\lhd_\sigma$ is a congruence w.r.t. application.
6. Finally prove that $[\![P']\!]^{D^\sigma} \lhd_\sigma P \iff P' \leq^{app}_\sigma P$ $(*)$.
The following two results are instrumental:
i) let $FV(M) \subseteq \{x_1, \ldots, x_n\}$; then

$$\forall i = 1, \ldots, n.\ d_i \lhd_\sigma P_i \implies [\![M]\!]^{D^\sigma}_{\rho[d_i/x_i]} \lhd_\sigma M[P_i/x_i],$$

ii)

$$d \sqsubseteq_{D^\sigma} d' \lhd_\sigma P' \leq^{app}_\sigma P \implies d \lhd_\sigma P.$$

Item ii) is proved using the induction-coinduction principle for $\lhd_\sigma$. To show implication $(\Rightarrow)$ in $(*)$ reason by coinduction, using the coinductive characterization of $\leq^{app}_\sigma$. The converse follows immediately from items i) and ii).
7. Deduce that $\leq^{app}_\sigma$ is a congruence w.r.t. application. Moreover, from $\approx^{app}_\sigma = \approx_\sigma$, using items i) and ii) above, deduce that D^σ is computationally adequate.

Logical Relations Method: Case $\sigma = h$. Since no closed λ-term is typable with 1, in order to define the relations $\trianglelefteq^n_h$, we need to consider the set of closed λ-terms extended with a constant 1 denoting the type 1. The relation $\trianglelefteq^n_h$ will be defined on $D^h_n \times \Lambda^0(\{1\})$. The constant 1 plays the role of a generic free variable; hence the extended evaluation $\Downarrow_h \subseteq \Lambda(\{1\}) \times Val_h(\{1\})$ is axiomatized by adding the following rule to those of Section 2: $\dfrac{}{1M_1 \ldots M_m \Downarrow_h 1M'_1 \ldots M'_m}\ m{\geq}0$

Definition 35. Define inductively the relations $\trianglelefteq^{n}_{h} \subseteq D^h_n \times \Lambda^0(\{1\})$ as:

$$\trianglelefteq^{0}_{h} = \{(\bot, P) \mid P \in \Lambda^0(\{1\})\} \cup \{(1,P) \mid \forall k \geq 0.\ \forall N_1 \ldots N_k \in \Lambda^0(\{1\}).\ PN_1 \ldots N_k \Downarrow_h\}$$

$$\trianglelefteq^{n+1}_{h} = \{(d,P) \in D^h_{n+1} \times \Lambda^0(\{1\}) \mid d =_{D^h} \bot \ \vee \ (d \neq_{D^h} \bot, 1 \ \wedge\ \forall (d', P') \in \trianglelefteq^{n}_{h}.\ (dd', PP') \in \trianglelefteq^{n}_{h}) \ \vee \ (d =_{D^h} 1 \ \wedge\ \forall k \geq 0.\ \forall N_1, \ldots, N_k \in \Lambda^0(\{1\}).\ PN_1 \ldots N_k \Downarrow_h)\}.$$

Lemma 36. *i)* $\forall d \in D^h_n,\ P \in \Lambda^0(\{1\}),\ d \trianglelefteq^{n}_{h} P \implies \forall m.\ d_m \trianglelefteq^{m}_{h} P$.
ii) $\triangleleft_h = \trianglelefteq_h$.

Proof. i) If $m \geq n$, then the thesis is immediate, by the definition of D^h_n. We are left to show that, for all $n \in \mathbf{N}$, $d_{n+1} \trianglelefteq^{n+1}_{h} P \implies d_n \trianglelefteq^{n}_{h} P$. This is proved by induction, using the well known facts: for all $d, d' \in D^h$, $(d_{n+1}d')_n = (d_{n+1}d'_n)_n = d_{n+1}d'_n = (dd'_n)_n$.
If $n = 0$, the thesis follows immediately by the definitions of $\trianglelefteq^{0}_{h}$ and $\trianglelefteq^{1}_{h}$. For $n+1 > 1$, suppose $\bot, 1 \neq d_{n+1} \trianglelefteq^{n+1}_{h} P$. Take $(d'_{n-1}, P') \in \trianglelefteq^{n-1}_{h}$. Using the equality above, $(d'_{n-1}, P') \in \trianglelefteq^{n}_{h}$, and hence $(d_{n+1}d'_{n-1}, PP') \in \trianglelefteq^{n}_{h}$. By induction hypothesis, $(d_n d'_{n-1}, PP') = ((d_{n+1}d'_{n-1})_{n-1}, PP') \in \trianglelefteq^{n-1}_{h}$. Hence $d_n \trianglelefteq^{n}_{h} P$.
ii) The thesis follows from item i). □

A crucial point in applying the logical relations method is to make sure that the least fixed point of the natural operator T_σ associated to the domain D^σ is $(\trianglelefteq_\sigma, \trianglelefteq_\sigma)$. To this end, we give:

Proposition 37. *i) Let $\mathcal{R}_h$ be defined as follows:*
$\mathcal{R}_h = \{((R^-, R^+), (\supseteq, \subseteq)) \mid R^-, R^+ \subseteq \bigcup_{n \in \mathbf{N}} (D^h_n \times \Lambda^0(\{1\})) \ \wedge$
$\trianglelefteq^{0}_{h} \subseteq R^-, R^+ \ \wedge\ A \cap R^- = \emptyset \ \wedge\ A \cap R^+ = \emptyset\}$,
where $A = \{(d,P) \in \bigcup_{n \in \mathbf{N}} (D^h_n \times \Lambda^0(\{1\})) \mid d =_{D^h} 1 \ \wedge\ \exists k \geq 0.\ \exists N_1, \ldots, N_k \in \Lambda^0(\{1\}).\ PN_1 \ldots N_k \not\Downarrow_h\}$. *Then $\mathcal{R}_h$ is a complete lattice.*
ii) The following operator $T_h : \mathcal{R}_h \to \mathcal{R}_h$ is well defined: $T_h(R^-, R^+) =$
$(\{(d,P) \mid d =_{D^h} \bot \ \vee\ (d \neq_{D^h} \bot, 1 \wedge P \Downarrow_h \ \wedge \forall (d', P') \in R^+.\ (dd', PP') \in R^-) \ \vee$
$(d =_{D^h} 1 \ \wedge\ \forall k \geq 0.\ \forall N_1, \ldots, N_k \in \Lambda^0(\{1\}). PN_1 \ldots N_k \Downarrow_h)\} \cap \bigcup_{n \in \mathbf{N}} (D^h_n \times \Lambda^0(\{1\}))$,
$\{(d,P) \mid d =_{D^h} \bot \ \vee\ (d \neq_{D^h} \bot, 1 \wedge P \Downarrow_h \ \wedge \forall (d', P') \in R^-.\ (dd', PP') \in R^+) \ \vee$
$(d =_{D^h} 1 \ \wedge\ \forall k \geq 0.\ \forall N_1, \ldots, N_k \in \Lambda^0(\{1\}). PN_1 \ldots N_k \Downarrow_h)\} \cap \bigcup_{n \in \mathbf{N}} (D^h_n \times \Lambda^0(\{1\})))$.

Theorem 38. *i) $(\trianglelefteq_h, \trianglelefteq_h)$ is the least fixed point of T_h.*
ii) The following induction-coinduction principle holds:

$$\frac{R^- \subseteq \pi_1(T_h(R^-, R^+)) \quad R^+ \supseteq \pi_2(T_h(R^-, R^+))}{R^- \subseteq \triangleleft_h \subseteq R^+}.$$

Proof. i) Using Lemma 36i), one can prove that $(\trianglelefteq_h, \trianglelefteq_h)$ is a fixed point of T_h. Moreover, one can check that, for all $n > 0$,
$T_h^{n+1}(\bigcup_{n\in\mathbf{N}}(D_n^h \times \Lambda^0(\{1\})) \setminus A, \trianglelefteq_h) = (\triangleleft_h^n \cap \bigcup_{n\in\mathbf{N}}(D_n^h \times \Lambda^0(\{1\})), \trianglelefteq_h^n)$.
This follows from the fact that, for $n > 0$,
$T(\triangleleft_h^n \cap \bigcup_{n\in\mathbf{N}}(D_n^h \times \Lambda^0(\{1\})), \triangleleft_h^n) = (\triangleleft_h^{n+1} \cap \bigcup_{n\in\mathbf{N}}(D_n^h \times \Lambda^0(\{1\})), \triangleleft_h^{n+1})$.
Hence the least fixed point of T_h is
$(\bigcap_{n\in\mathbf{N}} T_h^n(\bigcup_{n\in\mathbf{N}}(D_n^h \times \Lambda^0(\{1\})) \setminus A, \trianglelefteq_h), \bigcup_{n\in\mathbf{N}} T_h^n(\bigcup_{n\in\mathbf{N}}(D_n^h \times \Lambda^0(\{1\})), v)) =$
$(\triangleleft_h \cap \bigcup_{n\in\mathbf{N}}(D_n^h \times \Lambda^0(\{1\})), \trianglelefteq_h) = (\trianglelefteq_h, \trianglelefteq_h)$.
ii) The proof follows from Lemma 36ii) and item i). □

Theorem 39. $d \triangleleft_h P \iff \forall e \triangleleft_h Q.\ de \triangleleft_h PQ$.

Proof. ($\Rightarrow$) The proof follows from the definition of application in the inverse limit model D^h: for all $d, e \in D^h$, $de = \bigsqcup_{n\in\mathbf{N}} d_{n+1}e_n$. Suppose that $d \triangleleft_h P$ and $e \triangleleft_h Q$. We will prove that, for all n, $(de)_n \triangleleft_h PQ$. From the hypotheses and from the definition of $\triangleleft_h$, using the definition of application, we immediately get, for all n, k, $((\bigsqcup_{m\in\mathbf{N}} d_{m+1}e_m)_n)_k \trianglelefteq_h^k PQ$, i.e., from the definition of $\triangleleft_h$, we have, for all n, $(\bigsqcup_{m\in\mathbf{N}} d_{m+1}e_m)_n \triangleleft_h PQ$.
($\Leftarrow$) From the Definition of $\triangleleft_h$. □

Let $\leq_h^{app,1}$ be the natural extension of $\leq_h^{app}$ to $\Lambda^0(\{1\}) \times \Lambda^0(\{1\})$, i.e. put $M \leq_h^{app,1} N$ iff $\forall P_1 \ldots P_n \in \Lambda^0(\{1\}).\ (MP_1 \ldots P_n \Downarrow_h \Rightarrow NP_1 \ldots P_n \Downarrow_h)$.

Lemma 40. *For all $d, d' \in D^h$, and for all $P, P' \in \Lambda^0(\{1\})$,*

$$d \sqsubseteq_{D^h} d' \triangleleft_h P' \leq_h^{app,1} P \implies d \triangleleft_h P.$$

Proof. Apply the induction-coinduction principle to $R^+ = \trianglelefteq_h$, $R^- = \{(d, P) \mid \exists d' \in D^h\ \exists P' \in \Lambda^0(\{1\}).\ d \sqsubseteq_{D^h} d' \triangleleft_h P' \leq_h^{app,1} P\} \cap \bigcup_{n\in\mathbf{N}}(D_n^h \times \Lambda^0(\{1\}))$. □

The following crucial theorem can be easily shown by induction on M, using Theorem 39:

Theorem 41. *Let $M \in \Lambda(\{1\})$, with $FV(M) \subseteq \{x_1, \ldots, x_n\}$. Then*

$$\forall i = 1, \ldots, n.\ d_i \triangleleft_h P_i \implies [\![M]\!]^{D^h}_{\rho[d_i/x_i]} \triangleleft_h M[P_i/x_i].$$

Corollary 42. *For all $P \in \Lambda^0(\{1\})$,* $[\![P]\!]^{D^h} \triangleleft_h P$.

Lemma 43. $[\![P']\!]^{D^h} \triangleleft_h P \iff P' \leq_h^{app,1} P$.

Proof. The implication ($\Rightarrow$) can be proved by coinduction. In fact it is easy to check, using Corollary 42 and Theorem 39 ($\Rightarrow$), that $\{(P', P) \mid [\![P']\!]^{D^h} \triangleleft_h P\}$ is a $\Psi_h^{\leq}$-bisimulation. The proof of the implication ($\Leftarrow$) follows from Lemma 40, and from Corollary 42. □

Theorem 44. *i) $\leq_h^{app}$ is a congruence w.r.t. application.*
ii) D^h is computationally adequate.

Proof. i) From Theorem 39 and Lemma 43 it follows that $\leq_h^{app,1}$ is a congruence w.r.t. application. Then the thesis follows from the fact that $(\leq_h^{app,1})_{|\Lambda^0 \times \Lambda^0} = \leq_h^{app}$, which in turn is a consequence of Corollary 12.
ii) Using Corollary 42 and Lemma 40, one can show that, if $[M]^{D^h} \sqsubseteq_h [N]^{D^h}$, then $[M]^{D^h} \lhd_h N$, hence by Lemma 43 $M \leq_h^{app} N$. Using i) and Theorems 11, and 13, we have $M \leq_h N$. □

5 Final Remarks

In this paper we introduced two general semantical methodologies for deriving a coinductive characterization of the observational equivalence. But other techniques are also possible. We mention the following three:
1. *Plain induction* on *computation steps* of $\rightarrow_\sigma^*$. This, purely syntactical, direct approach easily applies to $\rightarrow_l$ (see [1]). With suitable extensions in order to take care of open terms, it applies also to $\sigma = h, n$. However, this direct approach is rather problematic for call-by-value strategies such as $\rightarrow_v$, $\rightarrow_p$, or non deterministic strategies like $\rightarrow_o$. For a subtle but complex proof by induction on computation for $\rightarrow_v$ see [15].
2. *Coinductive argument* on *congruence candidate relations*. This method was introduced by D.Howe for the lazy call-by-name strategy $\rightarrow_l$, and later generalized to a class of *lazy* strategies by-name and by-value, including $\rightarrow_v$ ([11]). In [14], we generalize and strengthen Howe's method so as to deal with strategies whose evaluation relations cannot be axiomatized only on closed terms.
3. Method based on the *Separability algorithm*. This method works only if a suitable "separability algorithm" (see e.g. [12, 2]) is available. I.e. if there exists an effective method which, given two non $\approx_\sigma$-equivalent terms, M, N, allows to define an applicative context $C[\ ]$ such that either $C[M] \Downarrow_\sigma$ and $C[M] \not\Downarrow_\sigma$, or viceversa. At present, such an algorithm exists only for $\approx_h$, $\approx_n$, and $\approx_p$. Hence the method works only for $\rightarrow_h$, $\rightarrow_n$, and $\rightarrow_p$.

References

1. S.Abramsky, L.Ong, *Full Abstraction in the Lazy Lambda Calculus,* Information and Computation, 105(2):159–267, 1993.
2. H.Barendregt, *The Lambda Calculus, its Syntax and Semantics,* North Holland, Amsterdam, 1984.
3. H.Barendregt, M.Coppo, M.Dezani-Ciancaglini, *A filter lambda model and the completeness of type assignment,* J.Symbolic Logic, 48(4):931–940, 1983.
4. M.Coppo, M.Dezani-Ciancaglini, M.Zacchi, *Type Theories, Normal Forms and D_∞-Lambda-Models,* Information and Computation, 72(2):85–116, 1987.
5. L.Egidi, F.Honsell, S.Ronchi Della Rocca, *Operational, denotational and logical Descriptions: a Case Study,* Fundamenta Informaticae, 16(2):149–169, 1992.
6. P.Freyd, *Algebraically complete categories,* A.Carboni et al. eds, Category Theory '90 Springer LNM, 1488:95–104, Como, 1990.

7. F.Honsell, M.Lenisa, *Some Results on Restricted λ-calculi*, MFCS'93 Conference Proceedings, A.Borzyszkowski et al. eds., Springer LNCS, 711:84–104, 1993.
8. F.Honsell, M.Lenisa, *Final Semantics for untyped λ-calculus*, M.Dezani et al. eds, TLCA'95 Springer LNCS, 902:249–265, Edinburgh, 1995.
9. F.Honsell, M.Lenisa, *A Semantical Analysis of an Effective Perpetual Strategy*, Draft, November 1996.
10. F.Honsell, S.Ronchi Della Rocca, *An approximation theorem for topological lambda models and the topological incompleteness of lambda calculus*, J. of Computer and System Sciences, 45(1):49-75, 1992.
11. D.Howe, *Proving Congruence of Bisimulation in Functional Programming Languages*, Information and Computation, 124(2):103–112, 1996.
12. M.Hyland, *A survey of some useful partial order relations on terms of the lambda-calculus*, C.Böhm ed., Springer LNCS, 37:83–95, 1975.
13. M.Lenisa, *Final Semantics for a Higher Order Concurrent Language*, CAAP'96 Conference Proceedings, H.Kirchner ed., Springer LNCS, 1059:102–118, 1996.
14. M.Lenisa, *The Congruence Candidate Method for Giving Coinductive Characterizations of Observational Equivalences in λ-calculi*, to appear in CAAP'97 Proc..
15. I.Mason, S.Smith, C.Talcott, *From Operational Semantics to Domain Theory*, Information and Computation to appear.
16. A.M.Pitts, *Computational Adequacy via 'Mixed' Inductive Definitions*, MFPS'93, Brookes et al. eds., Springer LNCS, 802:72–82, 1994.
17. A.M.Pitts, *A Note on Logical Relations Between Semantics and Syntax*, to appear in WoLLIC'96 Proc., J. of the Interest Group in Pure and Applied Logics.
18. A.M.Pitts, *Relational Properties of Domains*, Information and Computation, 127:66–90, 1996.
19. G.D.Plotkin, *Call-by-name, Call-by-value and the λ-calculus*, Theoretical Computer Science 1:125-159, 1975.

Schwichtenberg-Style Lambda Definability Is Undecidable

Jan Małolepszy, Małgorzata Moczurad, Marek Zaionc

Computer Science Department, Jagiellonian University,
Nawojki 11, 30-072 Krakow, Poland
E-mail {madry, maloleps, zaionc}@ii.uj.edu.pl

Abstract. We consider the lambda definability problem over an arbitrary free algebra. There is a natural notion of primitive recursive function in such algebras and at the same time a natural notion of a lambda definable function. We shown that the question: "For a given free algebra and a primitive recursive function within this algebra decide whether this function is lambda definable" is undecidable if the algebra is infinite. The main part of the paper is dedicated to the algebra of numbers in which lambda definability is described by the Schwichtenberg theorem. The result for an arbitrary infinite free algebra has been obtained by a simple interpretation of numerical functions as recursive functions in this algebra. This result is a counterpart of the distinguished result of Loader in which lambda terms are evaluated in finite domains for which undecidability of lambda definability is proved.

1 Simple Typed λ-Calculus

We shall consider a simple typed λ-calculus with a single ground type 0. The set *TYPES* is defined as follows: 0 is a type and if τ and μ are types then $\tau \to \mu$ is a type. By type $\tau^k \to \mu$ we mean the type $\tau \to (... \to (\tau \to \mu)...)$ with exactly k occurrences of τ. For $k = 0$, $\tau^0 \to \mu$ is μ.

For any type τ there is given a denumerable set of variables $V(\tau)$. Any type τ variable is a type τ term. If T is a term of type $\tau \to \mu$ and S is a type τ term, then TS is a term which has type μ. If T is a type μ term and x is a type τ variable, then $\lambda x.T$ is a term of type $\tau \to \mu$. The axioms of equality between terms have the form of $\beta\eta$ conversions and the convertible terms will be written as $T =_{\beta\eta} S$. A closed term is a term without free variables. For more detailed treatment of typed λ-calculus see [12].

2 Free Algebras

Algebra A given by a signature $S_A = (\alpha_1, ..., \alpha_n)$ has n constructors $c_1, ... c_n$ of arities $\alpha_1, ..., \alpha_n$, respectively. Expressions of the algebra A are defined recursively as a minimal set such that if $\alpha_i = 0$ then c_i is an expression and if $t_1, ..., t_{\alpha_i}$ are expressions then $c_i(t_1, ..., t_{\alpha_i})$ is an expression. We may assume that at least one α_i is equal to 0, otherwise the set of expressions is empty. We are going to investigate mappings $f : A^k \to A$. Any constructor can be seen as a function $c_i : A^{\alpha_i} \to A$. Let $\vec{x}$ be a list of expressions $x_1, ..., x_k$ and $f_1, ..., f_n$ be functions of arity $k + \alpha_i$, respectively. A class $\mathcal{X}$ of mappings is closed under primitive recursion if the $(k+1)$-ary function h defined by:

$$h(c_i(y_1, ..., y_{\alpha_i}), \vec{x}) = f_i(h(y_1, \vec{x}), ..., h(y_{\alpha_i}, \vec{x}), \vec{x})$$

for all $i \leq n$ belongs to $\mathcal{X}$ whenever functions $f_1, ..., f_n$ are in $\mathcal{X}$. In the case when $\alpha_i = 0$ the equation is reduced to $h(c_i, \vec{x}) = f_i(\vec{x})$. Let us distinguish the following operations: p_i^k is the k-ary projection which extracts i-th argument, i.e. $p_i^k(x_1, ..., x_k) = x_i$ and C_i^k is k-ary constant function when $\alpha_i = 0$ which maps constantly into expression c_i, i.e $C_i^k(x_1, ..., x_k) = c_i$. The class of primitive recursive functions over algebra A is defined as a minimal class containing constructors, projections, constant functions, and closed under composition and primitive recursion. The subclass G^A of primitive recursive functions is a minimal class containing projections, constructors, and closed under composition. The subclass F_k^A for $k \geq 0$ is a class of all $(k+p)$-ary functions f on A such that for all expressions $E_1, ..., E_p \in A$ the function p defined as $p(x_1, ..., x_k) = f(x_1, ..., x_k, E_1, ..., E_p)$ belongs to G^A. This includes a case when $k = 0$, which means that F_0^A is a set of constant nulary constructors. The class F_λ^A is a minimal class containing constructors, projections, constant functions, and closed under composition and the following limited version of primitive recursion: if $f_i \in F_\lambda^A \cap F_{\alpha_i}^A$ for $i \leq n$ then the function h obtained from f_i by primitive recursion belongs to F_λ.

We are going to give more attention to the algebra of natural numbers $\mathbb{N}$ based on the signature $(0, 1)$. Traditionally two constructors of the algebra $\mathbb{N}$ are called 0 (zero) and s (successor). The class of primitive recursive functions over $\mathbb{N}$ is the least one containing the constant zero function, successor, and closed under composition and the following form of primitive recursion:

$$h(0, \vec{x}) = f_1(\vec{x})$$
$$h(s(y), \vec{x}) = f_2(h(y, \vec{x})), \vec{x})$$

In addition, we will discuss finite algebras A_n based on the signature $(0, ..., 0)$. Algebra A_n consists of a finite number of expressions $c_1, ..., c_n$. It is easy to observe that every function on A_n is primitive recursive.

3 Representability

If A is an algebra given by a signature $S_A = (\alpha_1, ..., \alpha_n)$ then by τ^A we mean a type $(0^{\alpha_1} \to 0), \ldots, (0^{\alpha_n} \to 0) \to 0$. Assuming that at least one α_i is 0, we get that type τ^A is not empty. There is a natural 1-1 isomorphism between expressions of the algebra A and closed terms of type τ^A. If c_i is a nulary constructor in A then the term $\lambda x_1 ... x_n . x_i$ represents c_i. If $\alpha_i > 0$ and $t_1, ..., t_{\alpha_i}$ are expressions in A represented by closed terms $T_1, ..., T_{\alpha_i}$ of type τ^A, then expression $c_i(t_1, ..., t_{\alpha_i})$ is represented by the term $\lambda x_1 ... x_n . x_i(T_1 x_1 ... x_n) .... (T_{\alpha_i} x_1 ... x_n)$.

Thus, we have a 1-1 correspondence between closed terms of the type τ^A and A. In fact any rank 2 type τ represents some free algebra A. If additionally τ is non-empty (there are some closed terms of type τ) then the algebra A is non-empty. The unique (up to $\beta\eta$-conversion) term which represents an expression t is denoted by $\underline{t}$.

Definition 1. A function $h : A^n \to A$ is λ-definable by a closed term H of type $(\tau^A)^n \to \tau^A$ iff for all expressions $t_1, ..., t_n$

$$H\underline{t_1} \dots \underline{t_n} =_{\beta\eta} \underline{h(t_1, ..., t_n)}$$

It is easy to observe that any closed term H of the type $(\tau^A)^n \to \tau^A$ uniquely defines the function $h : A^n \to A$ as follows: if $t_1, ..., t_n$ are expressions then the value is an expression represented by the term $H\underline{t_1}, \dots, \underline{t_n}$. On the other hand, one function can be represented by many unconvertible terms.

For the algebra of natural numbers $\mathbb{N} = (1, 0)$ the type $\tau^{\mathbb{N}} = (0 \to 0) \to (0 \to 0)$ is called the Church numerals type. In this case the natural isomorphism identifying closed terms of the type $\tau^{\mathbb{N}}$ with expressions of the algebra $\mathbb{N}$ (non-negative integers) is such that for a number $n \in \mathbb{N}$ the $\underline{n}$ is the closed term $\lambda ux.u(...(ux)...)$ with exactly n occurrences of variable u.

Definition 2. Function $f : \mathbb{N}^k \to \mathbb{N}$ is λ-definable by a closed term F of type $\underline{\mathbb{N}}^{\,k} \to \underline{\mathbb{N}}$ iff for all numbers $n_1, ..., n_k$

$$F\underline{n_1}...\underline{n_k} =_{\beta\eta} \underline{f(n_1, ..., n_k)}$$

The following characteristics of λ-definable functions has been proved.

Theorem 3. *(Schwichtenberg [8] and Statman [9]) λ-definable functions on $\mathbb{N}$ are exactly compositions of 0, 1, addition, multiplication, sq and $\overline{sq}$ where*

$$sq(x) = \begin{cases} 0, & if\ x = 0 \\ 1, & if\ x \neq 0 \end{cases} \qquad \overline{sq}(x) = \begin{cases} 1, & if\ x = 0 \\ 0, & if\ x \neq 0 \end{cases}$$

Theorem 4. *(Zaionc [14] and Leivant [6]) λ-definable functions on algebra A are exactly functions from the class F_λ^A.*

Theorem 5. *(Zaionc [13]) Every function on a finite free algebra is λ-definable.*

A much stronger version of the above, covering also all functionals of finite type, is proved in [13].

4 Reconstruction of the Polynomials

The aim of this section is to show the possibility of reconstructing extended polynomials from the finite set of appropriate examples. This problem has been stated by Bharat Jayraman for the purpose of extracting programs written in the form of λ-terms from the finite number of cases of the behaviour of those programs (for example see [2], [3]). In this technique for program extraction Huet unification procedure is used. See also the programming by examples paradigm in [5], [4]. In the case of reconstructing polynomials some partial solution was obtained by Kannan Govindarajan in [1].

In this section we will demonstrate two theorems. The first one concerns standard polynomials (composition of addition and multiplication) of k variables. It is proved that only two examples are needed to determine the polynomial. For more general case concerning extended polynomials (composition of addition, multiplication, sq and $\overline{sq}$) it is proved that $2^{k+1}-1$ examples are necessary and sufficient for extraction.

Definition 6. By the set of polynomials we mean the least set of functions from $\mathbb{N}^k$ to $\mathbb{N}$ containing all projections, constant functions, addition and multiplication closed for composition.

By the set of extended polynomials we mean the least set of functions from $\mathbb{N}^k$ to $\mathbb{N}$ containing all projections, constant functions, addition, multiplication and additionally sq and $\overline{sq}$ (see Theorem 3 for definition), closed for composition.

Definition 7. An extended polynomial $f : \mathbb{N}^k \to \mathbb{N}$ is determined by the finite set of p examples $\{(\vec{x_1}, y_1), ..., (\vec{x_p}, y_p)\}$, where $\vec{x_i} \in \mathbb{N}^k$ and $y_i \in \mathbb{N}$ for all $i \leq p$, if

- $f(\vec{x_i}) = y_i$ for all $i \leq p$
- if for some extended polynomial $g : \mathbb{N}^k \to \mathbb{N}$ $g(\vec{x_i}) = y_i$ for all $i \leq p$ then $f = g$.

Definition 8. A finite set $\mathcal{X} \subset \mathbb{N}^k \times \mathbb{N}$ is called a determinant if some extended polynomial $f : \mathbb{N}^k \to \mathbb{N}$ is determined by $\mathcal{X}$.

Definition 9. We are going to use the standard normal form notation for polynomials with positive integer coefficients. Every polynomial $f : \mathbb{N}^k \to \mathbb{N}$ has a unique representation in the following normal form:

1. $k = 1$

$$f(x) = \sum_{i=0}^{m} \alpha_i x^i \ ,$$

where $\alpha_i \in \mathbb{N}$ are polynomial coefficients,

2. $k > 1$

$$f(\vec{x}, x_k) = \sum_{i=0}^{m} \alpha_i(\vec{x}) x_k^i \ ,$$

where $\alpha_i \in (\mathbb{N}^{k-1} \to \mathbb{N})$ are polynomials in normal form.

Lemma 10. *Given polynomial* $f(x_1,\ldots,x_k) = \sum_{i=0}^{m} \alpha_i(x_1,\ldots,x_{k-1})x_k^i$. *If a point* $(a_1,\ldots,a_k) \in \mathbb{N}^k$ *satisfies the following condition:*

$$\forall 0 \le i \le m: \ a_k > \alpha_i(a_1,\ldots,a_{k-1}) \ ,$$

then there is an algorithm to find out values $\alpha_i(a_1,\ldots,a_{k-1})$, $i = 1,\ldots,m$ *from the value of the polynomial at the point* $(a_1,\ldots,a_k)$.

Proof. Let $W = f(a_1,\ldots,a_k)$. Because $a_k > \alpha_i(a_1,\ldots,a_{k-1})$ for all $0 \le i \le m$, the values $\alpha_i(a_1,\ldots,a_{k-1})$ can be computed as the digits of W in the number system with base a_k. W is divided by a_k until 0 is reached, and the computed reminders are the values $\alpha_0(a_1,\ldots,a_{k-1}), \alpha_1(a_1,\ldots,a_{k-1}),\ldots$. □

Definition 11. For the given total computable function $f : \mathbb{N}^k \to \mathbb{N}$ we define numbers $D_f, E_f \in \mathbb{N}$ and function $M_f : \mathbb{N} \to \mathbb{N}$ by: $D_f \stackrel{\text{def}}{=} f(2,\ldots,2)$, $E_f \stackrel{\text{def}}{=} if\ D_f = 0\ then\ 1\ else\ \lceil \log_2 D_f \rceil$, $M_f(x) \stackrel{\text{def}}{=} D_f x^{E_f} + D_f$.

Lemma 12. *For every polynomial* $f : \mathbb{N}^k \to \mathbb{N}$ *the following holds:*

1. D_f *is greater or equal to the sum of all the coefficients of the polynomial* f *in the long normal form, and thus greater or equal to each coefficient of* f.
2. E_f *is greater or equal to the degree of* f *(the largest sum of the exponents in the long normal form of the polynomial).*
3. $\forall x_1,\ldots,x_k: \ f(x_1,\ldots,x_k) \le f(x_{max},\ldots,x_{max}) \le M_f(x_{max})$, *where* $x_{max} = \max_{1 \le i \le k}\{x_i\}$.

Proof. 1. Let α_j be the subsequent coefficients of f (coefficients of the monomials in the long normal form of f). Then

$$D_f = f(2,\ldots,2) = \sum_j \alpha_j 2^{e_j} \ge \sum \alpha_j \ge \alpha_i \ \forall i \ .$$

2. Assume by contradiction that E_f is smaller than the degree of f. It means that f has (in the form of sum of monomials) the component $\alpha x_1^{e_1} \cdots x_k^{e_k}$, such that $\alpha > 0$, $\sum_{i=1}^{k} e_i > E_f$. Then $D_f = f(2,\ldots,2) > 2^{E_f}$ which contradicts $D_f \le 2^{E_f}$ (from the definition of E_f).
3. Let α_j be the subsequent coefficients of f. The inequality

$$\forall x_1,\ldots,x_k: \ f(x_1,\ldots,x_k) \le f(x_{max},\ldots,x_{max})$$

follows from the fact that every polynomial is monotonic. The second inequality

$$\forall x: \ f(x,\ldots,x) \le M_f(x)$$

is a consequence of the following:

$$\forall x: \ f(x,\ldots,x) = \sum_j \alpha_j x^{e_j} \le \left(\sum_j \alpha_j\right) x^{E_f} + \left(\sum_j \alpha_j\right) \le D_f x^{E_f} + D_f = M_f(x) \ .$$

□

Lemma 13. *Given polynomial* $f : \mathbb{N}^k \to \mathbb{N}$. *If the point* $(a_1, \ldots, a_k) \in \mathbb{N}^k$ *satisfies two conditions:*

1. $a_1 > D_f$
2. $\forall 2 \le i \le k: \; a_i > M_f(a_{i-1})$

then there is an algorithm to extract values of all the coefficients of the polynomial f *from the value of* f *for* $(a_1, \ldots, a_k)$.

Proof. Induction on the number of polynomial arguments.

1. $k = 1$
   From the first condition and Lemma 12 it comes that f and a_1 satisfy the condition of Lemma 10. From Lemma 10, the values of all the coefficients of the polynomial f can be computed from the value of f for a_1.
2. $k > 1$
   Assume that the lemma holds for all $(k-1)$-argument polynomials. Represent f in the short normal form:

$$f(x_1, \ldots, x_k) = \sum_{i=0}^{m} \alpha_i(x_1, \ldots, x_{k-1}) x_k^i \; .$$

The second condition for $i = k$ implies that f and $(a_1, \ldots, a_k)$ satisfy the condition of Lemma 10:

$$\begin{aligned} a_k > M_f(a_{k-1}) \ge f(a_{k-1}, \ldots, a_{k-1}) \ge f(a_1, \ldots, a_{k-1}, a_{k-1}) = \\ = \textstyle\sum_{i=0}^{m} \alpha_i(a_1, \ldots, a_{k-1}) a_{k-1}^i \ge \alpha_j(a_1, \ldots, a_{k-1}) \;\; \forall 0 \le j \le m \; . \end{aligned}$$

We can see that the point $(a_1, \ldots, a_k) \in \mathbb{N}^k$ satisfies the condition of Lemma 10. Therefore the values of the polynomial α_j for $(a_1, \ldots, a_{k-1})$ can be obtained from the value of f at $(a_1, \ldots, a_k)$ using the algorithm proposed in Lemma 10. To compute the values of all the coefficients of the polynomial f it suffices to compute the values of all the coefficients of the polynomials α_j from the values of α_j at $(a_1, \ldots, a_{k-1})$. The polynomials α_j are $(k-1)$-argument polynomials, so the induction assumption can be used. It is now sufficient to show for each $j = 0, \ldots, m$ that α_j and $(a_1, \ldots, a_{k-1})$ satisfy the conditions of the lemma:

(a)

$$a_1 > D_{\alpha_j}$$

comes from:

$$a_1 > D_f = \sum_{i=0}^{m} D_{\alpha_i} 2^i \ge D_{\alpha_j} \; ,$$

(b)

$$\forall 2 \le i \le k-1: \; a_i > M_{\alpha_j}(a_{i-1})$$

comes from:

$$D_f \ge D_{\alpha_j} \;\Rightarrow\; E_f \ge E_{\alpha_j} \;\Rightarrow\; M_f(x) \ge M_{\alpha_j}(x)$$
$$\forall 2 \le i \le k-1: \; a_i > M_f(a_{i-1}) \ge M_{\alpha_j}(a_{i-1}) \; .$$

For each $j = 0, \ldots, m$, α_j and $(a_1, \ldots, a_{k-1})$ satisfy the conditions of Lemma 10, so the values of all the coefficients of the polynomials α_j (which are in fact coefficients of the polynomial f) can be computed from the values of α_j for $(a_1, \ldots, a_{k-1})$ and thus the values of all the coefficients of the polynomial f can be computed from the value of f for $(a_1, \ldots, a_k)$.

□

Theorem 14. *There exists an algorithm that, for an arbitrary k-argument ($k \in \mathbb{N}$) polynomial f, extracts the values of all the coefficients of f from the values of f at two vectors of arguments:*

1. $(2, \ldots, 2)$,
2. $(a_1, \ldots, a_k)$, *where $a_1, \ldots, a_k$ are obtained from the value $f(2, \ldots, 2)$ according to the algorithm presented in Lemma 13, i.e.:*

$$a_1 = D_f + 1$$
$$a_i = M_f(a_{i-1}) + 1 \text{ for } i = 2, \ldots, k \; .$$

Proof. The following algorithm can be used to compute the values of all the coefficients of f:

1. Compute the value of f for $(2, \ldots, 2)$.
2. Compute D_f, E_f and M_f.
3. Compute the arguments $a_1, \ldots, a_k$ satisfying the conditions of Lemma 13.
4. Compute the value of f for $(a_1, \ldots, a_k)$.
5. Using the method from Lemma 13 compute the values of the coefficients of f.

□

Lemma 15. *Let f be a polynomial such that $f \not\equiv 0$. There is no algorithm to extract the values of all the coefficients of f from the single value of the polynomial f.*

Proof. It can be easily seen that for an arbitrary k-argument ($k \ge 1$) polynomial $f \ne 0$ and an arbitrary set of arguments $(a_1, \ldots, a_k)$, there always exists such a polynomial g that: $g(a_1, \ldots, a_k) = f(a_1, \ldots, a_k)$ and $g \ne f$. Thus, the set $(a_1, \ldots, a_k)$ does not determine f, and the coefficients of f cannot be obtained from the value of f for $(a_1, \ldots, a_k)$. □

5 Reconstruction of the Extended Polynomials

Lemma 16. *For every $\alpha, \beta \in \mathbb{N}$ the following holds:*

$$\begin{aligned}
sq(sq(\alpha)) &= sq(\alpha)\\
sq(\overline{sq}(\alpha)) &= \overline{sq}(\alpha)\\
sq(\alpha \cdot \beta) &= sq(\alpha) \cdot sq(\beta)\\
sq(\alpha + \beta) &= sq(\alpha) \cdot \overline{sq}(\beta) + sq(\alpha) \cdot sq(\beta) + \overline{sq}(\alpha) \cdot sq(\beta)\\
\overline{sq}(sq(\alpha)) &= \overline{sq}(\alpha)\\
\overline{sq}(\overline{sq}(\alpha)) &= sq(\alpha)\\
\overline{sq}(\alpha \cdot \beta) &= \overline{sq}(\alpha) \cdot sq(\beta) + \overline{sq}(\alpha) \cdot \overline{sq}(\beta) + sq(\alpha) \cdot \overline{sq}(\beta)\\
\overline{sq}(\alpha + \beta) &= \overline{sq}(\alpha) \cdot \overline{sq}(\beta)\\
sq(\alpha) \cdot sq(\alpha) &= sq(\alpha)\\
sq(\alpha) \cdot \overline{sq}(\alpha) &= 0\\
\overline{sq}(\alpha) \cdot \overline{sq}(\alpha) &= \overline{sq}(\alpha)\\
sq(\alpha) \cdot \alpha &= \alpha\\
\overline{sq}(\alpha) \cdot \alpha &= 0\\
sq(\alpha) + \overline{sq}(\alpha) &= 1
\end{aligned}$$

Proof. All the properties above can be easily proved from the definition of sq and $\overline{sq}$ (see Theorem 3). □

Lemma 17. *Every extended polynomial $f : \mathbb{N}^k \to \mathbb{N}$ has a unique representation in the following normal form:*

$$f(x_1, \ldots, x_k) = \sum_{Z \in \mathcal{P}(\{1,\ldots,k\})} f_Z(x_1, \ldots, x_k) SQ_Z(x_1, \ldots, x_k) \ ,$$

where $SQ_Z(x_1, \ldots, x_k) \stackrel{\text{def}}{=} \prod_{i \in Z} sq(x_i) \prod_{j \notin Z} \overline{sq}(x_j)$ and f_Z are polynomials of the arguments x_i for $i \in Z$ (in the normal form).

Proof. (Existence) It can be seen from Lemma 16 that every extended polynomial can be represented as a sum of the products of the form:

$$a \cdot x_{i_1}^{e_1} \cdots x_{i_p}^{e_p} \cdot sq(x_{i_{p+1}}) \cdots sq(x_{i_q}) \cdot \overline{sq}(x_{i_{q+1}}) \cdots sq(x_{i_r}) \ ,$$

where $\forall \alpha : \ i_\alpha \in \{1, \ldots, k\}$, $\forall \alpha \neq \beta : \ i_\alpha \neq i_\beta$, $\forall i \in \{1, \ldots p\} : \ e_i \in \mathbb{N}$, $a \in \mathbb{N}$ and $r \leq k$. Thus every extended polynomial can be represented as a sum of the products of the form:

$$a \cdot x_{i_1}^{e_1} \cdots x_{i_p}^{e_p} \cdot SQ_Z(x_1, \ldots, x_k) \ ,$$

where $i_1, \ldots, ip \in Z$.

(Uniqueness) The above representation is unique for the given extended polynomial f, i.e. if there are two representations of that form of the extended polynomial f, the representations are identical (all the coefficients in that representations are the same). Assume that the extended polynomials $g = f$ have the normal forms:

$$f(x_1,\ldots,x_k) = \sum_{Z\in\mathcal{P}(\{1,\ldots,k\})} f_Z(x_1,\ldots,x_k)SQ_Z(x_1,\ldots,x_k)$$

$$g(x_1,\ldots,x_k) = \sum_{Z\in\mathcal{P}(\{1,\ldots,k\})} g_Z(x_1,\ldots,x_k)SQ_Z(x_1,\ldots,x_k)$$

$$\forall x_1,\ldots,x_k : f(x_1,\ldots,x_k) = g(x_1,\ldots,x_k) \Rightarrow$$

$$\forall x_1,\ldots,x_k \forall Z \in \mathcal{P}(\{1,\ldots,k\}) : f_Z(x_1,\ldots,x_k) = g_Z(x_1,\ldots,x_k) \ .$$

Since f_Z and g_Z are standard polynomials over the same variables in normal form, they are identical. Thus there is only one representation of f in the normal form. □

Example 1. The extended polynomial

$$f(x,y) = 2(xy+y^2)x + 3x^2(1+y)\overline{sq}(y) + xy$$

has the normal form:

$$f(x,y) = ((x+2x^2)y + (2x)y^2)SQ_{\{1,2\}}(x,y) + (3x^2)SQ_{\{1\}}(x,y) \ .$$

Lemma 18. *Let f be an extended polynomial in normal form:*

$$f(x_1,\ldots,x_k) = \sum_{Z\in\mathcal{P}(\{1,\ldots,k\})} f_Z(x_1,\ldots,x_k)SQ_Z(x_1,\ldots,x_k) \ .$$

Let $Y \in \mathcal{P}(\{1,\ldots,k\})$ be a set of indices and $(y_1,\ldots,y_k) \in \mathbb{N}^k$ be a point such that $y_i \neq 0 \Leftrightarrow i \in Y$. Then:

$$f(y_1,\ldots,y_k) = f_Y(y_1,\ldots,y_k) \ .$$

Proof. Straight from the definition of SQ_Z functions. □

Definition 19. For every computable total function $f : \mathbb{N}^k \to \mathbb{N}$ define the set $Det_f \text{ı} \mathbb{N}^{k+1}$ which consists of $2^{k+1}-1$ tuples:

- one tuple $(0,\ldots,0,f(0,\ldots,0))$
- for every non-empty set $Z\text{ı}\{1,\ldots,k\}$ two tuples
  1. $(y_1^Z,\ldots,y_k^Z,f(y_1^Z,\ldots,y_k^Z))$ where $y_i^Z = \begin{cases} 0, \text{ if } i \notin Z \\ 2, \text{ if } i \in Z \end{cases}$
  2. $(a_1^Z,\ldots,a_k^Z,f(a_1^Z,\ldots,a_k^Z))$ where $a_1^Z = D_{f_Z}+1$ and $a_i^Z = M_{f_Z}(a_{i-1}^Z)+1$

(see the definition of f_Z in Lemma 17 and D_{f_Z} and M_{f_Z} in Definition 11).

Theorem 20. *There is an algorithm that for an arbitrary k-argument ($k \in \mathbb{N}$) extended polynomial f computes the values of all the coefficients of f from the values of f at ($2^{k+1}-1$) points and at the same time computes the determinant set Det_f for f :*

1. *2^k sets $(y_1, \ldots, y_k)$, where $y_i \in \{0, 2\}$*
2. *$(2^k - 1)$ sets $(a_1^Z, \ldots, a_k^Z)$, where $Z \in \mathcal{P}(\{1, \ldots, k\})$, and $a_1^Z, \ldots, a_k^Z$ can be computed from the values of f for the sets of arguments $(y_1^Z, \ldots, y_k^Z)$, where*

$$y_i^Z = \begin{cases} 0, \text{ if } i \notin Z \\ 2, \text{ if } i \in Z \end{cases}$$

according to the conditions of Lemma 13, i.e.:

$$\begin{aligned} a_1^Z &= D_{fz} + 1 \\ a_i^Z &= M_{fz}(a_{i-1}^Z) + 1 \text{ for } i = 2, \ldots, k \ . \end{aligned}$$

Proof. The following algorithm can be used to compute the values of all the coefficients of f :

1. For $Z = \phi$
   (a) Compute the value of f for the set $(0, \ldots, 0)$.
   (b) This value is the value of the only coefficient of f_Z.
2. For every $Z \in \mathcal{P}(\{1, \ldots, k\})$, $Z \neq \phi$
   (a) Compute the value of f for the set $(y_1^Z, \ldots, y_k^Z)$, where

$$y_i^Z = \begin{cases} 0, \text{ if } i \notin Z \\ 2, \text{ if } i \in Z \ . \end{cases}$$

   (b) Compute D_{f_Z}, E_{f_Z} and M_{f_Z}.
   (c) Compute the arguments $a_1^Z, \ldots, a_k^Z$ satisfying the conditions of lemma 13.
   (d) Compute the value of f for the set $(a_1^Z, \ldots, a_k^Z)$.
   (e) Using the method from Lemma 13 compute the values of the coefficients of f_Z.

It is clear from the construction above that the set Det_f is a determinant set for f. □

Lemma 21. *There exist the k-argument extended polynomials whose coefficients cannot be computed from less than ($2^{k+1} - 1$) sets of arguments.*

Proof. Consider an arbitrary k-argument extended polynomial f such that:

$$f(x_1, \ldots, x_k) = \sum_{Z \in \mathcal{P}(\{1, \ldots, k\})} f_Z(x_1, \ldots, x_k) SQ_Z(x_1, \ldots, x_k) \ ,$$

where $\forall Z \in \mathcal{P}(\{1, \ldots, k\}) : f_Z \neq 0$. Assume that there exists a set E of less than ($2^{k+1} - 1$) examples determining polynomial f.

1. $(0,\ldots,0) \in E$. Then value of f_ϕ is not known and the values of f_ϕ coefficients cannot be computed.
2. $(0,\ldots,0) \notin E$. Then, for some $Y \in \mathcal{P}(\{1,\ldots,k\})$, there exists in E at most one example with the arguments $(y_1,\ldots,y_k)$ such that $y_i \neq 0 \Leftrightarrow i \in Y$. Thus at most one value of f_Y is known, so by Lemma 15 the values of f_Y coefficients cannot be computed.

□

Theorem 22. *There exists an algorithm that takes any computable total function $f : \mathbb{N}^k \to \mathbb{N}$ as its input (in the form of the Turing machine or in the form of the system of recursive equations etc.) and effectively produces the unique extended polynomial f^* such that values of f and f^* are identical on $2^{k+1}-1$ elements of the determinant set Det_f or prints an information that there is no extended polynomial f^* identical with f on the set Det_f.*

Proof. For the purpose of constructing f^* we consider the same algorithm as presented in Theorem 20. Define the set Det_f by examining the values of f at the set of $2^{k+1}-1$ tuples of arguments. By applying the algorithm described in Theorem 20 we extract all the coefficients of f^* which crosses the set Det_f. In fact only 2^k points (about half) from Det_f are used to compute the coefficients of the extended polynomial f^*. When extended polynomial f^* is established, determine if values of f^* are the same as f on Det_f. If the answer is *YES* then extended polynomial f^* is returned. If the answer is *NO* then our algorithm prints an information that there is no extended polynomial f^* which coincides with f on Det_f. Let us call $\mathcal{POLY}$ the procedure describe above. □

Example 2. Application of the procedure $\mathcal{POLY}$ to the function $f(x,y) = y^x + x + 1$.

Using construction from Theorem 22 we construct the following extended polynomial. For the set $Z_1 = \{1,2\}$ we get $f(2,2) = 7$, therefore $D_{fZ_1} = 7$, $E_{fZ_1} = 3$, $M_{fZ_1}(x) = 7x^3 + 7$. So $a_1 = 8$. Since $7 \times 8^3 + 7 = 3591$, we take $a_2 = 3592$. Consequently we need to compute $f(8,3592) = 3592^8 + 9$. So the polynomial we get is $f^*_{Z_1}(x,y) = y^8 + x + 1$. The same must be done for sets $Z_2 = \{1\}$ and for $Z_3 = \{2\}$. Finally we obtain that $f^*_{Z_2}(x,y) = x+1$ and $f^*_{Z_3}(x,y) = 2$ so the total configuration for the function $f^*(x,y)$ in normal form is

$$(y^8 + x + 1)sq(x)sq(y) + (x+1)sq(x)\overline{sq}(y) + 2\overline{sq}(x)sq(y) + \overline{sq}(x)\overline{sq}(y)$$

and the determinant set Det_f is

$$\{(0,0,1),(2,2,7),(8,3592,3592^8+9),(2,0,3),(4,0,5),(0,2,2),(0,3,2)\} .$$

Of course, the extended polynomial f^* and function f must agree on the following subset of Det_f

$$\{(0,0,1),(8,3592,3592^8+9),(4,0,5),(0,3,2)\} .$$

Since on the point $(2,2,7)$ from Det_f functions f and f^* do not agree, ($f(2,2) = 7$ but $f^*(2,2) = 259$) the result of the procedure $\mathcal{POLY}$ is to print a message that there is no extended polynomial f^* which coincides with f on Det_f.

6 Main Result

Theorem 23. *Decidability of two following problems is equivalent.*

1. *For a given total primitive recursive function* $f : \mathbb{N}^k \to \mathbb{N}$ *decide whether or not* $\forall \vec{x} \in \mathbb{N}^k\ f(x) = 0$.
2. *For a given primitive recursive function* $f : \mathbb{N}^k \to \mathbb{N}$ *decide whether or not* f *is an extended polynomial.*

Proof. ($1 \to 2$) Suppose the $f \equiv 0$ problem is decidable. Let $\mathcal{Q}$ be a procedure accepting it. We construct an algorithm to recognize if f is an extended polynomial. The algorithm takes the function f given by a set of recursive equations as its input. Function f is sent to the procedure $\mathcal{POLY}$ which generates the set Det_f and extended polynomial f^* in normal form which is identical with f on the set Det_f if such extended polynomial f^* exists. At this stage the extended polynomial f^* is given. Then next procedure writes a system of primitive recursive equations for the extended polynomial f^*, for function $x \dot{-} y = [if\ x > y\ then\ x - y\ else\ 0]$ and accordingly for the function $(f \dot{-} f^*) + (f^* \dot{-} f)$. The system of primitive recursive equations for the function $(f \dot{-} f^*) + (f^* \dot{-} f)$ is forwarded to the procedure $\mathcal{Q}$. If $\mathcal{Q}$ accepts $(f \dot{-} f^*) + (f^* \dot{-} f)$ then it means that $f \equiv f^*$, therefore f is an extended polynomial, otherwise our procedure does not accept f. If $\mathcal{POLY}$ returns an information that there is no extended polynomial f^* which coincides with f on Det_f, then positively f is not an extended polynomial. Otherwise f in normal form should be returned.
($2 \to 1$) Suppose the *extended polynomial* problem is decidable. Let $\mathcal{P}$ be a procedure accepting it. We are going to construct an algorithm to recognize the $f \equiv 0$ problem. The algorithm constructed takes a function f given by a set of recursive equations as its input. Function f is sent to the procedure $\mathcal{P}$ which recognizes if f is an extended polynomial. If $\mathcal{P}$ does not accept f then surely $f \not\equiv 0$ since the function constantly equal to 0 is an extended polynomial. Therefore our algorithm does not accept f. If $\mathcal{P}$ accepts f then f is an extended polynomial. In this case we send f to the procedure $\mathcal{POLY}$ to generate f^* as a normal form of f. Then we examine if all coefficients of f^* are zeros or not. Accordingly our algorithm accepts f or not. □

Theorem 24. *For a given primitive recursive function f it is undecidable to recognize whether or not the function is constantly equal to* 0.

Proof. The proof is by reducing the Post correspondence problem to the $f \equiv 0$ problem. For a given Post system $S = \{(x_1, y_1), ..., (x_k, y_k)\}$ over $\Sigma = \{a, b\}$, we construct a primitive recursive function $f_S(n) = [if\ n = 0\ then\ 0\ else\ g_S(n)]$, such that S has no solution if and only if f_S is constantly equal to zero. The function g_S decides if the number n is a code of the solution of S :

$$g_S(n) = con_m(\lambda, \lambda, n) \ ,$$

where $m = k+1$ and

$$con_m(x,y,n) = \begin{cases} 1, & \text{if } n=0, \text{ and } x=y \\ 0, & \text{if } n=0, \text{ and } x \neq y \\ con_m(S_1(n \bmod m) \circ x, S_2(n \bmod m) \circ y, (n \operatorname{div} m)), \\ \quad \text{if } n \neq 0 \ , \end{cases}$$

where $\circ$ stands for concatenation of words and S_1, S_2 are selectors defined by:

$$S_1(x) = \begin{cases} \lambda, & x=0 \\ x_1, & x=1 \\ \vdots \\ x_k, & x=k \\ \lambda, & x>k \end{cases} \qquad S_2(x) = \begin{cases} \lambda, & x=0 \\ y_1, & x=1 \\ \vdots \\ y_k, & x=k \\ \lambda, & x>k \end{cases}$$

All functions used (*mod*, *div*, $\circ$) and defined above (f_S, g_S, con_m, S_1 and S_2) are primitive recursive ones. The function con_m used here maps the digits d_i of a base-m representation of a natural number $n = \overline{d_1...d_p}$ into a series of Post pairs whose numbers are d_i. The base m is chosen as $k+1$, where k is the number of pairs in the Post system. Thus, each digit except zero represents a pair. Zero is simply skipped when scanning the digits. □

Example 3. Let us take a Post system $S = \{(aa, a), \ (ba, ab), \ (b, aba)\}$. For this system we define:

$$S_1(x) = \begin{cases} \lambda, & x=0 \\ aa, & x=1 \\ ba, & x=2 \\ b, & x=3 \\ \lambda, & x>3 \end{cases} \qquad S_2(x) = \begin{cases} \lambda, & x=0 \\ a, & x=1 \\ ab, & x=2 \\ aba, & x=3 \\ \lambda, & x>3 \end{cases}$$

The sequence 312 is not a solution of the system S and we have
$g_S(312) = con_{10}(\lambda, \lambda, 312) = con_{10}(ba, ab, 31) = con_{10}(aaba, aab, 3) = con_{10}(baaba, abaaab, 0) = 0 \Rightarrow f_S(312) = 0$.
But
$g_S(1231) = con_{10}(\lambda, \lambda, 1231) = con_{10}(aa, a, 123) = con_{10}(baa, abaa, 12) = con_{10}(babaa, ababaa, 1) = con_{10}(aababaa, aababaa, 0) = 1 \Rightarrow f_S(1231) = 1$.
Thus Post system S has a solution 1231.

Theorem 25. *For a given primitive recursive function f it is undecidable to recognize whether or not the function is λ-definable.*

Proof. By Theorem 3 λ-definable functions are just extended polynomials. By Theorem 23 decidability of identification of extended polynomials is equivalent with the decidability of identification of total 0 function. But according to Theorem 24 this is undecidable. □

7 Undecidability of λ-Definability in Free Algebras

In this chapter we are going to extend the result of Theorem 25 to λ-definability in any infinite free algebra (see Sect. 3 about λ-representability). The proof of undecidability of the λ-definability is done by a simple reduction of the numerical primitive recursive function to some primitive recursive function on given free algebra in a way that preserves λ-definability.

Let us assume that A is infinite. Therefore the signature $S_A = (\alpha_1, ..., \alpha_n)$ of A must contain at least one 0 and at least one positive integer. Otherwise A is either empty or finite. Without loss of generality we may assume that $\alpha_1 = 0$ and $\alpha_2 > 0$. Hence let us take the signature of A as $S_A = (0, \alpha_2, ..., \alpha_n)$ where $\alpha_2 > 0$. Let us name the constructors of A by ϵ of arity $\alpha_1 = 0$ and d of arity $\alpha_2 > 0$ and $c_3, \ldots, c_n$ of arities $\alpha_3, \ldots, \alpha_n$, respectively.

Definition 26. By $lm(t)$ we mean the length of the leftmost path the expression t in algebra A:

$$\begin{aligned} lm(\epsilon) &= 0 \\ lm(\ d(t_1, \ldots, t_{\alpha_2})\ ) &= 1 + lm(t_1) \\ lm(\ c_i(t_1, \ldots, t_{\alpha_i})\ ) &= 1 + lm(t_1)\ . \end{aligned}$$

Definition 27. Let us define the translation of numbers to terms of algebra A by associating with n the full α_2-ary tree n^A of height n:

$$\begin{aligned} 0^A &= \epsilon \\ (n+1)^A &= d(n^A, \ldots, n^A)\ . \end{aligned}$$

For any n it is obvious that $lm(n^A) = n$.

Definition 28. Let us define the primitive recursive function *tree* in algebra A by:

$$\begin{aligned} tree(\epsilon) &= \epsilon \\ tree(d(t_1, \ldots, t_{\alpha_2})) &= d(tree(t_1), \ldots, tree(t_1)) \\ tree(c_i(t_1, \ldots, t_{\alpha_i})) &= d(tree(t_1), \ldots, tree(t_1))\ . \end{aligned}$$

It is easy to observe that for any term t in A and for the number n which is the leftmost depth of the term t $tree(t) = n^A$ which means $tree(t) = lm(t)^A$.

Definition 29. Define the translation scheme for all primitive recursive functions on numbers into the primitive recursive functions on algebra A. We associate with each primitive recursive numerical function f the primitive recursive function f^A on algebra A with the same arity. The coding is carried out by structural induction.

Coding scheme:

- zero: for $f(\vec{x}) = 0$ we define $f^A(\vec{t}) = \epsilon$,
- successor: for $f(x) = x+1$ we define $f^A(t) = d(tree(t), \ldots, tree(t))$,
- projection: for $f(x_1, \ldots, x_k) = x_i$ we define $f^A(t_1, \ldots t_k) = tree(t_i)$,
- composition:
  if f is in the form $f(\vec{x}) = g(h_1(\vec{x}), \ldots, h_n(\vec{x}))$ and $g^A, {h_1}^A, \ldots, {h_n}^A$ are already defined then $f^A(\vec{t}) = g^A({h_1}^A(\vec{t}), \ldots, {h_n}^A(\vec{t}))$,
- primitive recursion: if f is given by $f(0, \vec{x}) = g(\vec{x})$ and $f(y+1, \vec{x}) = h(f(y, \vec{x}), \vec{x})$ and functions g^A, h^A are already defined then f^A is defined by the recursion scheme in algebra A as:

$$f^A(\epsilon, \vec{t}) = g^A(\vec{t})$$
$$f^A(d(t_1, \ldots, t_{\alpha_2}), \vec{t}) = h^A(f^A(t_1, \vec{t}), \vec{t})$$
$$f^A(c_i(t_1, \ldots, t_{\alpha_i}), \vec{t}) = h^A(f^A(t_1, \vec{t}), \vec{t}) \ .$$

Lemma 30. *For any primitive recursive function $f : \mathbb{N}^k \to \mathbb{N}$ and for all expressions $t_1, \ldots, t_k$*

$$f^A(t_1, \ldots t_k) = (f(lm(t_1), \ldots, lm(t_k))^A \ .$$

Proof. Structural induction on primitive recursive functions. □

Definition 31. Let τ^A be a type representing algebra A in typed λ-calculus and let $\tau^{\mathbb{N}}$ be the type of Church numerals. We are going to define λ-terms $CODE$ and $DECODE$ of types $\tau^A \to \tau^{\mathbb{N}}$ and $\tau^{\mathbb{N}} \to \tau^A$, respectively, by:

$$CODE = \lambda tux.tx(\lambda y_1 \ldots y_{\alpha_2}.uy_1)(\lambda y_1 \ldots y_{\alpha_3}.x) \ldots (\lambda y_1 \ldots y_{\alpha_n}.x)$$
$$DECODE = \lambda nxpy_3 \ldots y_n.n(\lambda y.p\underbrace{y \ldots y}_{\alpha_2})x \ .$$

It is obvious that for any number $n \in \mathbb{N}$ and any term t in algebra A:

$$CODE\underline{t} = \underline{lm(t)}$$
$$DECODE\underline{n} = \underline{n^A} \ .$$

Theorem 32. *For any primitive recursive function $f : \mathbb{N}^k \to \mathbb{N}$, f is λ-definable if and only if f^A is λ-definable.*

Proof. Let $f : \mathbb{N}^k \to \mathbb{N}$ be primitive recursive function, λ-definable by a term F. Define a term F^A of type $(\tau^A)^k \to \tau^A$ as:

$$\lambda t_1 \ldots t_k.DECODE(F(CODEt_1) \ldots (CODEt_k)) \ .$$

We are going to prove that f^A is λ-definable by F^A which means that $\underline{f^A(t_1, \ldots, t_k)} = F^A\underline{t_1} \ldots \underline{t_k}$ for all expressions $t_1, \ldots, t_k$.

$F^A\underline{t_1} \ldots \underline{t_k} =$

$$
\begin{aligned}
&= DECODE(F(CODE\underline{t_1})\dots(CODE\underline{t_k})) && \text{(definition)}\\
&= DECODE(F\ \underline{lm(t_1)}\dots\underline{lm(t_k)}) && \text{(equation in 31)}\\
&= DECODE(\underline{f(lm(t_1),\dots,lm(t_k))}) && \text{(assumption)}\\
&= \underline{f(lm(t_1),\dots,lm(t_k))^A} && \text{(equation in 31)}\\
&= \underline{f^A(t_1,\dots,t_k)} && \text{(Lemma 30).}
\end{aligned}
$$

Let us assume now that function $f^A : A^k \to A$ is λ-definable by the term F^A. Define the term F by:

$$\lambda n_1 \dots n_k.CODE(F^A(DECODE(n_1)\dots DECODE(n_k)))\ .$$

In order to prove λ-definability of f we need to show that $F\underline{n_1}\dots\underline{n_k} = \underline{f(n_1,\dots,n_k)}$ for all numbers $n_1,\dots,n_k$.

$$
\begin{aligned}
&F\underline{n_1}\dots\underline{n_k} =\\
&\quad = CODE(F^A(DECODE(\underline{n_1}))\dots(DECODE(\underline{n_k}))) && \text{(definition of F)}\\
&\quad = CODE(F^A(\underline{n_1{}^A})\dots(\underline{n_k{}^A})) && \text{(equation in 31)}\\
&\quad = CODE(\underline{f^A(n_1{}^A,\dots,n_k{}^A)}) && \text{(assumption)}\\
&\quad = CODE(\underline{f(lm(n_1{}^A),\dots,lm(n_k{}^A))^A}) && \text{(Lemma 30)}\\
&\quad = CODE(\underline{f(n_1,\dots,n_k)^A}) && \text{(equation in 27)}\\
&\quad = \underline{lm(f(n_1,\dots,n_k)^A)} && \text{(equation in 31)}\\
&\quad = \underline{f(n_1,\dots,n_k)} && \text{(equation in 27).}
\end{aligned}
$$

□

Theorem 33. *λ-definability in algebra A is undecidable if and only if A is infinite.*

Proof. ($\Leftarrow$) Suppose λ-definability in algebra A is decidable. Prove that λ-definability for primitive recursive numerical functions is decidable. Given primitive recursive function. Using coding from Definition 27 one can produce the system of primitive recursive functions in A. Employing the hypothetical procedure one can decide if the coded system of primitive recursive functions on A is λ-definable. Hence, using the observation from Theorem 32 determine the λ-definability of the primitive recursive numerical functions we begun with.
($\Rightarrow$) If A is finite then every function is λ-definable (see Theorem 3 and [13]), hence the problem is decidable. □

Acknowledgements

The authors wish to thank Piergiorgio Odifreddi, Daniel Leivant and especially Paweł Urzyczyn for their scientific support and the attention they gave to this work. Thanks are also due to Paweł Idziak for the interesting and valuable discussions.

References

1. K. Govindarajan. A note on Polynomials with Non-negative Integer Coefficients. Private communication; to appear in the *American Mathematical Monthly.*
2. J. Haas and B. Jayaraman. Interactive Synthesis of Definite Clause Grammars. In: K.R. Apt (ed.) *Proc. Joint International Conference and Symposium on Logic Programming,* Washington DC, MIT Press, November 1992, 541-555.
3. J. Haas and B. Jayaraman. From Context-Free to Definite-Clause Grammars: A Type-Theoretic Approach. *Journal of Logic Programming,* accepted to appear.
4. M. Hagiya. From Programming by Example to Proving by Example. In: T. Ito and A. R. Meyer (eds.), *Proc. Intl. Conf. on Theoret. Aspects of Comp. Software,* 387-419, Springer-Verlag LNCS 526.
5. M. Hagiya. Programming by Example and Proving by Example using Higher-order Unification. In: M. E. Stickel (ed.) *Proc. 10th CADE,* Kaiserslautern, Springer-Verlag LNAI 449, July 1990, 588-602.
6. D. Leivant. Functions over free algebras definable in the simple typed lambda calculus. *Theoretical Computer Science* **121** (1993), 309-321.
7. R. Loader. The Undecidability of λ-Definability. Private communication; will be published in forthcoming Alonzo Church Festschrift book.
8. H. Schwichtenberg. Definierbare Funktionen in λ-Kalkuli mit Typen. *Arch. Math. Logik Grundlagenforsch.* **17** (1975/76), 113–144.
9. R. Statman. The typed λ-calculus is not elementary recursive. *Theoretical Computer Science* **9** (1979), 73-81.
10. R. Statman. On the existence of closed terms in the typed λ-calculus. In: R. Hindley and J. Seldin, eds. *To H. B. Curry, Essays on Combinatory Logic, Lambda Calculus and Formalism,* Academic Press, London and New York, 1980.
11. R. Statman. Equality of functionals revisited. In: L.A. Harrington et al. (Eds.), *Harvey Friedman's Research on the Foundations of Mathematics,* North-Holland, Amsterdam, (1985), 331-338.
12. D.A. Wolfram. *The Clausual Theory of Types, Cambridge Tracts in Theoretical Computer Science* **21**, Cambridge University Press 1993.
13. M. Zaionc. On the λ-definable higher order Boolean operations. *Fundamenta Informaticæ* **XII** (1989), 181–190.
14. M. Zaionc. λ-definability on free algebras. *Annals of Pure and Applied Logic* **51** (1991), 279-300.

Outermost-Fair Rewriting

Femke van Raamsdonk

CWI, P.O. Box 94079, 1090 GB Amsterdam, The Netherlands

Abstract. A rewrite sequence is said to be outermost-fair if every outermost redex occurrence is eventually eliminated. O'Donnell has shown that outermost-fair rewriting is normalising for almost orthogonal first-order term rewriting systems. In this paper we extend this result to the higher-order case.

1 Introduction

It may occur that a term can be rewritten to normal form but is also the starting point of an infinite rewrite sequence. In that case it is important to know how to rewrite the term such that eventually a normal form is obtained. The question of how to rewrite a term can be answered by a strategy, which selects one or more redex occurrences in every term that is not in normal form. If repeatedly contracting the redex occurrences that are selected by the strategy yields a normal form whenever the initial term has one, the strategy is said to be *normalising*.

A classical result for λ-calculus with β-reduction is that the strategy selecting the leftmost redex occurrence is normalising. This is proved in [CFC58]. For orthogonal first-order term rewriting systems, O'Donnell has shown in [O'D77] that the parallel-outermost strategy, which selects all redex occurrences that are outermost to be contracted simultaneously, is normalising. This result is a consequence of a stronger result which is also proved in [O'D77], namely that every outermost-fair rewrite sequence eventually ends in a normal form whenever the initial term has one. A rewrite sequence is said to be outermost-fair if every outermost redex occurrence is eventually eliminated.

This paper is concerned with the question of how to find a normal form in a higher-order rewriting system, in which rewriting is defined modulo simply typed λ-calculus. We extend the result by O'Donnell to the higher-order case: we show that outermost-fair rewriting is normalising for almost orthogonal higher-order rewriting systems, that satisfy some condition on the bound variables. This condition is called full extendedness. As in the first-order case, an immediate corollary of the main result is that the parallel-outermost strategy is normalising for orthogonal higher-order rewriting systems that are fully extended.

Our result extends and corrects a result by Bergstra and Klop, proved in the appendix of [BK86], which states that outermost-fair rewriting is normalising for orthogonal Combinatory Reduction Systems. Unfortunately, the proof presented in [BK86] is not entirely correct.

The remainder of this paper is organised as follows. The next section is concerned with the preliminaries. In Section 3 the notion of outermost-fair rewriting is explained. In Section 4 the main result of this paper is proved, namely that outermost-fair rewriting is normalising for the class of almost orthogonal and fully extended higher-order rewriting systems. The present paper is rather concise in nature; for a detailed account the interested reader is referred to [Raa96].

2 Preliminaries

In this section we recall the definition of higher-order rewriting systems [Nip91, MN94], following the presentation in [Oos94, Raa96]. We further give the definitions of almost orthogonality and full extendedness. The reader is supposed to be familiar with simply typed λ-calculus with β-reduction (denoted by $\rightarrow_\beta$) and restricted $\overline{\eta}$-expansion (denoted by $\rightarrow_{\overline{\eta}}$); see for instance [Bar92, Aka93]. *Simple types*, written as $A, B, C, \ldots$ are built from base types and the binary type constructor $\rightarrow$. We suppose that for every type A there are infinitely many *variables of type* A, written as $x^A, y^A, z^A, \ldots$.

Higher-Order Rewriting Systems. The meta-language of higher-order rewriting systems, which we call the *substitution calculus* as in [Oos94, OR94, Raa96], is simply typed λ-calculus.

A *higher-order rewriting system* is specified by a pair $(\mathcal{A}, \mathcal{R})$ consisting of a rewrite alphabet and a set of rewrite rules over $\mathcal{A}$.

A *rewrite alphabet* is a set $\mathcal{A}$ consisting of simply typed *function symbols*. A *preterm* of type A over $\mathcal{A}$ is a simply typed λ-term of type A over $\mathcal{A}$. Preterms are denoted by $s, t, \ldots$. Instead of $\lambda x^A.s$ we will write $x^A.s$. We will often omit the superscript denoting the type of a variable. Preterms are considered modulo the equivalence relation generated by α, β and η. Every $\alpha\beta\eta$-equivalence class contains a $\beta\overline{\eta}$-normal form that is unique up to α-conversion. Such a representative is called a *term*. Terms are the objects that are rewritten in a higher-order rewriting system.

In the following, all preterms are supposed to be in $\overline{\eta}$-normal form. Note that the set of $\overline{\eta}$-normal forms is closed under β-reduction.

For the definition of a rewrite rule we first need to introduce the notion of rule-pattern, which is an adaptation of the notion of pattern due to Miller [Mil91]. A *rule-pattern* is a closed term of the form $x_1.\ldots.x_m.fs_1\ldots s_n$ such that every $y \in \{x_1, \ldots, x_m\}$ occurs in $fs_1\ldots s_n$, and it occurs only in subterms of the form $yz_1\ldots z_p$ with $z_1, \ldots, z_p$ the $\overline{\eta}$-normal forms of different bound variables not among $x_1, \ldots, x_m$. The function symbol f is called the *head-symbol* of the rule-pattern.

A *rewrite rule* over a rewrite alphabet $\mathcal{A}$ is defined as a pair of closed terms over $\mathcal{A}$ of the same type of the form $x_1.\ldots.x_m.s \rightarrow x_1.\ldots.x_m.t$, with $x_1.\ldots.x_m.s$ a rule-pattern. The head-symbol of a rewrite rule is the head-symbol of its left-hand side. Rewrite rules are denoted by $\mathtt{R}, \mathtt{R}', \ldots$.

The next thing to define is the rewrite relation of a higher-order rewriting system $(\mathcal{A}, \mathcal{R})$. To that end we need to introduce the notion of context. We use the symbol $\Box^A$ to denote a variable of type A that is supposed to be free. A *context of type* A is a term with one occurrence of $\Box^A$. A context of type A is denoted by $C\Box^A$ and the preterm obtained by replacing $\Box^A$ by a term s of type A is denoted by $C[s]$.

Now the *rewrite relation* of a higher-order rewriting system $(\mathcal{A}, \mathcal{R})$, denoted by $\rightarrow$, is defined as follows : we have $s \rightarrow t$ if there is a context $C\Box^A$ and a rewrite rule $l \rightarrow r$ in $\mathcal{R}$ with l and r of type A such that s is the β-normal form of $C[l]$ and t is the β-normal form of $C[r]$. That is, such a rewrite step is decomposed as

$$s\ {}_{\beta}^{!}\!\twoheadleftarrow C[l] \rightarrow C[r] \twoheadrightarrow_{\beta}^{!} t$$

where $\twoheadrightarrow_{\beta}^{!}$ denotes a β-reduction to β-normal form. Note that the rewrite relation is defined on terms, not on preterms. The rewrite relation is decidable because the left-hand sides of rewrite rules are required to be rule-patterns.

As an example, we consider untyped lambda-calculus with beta-reduction and eta-reduction in the format of higher-order rewriting systems. The rewrite alphabet consists of the function symbols $\mathsf{app} : 0 \rightarrow 0 \rightarrow 0$ and $\mathsf{abs} : (0 \rightarrow 0) \rightarrow 0$ with 0 the only base type. The rewrite rules are given as follows:

$$\begin{array}{rcl} z.z'.\mathsf{app}(\mathsf{abs}\,x.zx)z' & \rightarrow_{\mathsf{beta}} & z.z'.zz' \\ z.\mathsf{abs}(x.\mathsf{app}zx) & \rightarrow_{\mathsf{eta}} & z.z \end{array}$$

The rewrite step $\mathsf{app}(\mathsf{abs}\,x.x)y \rightarrow_{\mathsf{beta}} y$ is obtained as follows:

$$\mathsf{app}(\mathsf{abs}\,x.x)y\ {}_{\beta}^{!}\!\twoheadleftarrow (z.z'.\mathsf{app}(\mathsf{abs}\,x.zx)z')(x.x)y \rightarrow_{\mathsf{beta}} (z.z'.zz')(x.x)y \twoheadrightarrow_{\beta}^{!} y.$$

In the sequel, types won't be mentioned explicitly.

Residuals. In the remainder of this paper the notions of redex occurrence and residual will be important. For their definitions we need two auxiliary notions we suppose the reader is familiar with.

The first one is the notion of position, and an ordering $\preceq$ on positions. A *position* is a finite sequence over $\{0, 1\}$. We write positions as $\phi, \chi, \psi, \ldots$. There is an operator for concatenating positions that is denoted by juxtaposition and is supposed to be associative. The neutral element for concatenation of positions is the empty sequence denoted by ϵ.

The set of positions of a (pre)term, and the sub(pre)term of a (pre)term at position ϕ are defined as in λ-calculus. For instance, the set of positions of the term $f(x.x)a$ is $\{\epsilon, 0, 00, 01, 010, 1\}$. The subterm of $f(x.x)a$ at position 01 is $x.x$ and the subterm of $f(x.x)a$ at position 1 is a.

The ordering $\preceq$ on the set of positions is defined as follows: we have $\phi \preceq \chi$ if there exists a position ϕ' such that $\phi\phi' = \chi$. The strict variant of this order is obtained by requiring in addition that $\phi' \neq \epsilon$.

The second auxiliary notion needed for the definition of the residual relation of a higher-order rewriting system, is a *descendant relation* tracing positions along β-reductions. We illustrate this notion by an example. Consider the preterm $f((x.y.xx)ab)$ and its reduction to β-normal form $f((x.y.xx)ab) \twoheadrightarrow_{\beta}^{!} f(aa)$. The descendant of the position 0 in $f((x.y.xx)ab)$ is the position 0 in

$f(aa)$, the descendants of the position 101 in $f((x.y.xx)ab)$ are the positions 10 and 11 in $f(aa)$, and the descendant of the position 10000 in $f((x.y.xx)ab)$ is the position 1 in $f(aa)$.

A *redex occurrence* in a term s is a pair $(\phi, l \to r)$ consisting of a position in s and a rewrite rule such that ϕ is the descendant of the position of the head-symbol of l along the reduction $C[l] \twoheadrightarrow^!_\beta s$. A redex occurrence $(\phi, l \to r)$ as above is said to be *contracted* in the rewrite step $s \to t$ with $s \,{}^!_\beta\!\twoheadleftarrow C[l]$ and $C[r] \twoheadrightarrow^!_\beta t$. Redex occurrences are denoted by $u, v, w, \ldots$.

In the remainder of the paper it will often be essential to know which redex occurrence is contracted in a rewrite step. This is made explicit by writing $u : s \to t$ or $s \xrightarrow{u} t$, if the redex occurrence u is contracted in the rewrite step $s \to t$.

The ordering on positions induces an ordering on redex occurrences as follows. Let $(\phi, l \to r)$ and $(\phi', l' \to r')$ be redex occurrences in a term t. Suppose that $l = x_1.\ldots.x_m.fs_1.\ldots.s_n$ and $l' = x_1.\ldots.x_{m'}.f's'_1.\ldots.s'_{n'}$. Then $\phi = \phi_0\, 0^m$ and $\phi' = \phi'_0\, 0^{m'}$ for some ϕ_0 and ϕ'_0. Now an ordering on redex occurrences, also denoted by $\preceq$, is defined as follows: $(\phi, l \to r) \preceq (\phi', l' \to r')$ if $\phi_0 \preceq \phi'_0$. For example, if $x.fx \to_{R_1} x.gx$ and $a \to_{R_2} b$ are rewrite rules, then we have $(0, R_1) \preceq (1, R_2)$ in the term fa. For the intuition behind the ordering on redex occurrences, it might be helpful to think of $fs_1 \ldots s_m$ as $f(s_1, \ldots, s_m)$.

The *descendant relation* for a higher-order rewriting system is defined using the descendant relation for β as follows. Let $s \,{}^!_\beta\!\twoheadleftarrow C[l] \to C[r] \twoheadrightarrow^!_\beta s'$ be the decomposition of a rewrite step $s \to s'$. A position ϕ in s descends to a position ϕ' in s' if there is a position χ in the $C\Box$-part of $C[l]$, which is hence also a position in $C[r]$, such that χ descends to ϕ along $C[l] \twoheadrightarrow^!_\beta s$ and χ descends to ϕ' along $C[r] \twoheadrightarrow^!_\beta s'$. Note that a position in s which descends from a position in the l-part of $C[l]$ along $C[l] \twoheadrightarrow_\beta s$ does not have a descendant in s'.

As an example, we consider the higher-order rewriting system defined by the rewrite rule

$$x.fx \to x.gxx.$$

We have the rewrite step $h(fa) \to h(gaa)$ which is obtained as follows:

$$h(fa) \,{}^!_\beta\!\twoheadleftarrow h((x.fx)a) \to h((x.gxx)a) \twoheadrightarrow^!_\beta h(gaa).$$

The descendant of the position 0 in $h(fa)$ is the position 0 in $h(gaa)$, the descendant of the position 11 in $h(fa)$ are the positions 101 and 11 in $h(gaa)$ and the position 10 in $h(fa)$ doesn't have a descendant in $h(gaa)$.

In this paper we will be concerned with higher-order rewriting systems that have the following property: if $(\phi, l \to r)$ is a redex occurrence in s, and ϕ descends to ϕ' along the rewrite step $u : s \to s'$, then $(\phi', l \to r)$ is a redex occurrence in s'. The redex occurrence $(\phi', l \to r)$ is then said to be a *residual* of the redex occurrence $(\phi, l \to r)$. If u and v are redex occurrences in a term s, then the set of residuals of v after performing the rewrite step $u : s \to t$ is denoted by $\mathsf{Res}(s, u)(v)$. The residual relation is extended in a straightforward way to rewrite sequences consisting possibly of more than one step and sets of redex occurrences.

Almost Orthogonality and Full Extendedness. The main result of this paper is concerned with higher-order rewriting systems that are *almost orthogonal* and *fully extended.* We will now explain the notions of almost orthogonality and full extendedness.

For the definition of almost orthogonality we need the notion of left-linearity. A rule-pattern $x_1.\ldots.x_m.fs_1\ldots s_n$ is *linear* if every $y \in \{x_1,\ldots,x_m\}$ occurs at most once (and hence, by the definition of a rule-pattern, exactly once) in $fs_1\ldots s_n$. A rewrite rule is said to be *left-linear* if its left-hand side is linear, and a higher-order rewriting system is said to be *left-linear* if all its rewrite rules are left-linear. For example, the rewrite rule $x.fx \to x.gx$ is left-linear, but the rewrite rule $x.fxx \to x.gx$ is not.

A higher-order rewriting system is said to be *orthogonal* if it is left-linear and has no critical pairs. A higher-order rewriting system is said to be *weakly orthogonal* if it is left-linear and all its critical pairs are trivial. Almost orthogonality lies in between orthogonality and weak orthogonality. A higher-order rewriting system is said to be *almost orthogonal* if it is weakly orthogonal with the additional property that overlap between redex occurrences occurs only at the root of the redex occurrences. The notion of overlapping redex occurrences is defined below. The notion of critical pair is the usual one, as defined in [Hue80]. The formal definition for higher-order rewriting systems is not given in the present paper.

Definition 1. Let $u = (\phi 0^n, l \to r)$ and $u' = (\phi' 0^{n'}, l' \to r')$ with $l = x_1.\ldots.x_m.fs_1\ldots s_n$ and $l' = x_1.\ldots.x_{m'}.f's'_1\ldots s'_{n'}$ be redex occurrences in a term s.

1. The redex occurrence u *nests* the redex occurrence u' if $\phi' = \phi\chi\psi$ such that the subterm of $fs_1\ldots s_n$ at position $\chi 0^i$ is a variable $y \in \{x_1,\ldots,x_m\}$, and ψ is an arbitrary position.
2. The redex occurrences u and u' are *overlapping* if
   (a) they are different,
   (b) $\phi \preceq \phi'$ and u does not nest u', or $\phi' \preceq \phi$ and u' does not nest u.

Definition 2. 1. A higher-order rewriting system is said to be *weakly head-ambiguous* if for every term s and for every two overlapping redex occurrences $u = (\phi 0^n, l \to r)$ and $u' = (\phi' 0^{n'}, l' \to r')$ with $l = x_1.\ldots.x_m.fs_1\ldots s_n$ and $l' = x_1.\ldots.x_{m'}.f's'_1\ldots s'_{n'}$ in s, we have $u : s \to t$ and $u' : s \to t$, that is, u and u' define the same rewrite step, and moreover $\phi = \phi'$, and hence necessarily $n = n'$.

2. A higher-order rewriting system is said to be *almost orthogonal* if it is left-linear and weakly head-ambiguous.

The rewriting system defined by the rewrite rules

$$x.fx \to x.gx$$
$$a \to b$$

is orthogonal. The rewriting system defined by the rewrite rules

$$fa \to fb$$

$$a \to b$$

is weakly orthogonal but not almost orthogonal. The rewriting system for parallel or defined by the rewrite rules

$$x.fax \to x.a$$

$$x.fxa \to x.a$$

is almost orthogonal but not orthogonal.

We will use the following notation. By $u \between v$ we denote that the redex occurrences u and v are overlapping. Two overlapping redex occurrences are by definition different, so the relation $\between$ is not reflexive. In an almost orthogonal rewriting system, we have the following: if $u \between v$ and $v \between w$, then $u \between w$ or $u = w$. This implication does not hold in a weakly orthogonal rewriting system. Consider for instance the weakly orthogonal higher-order rewriting system defined by the following rewrite rules:

$$x.f(gx) \to_{\mathsf{R}_1} x.f(gb)$$

$$ga \to_{\mathsf{R}_2} gb$$

$$a \to_{\mathsf{R}_3} b$$

In the term $f(ga)$, we have $(0, \mathsf{R}_1) \between (10, \mathsf{R}_2)$ and $(10, \mathsf{R}_2) \between (11, \mathsf{R}_3)$ but not $(0, \mathsf{R}_1) \between (11, \mathsf{R}_3)$.

We denote by $u \parallel v$ that the redex occurrences u and v are not overlapping. The relation $\parallel$ is reflexive. It is extended in the obvious way to denote that a redex occurrence is not overlapping with a set of redex occurrences.

Finally, the notion of full extendedness will be needed. The definition is given in [HP96, Oos96a].

Definition 3. A rewrite rule $x_1. \ldots .x_m.s \to x_1. \ldots .x_m.t$ is said to be *fully extended* if every occurrence of $y \in \{x_1, \ldots, x_m\}$ in s has the $\overline{\eta}$-normal form of every bound variable in which scope it occurs as argument.

A higher-order rewriting system is *fully extended* if every rewrite rule of it is.

An example of a rewrite rule that is fully extended is the rule for **beta** in the higher-order rewriting system representing lambda-calculus. The rewrite rule for **eta** in the same system is not fully extended, because in $\mathsf{abs}(x.\mathsf{app}zx)$, the variable z does not have the variable x as an argument. Note that the traditional version of the eta-reduction rule contains a side-condition concerning the bound variable.

In Section 4 it will be explained which restrictions imposed on higher-order rewriting systems are necessary for outermost-fair rewriting to be normalising, and which are mainly there to make the proof work.

The Weakly Orthogonal Projection. The weakly orthogonal projection is defined by van Oostrom in [Oos94, p.49]. It is used to prove confluence of weakly orthogonal higher-order rewriting systems by developments. Here we recall the construction of the weakly orthogonal projection which, as the terminology indicates, is defined for all weakly orthogonal higher-order rewriting system. We will use it for the smaller class consisting of all almost orthogonal and fully extended higher-order rewriting systems.

The reader is supposed to be familiar with complete developments and the projection of a rewrite sequence over a rewrite step in the orthogonal case. A complete development of a set of (non-overlapping) redex occurrences $\mathcal{U}$ is denoted by $\mathcal{U} : s \mathrel{-\!\!\circ\!\!\!\rightarrow} t$ or by $s \overset{\mathcal{U}}{\mathrel{-\!\!\circ\!\!\!\rightarrow}} t$. If v is a redex occurrence in s, then the set of residuals of v after performing a complete development of $\mathcal{U}$ is denoted by $\mathsf{Res}(s,\mathcal{U})(v)$. A ***development rewrite sequence*** is obtained by concatenating complete developments.

Let $\mathcal{H}$ be a weakly orthogonal higher-order rewriting system. Consider a finite or infinite rewrite sequence

$$\tilde{\sigma} : s_0 \overset{\tilde{u}_0}{\rightarrow} s_1 \overset{\tilde{u}_1}{\rightarrow} s_2 \overset{\tilde{u}_2}{\rightarrow} s_3 \overset{\tilde{u}_3}{\rightarrow} \ldots$$

and a rewrite step

$$v : s_0 \rightarrow t_0.$$

Let $\mathcal{V}_0 = \{v\}$ and define for $m \geq 0$ the following:

$$u_m = \begin{cases} \tilde{u}_m \text{ if } \tilde{u}_m \parallel \mathcal{V}_m, \\ v_m \text{ if } \tilde{u}_m \nparallel v_m \text{ for some } v_m \in \mathcal{V}_m, \end{cases}$$

$$\mathcal{V}_{m+1} = \mathsf{Res}(s_m, u_m)(\mathcal{V}_m).$$

Since the rewriting system is weakly orthogonal, we have for every $m \geq 0$ that $u_m : s_m \rightarrow s_{m+1}$ if $\tilde{u}_m : s_m \rightarrow s_{m+1}$. Let σ be the rewrite sequence

$$\sigma : s_0 \overset{u_0}{\rightarrow} s_1 \overset{u_1}{\rightarrow} s_2 \overset{u_2}{\rightarrow} s_3 \overset{u_3}{\rightarrow} \ldots.$$

By construction, we have for every $m \geq 0$ that $u_m \parallel \mathcal{V}_m$. Note that $\mathcal{V}_m$ is the set of residuals of v in s_m. For every $m \geq 0$ we define

$$\mathcal{U}_m = \mathsf{Res}(s_m, \mathcal{V}_m)(u_m).$$

Let τ be the development rewrite sequence

$$\tau : t_0 \overset{\mathcal{U}_0}{\mathrel{-\!\!\circ\!\!\!\rightarrow}} t_1 \overset{\mathcal{U}_1}{\mathrel{-\!\!\circ\!\!\!\rightarrow}} t_2 \overset{\mathcal{U}_2}{\mathrel{-\!\!\circ\!\!\!\rightarrow}} t_3 \overset{\mathcal{U}_3}{\mathrel{-\!\!\circ\!\!\!\rightarrow}} \ldots.$$

In a diagram, the situation is depicted as follows:

$$\begin{array}{lccccccccc}
\tilde{\sigma} : & s_0 & \xrightarrow{\tilde{u}_0} & s_1 & \xrightarrow{\tilde{u}_1} & s_2 & \xrightarrow{\tilde{u}_2} & s_3 & \xrightarrow{\tilde{u}_3} & \cdots \\
 & & & & & & & & & \\
\sigma : & s_0 & \xrightarrow{u_0} & s_1 & \xrightarrow{u_1} & s_2 & \xrightarrow{u_2} & s_3 & \xrightarrow{u_3} & \cdots \\
 & \mathcal{V}_0 \downarrow & & \mathcal{V}_1 \downarrow & & \mathcal{V}_2 \downarrow & & \mathcal{V}_3 \downarrow & & \\
\tau : & t_0 & \xrightarrow{\mathcal{U}_0} & t_1 & \xrightarrow{\mathcal{U}_1} & t_2 & \xrightarrow{\mathcal{U}_2} & t_3 & \xrightarrow{\mathcal{U}_3} & \cdots
\end{array}$$

We use the following terminology. The rewrite sequence σ is said to be a ***simulation*** of the rewrite sequence $\tilde{\sigma}$. The development rewrite sequence τ is said to be the ***orthogonal projection*** of the rewrite sequence σ over the rewrite step $v : s_0 \to t_0$. The development rewrite sequence τ is said to be a ***weakly orthogonal projection*** of the rewrite sequence $\tilde{\sigma}$ over the rewrite step $v : s_0 \to t_0$.

In almost orthogonal higher-order rewriting systems, a simulation of a rewrite sequence constructed for the weakly orthogonal projection is unique. This is the case since if we have $u_m \nparallel v_m$ and $u_m \nparallel v'_m$ in the notation as above, then by almost orthogonality $v_m \nparallel v'_m$ or $v_m = v'_m$. Since $\mathcal{V}_m$ consists of non-overlapping redex occurrences, we have $v_m = v'_m$. In a weakly orthogonal higher-order rewriting system, a simulation of a rewrite sequence constructed for the weakly orthogonal projection is not necessarily unique.

We conclude this section with an example of the weakly orthogonal projection.

Example 1. Consider the rewriting system defined by the following rules:

$$x.fax \to_{\mathrm{R}_1} x.a$$
$$x.fxa \to_{\mathrm{R}_2} x.a$$
$$x.gx \to_{\mathrm{R}_3} x.hxx$$

It is almost orthogonal but not orthogonal. We construct the weakly orthogonal projection of the rewrite sequence

$$\tilde{\sigma} : g(faa) \xrightarrow{\tilde{u}_0} h(faa)(faa) \xrightarrow{\tilde{u}_1} ha(faa)$$

with $\tilde{u}_0 = (0, \mathrm{R}_3)$ and $\tilde{u}_1 = (0100, \mathrm{R}_2)$ over the rewrite step $v : g(faa) \to ga$ with $v = (100, \mathrm{R}_1)$. This yields the following:

$$\begin{array}{llllll}
\tilde{\sigma} : & g(faa) & \xrightarrow{\tilde{u}_0} & h(faa)(faa) & \xrightarrow{\tilde{u}_1} & ha(faa) \\
 & & & & & \\
\sigma : & g(faa) & \xrightarrow{u_0} & h(faa)(faa) & \xrightarrow{u_1} & ha(faa) \\
 & \mathcal{V}_0 \downarrow & & \mathcal{V}_1 \downarrow & & \mathcal{V}_2 \downarrow \\
\tau : & ga & \xrightarrow{\mathcal{U}_0} & haa & \xrightarrow{\mathcal{U}_1} & haa
\end{array}$$

with

$$\begin{array}{ll} u_0 = (0, \mathrm{R}_3) & \mathcal{U}_0 = \{(0, \mathrm{R}_3)\} \\ u_1 = (0100, \mathrm{R}_1) & \mathcal{U}_1 = \emptyset \end{array}$$

and

$$\mathcal{V}_0 = \{(100, \mathrm{R}_1)\}, \mathcal{V}_1 = \{(0100, \mathrm{R}_1), (100, \mathrm{R}_1)\}, \mathcal{V}_2 = \{(100, \mathrm{R}_1)\}.$$

3 Outermost-Fair Rewriting

A rewrite sequence is said to be outermost-fair, or in the terminology of [O'D77], eventually outermost, if every outermost redex occurrence is eventually eliminated. So a rewrite sequence is outermost-fair either if it ends in a normal form or if it is impossible to trace infinitely long an outermost redex occurrence. In order to formalise the notion of an outermost-fair rewrite sequence, the definition of an infinite outermost chain is given. First the definition of an outermost redex occurrence is presented.

Definition 4. A redex occurrence $(\phi, l \to r)$ in a term s is said to be *outermost* if for every redex occurrence $(\phi', l' \to r')$ in s we have the following: if $(\phi', l' \to r') \preceq (\phi, l \to r)$ then $\phi' = \phi$.

The term $g(faa)$ in the rewriting system of Example 1 contains one outermost redex occurrence, namely $(0, \mathsf{R}_3)$. The term faa in the same rewriting system contains two outermost redex occurrences, namely $(00, \mathsf{R}_1)$ and $(00, \mathsf{R}_2)$. This shows that, in an almost orthogonal higher-order rewriting system, outermost redex occurrences may be overlapping.

Definition 5. Let $\sigma : s_0 \xrightarrow{u_0} s_1 \xrightarrow{u_1} s_2 \xrightarrow{u_2} \ldots$ be an infinite rewrite sequence. An *infinite outermost chain in* σ is an infinite sequence of redex occurrences $w_m, w_{m+1}, \ldots$ such that

1. w_p is an outermost redex occurrence in s_p for every $p \geq m$,
2. w_p is a residual of w_{p-1} for every $p > m$.

Note that in the previous definition $w_p \parallel u_p$ for every $p \geq m$. Now the key definition of this paper can be given.

Definition 6. A rewrite sequence is said to be *outermost-fair* either if it ends in a normal form or if it is infinite and does not contain an infinite outermost chain.

The definition of an outermost-fair rewrite sequence is illustrated by the following example.

Example 2. Consider the higher-order rewriting system defined by the following rewrite rules:

$$x.fcx \to_{\mathsf{R}_1} x.fbx$$

$$b \to_{\mathsf{R}_2} c$$

$$a \to_{\mathsf{R}_3} a$$

1. The rewrite sequence

$$\sigma : fca \xrightarrow{u_0} fba \xrightarrow{u_1} fca \xrightarrow{u_2} \ldots$$

with $u_{2m} = (00, \mathrm{R}_1)$ and $u_{2m+1} = (01, \mathrm{R}_2)$ for every $m \geq 0$ is outermost-fair. Note that there is an infinite sequence of residuals starting in the first term of σ, namely $(1, \mathrm{R}_3), (1, \mathrm{R}_3), (1, \mathrm{R}_3), \ldots$. This infinite sequence of residuals is however not an infinite outermost chain, although infinitely many residuals in it are outermost.

2. The rewrite sequence

$$\sigma : faa \xrightarrow{u_0} faa \xrightarrow{u_1} faa \xrightarrow{u_2} \ldots$$

with $u_m = (01, \mathrm{R}_3)$ for every $m \geq 0$ is not outermost-fair since we have an infinite outermost chain, namely $(1, \mathrm{R}_3), (1, \mathrm{R}_3), (1, \mathrm{R}_3), \ldots$.

4 Outermost-Fair Rewriting is Normalising

In this section we present the proof of the main result of this paper, namely that outermost-fair rewriting is normalising for almost orthogonal and fully extended higher-order rewriting systems.

The Structure of the Proof. The structure of the proof is as follows. Suppose that the finite or infinite rewrite sequence

$$\tilde{\sigma} : s_0 \xrightarrow{\tilde{u}_0} s_1 \xrightarrow{\tilde{u}_1} s_2 \xrightarrow{\tilde{u}_2} s_3 \xrightarrow{\tilde{u}_3} \ldots$$

is outermost-fair. Let $v : s_0 \to t_0$ be a rewrite step and construct the weakly orthogonal projection of $\tilde{\sigma}$ over v as explained in Section 2. In a diagram:

$$\begin{array}{lccccccccc}
\tilde{\sigma} : & s_0 & \xrightarrow{\tilde{u}_0} & s_1 & \xrightarrow{\tilde{u}_1} & s_2 & \xrightarrow{\tilde{u}_2} & s_3 & \xrightarrow{\tilde{u}_3} & \ldots \\
\\
\sigma : & s_0 & \xrightarrow{u_0} & s_1 & \xrightarrow{u_1} & s_2 & \xrightarrow{u_2} & s_3 & \xrightarrow{u_3} & \ldots \\
 & {\scriptstyle \mathcal{V}_0}\big\downarrow & & {\scriptstyle \mathcal{V}_1}\big\downarrow & & {\scriptstyle \mathcal{V}_2}\big\downarrow & & {\scriptstyle \mathcal{V}_3}\big\downarrow & & \\
\tau : & t_0 & \xrightarrow{\mathcal{U}_0} & t_1 & \xrightarrow{\mathcal{U}_1} & t_2 & \xrightarrow{\mathcal{U}_2} & t_3 & \xrightarrow{\mathcal{U}_3} & \ldots
\end{array}$$

The proof consists of the following steps:

1. If $\tilde{\sigma}$ is outermost-fair, then σ is outermost-fair (Proposition 7).
2. If σ is outermost-fair, then τ is outermost-fair (Proposition 8).
3. If σ is outermost-fair and τ ends in a normal form t, then σ ends in t (Proposition 9).
4. If a term s has a normal form t then every outermost-fair rewrite sequence starting in s eventually ends in t (Theorem 10).

The notation in the proof will be such that the diagram above applies.

Some Observations. Three restrictions are imposed on the higher-order rewriting systems that we consider: rewrite rules must be left-linear, all critical pairs are trivial and overlap occurs only at the root of redex occurrences, and finally rewrite rules must be fully extended. Before we embark on the proof, we first analyse the rôle of these restrictions.

The restriction to left-linear systems is necessary, since outermost-fair rewriting may not be normalising in a higher-order rewriting system that is not left-linear. This is illustrated by the following example:

$$x.fxx \to_{\mathsf{R}_1} x.b$$
$$x.gx \to_{\mathsf{R}_2} x.gx$$
$$a \to_{\mathsf{R}_3} b$$

The rewrite sequence $f(ga)(gb) \to f(ga)(gb) \to \ldots$ in which alternately the outermost redex occurrence $(010, \mathsf{R}_2)$ and the outermost redex occurrence $(10, \mathsf{R}_2)$ are contracted, is outermost-fair but does not end in the normal form b, although we have $f(ga)(gb) \to f(gb)(gb) \to b$.

The restriction to fully extended systems is also necessary. This was pointed out to me by Vincent van Oostrom [Oos96b], who gave the following example. Consider the higher-order rewriting system defined by the following rewrite rules:

$$z.f(x.z) \to_{\mathsf{R}_1} z.a$$
$$z.gz \to_{\mathsf{R}_2} z.a$$
$$z.hz \to_{\mathsf{R}_3} z.hz$$

It is not fully extended because of the rewrite rule R_1: the variable z in $f(x.z)$ doesn't have the variable x as an argument. The rewrite sequence $f(x.h(gx)) \to f(x.h(gx)) \to \ldots$ in which in every step the outermost redex occurrence $(100, \mathsf{R}_3)$ is contracted, is outermost-fair but does not end in the normal form a, although we have $f(x.h(gx)) \to f(x.ha) \to a$. Note that $(0, \mathsf{R}_1)$ is *not* a redex occurrence in $f(x.h(gx))$ because of the occurrence of x at position 1011.

In both cases the problem is that an outermost-redex occurrence can be created by contracting a redex occurrence that is not outermost. In a higher-order rewriting system that is almost orthogonal and fully extended, an outermost redex occurrence can only be created by contracting a redex occurrence that is outermost itself.

Another important point is the elimination of outermost redex occurrences. In a higher-order rewriting system that is almost orthogonal and fully extended, an outermost redex occurrence can only be eliminated by contracting a redex occurrence that is also outermost itself. To be more precise, an outermost redex occurrence w can be eliminated in one of the following three ways:

1. by contracting w,
2. by contracting a redex occurrence that is overlapping with w,
3. by contracting a redex occurrence that creates a new outermost redex occurrence such that the residual of w is not outermost anymore.

It is quite easy to see that if a system is not left-linear, it can happen that an outermost redex occurrence is eliminated by contracting a redex occurrence that is not outermost.

Finally we discuss the restriction to higher-order rewriting systems that are weakly head-ambiguous. In a system that is weakly orthogonal but not almost orthogonal, it can happen that an outermost redex occurrence is created by contracting a redex occurrence that is not outermost, and it can also happen that an outermost redex occurrence is eliminated by contracting a redex occurrence that is not outermost. In both cases, the redex occurrence creating or eliminating an outermost redex occurrence is not necessarily outermost itself, but it is overlapping with an outermost redex occurrence. Consider for instance the higher-order rewriting system defined by the following rewrite rules:

$$fa \rightarrow_{R_1} fb$$
$$a \rightarrow_{R_2} b$$
$$b \rightarrow_{R_3} c$$

In the rewrite step $(1, R_2) : fa \rightarrow fb$ the outermost redex occurrence $(0, R_1)$ is eliminated although the redex occurrence $(1, R_2)$ is not outermost itself. Moreover, in the same rewrite step the outermost redex occurrence $(1, R_3)$ is created.

We conclude that the restrictions of left-linearity and fully extendedness are necessary for the result to hold; the restriction to critical pairs that are trivial and that have overlap only at the root of the redex occurrences is mainly there to make the proof work. It is imaginable that a proof can be given for the larger class of weakly orthogonal and fully extended higher-order rewriting systems. In fact, in an earlier version we erroneously claimed to prove normalisation of outermost-fair rewriting for weakly orthogonal systems. The restriction on critical pairs can probably not be relaxed much more: for left-linear higher-order rewriting systems that are *parallel-closed* as defined by Huet in Section 3.3 of [Hue80], outermost-fair rewriting is not normalising. This is illustrated by the higher-order rewriting system defined by the following rewrite rules:

$$a \rightarrow_{R_1} b$$
$$x.hx \rightarrow_{R_2} x.hx$$
$$g(hb) \rightarrow_{R_3} b$$

It is easy to see that it is left-linear and parallel-closed. The rewrite sequence $g(ha) \rightarrow g(ha) \rightarrow g(ha) \rightarrow \ldots$ in which in each rewrite step the redex occurrence $(10, R_2)$ is contracted, is outermost-fair but does not end in a normal form, although we have $g(ha) \rightarrow g(hb) \rightarrow b$.

The Proof. In the following, we consider an almost orthogonal and fully extended higher-order rewriting system $\mathcal{H}$.

Proposition 7. *Let $\tilde{\sigma}$ be an outermost-fair rewrite sequence issuing from s_0 and let $v : s_0 \rightarrow t_0$ be a rewrite step. Let σ be the simulation of $\tilde{\sigma}$ constructed for the weakly orthogonal projection of $\tilde{\sigma}$ over $v : s_0 \rightarrow t_0$. Then σ is outermost-fair.*

Proof. Let $\tilde{\sigma} : s_0 \overset{\tilde{u}_1}{\to} s_1 \overset{\tilde{u}_2}{\to} s_2 \overset{\tilde{u}_3}{\to} \ldots$ be an outermost-fair rewrite sequence and let σ be the simulation of $\tilde{\sigma}$ for the weakly orthogonal projection of $\tilde{\sigma}$ over some rewrite step $v : s_0 \to t_0$.

If $\tilde{\sigma}$ ends in a normal form, then σ also ends in a normal form, so we restrict attention to the case that $\tilde{\sigma}$ is infinite and does not contain an infinite outermost chain. It is sufficient to show the following: if an outermost redex occurrence w_m in s_m is eliminated in a rewrite step $\tilde{u}_m : s_m \to s_{m+1}$, then w_m is also eliminated in the rewrite step $u_m : s_m \to s_{m+1}$. There are two possibilities: either $u_m = \tilde{u}_m$ or $u_m \sharp \tilde{u}_m$. In the first case, the outermost redex occurrence w_m is clearly eliminated in the rewrite step $u_m : s_m \to s_{m+1}$. For the case that $u_m \sharp \tilde{u}_m$, we consider the different possibilities in which the redex occurrences w_m is eliminated in the rewrite step $\tilde{u}_m : s_m \to s_{m+1}$.

1. If $\tilde{u}_m = w_m$, then we have $u_m \sharp w_m$, and hence the outermost redex occurrence w_m is eliminated in the rewrite step $u_m : s_m \to s_{m+1}$.
2. If $\tilde{u}_m \sharp w_m$, then we have either $u_m = w_m$ or $u_m \sharp w_m$, and in both cases the outermost redex occurrence w_m is eliminated in the rewrite step $u_m : s_m \to s_{m+1}$.
3. Finally, the outermost redex occurrence w_m can be eliminated in the rewrite step $\tilde{u}_m : s_m \to s_{m+1}$ because an outermost redex occurrence is created above the residual of w_m. The same outermost redex occurrence is created by contracting u_m, so also in that case the outermost redex occurrence w_m is eliminated in the rewrite step $u_m : s_m \to s_{m+1}$. □

The proof of the following proposition requires some auxiliary definitions and results and is for lack of space omitted. The interested reader is referred to Proposition 6.2.11 in [Raa96]. Some of the ideas of the proof are also present in the proof of Proposition 9.

Proposition 8. *Let $\sigma : s_0 \overset{u_0}{\to} s_1 \overset{u_1}{\to} s_2 \overset{u_2}{\to} \ldots$ be an infinite rewrite sequence. Let $v : s_0 \to t_0$ be a rewrite step. Let $\tau : t_0 \overset{\mathcal{U}_0}{\twoheadrightarrow} t_1 \overset{\mathcal{U}_1}{\twoheadrightarrow} t_2 \overset{\mathcal{U}_2}{\twoheadrightarrow} \ldots$ be the orthogonal projection of the rewrite sequence σ over the rewrite step $v : s_0 \to t_0$. If τ contains an infinite outermost chain, then σ contains an infinite outermost chain.*

A direct consequence of Proposition 8 is that if a rewrite sequence is outermost-fair, then its orthogonal projection over some rewrite step is also outermost-fair.

Proposition 9. *Let $\sigma : s_0 \overset{u_0}{\to} s_1 \overset{u_1}{\to} s_2 \overset{u_2}{\to} \ldots$ be an outermost-fair rewrite sequence. Let $\tau : t_0 \overset{\mathcal{U}_0}{\twoheadrightarrow} t_1 \overset{\mathcal{U}_1}{\twoheadrightarrow} t_2 \overset{\mathcal{U}_2}{\twoheadrightarrow} \ldots$ be the orthogonal projection of σ over the rewrite step $v : s_0 \to t_0$. If τ ends in a normal form t then σ ends also in t.*

Proof. Let $\sigma : s_0 \overset{u_0}{\to} s_1 \overset{u_1}{\to} \ldots$ be an outermost-fair rewrite sequence. Let $\tau : t_0 \overset{\mathcal{U}_0}{\twoheadrightarrow} t_1 \overset{\mathcal{U}_1}{\twoheadrightarrow} \ldots$ be the orthogonal projection of σ over the rewrite step $v : s_0 \to t_0$. Suppose that τ ends in a normal form $t = t_m$. In a picture:

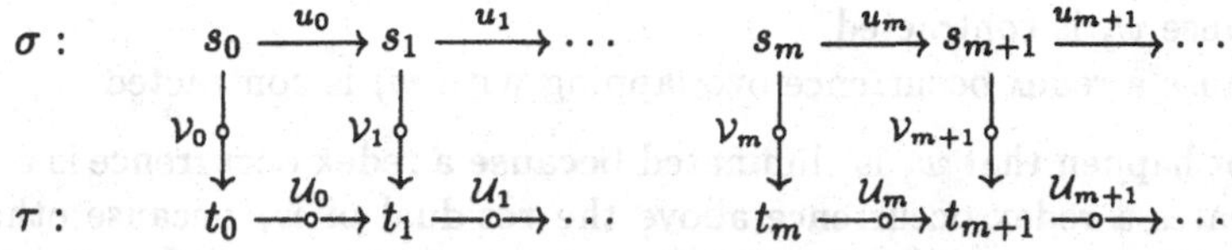

$$
\begin{array}{lccccccccccc}
\sigma: & s_0 & \xrightarrow{u_0} & s_1 & \xrightarrow{u_1} & \cdots & & s_m & \xrightarrow{u_m} & s_{m+1} & \xrightarrow{u_{m+1}} & \cdots \\
& \downarrow{\scriptstyle \mathcal{V}_0} & & \downarrow{\scriptstyle \mathcal{V}_1} & & & & \downarrow{\scriptstyle \mathcal{V}_m} & & \downarrow{\scriptstyle \mathcal{V}_{m+1}} & & \\
\tau: & t_0 & \xrightarrow{\mathcal{U}_0} & t_1 & \xrightarrow{\mathcal{U}_1} & \cdots & & t_m & \xrightarrow{\mathcal{U}_m} & t_{m+1} & \xrightarrow{\mathcal{U}_{m+1}} & \cdots
\end{array}
$$

We have $\emptyset = \mathcal{U}_m = \mathcal{U}_{m+1} = \ldots$, so $t = t_m = t_{m+1} = \ldots$. Now we define the following:

1. $t^0_m = s_m$,
2. $\mathcal{W}^0_m = \mathcal{V}_m$,
3. $\mathcal{V}^0_m$ is the set of all outermost redex occurrences of $\mathcal{W}^0_m$ in t^0_m.

For $i > 0$ we define:

1. t^{i+1}_m is the term obtained by performing a complete development of $\mathcal{V}^i_m$ in t^i_m,
2. $\mathcal{W}^{i+1}_m = \mathsf{Res}(t^i_m, \mathcal{V}^i_m)(\mathcal{W}^i_m)$,
3. $\mathcal{V}^{i+1}_m$ is the set of all outermost redex occurrences of $\mathcal{W}^{i+1}_m$ in t^{i+1}_m.

Let moreover p be the smallest natural number such that $\mathcal{V}^p_m = \emptyset$. We consider the following stepwise development of $\mathcal{V}_m$:

$$s_m = t^0_m \xrightarrow{\mathcal{V}^0_m} t^1_m \xrightarrow{\mathcal{V}^1_m} \ldots \xrightarrow{\mathcal{V}^{p-1}_m} t^p_m = t_m.$$

Note that indeed $t^p_m = t_m$ since t_m is a normal form. Projecting σ over this rewrite sequence yields the following:

$$
\begin{array}{ccccc}
t^0_m & \xrightarrow{u_m} & t^0_{m+1} & \xrightarrow{u_{m+1}} & \cdots \\
\downarrow{\scriptstyle \mathcal{V}^0_m} & & \downarrow{\scriptstyle \mathcal{V}^0_{m+1}} & & \\
t^1_m & \longrightarrow & t^1_{m+1} & \longrightarrow & \cdots \\
\vdots & & \vdots & & \\
t^{p-1}_m & \longrightarrow & t^{p-1}_{m+1} & \longrightarrow & \\
\downarrow{\scriptstyle \mathcal{V}^{p-1}_m} & & \downarrow{\scriptstyle \mathcal{V}^{p-1}_{m+1}} & & \\
t^p_m & \longrightarrow & t^p_{m+1} & \longrightarrow & \cdots
\end{array}
$$

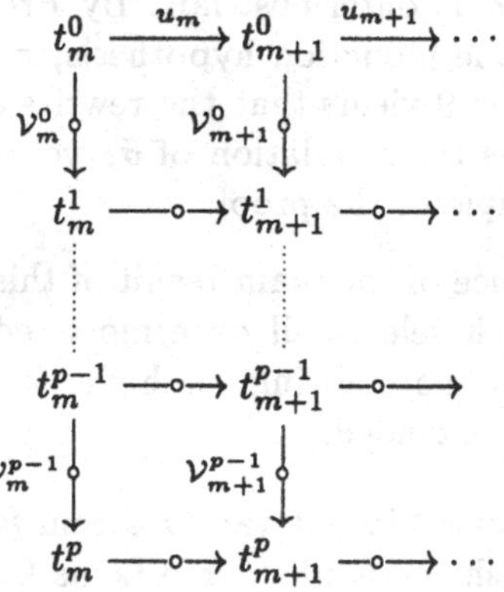

We have that $\mathcal{V}^i_j$ is a set of outermost redex occurrences in t^i_j for every $i \in \{0, \ldots, p-1\}$ and $j \geq m$.

Now we show that there exists an n such that $\mathcal{V}^0_n = \ldots = \mathcal{V}^{p-1}_n = \emptyset$. That is, eventually the rewrite sequences σ and τ coincide and σ also ends in t.

Let $j \geq m$ arbitrary. We define $f(j)$ to be the smallest number such that $\mathcal{V}^{f(j)}_j \neq \emptyset$. Then $\mathcal{V}^{f(j)}_j$ consists of outermost redex occurrences in s_j $(= t^0_j)$. Since the rewrite sequence σ is outermost-fair, a redex occurrence w_j in $\mathcal{V}^{f(j)}_j$ will eventually be eliminated. This elimination can happen in one of the following two ways:

1. because w_j is contracted,
2. because a redex occurrence overlapping with w_j is contracted.

It cannot happen that w_j is eliminated because a redex occurrence is contracted that creates a redex occurrence above the residual of w_j because otherwise t_j^p would not be a normal form, which is a contradiction since $t_j^p = t_j = t$. Hence there exists a j' such that $\mathcal{V}_{j'}^{f(j)} = \emptyset$.

This shows that eventually σ coincides with τ and hence σ ends in the normal form t. □

Finally we present the proof of the main result.

Theorem 10. *Let s_0 be a weakly normalising term. Every outermost-fair rewrite sequence starting in s_0 eventually ends in a normal form.*

Proof. Let s_0 be a weakly normalising term and consider an outermost-fair rewrite sequence $\tilde{\sigma} : s_0 \overset{\tilde{u}_0}{\to} s_1 \overset{\tilde{u}_1}{\to} s_2 \overset{\tilde{u}_2}{\to} \ldots$ starting in s_0. Let t be the normal form of s_0. We fix a rewrite sequence $\rho : s_0 \twoheadrightarrow t$ from s_0 to its normal form consisting of m rewrite steps. Now we prove by induction on m that $\tilde{\sigma}$ eventually ends in the normal form t.

If $m = 0$, then $s_0 = t$ and hence the statement trivially holds.

If $m > 0$, then the rewrite sequence ρ is of the form $s_0 \overset{v}{\to} t_0 \twoheadrightarrow t$. We construct the weakly orthogonal projection of the rewrite sequence $\tilde{\sigma}$ over the rewrite step $v : s_0 \to t_0$. Let σ be the simulation of $\tilde{\sigma}$ and let τ be the orthogonal projection of σ over v. By hypothesis $\tilde{\sigma}$ is outermost-fair. Hence by Proposition 7 the rewrite sequence σ is outermost-fair. By Proposition 8 this yields that τ is outermost-fair. By the induction hypothesis, τ ends in the normal form t. Applying now Proposition 9 yields that the rewrite sequence σ also ends in the normal form t. Since σ is the simulation of $\tilde{\sigma}$, we have that $\tilde{\sigma}$ also ends in the normal form t. This completes the proof. □

An immediate consequence of the main result of this paper is that the parallel-outermost strategy, which selects all outermost redex occurrences to be contracted simultaneously, is normalising for higher-order rewriting systems that are orthogonal and fully extended.

Acknowledgements. I thank Vincent van Oostrom for corrections and inspiring discussions. I further wish to thank Pierre-Louis Curien, Zurab Khasidashvili, Jan Willem Klop, Aart Middeldorp and the anonymous referees for comments and suggestions. The diagrams are made using the package Xy-pic of Kristoffer H. Rose.

References

[Aka93] Yohji Akama. On Mints' reduction for ccc-calculus. In M. Bezem and J.F. Groote, editors, *Proceedings of the International Conference on Typed Lambda Calculi and Applications (TLCA '93)*, pages 1–12, Utrecht, The Netherlands, March 1993. Volume 664 of Lecture Notes in Computer Science.

[Bar92] H.P. Barendregt. Lambda calculi with types. In S. Abramsky, Dov M. Gabbay, and T.S.E Maibaum, editors, *Handbook of Logic in Computer Science*, volume 2, pages 117–310. Oxford University Press, New York, 1992.

[BK86] J.A. Bergstra and J.W. Klop. Conditional rewrite rules: Confluence and termination. *Journal of Computer and System Sciences*, 32:323–362, 1986.

[CFC58] Haskell B. Curry, Robert Feys, and William Craig. *Combinatory Logic, volume I*. Studies in Logic and the Foundations of Mathematics. North-Holland Publishing Company, Amsterdam, 1958.

[HP96] Michael Hanus and Christian Prehofer. Higher-order narrowing with definitional trees. In Harald Ganzinger, editor, *Proceedings of the 7th International Conference on Rewriting Techniques and Applications (RTA '96)*, volume 1103 of *Lecture Notes in Computer Science*, pages 138–152, New Brunswick, USA, 1996.

[Hue80] Gérard Huet. Confluent reductions: Abstract properties and applications to term rewriting systems. *Journal of the Association for Computing Machinery*, 27(4):797–821, October 1980.

[Mil91] Dale Miller. A logic programming language with lambda-abstraction, function variables, and simple unification. *Journal of Logic and Computation*, 1(4):497–536, 1991.

[MN94] Richard Mayr and Tobias Nipkow. Higher-order rewrite systems and their confluence. Technical Report TUM-I9433, Technische Universität München, August 1994.

[Nip91] Tobias Nipkow. Higher-order critical pairs. In *Proceedings of the sixth annual IEEE Symposium on Logic in Computer Science (LICS '91)*, pages 342–349, Amsterdam, The Netherlands, July 1991.

[O'D77] Michael J. O'Donnell. *Computing in Systems Described by Equations*, volume 58 of *Lecture Notes in Computer Science*. Springer Verlag, 1977.

[Oos94] Vincent van Oostrom. *Confluence for Abstract and Higher-Order Rewriting*. PhD thesis, Vrije Universiteit, Amsterdam, March 1994. Available at `http://www.cs.vu.nl/~oostrom`.

[Oos96a] Vincent van Oostrom. Higher-order families. In Harald Ganzinger, editor, *Proceedings of the 7th International Conference on Rewriting Techniques and Applications (RTA '96)*, volume 1103 of *Lecture Notes in Computer Science*, pages 392–407, New Brunswick, USA, 1996.

[Oos96b] Vincent van Oostrom. Personal communication, 1996.

[OR94] Vincent van Oostrom and Femke van Raamsdonk. Weak orthogonality implies confluence: the higher-order case. In A. Nerode and Yu.V. Matiyasevich, editors, *Proceedings of the Third International Symposium on Logical Foundations of Computer Science (LFCS '94)*, volume 813 of *Lecture Notes in Computer Science*, pages 379–392, St. Petersburg, July 1994.

[Raa96] Femke van Raamsdonk. *Confluence and Normalisation for Higher-Order Rewriting*. PhD thesis, Vrije Universiteit, Amsterdam, May 1996. Available at `http://www.cwi.nl/~femke`.

Pomset Logic:
A Non-commutative Extension
of Classical Linear Logic

Christian Retoré

retore@loria.fr http://www.loria.fr/~retore
Projet Calligramme, INRIA-Lorraine & CRIN-C.N.R.S.
B.P. 101 54 602 Villers lès Nancy cedex France

Abstract. We extend the multiplicative fragment of linear logic with a non-commutative connective (called `before`), which, roughly speaking, corresponds to sequential composition. This lead us to a calculus where the conclusion of a proof is a Partially Ordered MultiSET of formulae.
We firstly examine coherence semantics, where we introduce the `before` connective, and ordered products of formulae. Secondly we extend the syntax of multiplicative proof nets to these new operations.
We then prove strong normalisation, and confluence.
Coming back to the denotational semantics that we started with, we establish in an unusual way the soundness of this calculus with respect to the semantics. The converse, i.e. a kind of completeness result, is simply stated: we refer to a report for its lengthy proof.
We conclude by mentioning more results, including a sequent calculus which is interpreted by both the semantics and the proof net syntax, although we are not sure that it takes all proof nets into account.
The relevance of this calculus to computational linguistics, process calculi, and semantics of imperative programming is briefly explained in the introduction.

Introduction

Motivation. Linear logic [10, 36, 12] succeeds in modelling computational phenomena because it is both a neat logical system and a resource sensible logic. This explains its relevance to various areas like: (1) process calculi and concurrency, e.g. [1, 4, 18, 19, 25], (2) functional programming, e.g. [1, 14], (3) logic programming, e.g. [17, 25].

In some situations *non-commutative* features would be welcome, to handle phenomena like: prefixing or sequential composition in area (1), strategy optimisation in areas (2,3). In other areas these non-commutative features are even necessary to obtain a logical model: for instance if one wants a logical description of state manipulation as appearing in the semantics of imperative programming [35] or if one wants a logical system corresponding to a categorial grammar in computational linguistics [20, 24].

One possible direction consists in leaving out the exchange rule, i.e. in working with non-commutative connectives, as in [2, 3], but it is then highly difficult to also include commutative features, which are needed too.

Principles of the calculus. Here we *extend* multiplicative linear logic with a non commutative connective which is the only possible one according to coherence semantics. This connective, called **before** $<$, is non-commutative ($A < B \not\equiv B < A$), self-dual ($(A < B)^\perp \equiv A^\perp < B^\perp$), and associative. Furthermore, its semantical definition suggests considering n-ary connectives defined by partial orders.

This led us to consider in [28, pp. 85–139], a calculus where the connectives are linear multiplicative conjunction $\otimes$, disjunction $\wp$, and **before** $<$, and where the conclusion of a proof is a partially ordered multiset of formulae. This article is devoted to the syntax of proof nets for this calculus, which is a natural extension of multiplicative proof nets enriched with the *mix* rule [8, pp. 277–278].

Results. We first describe the semantical construction which gave rise to this calculus, and next we turn our attention to the proof net syntax.

As usual, we first define proof structure, and only the proof structures enjoying a global criterion are called proof nets.

The correctness criterion simply extends the usual one, in the style of [7, pp. 193–194], but we give here a formulation which is new with respect to [28, pp. 93–100], with edge bicoloured graphs, i.e. in the style of [32, p. 8]. It is thus immediate that this criterion may be checked in polynomial time: this makes the proof net syntax a sensible syntax by itself.

We then define cut-elimination, as three local graph rewriting rules. Once we have proved that these rules preserve the correctness of the proof net, we very easily get *strong normalisation* and *confluence*.

Next we provide a denotational semantics for this calculus using the semantics defined in the beginning, and prove that each proof admits a non-trivial semantics preserved under cut-elimination. This is done by an extension of the experiment method of [10, pp. 57–60]. In other words, we show that our calculus is sound w.r.t. this semantics, seizing the opportunity of a method which does not rely on an inductive definitions of the proofs. We state a kind of converse, i.e. a kind of completeness result with respect to this semantics, but the proof which can be found in [29, pp. 19–22], is too lengthy to be given here — as opposed to the usual case which is very simple, see [33].

Finally we briefly explain that the partial orders which correspond to the n-ary connectives definable from the binary connectives exactly correspond to series parallel orders — see e.g. [37]. The formulae corresponding to series-parallel orders do not involve the **times** connective, and the **before** connective corresponds to series composition, while the **par** connective corresponds to parallel composition.

We conclude by giving some more results that we obtained on this calculus, mainly in [28, pp. 85–139], and then raise the difficult question of finding a sequent calculus *exactly* corresponding to these proof nets — a sound one, found in [28, pp. 111–122], is briefly given, but we are unable to show that it takes all the proof nets into account.

Relation to other studies in linear logic. Regarding non-commutative linear logic, we succeed in having both the usual commutative connectives **par** and **times** together with a non-commutative connective, and to endow our calculus with a simple denotational semantics.

But this calculus is also a new approach for dealing with n-ary connectives, here defined by partial orders. Regarding the unrelated attempt of [9, 7], we succeed in giving the connectives we define a denotational semantics, and a sequent calculus — even if it is not a complete one.

In fact, with respect to the two aforementioned trends, our main success is to provide a simple computational meaning to our n-ary connectives and non-commutative features. Indeed, the cuts, i.e. the computation to be performed, are also involved in the partial order, and therefore a concurrent strategy is described within the syntax itself. This fits in with the computational interpretation of linear logic of [1] and is the starting point of the use [4, 5] and [26, 27] made of this calculus.

Computational meaning of before and of partial orders. As usual cut links may be viewed as particular final **times** links, which allows us to make them appear in the partial order. This partial order may thus be viewed as a strategy for computing the proof net, which is described within the syntax itself. This is a concurrent strategy, which simply consists in first evaluating the cuts (i.e. the computations to be performed) according to this order.

Notice that our calculus relates to true concurrency rather than to CCS-like calculi where $(P|Q) = P.Q + Q.P$ using the notation of [23].

Also notice that a cut between $A \wp B$ and $A^\perp \otimes B^\perp$ reduces to a cut between A and $A^\perp$ and a cut between B and $B^\perp$ that can be done in *parallel*, while a cut between $A < B$ and $A^\perp < B^\perp$reduces to a cut between A and $A^\perp$ which is to be computed first and a cut between B and $B^\perp$ which is to be computed next — the identity $(A < B)^\perp \equiv A^\perp < B^\perp$ *without swap*, as opposed to [2, 3], is needed to allow such an interpretation.

That is the reason why $A < B$ is to be intuitively understood as *sequential composition*, as shown in [4].

In the plain logical calculus the order only describes a strategy, since we can forget (part of) it and still obtain a proof net. Nevertheless when there are proper axioms modelling some "real world" constraints, as the order appearing in these axioms merges with the order introduced by the proof net, we can no more forget parts of the order, and we are thus able to model temporal constraints, like "A ought to be done before B".

Applications. This work has already been used, on the basis of [28], to provide some solutions or insights to the aforementioned fields:

In [4, 5] the authors consider proof nets as processes, and show that the correctness of the proof net corresponds to freedom from deadlock. The connectives are interpreted as follows: $\wp$ parallel composition, $\otimes$ internal choice, and $<$ sequential composition.

In [26, 27] the author makes use of the calculus to model by local means state change in imperative programming. For instance, a buffer of type A has the type $!(A^{\perp} < A)$: thus it forces that it *first* get something of type A (write), and *then* produce something of type A (read).

In [15] the author suggests that in the *proof search as computation* paradigm the order we have on the formulae, should force some subgoals to be proved before some others.

In [21, 34], this calculus is applied to categorial grammars — see e.g. [20, 24]. We seize the opportunity that this calculus is able to handle *partial* orders instead of *linear* orders [2, 3, 20] to provide a logical treatment of linguistics phenomena hitherto absent from categorial grammars, like relatively free word order, gapping, head wrapping, as explained in [21, 34]. In such a grammar, the lexicon associates each word with a partial proof net, which contains an axiom labelled with the word. To analyse a sentence we must first make a complete proof net with the parts corresponding to the words of the sentence, i.e. check if the criterion holds (the consumption of the valencies is correct), and then check whether the induced order is included in the word order of the sentence (the order of the words is correct). Just to mention one example, we are thus able to model correctly French perception verbs: in this grammar, both *Pierre entend Marie chanter* et *Pierre entend chanter Marie* are recognised as correct sentences, but, and that is the most important, they both come out with the *same* analysis.

1 A guideline: coherence semantics

General framework Coherence spaces are a denotational semantics tightly related to linear logic, as explained in [10, 36, 12], and in particular to its proof net syntax [10, 33, 29]. Actually linear logic was even discovered through this semantics, which belongs to the world of stable semantics, introduced in [6] and which incorporates more computational behaviour than plain Scott semantics.

A **coherence space** A is a simple graph, i.e. a set endowed with a symmetric and anti-reflexive relation. The vertices are called **tokens**, and their set is called the **web** of the coherence space A, denoted by $|A|$. Given two tokens $a, a' \in |A|$, write $a \frown a'[A]$ for a and a' are adjacent, or **strictly coherent**, and $a \smile a'[A]$ for a and a' are neither adjacent nor equal, i.e. are **strictly incoherent**.

A **clique** of a coherence spaces is a set of pairwise adjacent vertices or coherent tokens. A **linear morphism** or **linear map** from a coherence space A to a coherence space B, is a relation $\ell \in |A| \times |B|$ satisfying:

$$\forall (a,b), (a',b') \in \ell \quad (a = a' \Rightarrow (b = b' \vee b \frown b'[B])) \wedge (a \frown a'[A] \Rightarrow b \frown b'[B])$$

Linear morphisms compose as relations, and coherence spaces with linear morphisms form a category. Let us write $A \equiv B$ whenever there exists a canonical linear morphism from A to B and one from B to A, one being the inverse of the other.

The interpretation of the different levels of linear logic within coherent spaces proceeds as follows:

Syntax	Semantics
formula F	**coherence space** also denoted by F
propositional variable α	arbitrary coherence space α
(n-ary) connective	(n-ary) operation on coherence spaces
proof of a formula F	**clique** of the corresponding coherence space F
proof of a sequent $A \vdash B$	linear morphism from A to B
normalisation of a proof	**equality** of the corresponding clique(s)

Figure 1

As usual a proof of a sequent is interpreted as the proof derived from it by replacing the left and right commas with the corresponding connectives, i.e. a proof of $A_1 \otimes \cdots \otimes A_n \vdash B_1 \wp \cdots \wp B_p$. Furthermore, given two coherence spaces A and B, there exists a coherence space $A \multimap B = A^\perp \wp B$, to be defined next, whose cliques correspond to linear morphisms from A to B.

Introducing before as *the* non-commutative multiplicative connective
As shown in Figure 1, the first step towards such a semantics is to interpret the formulae.

Once the interpretation of propositional variables is set, we just need to have an n-ary operation (more precisely, an n-ary functor) interpreting each n-ary connective. For instance linear negation, the unary connective $(\ldots)^\perp$, is the idempotent contravariant functor defined by:

$A^\perp$ is defined by $|A^\perp| = |A|$ and $a \frown a'[A^\perp]$ whenever $a \smile a'[A]$

$\ell^\perp = \{(b,a) \in |B| \times |A| \ / \ (a,b) \in \ell\}$

Among the connectives, the ones that map n coherence spaces $A_1, \ldots, A_n$ on a coherence space whose web is the Cartesian product $|A_1| \times \cdots |A_n|$ are said to be **multiplicative** connectives. So linear negation is a unary multiplicative connective. A connective is said to be positive whenever as a functor it is covariant in all its argument. With the help of the (contravariant) linear negation, they are the basic connectives from which the others may be defined as short hands.

Let $\heartsuit$ be a positive binary multiplicative connective. To define it we must specify according to the coherence of a and a' (in A), and to the one of b and b' (in B) the coherence of two pairs (a,b) and (a',b'). This can be pictured in a 3×3 array, but if $\heartsuit$ is positive, all the 9 cases are *a priori* filled in, but two, which are $\smile\heartsuit\frown$ and $\frown\heartsuit\smile$ (Figure 2). From this array, we observe that there only exist four binary multiplicative connectives positive in both their arguments, the two commutative ones being well-known, as shown in Figure 3.

$A\heartsuit B$		B		
		$\smile$	$=$	$\frown$
A	$\smile$	$\smile$	$\smile$	$\smile\heartsuit\frown$
A	$=$	$\smile$	$=$	$\frown$
A	$\frown$	$\frown\heartsuit\smile$	$\frown$	$\frown$

Figure 2

The multiplicative (positive) binary connectives				
$\smile\heartsuit\frown$	$\frown\heartsuit\smile$	commutative	notation	name
$\frown$	$\frown$	yes	$A \wp B$	**par**
$\smile$	$\smile$	yes	$A \otimes B$	**times**
$\frown$	$\smile$	no	$A < B$	**before**
$\smile$	$\frown$	no	$B < A$	"reverse **before**"

Figure 3

Regarding the two commutative connectives, they are known to be associative and to enjoy the De Morgan laws, which turn one into the other: $(A \otimes B)^{\perp} \equiv A^{\perp} \wp B^{\perp}$ and $(A \wp B)^{\perp} \equiv A^{\perp} \otimes B^{\perp}$. What are the corresponding properties of **before**, and what is its relation to its commutative companions? A mere computation shows that:

Proposition 1. *The* **before** *connective is:*

- non-commutative $A < B \not\equiv B < A$
- self dual $(A < B)^{\perp} \equiv A^{\perp} < B^{\perp}$
- associative $A < (B < C) \equiv (A < B) < C$
- *regarding linear implication it lies* in between **par** and **times**, *i.e. there exists a canonical linear morphism from $A \otimes B$ to $A < B$ and one from $A < B$ to $A \wp B$. — the relation(s) defining these linear maps simply being the identity relation:* $\{((a, b), (a, b))/(a, b) \in |A \times B|\}$.

Notice that coherence according to **before** may be defined in a lexicographic manner: $(a, b) \frown (a', b')[A < B]$ iff $(a \frown a'[A] \wedge b = b') \vee b \frown b'[B]$. This suggests introducing the following n-ary multiplicative connectives: let $(A_i)_{(i \in I)}$ be a family of coherence spaces ordered by a partial order u, written $A_k < A_l[\mathrm{u}]$. We define the **ordered product** of this ordered family as the coherence space $\prod_{\mathrm{u}} A_i$ whose web is the Cartesian product of the webs of the A_i's — $|\prod_{\mathrm{u}} A_i| = |A_1| \times \ldots \times |A_n|$ — and whose strict coherence is defined by:

$$(a_1, \ldots, a_n) \frown (a'_1, \ldots, a'_n) \left[\textstyle\prod_{\mathrm{u}} A_i\right] \text{ iff } \exists i \,.\, a_i \frown a'_i[A_i] \wedge \forall A_j > A_i[\mathrm{u}]\ a_j = a'_j$$

Notice that $\prod_{\varnothing} A, B \equiv A \wp B$ while $\prod_{A<B} A, B \equiv A < B$.

The next sections present proofs whose semantical interpretations will be cliques of these ordered products of coherence spaces.

2 Language and sequents

Without any further structure on the conclusions of a proof, or sequent, there is no way to introduce this non-commutative connective **before**— there are only two possible multiplicative rules, which are the usual rules for **times** and **par** of linear logic.

As it is suggested by the semantics above to work with partially ordered multiset of formulae, let us look for a logical calculus whose conclusions will be ordered multisets of formulae of $\mathcal{F} ::= \mathcal{P} \mid \mathcal{F}^{\perp} \mid \mathcal{F} \wp \mathcal{F} \mid \mathcal{F} < \mathcal{F} \mid \mathcal{F} \otimes \mathcal{F}$, where $\mathcal{P}$ is a set of propositional variables.

On the set $\mathcal{F}$ of formulae, we have the following De Morgan laws:
$(A^{\perp})^{\perp} \equiv A \quad (A \wp B)^{\perp} \equiv A^{\perp} \otimes B^{\perp} \quad (A < B)^{\perp} \equiv A^{\perp} < B^{\perp} \quad (A \otimes B)^{\perp} \equiv A^{\perp} \wp B^{\perp}$.

Therefore, as is usual in the theory of proof nets, we shall only consider formulae up to De Morgan equivalence. Indeed, each formula of $\mathcal{F}$ has a unique representative in the following set of formulae:

$$\mathcal{M} ::= \mathcal{P} \mid \mathcal{P}^{\perp} \mid \mathcal{M} \wp \mathcal{M} \mid \mathcal{M} < \mathcal{M} \mid \mathcal{M} \otimes \mathcal{M}$$

Consequently, we shall only consider formulae of $\mathcal{M}$, and $F^{\perp}$ should be understood as the unique formula F' of $\mathcal{M}$ which is equivalent to the formula $(F)^{\perp}$ of $\mathcal{F}$.

Another familiar property underlined by proof nets, is that a cut may be viewed as a **times** $K \otimes K^{\perp}$ between the two dual formulae K and $K^{\perp}$ that vanish in a cut rule.[1] Here we shall use this view of cuts as **times** formulae to keep a track of these cuts: thus the partial order on the formulae of a sequent may also involves the cuts, i.e. the computations to be performed.

The conclusion of a proof will be a partially ordered multi-set of formulae and cuts written:

$$\vdash A_1, \ldots, A_n, G_1^{\bullet}, \ldots, G_p^{\bullet}[u] \quad \text{with:}$$

- $A_1, \ldots, A_n$ being formulae of $\mathcal{M}$
- $G_1^{\bullet} = X_1 \otimes X_1^{\perp}, \ldots, G_p^{\bullet} = X_p \otimes X_p^{\perp}$, being cuts — where $\forall i \in [1,p]\ X_i \in \mathcal{M}$.
- u being an order on the multiset $\{A_1, \ldots, A_n, G_1^{\bullet}, \ldots, G_p^{\bullet}\} \subset \mathcal{M}$

We now define a proof syntax dealing with such conclusions, in the proof net style, because this calculus is a simpler extension of the proof net syntax than of the sequent syntax.

3 Ordered proof nets

In this section we introduced ordered proof nets in the framework developed in [32] for the usual multiplicative calculus, by extending the definition of a R&B-graph to the directed case.

We first present ordered proof structures from a kind of sub-formula trees, R&B-trees — this is the most intuitive definition — , and then define them à la Girard with links, — and that will be more convenient for the proofs in the next sections.

Directed R&B-graphs A **R&B-graph** $G = (V; B, R)$ is an edge-bi-coloured graph such that:

[1] To be more precise, we should apply some second order existential quantification to this **times** formula to obtain $\exists X.X \otimes X^{\perp}$ which is equivalent to $\perp$.

- V is a set of **vertices**
- R is a set of ordered pairs $(x, y) \in V^2$ such that $x \neq y$ — in case we have both (x, y) and (y, x) in R we speak of the **R-edge** $x - y$; in case we have $(x, y) \in R$ and $(y, x) \notin R$ we speak of the **R-arc** $x \to y$.
- B is a set of ordered pairs $(x, y) \in V^2$ such that
  - $x \neq y$ — no loop
  - $(x, y) \in B \Rightarrow (y, x) \in B$ — B is a set of edges, as opposed to arcs, and the **B-edge** $\{(x, y), (y, x)\}$ will be simply denoted by $x - y$
  - $\forall x \exists! y \ (x, y) \in B$ — B is a perfect matching of the full graph including B and R edges

These R&B-graphs clearly can be pictured as edge-bicoloured graphs with vertices V in which B-edges are Bold (or Blue), an R-arcs or R-edges are Regular (or Red). In a R&B-graph, an **alternating path** p of length n from x_0 to x_n is a sequence of n consecutive arcs $p = (x_0, x_1)(x_1, x_2) \cdots (x_{n-1}, x_n)$ alternatively in B and in R, that is to say $(x_{i-1}, x_i) \in R \Rightarrow (x_i, x_{i+1}) \in B$ and $(x_{i-1}, x_i) \in B \Rightarrow (x_i, x_{i+1}) \in R$. An alternating path is said to be **elementary**, whenever no vertex appears more than twice in its formal expression. In this case we speak of an **æ-path**. In case $x_0 = x_n$ and n is even we speak of an alternating elementary circuit, **æ-circuit** for short.

Ordered proof nets as directed R&B-graphs

Definition 2. Given a formula C, we defines its R&B-trees as edge bicoloured graphs.[2] For every formula C, the discrete graph consisting of a vertex C° is a R&B-tree of C, with root C, and leaf C. If C is a non atomic formula, namely $C = A * B$ with $* \in \{\wp, \otimes, <\}$, and if $T(A)$ and $T(B)$ are respectively R&B-trees of A and B, then $T(A) * T(B)$, defined in Figure 4, is a R&B-tree of C, with root C, and both the leaves of $T(A)$ and $T(B)$ as leaves.

C°	$T(A) \wp T(B)$	$T(A) < T(B)$	$T(A) \otimes T(B)$
$\overset{\circ}{C}$	$A \wp B$	$A < B$	$A \otimes B$

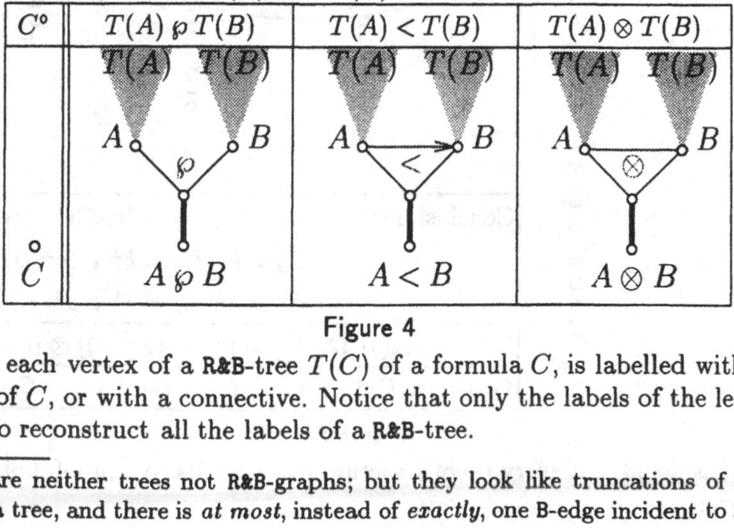

Figure 4

Thus each vertex of a R&B-tree $T(C)$ of a formula C, is labelled with a subformula of C, or with a connective. Notice that only the labels of the leaves are needed to reconstruct all the labels of a R&B-tree.

---

[2] They are neither trees not R&B-graphs; but they look like truncations of the subformula tree, and there is *at most*, instead of *exactly*, one B-edge incident to a vertex.

Definition 3. A proof structure with $C_1, \ldots, C_n, G_1^\bullet, \ldots, G_p^\bullet$ as conclusions and cuts — thus $G_i = X_i \otimes X_i^\perp$ — with order u on its conclusions and cuts is a R&B-graph which consists of:

- a family of R&B-trees $T(C_1), \ldots, T(C_n), T(X_1) \otimes T(X_1^\perp), \ldots, T(X_p) \otimes T(X_p^\perp)$ where $T(X)$ denotes a R&B-tree of X. This part represents the syntactic forest of the sequent $C_1, \ldots, C_n, X_1 \otimes X_1^\perp, \ldots, X_p \otimes X^\perp$
- a family of B-edges, called *axioms*, each of them linking two leaves being the negation one of the other, in such a way that for each leaf there exists exactly one axiom incident to it.
- a family of R-*arcs*, representing the *order*, i.e. there is one such R-arc from a conclusion or cut X to another Y whenever $X < Y[\mathrm{u}]$
- a special mark, •, on the roots of the $T(G_i)$ to make a distinction between a cut and a conclusion $X \otimes X^\perp$ which is not considered as a cut.

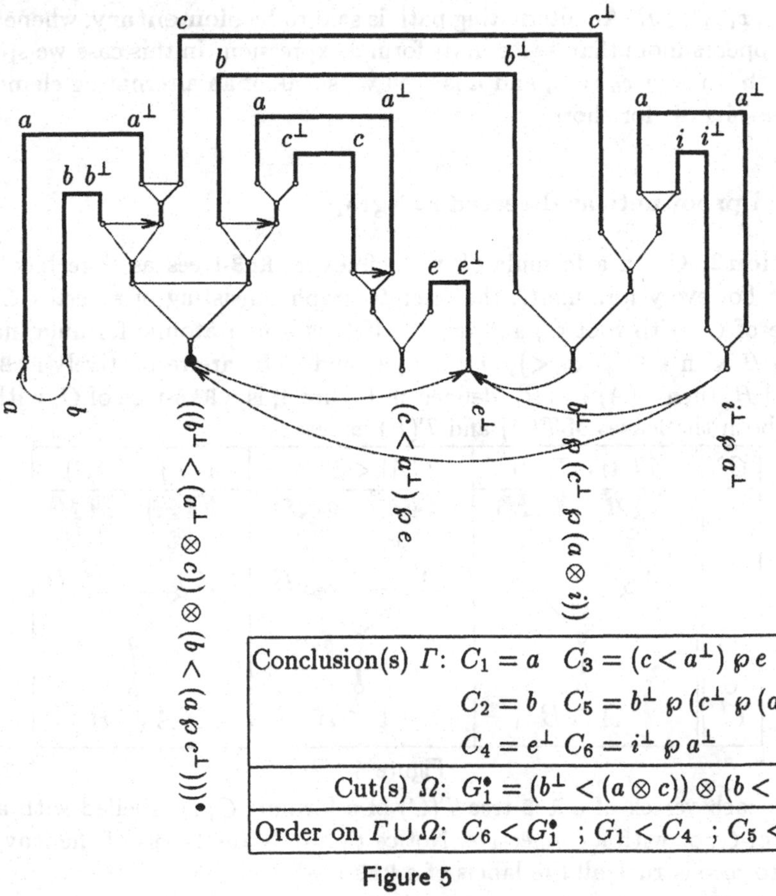

Conclusion(s) Γ:	$C_1 = a \quad C_3 = (c < a^\perp) \wp e$ $C_2 = b \quad C_5 = b^\perp \wp (c^\perp \wp (a \otimes i))$ $C_4 = e^\perp \quad C_6 = i^\perp \wp a^\perp$
Cut(s) Ω:	$G_1^\bullet = (b^\perp < (a \otimes c)) \otimes (b < (a \wp c))$
Order on $\Gamma \cup \Omega$:	$C_6 < G_1^\bullet \;;\; G_1 < C_4 \;;\; C_5 < C_4$

Figure 5

Now, as usual, not every proof structure corresponds to a proof, but only the proof nets:

Definition 4. A **proof net** is a proof structure which contains no æ-circuit. [3]

The example of figure 5 is a proof net but if the conclusion $C_5 = b^\perp \wp (c^\perp \wp (a \otimes i))$ were to be replaced with $C_5' = b^\perp < (c^\perp \wp (a \otimes i))$, there would be an æ-circuit: $[b] -_{\mathtt{B}} [b^\perp] \rightarrow_{\mathtt{R}} [<] -_{\mathtt{B}} [b^\perp < (c^\perp \wp (a \otimes i))] -_{\mathtt{B}} [\wp] -_{\mathtt{R}} [c^\perp] -_{\mathtt{B}} [c] -_{\mathtt{R}} [\otimes] -_{\mathtt{B}} [a^\perp \otimes c] -_{\mathtt{R}} [<] -_{\mathtt{R}} [b^\perp < (a^\perp \otimes c)] -_{\mathtt{R}} [b < (a \wp c^\perp)] -_{\mathtt{B}} [<] -_{\mathtt{R}} [b]$

Here is an important proposition, albeit easy, which means that the proof net syntax is a sensible syntax by itself:

Proposition 5. *There is a polynomial (cubic) algorithm which checks whether a proof structure is a proof net.*

Proof. Let X and Y be two vertices in a given proof structure. Checking whether there exists an æ-path from X to Y starting with its unique incident B-edge is a standard breadth search algorithm. Each B-edge is visited once in each direction, and it is thus quadratic in twice the number of B-edges, i.e. in the number of vertices (the B-edges are a perfect matching of the graph). If we take $Y = X$ and repeat this for any vertex X, we get a cubic algorithm which checks the absence or presence of æ-circuit.

Let us now define the **links** of the ordered proof structures, i.e. the bricks they are made of, and their premises and conclusions. A $*$-link, with $* \in \{\otimes, \wp, <\}$ is the **R&B**-graph on four vertices A, B, $*$, $A * B$, which appears in figure 4; A and B are said to be its premises, and $A * B$ is said to be its conclusion. An axiom link is an axiom, i.e. a B-edge whose end vertices are A and $A^\perp$. This link has no premise and two conclusions, namely A and $A^\perp$. An ordered proof structure may also be defined as a set of links such that each formula is the conclusion of exactly one link, and the premise of at most one conclusion, plus a family of R-arcs which is an order between conclusions, i.e. between the formulae which are not the premise of any link.

4 Cut elimination

We now define cut-elimination as a local graph rewriting system which turns a proof net into a proof net, in such a way that *the restriction of the order to the conclusions is preserved under cut elimination.* A cut is a `times` link between two dual formulae K and $K^\perp$. Each of the formulae K and $K^\perp$ is the conclusion of a unique link, say k and $k^\perp$. There are three elementary steps of cut elimination to be described, according to the nature of the links k and $k^\perp$:

AX/? k or $k^\perp$ is an **axiom** link

BF/BF k and $k^\perp$ are **before** links

TS/PAR k is a **times** link and $k^\perp$ a **par** link.

[3] Beware that the adjective *elementary* in *alternating elementary circuit* is necessary. For instance, there exist proof nets of the usual multiplicative calculus, hence of this extension that we are presently defining, which contain alternating circuits.

The AX/? **elementary step** Here is the picture of this elementary step. The vertex X_i (resp. Y_j) is a conclusion or a cut which is, according to the order between conclusions and cuts, below (resp. above) the cut we are reducing and there may be several such conclusions X_i (resp. Y_j). We first suppress the **`axiom`** link and then the **`cut/times`** link and its incident R-arcs. We then identify the vertices labelled with A.
The order on the conclusions and cuts in the reduct simply is the restriction of the order in the redex to the remaining conclusions and cuts.

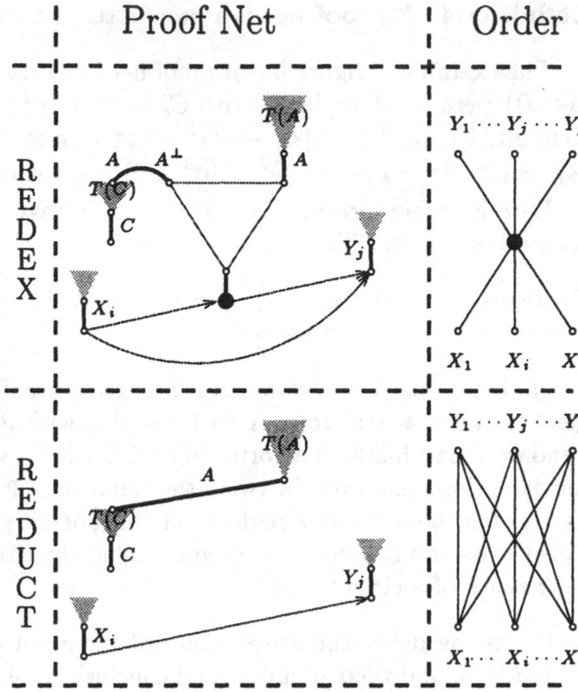

The BF/BF **elementary step** Here is the picture of this elementary step. The vertex X_i (resp. Y_j) is a conclusion or a cut which is, according to the order on conclusions and cuts, below (resp. above) the cut we are reducing, and there may be several such conclusions X_i (resp. Y_j).
In the order, the cut $\bullet$ is split into two cuts $\bullet_1$ and $\bullet_2$ with $\bullet_1 < \bullet_2$, which occupy the place of $\bullet$.

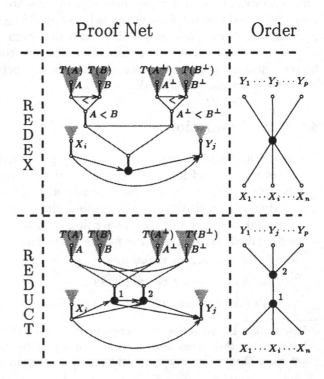

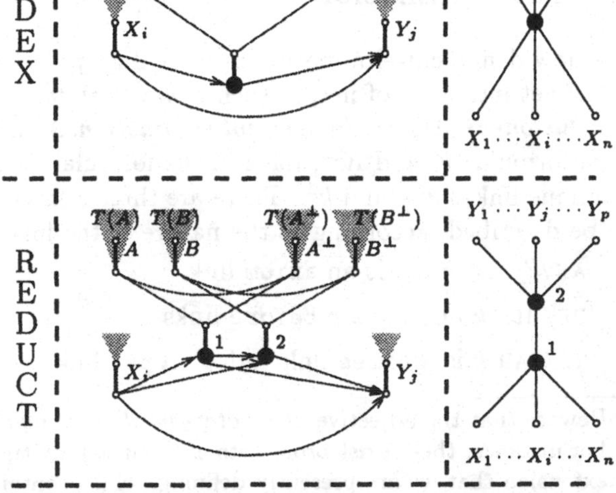

The TS/PAR elementary step Here is the picture of this elementary step. The vertex X_i (resp. Y_j) is a conclusion or a cut which is, according to the order on conclusions and cuts, below (resp. above) the cut we are reducing, and there may be several such conclusions X_i (resp. Y_j).
In the order, the cut $\bullet$ is split into two cuts $\bullet_1$ and $\bullet_2$, unordered, which occupy the place of $\bullet$.

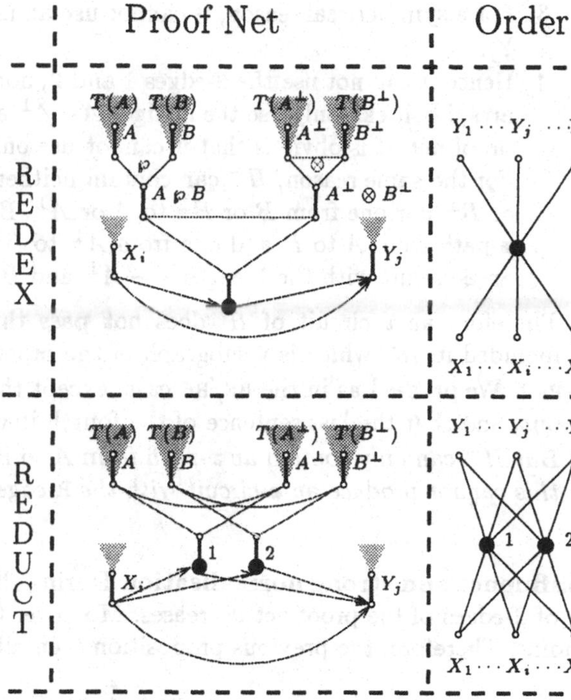

Proposition 6. *The elementary steps* AX/?, TS/PAR, *and* BF/BF *preserve the absence of* æ*-circuit, as well as the order between the remaining conclusions and cuts.*

Proof. The last part of the statement is obvious for all elementary steps, from the Hasse diagrams of the orders of the redex and the reduct.

Let us call Π the original proof net and Π' its reduct. We now show that if Π contains no æ-circuit, so does Π', for each elementary step.

AX/? Observe that any æ-path in the reduct defines an æ-path (whose endings have the same colour) in the reduct.

BF/BF Let Π^- be the part of Π which is common to Π', and let π' the subgraph of Π' consisting in the two B-edges 1 and 2 incident with $\bullet^1$ and $\bullet^2$, and their adjacent R-edges. We assume there is an æ-circuit in Π'.

1. If it uses the R-arc $\bullet^1 \rightarrow \bullet^2$, then it uses:
   - (either the R-edge from A to 1 or the R-edge from $A^\perp$ to 1), and
   - the B-edge 1, the R-arc $\bullet^1 \rightarrow \bullet^2$, the B-edge 2, and (either the R-edge from B to 2 or the R-edge from $B^\perp$ to 2).

   Because Π is a proof net, Π^- can contain neither any æ-path from B to A, nor one from $B^\perp$ to $A^\perp$, nor one from B to $A^\perp$, nor one from $B^\perp$ to A. Hence there is no æ-circuit using $\bullet^1 \rightarrow \bullet^2$.
2. If it uses an R-edge from some X_i to $\bullet^1$ or $\bullet^2$, there should be in Π^- an æ-path from a vertex of π' to X, but this is impossible because Π is a proof net.

3. For a symmetrical reason, it cannot use an R-edge from $\bullet^1$ or $\bullet^2$ to some Y_j.
4. Hence it can not use the B-edges 1 and 2, nor their adjacent R-edges and arcs, i.e. it can only use the R-edges $A - A^\perp$ and $B - B^\perp$. Because Π is a proof net, it is obvious that it cannot use only one of these two R-edges. For the same reason, Π^- can contain neither a path from A or $A^\perp$ to B or $B^\perp$ nor one from B or $B^\perp$ to A or $A^\perp$. But Π^- can only contain an æ-path from A to B and one from $A^\perp$ to $B^\perp$, and this cannot produce an æ-circuit with the R-edges $A - A^\perp$ and $B - B^\perp$.

Therefore an æ-circuit of Π' does not pass through π', hence should be included in Π^- which is a subgraph of the proof net Π, contradiction.

TS/PAR We proceed as in the BF/BF case, except that the first item can be left out, and that the last sentence of the fourth item should be replaced with: *But Π^- can only contain an æ-path from A to B and one from B to A, and this cannot produce an æ-circuit with the* R-*edges $A - A^\perp$ and $B - B^\perp$.*

Confluence and strong normalisation During all elementary steps, the number of B-edges of the proof net decreases. Moreover the redex configurations are disjoint. Therefore, the previous proposition 6 entails the following:

Theorem 7. *The calculus of ordered proof nets enjoys strong normalisation and confluence: a proof net with conclusions and cuts $F_1, \ldots, F_n, G_1^\bullet, \ldots, G_p^\bullet$ ordered by* u *reduces to a cut free proof net with conclusions $F_1, \ldots, F_n$ ordered by* $\mathrm{u}|_{F_1,\ldots,F_n}$.

5 Denotational semantics for this calculus

We compute the semantics $\| \Pi \|$ of a proof net Π by extending the method of experiments of [10, pp. 57–60] or of [29, 33] to this calculus.

We assume that we have an interpretation, i.e. that any atomic formula is associated with a coherence space. Consequently, to each formula F is associated a coherence space also denoted by F, by means of the interpretation of the connectives given in the first section.

Definition 8. Let Π be proof net with conclusion$\vdash F_1, \ldots, F_p, F_1^\bullet, \ldots, F_n^\bullet[\mathrm{u}]$. An **experiment** of Π is a labelling of the sub-formulae appearing in the proof net satisfying:

- the label of a sub-formula F is a token u of the corresponding coherence space F ($u \in |F|$).
- the conclusion $A * B$ with $* \in \{\wp, \otimes, <\}$ of a link $\{\wp, \otimes, <\}$ has the label (t_1, t_2) iff its two premises A and B respectively have the labels t_1 and t_2
- the two conclusions A and $A^\perp$ of an axiom link have the same label — this makes sense because $|A| = |A^\perp|$.

Hence an experiment of Π is completely defined by associating a point of the web $|A| = |A^{\perp}|$ to every axiom $A - A^{\perp}$ of Π.

The experiment is said to **succeed** whenever: the label (t_1, t_2) of a cut $F_j^{\bullet} = E \otimes E^{\perp}$ satisfies $t_1 = t_2$.

Finally, the **semantics** of Π is the set $\|\Pi\|$ of the tuples $(t_1, ..., t_p)$ such that there exists a successful experiment according to which the conclusions of the proof net, i.e. $F_1, ..., F_p$, are respectively labelled $t_1, ..., t_p$.

Proposition 9. *Let Π be a proof net with conclusion $\vdash F_1, ..., F_p, F_1^{\bullet}, ..., F_n^{\bullet}[\mathrm{u}]$, let $\mathfrak{v} = \mathrm{u}|_{F_1,...,F_p}$. Then $\|\Pi\|$ is a clique of $\prod_{\mathfrak{v}} F_1, ..., F_p$.*

Proof. We assume that two experiments yields two strictly incoherent tuples $(t_1, ..., t_p)$ and $(t'_1, ..., t'_p)$ of $\prod_{\mathfrak{v}} F_1, ..., F_p$, and show that Π contains an æ-circuit. The method consists in extending an æ-path, until we obtain an æ-circuit.

We write $A : \frown$ (resp. $\smile$) whenever the labels of A according to the two experiments are strictly coherent (resp. incoherent) in A.

The æ-path to be constructed has a marking **up** or **down**, and throughout its construction fulfils the following requirement:

1. the path always ends on a sub-formula of a conclusions or cut, which may be a conclusion but not a cut.
2. if the mark is **up** (resp. **down**) the path ends on a formula $F : \smile$ (resp. $F : \frown$)
3. when the path uses an edge of a **par** or **before** branching from a premise to the conclusion (resp. from the conclusion to a premise), the two experiments are strictly coherent (resp. strictly incoherent) both in the premise and the conclusion.

We start with the empty path $t_i \smile t'_i : F_i$, and marking **up** — our assumption makes sure there is such an F_i. Assuming we already have built a path satisfying (1), (2) and (3), we review all its possible endings and markings.

Here are three cases, the others being similar — see [29, pp. 24–29], for a complete description of all the cases:

- *The path ends on the A premise of $A < B$ with marking* **down** So $A : \frown$ (2). If $B : \smile$ we extend our path using the R-arc from A to B and put the marking **up**. If $B : \frown$, if the path already used the R-edge $A < B - B$ it used it from $A < B$ to B (3), and using the arc $A \to B$ we have an æ-circuit. Otherwise we extend our æ-path with the R-edge $A - A < B$ and keep the marking **down**.
- *The path ends in a conclusion, with marking* **down**. So $A_q : \frown$ (2), and there is an arc from A_q to some $A_r : \smile$. Indeed, if $(a_1, ..., a_p) \smile (a'_1, ..., a'_p)$ $[\prod_{\mathfrak{v}} A_i]$ and $a_k \frown a'_k$ then $\exists A_l > A_k[\mathrm{u}]\ a_l \smile a'_l[A_l]$. We extend our æ-path using this arc, and put the marking **up**.
- *The path ends on the premise E of a cut $F_s^{\bullet} = (E \otimes E^{\perp})^{\bullet}$, with marking* **down**. So $E : \frown$ (2), and, because the experiments succeed, $E^{\perp} : \smile$. We extend the elementary alternating path with the R-edge $E - E^{\perp}$, with marking **up** — so (1) is still fulfilled.

It is easily seen that (1), (2) and (3) are preserved while extending the path.

Theorem 10. *Every proof net Π has a non-trivial semantics preserved by cut elimination.*

Proof. When Π is a cut-free proof net, any of its experiments succeeds, so the square of the cardinal of $||\Pi||$ is the product of the cardinals of the webs corresponding to leaves. So the semantics of a cut-free proof net is always non-trivial.

Let Π' be a proof net obtained from Π by one of the elementary cut-elimination steps; the *successful* experiments of Π and Π' clearly are in a one-to-one correspondence. Therefore we can speak of a denotational semantics: $||\Pi|| = ||\Pi'||$.

The theorem follows from these two remarks.

In [29] we have established the "strong" converse which follows. It expresses a kind of completeness of coherence semantics w.r.t. ordered proof nets. The proof, which is very simple for the usual multiplicative calculus enriched with the *mix* rule, see [33], is, in the case of ordered proof structures and nets too lengthy to be given here; one should refer to [29, pp. 19–22].

Theorem 11. *There exits a four-token coherence space Z such that, when we interpret each atomic formula by Z, a proof structure Π is a proof net iff $||\Pi||$ is a clique.*

6 Other results on this calculus

We mention here some more results which may be of interest to the reader.

η-rule The η-rule holds for this calculus [28, p. 104]: thus the axiom links may be restricted to atomic formulae.

Extension to full linear logic There is no difficulty in extending this calculus to the modalities "?" and "!" and additive connectives, using proof nets with boxes [10, pp. 43–46] for "!" and "&". The order outside a !-box is the same as inside. The order on both proof nets included in a &-box is asked to be the same, and is the one on the outside of the &-box.

Relation to the usual multiplicative calculus enriched with the mix *rule* This calculus is a faithful extension of proof nets with *mix* [8, pp. 277–278]. Firstly the ordered proof nets without any `before` link exactly are the *mix* proof net. Secondly, in a proof net including some `before` links, each `before` link may be turned into a `par` or `times` link in order to get a correct *mix* proof net. It is trivial for cut-free proof nets, turning `before` links into `par` links, but it is trickier for non-cut-free proof nets, see [28, pp. 104–106].

A modality corresponding to **before** Answering with respect to **before** a question of [11, p. 257], we found, in the category of coherence spaces a self-dual modality which enables contraction w.r.t. **before** on both sides [30, 31]. Its syntax, now under study, is intended for a constructive treatment of classical logic.

7 The problem of finding a sequent calculus

Finding a sequent calculus corresponding to proof nets dealing with n-ary connectives is not easy — e.g. for the n-ary connectives of [7, pp. 196–197] there is no sequent calculus at all.

Because of the De Morgan laws, we can limit ourselves to a calculus of right handed sequent. For this pomset logic, we only have an imperfect solution, the calculus of ordered sequents presented in [28, pp. 111–122]. The axiom is $\vdash A, A^{\perp}[\emptyset]$ — thus proofs start with an empty order on the conclusions and cuts — and the other rules are the following:

$$\frac{\vdash \Gamma, A, B[\mathfrak{u}]}{\vdash \Gamma, A \wp B[\mathfrak{u}']}\wp \qquad \frac{\vdash \Gamma, A, B[\mathfrak{u}]}{\vdash \Gamma, A < B[\mathfrak{u}']}< \qquad \frac{\vdash \Gamma[\mathfrak{u}] \quad \vdash \Delta[\mathfrak{v}]}{\vdash \Gamma, \Delta[\mathfrak{w}]}\,mix \qquad \frac{\vdash \Gamma, A[\mathfrak{u}] \quad \vdash \Delta, B[\mathfrak{v}]}{\vdash \Gamma, \Delta, A \otimes B[\mathfrak{w}]}\otimes$$

In the $\wp$- and $<$-rule the order $\mathfrak{u}'$ is obtained from $\mathfrak{u}$ by identifying A and B, to a formula respectively called $A \wp B$ or $A < B$. But, for applying the $\wp$-rule, A and B are asked to have exactly the same predecessors and successors, while, for applying a $<$-rule, A is asked to be the only predecessor of B and B is asked to be the only successor of A.

In the *mix* and $\otimes$-rule, $\mathfrak{w}$ is any order such that $\mathfrak{w}|_{\bar{\mathfrak{u}}} = \mathfrak{u}$ and $\mathfrak{w}|_{\bar{\mathfrak{v}}} = \mathfrak{v}$ (where $\bar{\mathfrak{u}}$ denotes the domain of the order $\mathfrak{u}$, e.g. Γ, A in the case of the $\otimes$-rule) which satisfies the following property:

$$\forall U, U' \in \bar{\mathfrak{u}}\ \forall V, V' \in \bar{\mathfrak{v}} \quad U{<}V'[\mathfrak{w}] \wedge V{<}U'[\mathfrak{w}] \ \Rightarrow\ U{<}U'[\mathfrak{u}] \vee V{<}V'[\mathfrak{v}]$$

— in the $\otimes$-rule case, one should read $A \otimes B$ as A (resp. B) when it is considered as a formula in $\bar{\mathfrak{u}}$ (resp. $\bar{\mathfrak{v}}$). Furthermore, for the $\otimes$-rule, if $U{<}V[\mathfrak{w}]$ (resp. $U{>}V[\mathfrak{w}]$) with $U \in \bar{\mathfrak{u}}$, $V \in \bar{\mathfrak{v}}$ then $U{<}A[\mathfrak{u}] \vee B{<}V[\mathfrak{v}]$, (resp. $U{>}A[\mathfrak{w}] \vee B{>}V[\mathfrak{w}]$).

The proofs of this sequent calculus translate into ordered proof *nets* and are interpreted by coherence semantics; it is even showed to be the largest "standard" sequent calculus with these properties. Nevertheless, we are unable to prove that every ordered proof net does actually correspond to a proof of this sequent calculus. The techniques of [32] give the hope of a better solution, and we will present an outcome of this study, which enlightens the meaning of **before**, and then explain where the difficulty lies.

Definable connectives: series parallel orders A first interesting step is to look at the definable n-ary connectives, i.e. the ones which behaves like formulae. Let us explain what this means by analogy with the usual multiplicative calculus, whose only connectives are $\wp$ and $\otimes$. Given any proof structure Π with conclusions $A_1, \ldots, A_p, B_1, \ldots, B_n$, there is a formula $\mathcal{F}(A_1, \ldots, A_p)$, namely $A_1 \wp \cdots \wp F_p$ such that Π is a proof net iff Π' is, where Π' is the proof structure with the same axiom links but where the conclusions $A_1, \ldots, A_n$ are replaced by the single conclusion $\mathcal{F}(A_1, \ldots, A_n)$ — Π' obtained from Π by writing the (R&B) formula tree of $\mathcal{F}$ below the conclusions A_i. Moreover, the formula $\mathcal{F}$ is unique up to the commutativity and associativity of $\wp$.

Here, assuming the order on the A_i's is u, and that no A_i is related to a B_j by the order, we have a similar result, which enlightens the meaning of the connectives. In [28, pp. 106–109], we have shown that:

- The formula $\mathcal{F}$ does not use the **times** connective.
- The possibility of writing a **par** link with A_1 and A_2 as hypotheses exactly corresponds to A_i and A_j having the same predecessors and successors.
- The possibility of writing a **before** link with A_1 and A_2 as hypotheses exactly corresponds to A_i being the only predecessor of A_j and A_j the only successor of A_i
- In both cases the order which has to be taken on $A_1 \wp A_2, A_3, \ldots, A_p$ or $A_1 < A_2, A_3, \ldots, A_p$ is obtained by identifying A_1 and A_2.

So the question turns out to be: when does u reduce to a single point by these contractions? We have proved that it is the case iff u is an N-free order.[4] The sequence of contractions, i.e. the way of writing $\mathcal{F}$ from $\wp$ and $<$, is unique up to the commutativity of $\wp$ and to the associativity of $\wp$ and $<$.

But the N-free orders are known to be series parallel orders [37, pp. 310–311], i.e. to belong to the smallest class of orders closed under disjoint union and ordinal sum, and reading our formula from outermost connectives to innermost ones we obtain the decomposition of u as a series parallel order, $\wp$ corresponding to parallel composition, and $<$ to serial composition.

The funny thing about it is that these orders were introduced to model concurrency constraints, which exactly correspond to the inner strategy described by the order. Nevertheless our calculus is able to deal with *any* constraint order, although the constraints expressed by a non-series-parallel order are harder to understand.

Towards a complete sequent calculus Restricting ourselves to definable connectives seems a sensible approach, but the connective **times** does not fit in this setting. This is why we are now thinking of a more general calculus, dealing with a class of relations which include series-parallel orders and series parallel graphs: thus the connective **times**, corresponding to undirected series composition, fits in. This leaves the ordered proof nets almost unchanged, but allows a unary rule for the connective **times**, ruled by the relation.

This approach looks very promising. For the usual multiplicative calculus, it has already brought some results, [32]. But this non-commutative case, which involves directed graphs, is more difficult to handle than the usual one for which numerous proofs are known [10, 7, 8, 32]. Indeed, connectivity questions are much more complex for directed graphs (see e.g. [16]) and, even in [22], there is

[4] The Hasse diagram of the restriction of u to four points never is N.

nothing about matchings for directed graphs... Furthermore, we are not simply looking for a bridge, as in the usual calculus, but for an edge cut-set with a given property.

Roughly speaking, we are looking for an inductive definition of the graphs with a perfect matching which possesses no æ-circuit; when the graphs are not directed it is done [32], but otherwise we just obtained some partial results, and we think it is too early to tell more than these hints.

Acknowledgements. Thanks to Jean-Yves Girard for his suggestions and encouragements, to Jacques van de Wiele for helpful advice, and to Maurice Pouzet who pointed to my attention the already existing work on series-parallel orders. Many thanks to Roger Hindley, François Lamarche and the referees for their numerous corrections.

References

1. S. Abramsky. Computational interpretations of linear logic. *Theoretical Computer Science*, 111:3–57, 1993.
2. V. M. Abrusci. Phase semantics and sequent calculus for pure non-commutative classical linear logic. *Journal of Symbolic Logic*, 56(4):1403–1451, December 1991.
3. V. M. Abrusci. Non- commutative proof nets. In [13], pages 271–296, 1995.
4. A. Asperti. A linguistic approach to dead-lock. Technical Report LIENS 91-15, Dép. Maths et Info, Ecole Normale Supérieure, Paris, October 1991.
5. A. Asperti and G. Dore. Yet another correctness criterion for multiplicative linear logic with mix. In A. Nerode and Yu. Matiyasevich, editors, *Logical Fundations of Computer Science*, volume 813 of *Lecture Notes in Computer Science*, pages 34–46, St. Petersburg, July 1994. Springer Verlag.
6. G. Berry. *Modèles complètements adéquats et stables des lambda calculs typés.* Thèse d'état, spécialité sciences mathématiques, Université Paris 7, 1979.
7. V. Danos and L. Regnier. The structure of multiplicatives. *Archive for Mathematical Logic*, 28:181–203, 1989.
8. A. Fleury and C. Retoré. The mix rule. *Mathematical Structures in Computer Science*, 4(2):273–285, June 1994.
9. J.-Y. Girard. Multiplicatives. *Rendiconti del Seminario dell'Universitá é Politecnico Torino*, pages 11–33. October 1986. Special issue on Logic and Computer Science.
10. J.-Y. Girard. Linear logic. *Theoretical Computer Science*, 50(1):1–102, 1987.
11. J.-Y. Girard. A new constructive logic: classical logic. *Mathematical Structures in Computer Science*, 1(3):255–296, November 1991.
12. J.-Y. Girard. Linear logic: its syntax and semantics. In [13], pages 1–42, 1995.
13. J.-Y. Girard, Y. Lafont, and L. Regnier, editors. *Advances in Linear Logic*, volume 222 of London Mathematical Society Lecture Notes. Cambridge University Press, 1995.
14. G. Gonthier, J.-J. Lévy and M. Abadi. The geometry of optimal lambda reduction. In *Principles Of Programming Languages*, pages 15–26, 1992.
15. A. Guglielmi. Concurrency and plan generation in a logic programming language with a sequential operator. In P. van Hentenryck, editor, *International Conference on Logic Programming*, pages 240–254, Genova, 1994. M.I.T. Press.
16. Y. O. Hamidoune. Connexité dans les graphes orientés. *Journal of Combinatorial Theory (B)*, 30:1–11, 1981.

17. J. S. Hodas and D. Miller. Logic programming in a fragment of intuitionistic linear logic. *Information and computation*, pages 327–365, 1994.
18. M. Kanovitch. Linear logic as a logic of computation. *Annals of pure and applied logic*, 67:183–212, 1994.
19. N. Kobayashi and A. Yonezawa. Higher-order concurrent linear logic programming. In *Theory and Practice of Parallel Programming*, volume 907 of *LNCS*, pages 137–166. Springer Verlag, 1995.
20. J. Lambek. The mathematics of sentence structure. *American mathematical monthly*, pages 154–170, 1958.
21. A. Lecomte and C. Retoré. Pomset logic as an alternative categorial grammar. In G. Morrill and R. Oehrle, editors, *Formal Grammar*, pages 181–196, FoLLI.
22. L. Lovàsz and M.D. Plummer. *Matching Theory*, volume 121 of *Mathematics Studies*. North-Holland, 1986. Annals of Discrete Mathematics 29.
23. R. Milner. *Communication and Concurrency*. International series in computer science. Prentice Hall, 1989.
24. G. V. Morrill. *Type Logical Grammar*. Kluwer Academic Publishers, Dordrecht and Hingham, 1994.
25. G. Perrier. Concurrent Programming as Proof Net Construction. Research Report 96-R-132, CRIN-CNRS, Nancy, September 1996.
26. U. S. Reddy. A linear logic model of state. Electronic manuscript, University of Illinois (anonymous FTP from cs.uiuc.edu), October 1993.
27. U. S. Reddy. Global state considered unnecessary: An introduction to object-based semantics. *Journal of Lisp and Symbolic Computation*, 1996. (to appear).
28. C. Retoré. *Réseaux et Séquents Ordonnés*. Thèse de Doctorat, spécialité Mathématiques, Université Paris 7, février 1993.
29. C. Retoré. On the relation between coherence semantics and multiplicative proof nets. Rapport de Recherche 2430, INRIA, décembre 1994.
30. C. Retoré. A self-dual modality for "before" in the category of coherence spaces and in the category of hypercoherences. Rapport de Recherche 2432, INRIA, décembre 1994.
31. C. Retoré. Une modalité autoduale pour le connecteur "précède". In Pierre Ageron, editor, *Catégories, Algèbres, Esquisses et Néo-Esquisses*, Publications du Département de Mathématiques, Université de Caen, pages 11–16, septembre 1994.
32. C. Retoré. Perfect matchings and series-parallel graphs: multiplicative proof nets as R&B-graphs. In J.-Y. Girard, M. Okada, and A. Scedrov, editors, *Linear'96*, volume 3 of *Electronic Notes in Computer Science*. Elsevier, 1996. (available from http://pigeon.elsevier.nl/mcs/tcs/pc/menu.html).
33. C. Retoré. A semantical characterisation of the correctness of a proof net. *Mathematical Structures in Computer Science*. (to appear)
34. I. Schena. Pomset logic and discontinuity in natural language. In C. Retoré, editor, *Logical Aspects of Computational Linguistics*, pages 7–12. CRIN-C.N.R.S. & INRIA-Lorraine, 1996. (to appear in LNCS/LNAI series, Springer-Verlag).
35. R. D. Tennent. *Semantics of programming languages*. Prentice-Hall International, London, 1991.
36. A. S. Troelstra. *Lectures on Linear Logic*, volume 29 of *Center for the Study of Language and Information Lecture Notes*. Cambridge University Press, 1992.
37. J. Valdes, R.E. Tarjan, and E. L. Lawler. The recognition of Series-Parallel digraphs. *SIAM Journal of Computing*, 11(2):298–313, May 1982.

Computational Reflection in the Calculus of Constructions and its Application to Theorem Proving

Harald Rueß

Universität Ulm
Fakultät für Informatik
D-89069 Ulm, Germany
ruess@informatik.uni-ulm.de

Abstract. This paper describes a computational reflection mechanism for the *calculus of constructions*. In this framework it is possible to encode functions that operate on *syntactic* representations on the meta-level and to verify *semantic* relations between the object-level denotations of the source and the target of meta-functions. Moreover, it is shown how computational reflection can easily be integrated with existing proof development systems based on refinement methods in order to extend theorem proving capabilities in a sound way.

1 Introduction

A *self-referential* system is able to refer to (parts of) itself, and consists of a base system, the so-called *object-level*, an internal representation of (parts of) itself, the *meta-level*, and a *reflection* mechanism that expresses certain relationships between the object-level and its corresponding meta-level encoding. In his seminal paper, Weyhrauch [22] has already noted that self-referential theorem proving systems can be used to replace object-level deduction with meta-level computation, which, in many cases, may be more efficient. The main problem with this approach, however, is to guarantee that the introduction of new meta-level functions does not affect soundness.

Building on Weyhrauch's idea, various self-referential systems have been developed for soundly extending theorem proving systems [2,11,9,1,10,19,20], where a theorem prover is said to be *soundly extensible* if a meta-language can be employed to extend the reasoning capabilities and if these extensions can be shown to be correct with respect to some given correctness criterion. Usually, self-referential systems for soundly extending theorem provers are *conservative extensions* of their underlying base-level systems. Whereas conservative extensions can not give real *reflection principles* [12], they allow a single system to simulate a large amount of meta-reasoning by meta-functions and give the assurance that the resulting system remains consistent.

Basically, there have been two different approaches to obtain self-referential systems for sound extensions; namely *computational* and *deductive* reflection.

Deductive reflection [22,20] requires an embedding of the inference rules within the underlying logic and transitions between these meta-level encodings and the object-level are established by *deductive reflection rules.* In the *computational reflection* [2,9] paradigm the connection between meta-level encodings and their corresponding entities on the object-level is established by a *self-interpreter* that associates meta-level representations with the object-level values they denote. In this way, computational reflection systems avoid the complexity of embedding inference rules of the object-level. Compared with deductive reflection systems, however, the applicability of computational reflection is limited, since notions like provability or derivability are not encoded on the meta-level. The computational reflection frameworks described in [2,9], for example, are limited to establish equality of the (object-level) denotations of (meta-level) sources and targets of meta-functions.

Computational reflection is especially simple in an untyped setting [15], while the task of constructing self-interpreters for typed calculi is usually considerably more complicated: in any reasonable type system there are non well-typed terms, and, consequently, self-interpreters are partial functions that only work on representations of well-typed object-level entities. A self-interpreter for a typed language must therefore not only compute an object-level entity but also a type for this entity. Thus, the simplicity of computational reflection for untyped base calculi seems to be lost in a typed setting.

This is especially troublesome in a type-theoretic setting, where the notions of type-checking and proof-checking coincide. A large portion of Howe's computational reflection mechanism [9] for a fragment of NUPRL, for example, consists of type checking routines. Besides practical considerations of going through the whole work of encoding type-checking, there are also theoretical limitations of encoding type-checking for a type theory within itself: a crucial part of type-checking in a type-theoretic setting involves normalization and Girard et al. [5], for example, show that, using an impredicative encoding of the natural numbers, it is impossible to represent normalization in type-theories like *System F*.

Thus, it seems to be a better idea to already include type information into the representations and use representation types that only include representations of well-typed entities. Pfenning and Lee [18] use such an encoding to develop an interpreter for F^2 in F^3. On the negative side, however, they conclude that evaluation is just about the only meta-function definable on such a representation, since the choice of only representing well–typed objects forces representations by means of *higher-order abstract syntax*: both object-level variables and abstractions are implicitly represented by itself.

In the following, we combine and extend the ideas developed in [2,9,18] and construct a powerful computational reflection mechanism for a variant of the *calculus of constructions.* This reflection mechanism both

- preserves the simplicity of the original idea of computational reflection in a type-theoretic setting and
- opens up a wide range of applications for sound extensions of theorem provers by means of computational reflection.

The increased applicability is mainly due to the flexibility of choosing various forms of correctness criteria and, in the presence of higher type universes, to the fact that not only ordinary term expressions but also entities like formulae can be represented and manipulated on the meta-level.

The formal background of the calculus of constructions and refinement proofs is provided in Sections 2 and 3, respectively, and the computational reflection mechanism for a quantifier-free fragment of the calculus of constructions is developed in Sections 4 and 5. The restriction to this quantifier-free fragment is mainly due to presentation purposes and to the fact that many interesting meta-functions for theorem proving can already be defined in this fragment. An encoding of rewrites as verifiable meta-functions is provided in Section 6 and in Section 7 it is demonstrated how computational reflection (and rewrites) are applied in the refinement process. For lack of space we only sketch the constructions and proofs concerning the example rewrites in Section 6; an in-depth description of these developments, however, can be found in [19]. Finally, in Section 8, we extend our meta-level representations and the computational reflection function as described in Sections 4 and 5 to also include λ- and Π-abstractions and universe constants.

2 Formal Background

The *calculus of constructions* (CC) [4] forms the starting point of this investigation of computational reflection. It can be viewed as a unification of impredicative quantification in *System F* with dependent types. In its pure form, the dependent product $\Pi x : A.\ B$ is the only type constructor of the calculus of constructions.[1] Furthermore, abstraction is of the form $\lambda\, x : A.\ M$ and application is written as $M(N)$.[2] The treatment of rules is based on the notion of *type judgements*. Judgements are of the form $\Gamma \vdash M : A$ and express the fact that in context Γ term M is of type A. Here, a context is defined as a sequence of declarations $(x : A)$ and definitions $(x : A\ :=\ M)$. A context is said to be *valid* if every type A is well-formed in the preceding context. Terms that define types are collected in yet other types, the so-called universes *Prop*, *Type*$_i$ ($i \in \mathbb{N}_0$). An extension of the calculus of constructions with cumulative type universes has been introduced by Coquand and Huet [4], Luo [13] analyzed its meta-theory, and Harper and Pollack [7] studied *type checking* and *well-typedness* problems for variants of this calculus. One of Harper and Pollack's extensions to this calculus includes

[1] Whenever variable x does not occur free in B, product type $\Pi x : A.\ B$ is also denoted by $A \to B$, and $\to$ is right-associative. Variables and terms are usually denoted as x,y,z and M, N, A, B, respectively, while $M\,[x := N]$ denotes substitution of N for free occurrences of x in M. Bindings of the form $x?A$ indicate parameters that can be omitted in function application. Given $I\ :=\ \lambda\, A?\,\mathit{Type}, x : A.\ x$, for example, one may write $I(3)$ instead of $I(\mathit{nat})(3)$, since the argument *nat* can be inferred from the type of 3. Sometimes, however, one may want to make hidden applications explicit, and in these cases we write $I\{\mathit{nat}\}(3)$.

[2] Application associates to the left.

Formation:

$$nat : SET$$

Introduction:

$$0 : nat$$
$$succ : nat \rightarrow nat$$

Elimination:

$$\begin{aligned} elim_{nat} \ : \ & \Pi C : (nat \rightarrow Type \rightarrow C(0). \\ & (\Pi x : nat.\ C(x) \rightarrow C(succ(x))) \rightarrow \\ & (\Pi n : nat.\ C(n)) \end{aligned}$$

Equality: for $E \ := \ elim_{nat}(f_0)(f_{succ})$:

$$E(0) \simeq f_0$$
$$E(succ(x)) \simeq f_{succ}(x)(E(x)),$$

Fig. 1. Natural Numbers

an anonymous type universe *Type* as a means of implementing the *typical ambiguity* convention. Explicit universe levels may be omitted by using instead the anonymous universe, with the understanding that such a term stands for some consistent replacement of the anonymous universe by specific universes. Finally, $M_1 \equiv M_2$ and $M_1 \simeq M_2$ denote syntactic equality and the conversion relation generated by β-reduction, respectively.

Using the principle of *propositions-as-types*, the dependent product can be interpreted both as the type of dependent functions with domain A and codomain $B\,[x := N]$, with N the argument of the function at hand, and as the logical formula $\forall\, x : A.\ B$. In the logical interpretation $\lambda\, x : A.\ M$ is interpreted as a proof object for the formula $\forall\, x : A.\ B$, and, using a constructive interpretation of formulae, $\forall\, x : A.\ B$ is only valid if there exists a proof object that *inhabits* this formula. In this way it is possible to encode the usual logical notions ($\forall$, $\exists$, $\wedge$, $\vee$, $\Rightarrow$, ...) together with a natural-deduction style calculus for higher-order constructive logic with Leibniz equality (=) in CC (see, for example, [3]).

In a similar way, inductive data-types can be encoded in CC by means of impredicative quantification. For the well-known imperfections of these encodings, however, we prefer to extend CC for each data-type with notions related to formation, introduction, elimination, and equality rules (e.g. [16]). Figure 1, for example, lists the corresponding rules for the natural numbers; here and in the following we use the definition $SET \ := \ Type$. A homomorphic functional [21] (or *catamorphism*) hom_T on the inductive data-type T is a special case of elimination and is used in the following to encode functions with source of type T by means of (higher-order) primitive recursion. With the homomorphic functional

$$hom_{nat} \;:\; \Pi C?SET.\; C \to (C \to C) \to nat \to C$$

on *nat*, for example, addition on natural numbers, can be defined as

$$add : \; nat \to nat \to nat \;:=\; \lambda\, m : nat.\; hom_{nat}(m)(succ)$$

A prominent role throughout this paper plays the data-type *trm* : *SET* as provided in Figure 3, since it is used as a representation type for quantifier-free terms of CC. It introduces constructors for representing variables and application.

3 Proof Development by Refinement

A traditional method of searching for proofs (and programs) is to work backwards from goals to subgoals. In type-theoretic development tools such as LEGO [14] one typically starts with an *unknown* construction (or *meta-variable*) $?_p$ for solving goal A in context Γ:

$$\Gamma \vdash ?_p : A$$

A *refinement step* consists of reducing the task of constructing $?_p$ such that $\Gamma \vdash ?_p : A$ to the task of constructing terms $?_{p_1}, \ldots, ?_{p_n}$ such that:

$$\Gamma_1 \vdash ?_{p_1} : A_1, \ldots, \Gamma_n \vdash ?_{p_n} : A_n$$

A justification of this refinement step involves a theorem

$$H \;:\; \Pi x_1 : A_1, \ldots, x_n : A_n.\; A'$$

such that $H(?_{p_1}) \ldots (?_{p_n}) \;:\; A$. In the general case, refinement involves higher-order unification, and a subgoal A_k may depend on unknown constructions $?_{p_l}$ for $l < k$. The refinement or resolution process stops when a term p of the problem at hand is constructed that does not contain any unknown.

4 Representation Types

The first step towards our computational reflection architecture consists in an embedding of (parts of) CC within itself. More precisely, representation types for contexts and quantifier-free terms of CC are defined. Each context Γ is represented by an enumeration type, say cxt_Γ, that comprises a constructor term for every constant – defined or declared – in the context Γ, and each constructor is parameterized over the type of the constant it represents. Furthermore, associations between representations of constants and denotations are accomplished by means of a function $reflect_\Gamma$. More precisely:

Formation:

$$cxt_\Gamma \ : \ SET \rightarrow SET$$

Introduction:

$$\ulcorner x_0 \urcorner \ : \ cxt_\Gamma(A_0) \quad \ldots \quad \ulcorner x_n \urcorner \ : \ cxt_\Gamma(A_n)$$

Elimination:

$$\begin{aligned} elim_\Gamma : \Pi C : (\Pi A?SET.\ cxt_\Gamma(A)) \rightarrow Type. \\ C\{A_0\}(\ulcorner x_0 \urcorner) \rightarrow \ldots C\{A_n\}(\ulcorner x_n \urcorner) \\ \rightarrow \Pi A : SET,\ M : cxt_\Gamma(A).\ C(M) \end{aligned}$$

Equality: for $i = 1, \ldots n$:

$$elim_\Gamma(C)(f_0)\ldots(f_n)(\ulcorner x_i \urcorner) \simeq f_i$$

Fig. 2. Representation Type for Contexts

Definition 1. Let $\Gamma \ := \ x_0 : A_0, \ldots, x_n : A_n$ be a valid context where some of the x_i may be defined constants. This context Γ determines a data-type cxt_Γ that serves as a *representation type* for the context Γ; see Figure 2. Here $\ulcorner . \urcorner$ is an arbitrary injective naming function. Furthermore, define a function that reflects $\ulcorner x_i \urcorner$, for $i = 0, \ldots, n$, back to x_i.

$$\begin{aligned} reflect_\Gamma \ &: (\Pi A?SET.\ cxt_\Gamma(A) \rightarrow A) \ := \\ &elim_\Gamma(\lambda A : SET, \_ : cxt_\Gamma(A).\ A)(x_0)\ldots(x_n) \end{aligned}$$

Thus, a representation type cxt_Γ and a corresponding reflection function $reflect_\Gamma$ are defined for each valid context Γ. The main point about this definition is expressed by the fact that reflection really reflects representations of constants to the values they denote.

Lemma 2. *Let Γ be a valid context; then for all $c : cxt_\Gamma$:*

$$reflect_\Gamma(\ulcorner c \urcorner) \simeq c$$

The proof follows immediately form Definition 1 and the equality rule in Figure 2. For the remainder of this paper we assume all free occurrences of

$$cxt : SET \rightarrow SET$$
$$reflect_{cxt}(A) : cxt(A) \rightarrow A$$

to be implicitly quantified (here, $A?SET$). Implicit quantifications are made explicit in systems like LEGO by *discharging* [14] declared constants. In the following, it is sometimes necessary to make implicit quantifications explicit to improve readability.

Formation:

$$trm : SET \to SET$$

Introduction:

$$rep : \Pi A?SET.\ cxt(A) \to trm(A)$$

$$\begin{array}{l} app : \Pi A?SET,\ B : A \to SET,\ a : A. \\ \quad trm(\Pi x : A.\ B(x)) \to trm(A) \\ \quad\quad \to trm(B(a)) \end{array}$$

Elimination:

$$\begin{array}{l} elim_{trm}\ : \\ \Pi C : (\Pi A?SET.\ trm(A) \to Type). \\ \ (\Pi A?SET,\ c : cxt(A).\ C(rep(c))) \to \\ \ (\Pi A?SET, B : A \to SET, a : A, M : trm(\Pi x : A.\ B(x)), N : trm(A). \\ \quad C(M) \to C(N) \to C(app(B)(a)(M)(N))) \\ \quad \to \Pi A?SET.\ M : trm(A).\ C(M) \end{array}$$

Equality: for $E := elim_{trm}(f_{ref})(f_{app})$:

$$E(rep(c)) \simeq f_{rep}(c)$$

$$E(app(B)(a)(M)(N) \simeq f_{app}(B)(a)(M)(N)(E(M))(E(N))$$

Fig. 3. Representation Type for Terms

Next, we describe the representation of CC terms. In order to simplify the presentation of the encodings, we chose to only encode the quantifier-free part of CC. Moreover, the inclusion of type information in the representation forces the use of higher-order abstract syntax, and for the encoding of λ-abstraction by

$$lam(A)(B)\ :\ (\Pi x : A.\ trm(B(x))) \to trm(\Pi x : A.\ B(x))$$

with $A?SET$ and $B : A \to SET$, there seems to be no easy way to inspect and manipulate bound variables in the body of representations of λ-abstractions. Consequently, terms with bound variables are not represented on the meta-level in the following. As shown in Section 8 and in [19], however, the basic techniques extend to the case of representing λ-abstraction and, for the presence of cumulative universes, even product types.

A term M of type A is represented as an object of the representation type $trm(A)$ as proposed by Pfenning and Lee [18]. This consideration leads to the definition of the representation type $trm(A)$ for the quantifier-free fragment of CC in Figure 3. Here, terms $rep(x)$ are intended to represent variables x while the constructor *app* builds representations of applications. Note also that the constructor *app* in Figure 3 includes an extraneous argument $a : A$ that is necessary to express the range type $trm(B(a))$ of representations of dependent applications.

In order to express functions on the representation type $trm(A)$ one encodes a recursor as a "non-dependent" version of the term elimination $elim_{trm}$ (see Figure 3) in the usual way. This results in a defined constant hom_{trm} of type:

$$\begin{array}{l} \Pi\, C : SET \to SET. \\ \quad (\Pi\, A?SET.\ cxt(A) \to C(A)) \to \\ \quad (\Pi\, A?SET,\ B : A \to SET,\ a : A,\ C(\Pi x : A.\ B(x)),\ C(A). \\ \qquad C((\lambda\, x : A.\ B(x))a)) \to \\ \quad \Pi\, A?SET.\ trm(A).\ C(A) \end{array}$$

5 Computational Reflection

Using the homomorphic functional hom_{trm} on the inductive datatype *trm*, it is straightforward to encode a function *reflect* that relates a representation of some term of type A, i.e. an entity of type $trm(A)$, with a corresponding object-level denotation of type A; this function constitutes *computational reflection.*

Definition 3.

$$\begin{array}{l} reflect\ : (\Pi\, A?SET.\ trm(A) \to A)\ := \\ \quad hom_{trm}(\lambda\, A?SET.\ A) \\ \qquad (\lambda\, A?SET,\ c : cxt(A).\ reflect_{cxt}(c)) \\ \qquad (\lambda\, A?SET,\ B : A \to SET,\ N : A,\ M : (\Pi x : A.\ B(x)), \_ : A. \\ \qquad\quad M(N)) \end{array}$$

As stated above, the constants *cxt* and $reflect_{cxt}$ are assumed to be implicit parameters of *reflect*. Given a certain context Γ, these constants are assumed to be instantiated with cxt_Γ and $reflect_\Gamma$ as generated from context Γ according to Definition 1. Also, following [9], the computational *reflect*ion function above is called to be a *partial reflection* function, for its encoding is purely definitional.

Computational reflection permits getting rid of the extraneous argument in the representation of function application by overloading the constructor *app* for representing applications ($A?SET$, $B : A \to SET$):

$$\begin{array}{rcl} app(A)(B) & := & \lambda\, M : trm(\Pi x : A.\ B(x)),\ N : trm(A). \\ & & app\{A\}(B)(reflect(N))(M)(N) \end{array}$$

The following examples also use representations of non-dependent application ($A, B?SET$):

$$\begin{array}{rcl} app_{nd}(A)(B) & := & \lambda\, M : trm(A \to B),\ N : trm(A). \\ & & app\{A\}(\lambda\_ : A.\ B)(M)(N) \end{array}$$

Next, consider the characteristic equations for *reflect*.

Lemma 4. *Let $A ? SET$, $B : A \to SET$, $c : cxt(A)$, $M : trm(\Pi x : A.\ B(x))$, $N : trm(A)$; then:*

$$\begin{aligned} reflect(rep(c)) &\simeq reflect_{cxt}(c) \\ reflect(app(B)(M)(N)) &\simeq reflect(M)(reflect(N)) \end{aligned}$$

The proof of this fact follows from Definition 3 and the equality rules for *trm* in Figure 3. In the next step, quotation for the quantifier-free part of CC is defined.

Definition 5. Let $\Gamma := x_0 : A_0, \ldots, x_n : A_n$ be a valid context (with x_i either declared or defined), and L be a well-typed term in context Γ; then:

- If $L \equiv x_i$ for some $i = 0, \ldots, n$ then $\ulcorner L \urcorner := rep(\ulcorner x_i \urcorner)$.
- If L is of the form $M(N)$ with $M : \Pi x : A.\ B(x)$ and $N : A$ then define $\ulcorner L \urcorner := app(B)(\ulcorner M \urcorner)(\ulcorner N \urcorner)$.
- Otherwise quotation is undefined.

A term L for which quotation is defined is referred to as being reifable.

Finally, and most importantly, Theorem 6 states that *reflect* is an interpreter on the given encoding. This fact is proven by induction on the structure of reifable terms.

Theorem 6. *Let Γ be a valid context, M be a reifable term of type A in context Γ; then:*

$$reflect(cxt_{\Gamma})(reflect_{\Gamma})(\ulcorner M \urcorner) \simeq M$$

6 Rewrites as Meta-Functions

A *meta-function* is an ordinary CC function that operates on meta-level entities of type $trm(A)$. Such a function is able to both inspect and manipulate the syntactic form of a term. Now, computational reflection permits relating the source and the target of meta-functions on the object-level. Prominent examples of meta-functions are rewrites and, more generally, simplifications like the *cancel* function in [2].

Intuitively, a rewrite is a function that maps a representation of a term of type A to another representation of some other term of type A such that the denotations of the source and the target are equal.

Definition 7. Let $A : SET$, $f : rewrite(A)$, and $M : A$; then:

$$\begin{aligned} rewrite(A) &:= trm(A) \to trm(A) \\ correct(f)(M) &:= reflect(M) =_A reflect(f(M)) \end{aligned}$$

This (defined) correctness criterion corresponds directly to the built-in correctness criteria in [2,9]. In our case, however, this is just one instance of a correctness criterion for meta-functions, and, depending on the domain of application, appropriate correctness criteria may be defined (for example, for meta-functions manipulating formulae or meta-functions supporting refinement proofs of programs).

In the following, we sketch the definition and correctness proof of a simple rewrite that replaces (a representation of) $x \circ e$ with (a representation of) x. These developments are assumed to be — besides cxt and $reflect_{cxt}$ — implicitly quantified over the constants in the following context.

$$\begin{aligned} G &: SET \\ \circ &: G \to G \to G \\ e &: G \\ Ax &: \Pi x : G.\, x \circ e = x \\ \circ' &: cxt(G \to G \to G) \\ e' &: cxt(G) \\ H_1 &: reflect_{cxt}(\circ') = \circ \\ H_2 &: reflect_{cxt}(e') = e \end{aligned}$$

Thus, $\circ$ is a binary operation over G, e is a right-neutral element of $\circ$, and the constants $\circ'$, e' are assumed to be the representations of $\circ$ and e, respectively. In our applications of reflection in Section 7, the parameters $\circ'$ and e' will be instantiated with quotations corresponding to actual parameters for $\circ$ and e.

Now we have collected all the ingredients to define the following example rewrite.[3]

$$\begin{aligned} & f_\circ(\circ')(e') \ : rewrite(G) \ := \\ & \quad \lambda M : trm(G).\ \ hom_{maybe}(\lambda\, y : trm(G).\ app_{nd}(app_{nd}(\circ')(y))(e')) \\ & \qquad\qquad\qquad\qquad (M) \\ & \qquad\qquad\qquad\qquad (match(M)) \end{aligned}$$

This is a correct rewrite, since one can find a term $corr\_f_\circ$ such that

$$corr\_f_\circ \ : \ correct(f_\circ)$$

Rewrites are combined using the sequencing (*then*) and repetition (*repeat*) combinators in Figure 4. Of course, other combinators like the usual *try*, *orelse*, or *andthen* can easily be added to this framework. In a similar way, one encodes the rewrite $sub(f) : rewrite(A)$ that applies some correct rewrite f to the sub-terms of a representation term, and prove its correctness

$$sub\_corr(f) \ : \ correct(f) \Rightarrow correct(sub(f))$$

[3] Here, $maybe(G)$ is assumed to be an inductive data-type with constructors $yes(y)$ and no; $hom_{maybe}(f_{yes})(f_{no})(c)$ evaluates to $f_{yes}(y)$ if $c = yes(y)$ and to f_{no} if $c = no$. Furthermore, $match$ is assumed to be defined such that $match(M) \simeq yes(y)$ if and only if $M \equiv y \circ e'$.

Let $A?SET$, $f, g : rewrite(A)$, $n : nat$; then:

$$then(f)(g) \ : rewrite(A) \ := \ \lambda M : trm(A).\ g(f(M))$$

$$then\_corr(f)(g) \ : \ correct(f) \Rightarrow correct(g) \Rightarrow correct(then(f)(g))$$

$$repeat(n)(f) \ : rewrite(A) \ := \ hom_{nat}(id)(\lambda g : rewrite(A).\ then(f)(g))(n)$$

$$correct_{repeat}(n)(f) \ : \ correct(f) \Rightarrow correct(repeat(n)(f))$$

Fig. 4. Rewrite Combinators

It is a simple matter to define powerful rewrites from simpler ones and the rewrite combinators above. Consider, for example the definition of a rewrite

$$\begin{aligned} &f(+')(*')(0')(1') : rewrite(nat) \ := \\ &\quad repeat(42)(then(sub(f_\circ(+')(0')))(sub(f_\circ(*')(1')))) \end{aligned}$$

where $+', *'$ are of type $cxt(nat \rightarrow nat \rightarrow nat)$ and $0', 1'$ are of type $cxt(nat)$.

Here, $f_\circ(+')(0')$ and $f_\circ(*')(1')$ are instances of the rewrite $f_\circ$ above: they simply replace (representations of) $x + 0$ and $x * 1$ with x, respectively. In order to establish correctness of rewrite f, one needs the assumptions

$$\begin{aligned} H_1 : reflect_{cxt}(+') = + \ , \ H_2 : reflect_{cxt}(*') = * \\ H_3 : reflect_{cxt}(0') = 0 \ \ , \ H_4 : reflect_{cxt}(1') = 1 \end{aligned}$$

Now, the proof of correctness

$$corr\_f(cxt)(reflect_{cxt})(+')(*')(0')(1')(H_1)\ldots(H_4) \ : \ correct(cxt)(reflect_{cxt})(f)$$

for rewrite f follows directly from the correctness results for f_o, *then*, *repeat*, and *sub* by simply "chaining" the corresponding proof terms together. Note that the implicit arguments have been made explicit here, since this result is used in the sequel.

7 Computational Reflection in the Refinement Process

In the previous section, notions of rewriting have been encoded on the meta-level. There was, however, no indication of how to actually apply rewrites and other meta-functions in the refinement process.

Application of computational reflection in refinement proofs proceeds in three subsequent steps:

- **Reification:** given the current context Γ, one first computes cxt_Γ and $reflect_\Gamma$ according to Definition 1. Then, selected sub-terms P_i of the source goal are replaced by $reflect(cxt_\Gamma)(reflect_\Gamma)(\ulcorner P_i \urcorner)$. For Theorem 6 and the convertibility rule of CC,[4] this modified goal is inhabited by a proof term p if and only if the source goal is inhabited by p.
- **Refinement:** refinement with a selected meta-theorem; i.e. a theorem that mentions the computational reflection function *reflect*.
- **Reflection:** replacement of the meta-level representations in the resulting subgoals by their denotations on the object-level. This process simply involves normalization of the computational reflection function.

Obviously, it is possible to construct a proof term for the source goal from constructions of the resulting subgoals.

Next, we exemplify the application of a meta-theorem for the example rewrite f defined in Section 6. The meta-theorem

$$\begin{array}{l} rewrite\_on\_relation : \\ \quad \Pi\, cxt,\ reflect_{cxt}. \\ \qquad \text{LET } corr := correct(cxt)(reflect_{cxt}),\ refl := reflect(cxt)(reflect_{cxt}) \text{ IN} \\ \qquad\quad \Pi A, B?SET,\ R : A \to B \to Prop. \\ \qquad\qquad \Pi g_1 : rewrite(A).\ corr(g_1) \Rightarrow \Pi g_2 : rewrite(B).\ corr(g_2) \Rightarrow \\ \qquad\qquad\quad \Pi M_1, M_2 : trm(A).\ R(refl(g_1(M_1)))(refl(g_2(M_2))) \\ \qquad\qquad\qquad \Rightarrow R(refl(M_1))(refl(M_2)) \end{array}$$

for example, can be used to simplify arguments of a binary relation by means of rewrites g_1, g_2. This meta-theorem can easily be proven in a refinement style.

Now, assume the "toy" problem of finding a proof term for the unknown $?_p$ such that[5]

$$\underbrace{\Delta,\ x : nat}_{\Gamma} \vdash\ ?_p\ :\ \underbrace{x * 1}_{lhs}\ =\ \underbrace{(x + 0) * (1 + 0)}_{rhs}$$

In a first step one instantiates the computational reflection function with the representation type for the context Γ to obtain the function $reflect(cxt_\Gamma)(reflect_\Gamma)$ and reifies the left and the right hand side of the equation; for Theorem 6 and the convertibility rule of CC it suffices to construct a term $?_p$ for the goal:

$$\Gamma \vdash\ ?_p\ :\ reflect(cxt_\Gamma)(reflect_\Gamma)(\ulcorner lhs \urcorner) = reflect(cxt_\Gamma)(reflect_\Gamma)(\ulcorner rhs \urcorner)$$

Refinement using the (meta-) theorem $rewrite\_on\_relation(cxt_\Gamma)(reflect_\Gamma)$ yields the subgoals (in context Γ)

$?_{g_1} : rewrite(nat)$

$?_{c_1} : correct(cxt_\Gamma)(reflect_\Gamma)(?_{g_1})$

$?_{g_2} : rewrite(nat)$

---

[4] If $\Gamma \vdash M : A$ and $A \simeq A'$ then $\Gamma \vdash M : A'$.

[5] It is also assumed that x and $*$ are defined in context Δ in such a way that neither $x + 0$ nor $x * 1$ β-reduce to x.

$$?_{c_2} : correct(cxt_\Gamma)(reflect_\Gamma)(?_{g_2})$$
$$?_H : reflect(cxt_\Gamma)(reflect_\Gamma)(?_{g_1}(\ulcorner lhs \urcorner)) = reflect(cxt_\Gamma)(reflect_\Gamma)(?_{g_2}(\ulcorner rhs \urcorner))$$

Next, we choose the rewrite $f(\ulcorner + \urcorner)(\ulcorner * \urcorner)(\ulcorner 0 \urcorner)(\ulcorner 1 \urcorner)$ from Section 6 for the unknowns $?_{g_1}$ and $?_{g_2}$, and normalization yields the simplified subgoal

$$?_H : x = x$$

which obviously holds. The only thing left is to construct proof terms for the unknowns $?_{c_1}$ and $?_{c_2}$. This involves refinement of the proof term *corr_f* from Section 6, and the subgoals resulting from the assumptions H_1 through H_4 of *corr_f* can be proven immediately. This finishes the construction of a proof term by refinement and computational reflection for this toy example.

This example demonstrates that proof steps involving simplification on the meta-level can readily be carried out in development tools such as LEGO [14] using the `Refine` command. We were not able, however, to encode, apply, and prove the correctness of more advanced meta-functions like the *cancel* function [2] at reasonable cost; this is mainly due to the relative slowness of evaluation and the limited amount of automated deduction techniques currently available in the LEGO system.

One of the main points is that meta-functions like *f* are applicable for all contexts for which the prerequisites of the specified correctness criterion can be shown to hold. Also, the correctness of a simplification needn't necessarily be established before executing this simplification but may well be expressed as a proof obligation that is being proven in the future. Moreover, the way computational reflection is applied in the refinement process is not fixed: the meta-theorem *rewrite_on_relation* is just one particular example of a meta-theorem that involves computational reflection.

It may well be argued that goals like the above can be solved as easily and also "in-one-step" using an appropriate proof-search tactic [6]. While this claim is certainly true, there are some subtle –but important – differences to formal meta-functions in the framework developed herein. Tactics basically search for proofs in terms of primitive or derived inference rules, while formal meta-functions in reflective frameworks only work on the syntactic categories of the problem at hand; justifications of these algorithms are proven *once and for all*, and are simply instantiated for each problem. Of course, applications of formal meta-functions by means of computational reflection can and should be used as single steps in proof search tacticals.

The advantage of using formal meta-functions to soundly add theorem proving capabilities are manifold: first, abstraction from primitive (and derived) inference rules is possible, and, second, formal meta-functions have the potential of being more efficient than tactics. Whereas an in-depth discussion of the relative merits of formal meta-functions and tactics is well beyond the scope of this paper (for such comparisons see, for example, [8,20]), it should be noted that it is currently unclear if the efficiency required to complete the large-scale verifications described, for example, in [17] at reasonable cost can be achieved using an LCF-like tactics approach. Third, meta-functions are expressed in the same language as ordinary functions and can be developed and proven to be correct using the very same techniques. Thus, there is the potential to bootstrap the kernel system and use meta-functions to develop yet other meta-functions to soundly extend the core prover.

8 A Self-Interpreter for CC

So far we have restricted ourselves to a computational reflection framework for the quantifier-free fragment of CC, since this fragment is sufficient to encode a large number of useful meta-functions for theorem-proving. The basic ideas, however, extend readily to support computational reflection for λ-abstractions and, for the presence of cumulative universes, even for Π-types and universes. For an implementation of computational reflection in LEGO see Appendix A. With the exception of the universes *Prop* and $Type_i$, which are represented by entities built from a single constructor *con*, there is a constructor of the representation type $trm(A)$ for all the term constructors of CC.

$$\begin{aligned}
con &: trm(Type) \\
rep &: \Pi A ? Type.\ A \rightarrow trm(A) \\
lam &: \Pi A ? Type,\ B : A \rightarrow Type.\ (\Pi x : A.\ trm(B(x))) \rightarrow trm(\Pi x : A.\ B(x)) \\
pi &: \Pi A ? Type,\ B : A \rightarrow Type.\ trm(Type) \\
app &: \Pi A ? Type,\ B : A \rightarrow Type,\ a : A. \\
&\qquad trm(\Pi x : A.\ B(x)) \rightarrow trm(A) \rightarrow trm(B(a))
\end{aligned}$$

Using the representation type in Appendix A it is straightforward to represent, for example, the polymorphic identity $I := \lambda A : Type, x : A.\ x$ by

$$\begin{aligned}
I' :\ & trm(\Pi A : Type.\ A \rightarrow A) := \\
& lam(\lambda A : Type.\ A \rightarrow A)(\lambda A : Type.\ lam(\lambda_ : A.\ A)(\lambda x : A.\ rep(x)))
\end{aligned}$$

Since type information pertains to encodings, it is straightforward to define the function *reflect* in Appendix A that maps entities of $trm(A)$ to their denotation on the object level. Again, computational reflection can be used to get rid of the extraneous argument in the constructor *app* for representing applications. Using this new *app* constructor, one may, for example, represent the application $I(A)(a)$, for $A : Type$, by

$$\begin{aligned}
IAa' : trm(A) := \ & app\{A\}(\lambda_ : A.\ A) \\
& \quad (app\{Type\}(\lambda x : Type.\ x \rightarrow x)(I')(rep(A))) \\
& \quad (rep(a))
\end{aligned}$$

Now, using the representations above, one gets the expected conversions:

$$\begin{aligned}
reflect(I') &\simeq \lambda A : Type.\ x : A.\ x \\
reflect(IAa') &\simeq a
\end{aligned}$$

9 Related Work and Conclusions

A computational reflection mechanism for the calculus of constructions has been developed in this paper. This approach avoids to encode type-checking on the meta-level, since type information is already included into the representations types. Consequently, only well-typed terms can be represented and a reflection function connecting representations with the terms they denote becomes total. The resulting reflection mechanism can be readily applied in the proof-refinement process.

Our work differs from the work reported in [2] in that the entire mechanism is constructed within the theory; more precisely, both the meta-level encodings and the reflection function are definitional extensions of the base system. Another important difference stems from the higher-order nature of type theories such as CC. It allows for abstractions that drastically reduce the burden of verifying correctness of new procedures. These improvements also apply to the computational reflection mechanism developed by Howe [9]. The basic difference to Howe's approach, however, lies in the style of encoding, since Howe represents terms as abstract syntax and our encodings include type information. The consequences to this are manifold. First, it is straightforward to encode a self-interpreter that associates meta-level representations with the entities they denote. The second consequence of only representing well-typed terms concerns the fact that meta-functions may only construct a representation of another well-typed term; otherwise the meta-function is not even type-correct. This is in contrast with Howe's approach where the condition of producing representations of well-typed terms is included in the correctness condition. Third, for the presence of type universes, one may not only modify simple term expressions but also formulae using the same mechanism. Fourth, our computational reflection mechanism does not fix a single notion of correctness for meta-functions like in [2] or [9]; instead, notions of correctness that are appropriate for specific domains of applications can be defined. In the case of syntactically manipulating formulae, for example, an appropriate correctness criterion may be that the manipulated formula logically implies the original formula. Fifth, variations of this computational reflection mechanism can be formalized easily in other type theories. Altogether, the simplicity and flexibility of the computational reflection mechanism described in this paper opens up a wide range of possibilities for applying computational reflection to the task of constructing proofs (and programs) in a refinement style.

Acknowledgments: The author would like to thank F. von Henke for supporting this work, the anonymous referees for their helpful comments, and H. Pfeifer for proof-reading.

References

1. S.F. Allen, R.L. Constable, D.J. Howe, and W.E. Aitken. The Semantics of Reflected Proof. In *Proc. 5th Annual IEEE Symposium on Logic in Computer Science*, pages 95–105. IEEE CS Press, 1990.
2. R.S. Boyer and J.S. Moore. Metafunctions: Proving them Correct and Using them Efficiently as New Proof Procedures. In R.S. Boyer and J.S. Moore, editors, *The Correctness Problem in Computer Science*, chapter 3. Academic Press, 1981.
3. T. Coquand and G. Huet. Constructions: a Higher-Order Proof System for Mechanizing Mathematics. In B. Buchberger, editor, *EUROCAL'85: European Conference on Computer Algebra*, volume 203 of *Lecture Notes in Computer Science*, pages 151–184. Springer-Verlag, 1985.
4. T. Coquand and G. Huet. The Calculus of Constructions. *Information and Computation*, 76(2/3):95–120, 1988.
5. J.-Y. Girard, Y. Lafont, and P. Taylor. *Proofs and Types*, volume 7 of *Cambridge Tracts in Theoretical Computer Science*. Cambridge University Press, 1989.
6. M. J. Gordon, A. J. R. Milner, and C. P. Wadsworth. *Edinburgh LCF: a Mechanized Logic of Computation*, volume 78 of *Lecture Notes in Computer Science*. Springer-Verlag, Berlin, 1979.
7. R. Harper and R. Pollack. Type Checking, Universal Polymorphism, and Type Ambiguity in the Calculus of Constructions. In *TAPSOFT'89, volume II*, Lecture Notes in Computer Science, pages 240–256. Springer-Verlag, 1989.
8. J. Harrison. Metatheory and Reflection in Theorem Proving: A Survey and Critique. Technical Report CRC-053, SRI Cambridge, UK, 1995. See `http://www.cl.cam.ac.uk/ftp/hvg/papers`.
9. D.J. Howe. Computational Metatheory in Nuprl. In *Proc. 9th International Conference on Automated Deduction*, volume 310, pages 238–257. Springer-Verlag Lecture Notes in Computer Science, 1988.
10. D.J. Howe. Reflecting the Semantics of Reflected Proof. In P. Aczel, H. Simmons, and S. Wainer, editors, *Proof Theory*, pages 227–250. Cambridge University Press, 1992.
11. T.B. Knoblock and R.L. Constable. Formalized Metareasoning in Type Theory. In *Proceedings of LICS*, pages 237–248. IEEE, 1986. Also available as technical report TR 86-742, Department of Computer Science, Cornell University.
12. G. Kreisel and A. Lévy. Reflection Principles and their Use for Establishing the Complexity of Axiomatic Systems. *Zeitschrift für math. Logik und Grundlagen der Mathematik*, Bd. 14:97–142, 1968.
13. Z. Luo. $CC^{\omega}_{\subset}$ and its Metatheory. Technical Report ECS-LFCS-88-57, Laboratory for the Foundations of Computer Science, Edinburgh University, July 1988.
14. Z. Luo and R. Pollack. The Lego Proof Development System: A User's Manual. Technical Report ECS-LFCS-92-211, University of Edinburgh, 1992.
15. T. Mogensen. Efficient Self-Interpretation in Lambda Calculus. *J. Functional Programming*, 2(3):345–364, 1992.
16. B. Nordström, K. Petersson, and J.M. Smith. *Programming in Martin-Löf's Type Theory*. Number 7 in International Series of Monographs on Computer Science. Oxford Science Publications, 1990.
17. S. Owre, J. Rushby, N. Shankar, and F. von Henke. Formal Verification for Fault-Tolerant Architectures: Prolegomena to the Design of PVS. *IEEE Transactions on Software Engineering*, 21(2):107–125, February 1995.

18. F. Pfenning and P. Lee. Metacircularity in the Polymorphic λ-Calculus. *Theoretical Computer Science*, 89:137–159, 1991.
19. H. Rueß. *Formal Meta-Programming in the Calculus of Constructions*. PhD thesis, Universität Ulm, 1995.
20. H. Rueß. Reflection of Formal Tactics in a Deductive Reflection Framework. In M.A. McRobbie and J.K.Slaney, editors, *Automated Deduction - CADE-13*, volume 1104 of *Lecture Notes in Computer Science*. Springer-Verlag, 1996.
21. F. W. von Henke. An Algebraic Approach to Data Types, Program Verification, and Program Synthesis. In *Mathematical Foundations of Computer Science, Proceedings*, volume 45 of *Lecture Notes in Computer Science*. Springer-Verlag, 1976.
22. R. W. Weyhrauch. Prolegomena to a Theory of Mechanized Formal Reasoning. *Artificial Intelligence*, 13(1):133–170, 1980.

A *Lego*-Code for Self-Interpreter

```
Init CC';                                   (* Only Old Lego Version! *)

[T:Type->Type];

  contype == Type -> (T Type);

  reptype == {A:Type}A->(T A);

  lamtype == {A:Type}{B:A->Type}
                ({x:A}(T (B x)))->(T ({x:A}(B x)));

  pitype  == {A:Type}{B:A->Type}(T Type);

  apptype == {A:Type}{B:A->Type}{a:A}
                (T ({x:A}(B x)))->(T A)->(T (B a));
Discharge T;

trm[A:Type] ==
   {T:Type->Type}(contype T) -> (reptype T) ->
                 (lamtype T) -> (pitype T) -> (apptype T) ->
                    (T A);

trm_hom [T:Type->Type][f1: contype T][f2: reptype T][f3: lamtype T]
        [f4: pitype T][f5: apptype T][A|Type][t:(trm A)]: (T A)
  == (t T f1 f2 f3 f4 f5);

reflect [A|Type] ==                         (* Partial Reflection *)
     trm_hom ([T:Type]T)
        ([c:Type]c)
        ([A:Type][x:A]x)
        ([A:Type][B:A->Type][N:{x:A}(B x)]N)
        ([A:Type][B:A->Type]{x:A}(B x))
        ([A:Type][B:A->Type][a:A][M:{x:A}(B x)][N:A](M a));
```

Names, Equations, Relations: Practical Ways to Reason about *new*

Ian Stark

BRICS*, Department of Computer Science, University of Aarhus, Denmark

Abstract. The nu-calculus of Pitts and Stark is a typed lambda-calculus, extended with state in the form of dynamically-generated *names*. These names can be created locally, passed around, and compared with one another. Through the interaction between names and functions, the language can capture notions of scope, visibility and sharing. Originally motivated by the study of references in Standard ML, the nu-calculus has connections to other kinds of local declaration, and to the mobile processes of the π-calculus.
This paper introduces a logic of equations and relations which allows one to reason about expressions of the nu-calculus: this uses a simple representation of the private and public scope of names, and allows straightforward proofs of contextual equivalence (also known as observational, or observable, equivalence). The logic is based on earlier operational techniques, providing the same power but in a much more accessible form. In particular it allows intuitive and direct proofs of all contextual equivalences between first-order functions with local names.

1 Introduction

Many convenient features of programming languages today involve some notion of *generativity*: the idea that an entity may be freshly created, distinct from all others. This is clearly central to object-oriented programming, with the dynamic creation of new objects as instances of a class, and the issue of object identity. In the study of concurrency, the π-calculus [14] uses dynamically-generated names to describe the behaviour of mobile processes, whose communication topology may change over time. In functional programming, the language Standard ML [15] extends typed lambda-calculus with a number of features, of which mutable reference cells, exceptions and user-declared datatypes are all generative; as are the structures and functors of the module system. More broadly, the concept of lexical scope rests on the idea that local identifiers should always be treated as fresh, distinct from any already declared.

Such dynamic creation occurs at a variety of levels, from the run-time behaviour of Lisp's *gensym* to resolving questions of scope during program linking. Generally, the intention is that its use should be intuitive or even transparent to

---

* Basic Research in Computer Science, a centre of the Danish National Research Foundation.

the programmer. Nevertheless, for correct implementation and sound design it is essential to develop an appropriate abstract understanding of what it means to be *new*.

The *nu-calculus* was devised to explore this common property of generativity, by adding *names* to the simply-typed lambda-calculus. Names may be created locally, passed around, and compared with one another, but that is all. The language is reviewed in Section 2; a full description is given by Pitts and Stark in [22,23], with its operational and denotational semantics studied at some length in [26,27]. Central to the nu-calculus is the use of *name abstraction*: the expression $\nu n.M$ represents the creation of a fresh name, which is then bound to n within the body of M. So, for example, the expression

$$\nu n.\nu n'.(n = n')$$

generates two new names, bound to n and n', and compares them, finally returning the answer *false*. Functions may have local names that remain private and persist from one use of the function to the next; alternatively, names may be passed out of their original scope and can even outlive their creator. It is precisely this mobility of names that allows the nu-calculus to model issues of locality, privacy and non-interference.

Two expressions of the nu-calculus are *contextually equivalent*[1] if they can be freely exchanged in any program: there is no way in the language itself to distinguish them. Contextual equivalence is an excellent property in principle, but in practice often hard to work with because of the need to consider all possible programs. As a consequence a number of authors have made considerable effort, in various language settings, to develop convenient methods for demonstrating contextual equivalence.

Milner's context lemma [13], Gordon's 'experiments' [5], and the 'ciu' theorems of Mason and Talcott [10,28], provide one such approach. These show that instead of all program contexts, it is sufficient to consider only those in some particular form. For the nu-calculus, a suitable context lemma is indeed available [26, §2.6] and states that one need only consider so-called 'argument contexts'. However even this reduced collection of contexts is still inconveniently large, a problem arising from the imperative nature of name creation.

Alternatively, one can look for relations that imply contextual equivalence but are easier to work with. One possibility is to define such relations directly from the operational semantics of the language, as with the *applicative bisimilarity* variously used by Abramsky [1], Howe [8], Gordon [5], and others. Denotational semantics provides another route: if two expressions have equal interpretation in some adequate model, then they are contextually equivalent. For the nu-calculus, such operational methods are developed and refined in [22,23], while categorical models are presented in [27]. Both approaches are treated at length in [26].

In principle, methods such as these do give techniques for proving contextual equivalences. In practice however, they are often awkward and can require rather

[1] The same property is variously known as {operational/observational/observable} {equivalence/congruence}.

detailed mathematical knowledge. The contribution of this paper is to take two existing operational techniques, and extract from them a straightforward logic that allows simple and direct reasoning about contextual equivalence in the nu-calculus.

The first operational technique, ***applicative equivalence***, gives rise to an equational logic with assertions of the form

$$s, \Gamma \vdash M_1 =_\sigma M_2 \ .$$

If such an assertion can be proved using the rules of the logic, then it is certain that expressions M_1 and M_2 are contextually equivalent (here s and Γ list the free names and variables respectively). This equational scheme is simple, but not particularly complete: it is good for reasoning in the presence of names, but not so good at reasoning about names themselves.

The technique of ***operational logical relations*** refines this by considering just how different expressions make use of their local names. The corresponding logic is one of relational reasoning, with assertions of the form

$$\Gamma \vdash M_1 \ R_\sigma \ M_2 \ .$$

Here R is a relation between the free names of M_1 and M_2 that records information on their privacy and visibility. This logic includes the equational one, and is considerably more powerful: it is sufficient to prove all contextual equivalences between expressions of first-order function type.

It is significant that these schemes both build on existing methods; all the proofs of soundness and completeness work by transferring corresponding properties from the earlier operational techniques. For the completeness results in particular this is a considerable saving in proof effort. Such incremental development continues a form of 'technology transfer' from the abstract to the concrete: these same operational techniques were in turn guided by a denotational semantics for the nu-calculus based on categorical monads.

The layout of the paper is as follows: Section 2 reviews the nu-calculus and gives some representative examples of contextual equivalence; Section 3 describes the techniques of applicative equivalence and operational logical relations; Section 4 explains the new logic for equational reasoning; Section 5 extends this to a logic for relational reasoning; and Section 6 concludes.

Related Work

The general issue of adding effects to functional languages has received considerable attention over time, and there is a substantial body of work concerning operational and denotational methods for proving contextual equivalence. A selection of references can be found in [20,28], for example. However, not so many practical systems have emerged for reasoning about expressions and proving actual examples of contextual equivalence.

Felleisen and Hieb [2] present a calculus for equational reasoning about state and control features. This extends β_v-interconvertibility and is similar to the equational reasoning of this paper, in that it is correct and convenient for proving contextual equivalence, but not particularly complete.

Mason and Talcott's logic for reasoning about destructive update in Lisp [11] is again similar in power to our equational reasoning. Moreover, our underlying operational notion of applicative equivalence corresponds quite closely to Mason's 'strong isomorphism' [9]. Further work [12] adds some particular reasoning principles that resemble aspects of our relational reasoning, but can only be applied to first-order functions; by contrast, our techniques remain valid at all higher function types. In a similar vein, the 'variable typed logic of effects' (VTLoE) of Honsell, Mason, Smith and Talcott [7] is an operationally-based scheme for proving certain assertions about functions with state.

The 'computational metalanguage' of Moggi [16] provides a general method for equational reasoning about additions to functional languages. Its application to the nu-calculus is discussed in [26, §3.3], where it is shown to correspond closely to applicative equivalence. Related to this is 'evaluation logic', a variety of modal logic that can express the possibility or certainty of certain computational effects [17,21]. Moggi has shown how a variety of program logics, including VTLoE, can be expressed within evaluation logic [18].

Although the nu-calculus may appear simpler than the languages considered in the work cited, the notion of generativity it highlights is still of real significance. Moreover, the relational logic presented here goes beyond all of the above in the variety of contextual equivalences it can prove: we properly capture the subtle interaction between local declarations and higher-order functions.

2 The Nu-Calculus

A full description of the nu-calculus can be found in [26,27]; this section gives just a brief overview. The language is based on the simply-typed lambda-calculus, with a hierarchy of function types $\sigma \to \sigma'$ built over ground types o of *booleans* and ν of *names*. Expressions have the form

$$\begin{aligned} M \ ::= \ & x \mid n \mid \mathit{true} \mid \mathit{false} \mid \mathit{if}\ M\ \mathit{then}\ M\ \mathit{else}\ M \\ \mid \ & M = M \mid \nu n.M \mid \lambda x{:}\sigma.M \mid MM\,. \end{aligned}$$

Here x and n are typed variables and names respectively, taken from separate infinite supplies. The expression '$M = M'$' tests for equality between two names. Name abstraction $\nu n.M$ creates a fresh name bound to n within the body M; during evaluation, names may outlive their creator and escape from their original scope. We implicitly identify expressions which only differ in their choice of bound variables and names (α-conversion). A useful abbreviation is *new* for $\nu n.n$; this is the expression that generates a new name and then immediately returns it.

Expressions are typed according to the rules in Figure 1. The type assertion

$$s, \Gamma \vdash M : \sigma$$

says that in the presence of s and Γ the expression M has type σ. Here s is a finite set of names, Γ is a finite set of typed variables, and M is an expression with free names in s and free variables in Γ. The symbol $\oplus$ represents disjoint union, here in $s \oplus \{n\}$ and $\Gamma \oplus \{x : \sigma\}$. We may omit Γ when it is empty.

$$\frac{}{s, \Gamma \vdash x : \sigma}\ (x : \sigma \in \Gamma) \qquad \frac{}{s, \Gamma \vdash n : \nu}\ (n \in s) \qquad \frac{}{s, \Gamma \vdash b : o}\ (b = \mathit{true}, \mathit{false})$$

$$\frac{s, \Gamma \vdash B : o \quad s, \Gamma \vdash M : \sigma \quad s, \Gamma \vdash M' : \sigma}{s, \Gamma \vdash \mathit{if}\ B\ \mathit{then}\ M\ \mathit{else}\ M' : \sigma} \qquad \frac{s, \Gamma \vdash N : \nu \quad s, \Gamma \vdash N' : \nu}{s, \Gamma \vdash (N = N') : o}$$

$$\frac{s \oplus \{n\}, \Gamma \vdash M : \sigma}{s, \Gamma \vdash \nu n.M : \sigma} \qquad \frac{s, \Gamma \oplus \{x : \sigma\} \vdash M : \sigma'}{s, \Gamma \vdash \lambda x{:}\sigma.M : \sigma \to \sigma'} \qquad \frac{s, \Gamma \vdash F : \sigma \to \sigma' \quad s, \Gamma \vdash M : \sigma}{s, \Gamma \vdash FM : \sigma'}$$

Fig. 1. Rules for assigning types to expressions of the nu-calculus

An expression is in *canonical form* if it is either a name, a variable, one of the boolean constants *true* or *false*, or a function abstraction. These are to be the *values* of the nu-calculus, and correspond to weak head normal form in the lambda-calculus. An expression is *closed* if it has no free variables; a closed expression may still have free names. We define the sets

$$\begin{aligned} \mathrm{Exp}_\sigma(s, \Gamma) &= \{ M \mid s, \Gamma \vdash M : \sigma \} \\ \mathrm{Can}_\sigma(s, \Gamma) &= \{ C \mid C \in \mathrm{Exp}_\sigma(s, \Gamma),\ C \text{ canonical} \} \\ \mathrm{Exp}_\sigma(s) &= \mathrm{Exp}_\sigma(s, \emptyset) \\ \mathrm{Can}_\sigma(s) &= \mathrm{Can}_\sigma(s, \emptyset) \end{aligned}$$

of expressions and canonical expressions, open and closed.

The operational semantics of the nu-calculus is specified by the inductively defined *evaluation relation* given in Figure 2. Elements of the relation take the form

$$s \vdash M \Downarrow_\sigma (s')C$$

where s and s' are disjoint finite sets of names, $M \in \mathrm{Exp}_\sigma(s)$ and $C \in \mathrm{Can}_\sigma(s \oplus s')$. This is intended to mean that in the presence of the names s, expression M of type σ evaluates to canonical form C and creates fresh names s'. We may omit s or s' when they are empty.

Evaluation is chosen to be left-to-right and call-by-value, after Standard ML; it can also be shown to be deterministic and terminating [26, Theorem 2.4].

As an example of evaluation, consider the judgement

$$\vdash (\lambda x{:}\nu.(x = x))(\nu n.n) \Downarrow_o (n)\mathit{true}\ .$$

$$\text{(CAN)} \qquad \frac{}{s \vdash C \Downarrow_\sigma C} \quad C \text{ canonical}$$

$$\text{(COND1)} \qquad \frac{s \vdash B \Downarrow_o (s_1)\mathit{true} \qquad s \oplus s_1 \vdash M \Downarrow_\sigma (s_2)C}{s \vdash \mathit{if}\ B\ \mathit{then}\ M\ \mathit{else}\ M' \Downarrow_\sigma (s_1 \oplus s_2)C}$$

$$\text{(COND2)} \qquad \frac{s \vdash B \Downarrow_o (s_1)\mathit{false} \qquad s \oplus s_1 \vdash M' \Downarrow_\sigma (s_2)C'}{s \vdash \mathit{if}\ B\ \mathit{then}\ M\ \mathit{else}\ M' \Downarrow_\sigma (s_1 \oplus s_2)C'}$$

$$\text{(EQ1)} \qquad \frac{s \vdash N \Downarrow_\nu (s_1)n \qquad s \oplus s_1 \vdash N' \Downarrow_\nu (s_2)n}{s \vdash (N = N') \Downarrow_o (s_1 \oplus s_2)\mathit{true}} \quad n \in s$$

$$\text{(EQ2)} \qquad \frac{s \vdash N \Downarrow_\nu (s_1)n \qquad s \oplus s_1 \vdash N' \Downarrow_\nu (s_2)n'}{s \vdash (N = N') \Downarrow_o (s_1 \oplus s_2)\mathit{false}} \quad n, n' \text{ distinct}$$

$$\text{(LOCAL)} \qquad \frac{s \oplus \{n\} \vdash M \Downarrow_\sigma (s_1)C}{s \vdash \nu n.M \Downarrow_\sigma (\{n\} \oplus s_1)C} \quad n \notin (s \oplus s_1)$$

$$\text{(APP)} \qquad \frac{\begin{array}{c} s \vdash F \Downarrow_{\sigma \to \sigma'} (s_1)\lambda x{:}\sigma.M' \qquad s \oplus s_1 \vdash M \Downarrow_\sigma (s_2)C \\ s \oplus s_1 \oplus s_2 \vdash M'[C/x] \Downarrow_{\sigma'} (s_3)C' \end{array}}{s \vdash FM \Downarrow_{\sigma'} (s_1 \oplus s_2 \oplus s_3)C'}$$

Fig. 2. Rules for evaluating expressions of the nu-calculus

First the argument $\nu n.n$ (or *new*) is evaluated, returning a fresh name bound to n. This is in turn bound to the variable x, and the body of the function compares this name to itself, giving the result *true*. Compare this with

$$\vdash (\nu n'.\lambda x{:}\nu.(x = n'))(\nu n.n) \Downarrow_o (n', n)\mathit{false}\ .$$

Here the evaluation of the function itself creates a fresh name, bound to n'; the argument provides another fresh name, and the comparison then returns *false*.

Repeated evaluation of a name abstraction will give different fresh names. Thus the two expressions

$$\nu n.\lambda x{:}o.n \qquad \text{and} \qquad \lambda x{:}o.\nu n.n$$

behave differently: the first evaluates to the function $\lambda x{:}o.n$, with every subsequent application returning the private name bound to n; while the second gives a different fresh name as result each time it is applied. The expressions are distinguished by the program

$$(\lambda f : o \to \nu\,.\,(f\mathit{true} = f\mathit{true}))\,\langle\!\langle - \rangle\!\rangle$$

which evaluates to *true* or *false* according to how the hole $\langle\!\langle - \rangle\!\rangle$ is filled.

This leads us to the notion of *program context*. A formal definition is given in [26, §2.4]; here we simply note that the form $P\langle\!\langle - \rangle\!\rangle$ represents a program P with some number of holes $\langle\!\langle - \rangle\!\rangle$, and in $P\langle\!\langle (\vec{x})M \rangle\!\rangle$ these are filled by an expression M whose free variables are in the list $\vec{x}$. There is an arrangement to capture these free variables, and the completed program is a closed expression of boolean type.

Definition 1 (Contextual Equivalence). If $M_1, M_2 \in \mathrm{Exp}_\sigma(s,\Gamma)$ then we say that they are *contextually equivalent*, written

$$s,\Gamma \vdash M_1 \approx_\sigma M_2$$

if for all closing program contexts $P\langle\!\langle - \rangle\!\rangle$ and boolean values $b \in \{\mathit{true}, \mathit{false}\}$,

$$(\exists s_1 \,.\, s \vdash P\langle\!\langle(\vec{x})M_1\rangle\!\rangle \Downarrow_o (s_1)b) \quad \Longleftrightarrow \quad (\exists s_2 \,.\, s \vdash P\langle\!\langle(\vec{x})M_2\rangle\!\rangle \Downarrow_o (s_2)b).$$

That is, $P\langle\!\langle - \rangle\!\rangle$ always evaluates to the same boolean value, whether the hole is filled by M_1 or M_2. If both s and Γ are empty then we write simply $M_1 \approx_\sigma M_2$.

This is in many ways the right and proper notion of equivalence between nu-calculus expressions. However the quantification over all programs makes it inconvenient to demonstrate directly; as discussed in the introduction, the purpose of this paper is to present simple methods for reasoning about contextual equivalence without the need to consider contexts or even evaluation.

Examples.

Up to contextual equivalence unused names are irrelevant, as is the order in which names are generated:

$$s,\Gamma \vdash \quad \nu n.M \approx_\sigma M \qquad\qquad n \notin \mathit{fn}(M) \tag{1}$$

$$s,\Gamma \vdash \nu n.\nu n'.M \approx_\sigma \nu n'.\nu n.M \;. \tag{2}$$

Evaluation respects contextual equivalence:

$$s \vdash M \Downarrow_\sigma (s')C \quad \Longrightarrow \quad s \vdash M \approx_\sigma \nu s'.C \tag{3}$$

where $\nu s'.C$ abbreviates multiple name abstractions. A variety of equivalences familiar from the call-by-value lambda-calculus also hold. For instance Plotkin's β_v-rule [24]: if $C \in \mathrm{Can}_\sigma(s,\Gamma)$ and $M \in \mathrm{Exp}_{\sigma'}(s, \Gamma \oplus \{x : \sigma\})$ then

$$s,\Gamma \vdash \ (\lambda x{:}\sigma.M)C \approx_{\sigma'} M[C/x]. \tag{4}$$

Names can be used to detect that general β-equivalence fails, as with

$$(\lambda x{:}\nu.x = x)\mathit{new} \not\approx_o (\mathit{new} = \mathit{new}) \tag{5}$$

which evaluate to *true* and *false* respectively. More interestingly, distinct expressions may be contextually equivalent if they differ only in their use of 'private' names:

$$\nu n.\lambda x{:}\nu.(x = n) \approx_{\nu \to o} \lambda x{:}\nu.\mathit{false}\;. \tag{6}$$

Here the right-hand expression is the function that always returns false; while the left-hand expression evaluates to a function with a persistent local name n, that it compares against any name supplied as an argument. Although these function bodies are quite different, no external context can supply the private name bound to n that would distinguish between them; hence the original expressions are in fact contextually equivalent.

A range of further examples can be found in earlier work on the nu-calculus [22,23,26,27].

3 Operational Reasoning

This section describes two operational techniques for demonstrating contextual equivalences in the nu-calculus. ***Applicative equivalence*** captures much of the general behaviour of higher-order functions and their evaluation, while the more sophisticated ***operational logical relations*** highlight the particular properties of name privacy and visibility. Both are discussed in more detail in [22] and [26], which also give proofs of the results below.

Definition 2 (Applicative Equivalence). We define a pair of relations $s \vdash C_1 \sim_\sigma^{can} C_2$ for $C_1, C_2 \in \mathrm{Can}_\sigma(s)$ and $s \vdash M_1 \sim_\sigma^{exp} M_2$ for $M_1, M_2 \in \mathrm{Exp}_\sigma(s)$ inductively over the structure of the type σ, according to:

$$s \vdash b_1 \sim_o^{can} b_2 \iff b_1 = b_2$$

$$s \vdash n_1 \sim_\nu^{can} n_2 \iff n_1 = n_2$$

$$s \vdash \lambda x{:}\sigma.M_1 \sim_{\sigma\to\sigma'}^{can} \lambda x{:}\sigma.M_2 \iff \forall s', C \in \mathrm{Can}_\sigma(s \oplus s') . \; s \oplus s' \vdash M_1[C/x] \sim_{\sigma'}^{exp} M_2[C/x]$$

$$s \vdash M_1 \sim_\sigma^{exp} M_2 \iff \exists s_1, s_2, C_1 \in \mathrm{Can}_\sigma(s \oplus s_1), C_2 \in \mathrm{Can}_\sigma(s \oplus s_2) . \; s \vdash M_1 \Downarrow_\sigma (s_1)C_1 \;\&\; s \vdash M_2 \Downarrow_\sigma (s_2)C_2 \;\&\; s \oplus (s_1 \cup s_2) \vdash C_1 \sim_\sigma^{can} C_2 .$$

The intuition here is that functions are equivalent if they give equivalent results at possible arguments; while expressions in general are equivalent if they evaluate to equivalent canonical forms.

It is immediate that $\sim_\sigma^{exp}$ coincides with $\sim_\sigma^{can}$ on canonical forms; we write them indiscriminately as $\sim_\sigma$ and call the relation *applicative equivalence.*[2] We can extend the relation to open expressions: if $M_1, M_2 \in \mathrm{Exp}_\sigma(s, \Gamma)$ then we define

$$s, \Gamma \vdash M_1 \sim_\sigma M_2 \iff \forall s', C_i \in \mathrm{Can}_{\sigma_i}(s \oplus s') \quad i = 1, \ldots, n . \; s \oplus s' \vdash M_1[\vec{C}/\vec{x}] \sim_\sigma M_2[\vec{C}/\vec{x}]$$

where $\Gamma = \{x_1 : \sigma_1, \ldots, x_n : \sigma_n\}$.

Applicative equivalence is based on similar 'bisimulation' relations of Abramsky [1] and Howe [8] for untyped lambda-calculus, and Gordon [6] for typed lambda-calculus. It is well behaved and suffices to prove contextual equivalence:

Theorem 3. *Applicative equivalence is an equivalence, a congruence, and implies contextual equivalence.*

The proof of this centres on the demonstration that applicative equivalence is a congruence, *i.e.* it is preserved by all the rules for forming expressions of the

[2] This is a different relation to the applicative equivalence of [22, Def. 13] and [23, Def. 3.4] which (rather unfortunately) turns out not to be an equivalence at all.

nu-calculus. That it implies contextual equivalence follows from this without difficulty; details are in [26, §2.7].

Applicative equivalence verifies examples (1)–(4) above, and numerous others: a range of contextual equivalences familiar from the standard typed lambda-calculus, and all others that make straightforward use of names. What it cannot capture is the notion of privacy that lies behind example (6); where equivalence relies on a particular name remaining secret.

To address the distinction between private and public uses of names, we introduce the idea of a *span* between name sets. A span $R : s_1 \rightleftharpoons s_2$ is an injective partial map from s_1 to s_2; this is equivalent to a pair of injections $s_1 \leftarrowtail R \rightarrowtail s_2$, or a relation such that

$$(n_1, n_2) \in R \quad \& \quad (n_1', n_2') \in R \quad \Longrightarrow \quad (n_1 = n_1') \iff (n_2 = n_2')$$

for $n_1, n_1' \in s_1$ and $n_2, n_2' \in s_2$. The idea is that for any span R the bijection between $\mathrm{dom}(R) \subseteq s_1$ and $\mathrm{cod}(R) \subseteq s_2$ represents matching use of 'visible' names, while the remaining elements not in the graph of R are 'unseen' names. The identity relation $id_s : s \rightleftharpoons s$ is clearly a span; and if $R : s_1 \rightleftharpoons s_2$ and $R' : s_1' \rightleftharpoons s_2'$ are spans on distinct name sets, then their disjoint union $R \oplus R' : s_1 \oplus s_1' \rightleftharpoons s_2 \oplus s_2'$ is also a span. Starting from spans, we now build up a collection of relations between expressions of higher types.

Definition 4 (Logical Relations). If $R : s_1 \rightleftharpoons s_2$ is a span then we define relations

$$R_\sigma^{can} \subseteq \mathrm{Can}_\sigma(s_1) \times \mathrm{Can}_\sigma(s_2)$$
$$R_\sigma^{exp} \subseteq \mathrm{Exp}_\sigma(s_1) \times \mathrm{Exp}_\sigma(s_2)$$

by induction over the structure of the type σ, according to:

$$b_1 \; R_o^{can} \; b_2 \iff b_1 = b_2$$

$$n_1 \; R_\nu^{can} \; n_2 \iff n_1 \, R \, n_2$$

$$\begin{aligned}
&(\lambda x{:}\sigma.M_1) \; R_{\sigma\to\sigma'}^{can} \; (\lambda x{:}\sigma.M_2) \iff \\
&\qquad \forall R' : s_1' \rightleftharpoons s_2',\ C_1 \in \mathrm{Can}_\sigma(s_1 \oplus s_1'),\ C_2 \in \mathrm{Can}_\sigma(s_2 \oplus s_2') \,. \\
&\qquad\quad C_1 \; (R \oplus R')_\sigma^{can} \; C_2 \implies M_1[C_1/x] \; (R \oplus R')_{\sigma'}^{exp} \; M_2[C_2/x]
\end{aligned}$$

$$\begin{aligned}
&M_1 \; R_\sigma^{exp} \; M_2 \iff \\
&\qquad \exists R' : s_1' \rightleftharpoons s_2',\ C_1 \in \mathrm{Can}_\sigma(s_1 \oplus s_1'),\ C_2 \in \mathrm{Can}_\sigma(s_2 \oplus s_2') \,. \\
&\qquad\quad s_1 \vdash M_1 \Downarrow_\sigma (s_1')C_1 \;\; \& \;\; s_2 \vdash M_2 \Downarrow_\sigma (s_2')C_2 \;\; \& \;\; C_1 \; (R \oplus R')_\sigma^{can} \; C_2 .
\end{aligned}$$

This definition differs somewhat from that for applicative equivalence. Functions are now related if they take related arguments to related results; and expressions in general are related if some span can be found between their respective local names that will relate their canonical forms.

The operational relations R_σ^{can} and R_σ^{exp} coincide on canonical forms, and we may write them as R_σ^{opn} indiscriminately. We can extend the relations to open expressions: if $M_1 \in \mathrm{Exp}_\sigma(s_1, \Gamma)$ and $M_2 \in \mathrm{Exp}_\sigma(s_2, \Gamma)$ then define

$$\begin{array}{rcl} \Gamma \vdash M_1 \; R^{opn}_\sigma \; M_2 & \iff & \forall R' : s'_1 \rightleftharpoons s'_2, \\ & & \quad C_{ij} \in \mathrm{Can}_{\sigma_j}(s_i \oplus s'_i) \quad i = 1,2 \quad j = 1,\dots,n \,. \\ & & \quad (\,\&^n_{j=1} \,.\, C_{1j} \; (R \oplus R')^{can}_{\sigma_j} \; C_{2j}\,) \\ & & \quad \implies M_1[\vec{C}_1/\vec{x}] \; (R \oplus R')^{exp}_\sigma \; M_2[\vec{C}_2/\vec{x}] \end{array}$$

where $\Gamma = \{x_1 : \sigma_1, \dots, x_n : \sigma_n\}$.

The intuition is that if $\Gamma \vdash M_1 \; R^{opn}_\sigma \; M_2$ for some $R : s_1 \rightleftharpoons s_2$ then the names in s_1 and s_2 related by R are public and must be treated similarly by M_1 and M_2, while those names not mentioned in R are private and must remain so.

Theorem 5. *For any expressions $M_1, M_2 \in \mathrm{Exp}_\sigma(s, \Gamma)$:*

$$\Gamma \vdash M_1 \; (id_s)^{opn}_\sigma \; M_2 \quad \implies \quad s, \Gamma \vdash M_1 \approx_\sigma M_2 .$$

If σ is a ground or first-order type of the nu-calculus and Γ is a set of variables of ground type, then the converse also holds:

$$s, \Gamma \vdash M_1 \approx_\sigma M_2 \quad \implies \quad \Gamma \vdash M_1 \; (id_s)^{opn}_\sigma \; M_2 .$$

Proposition 6. *Logical relations subsume applicative equivalence: whenever we have $s, \Gamma \vdash M_1 \sim_\sigma M_2$ then also $\Gamma \vdash M_1 \; (id_s)^{opn}_\sigma \; M_2$.*

Thus logical relations can be used to demonstrate contextual equivalence, extending and significantly improving on applicative equivalence. They are not quite sufficient to handle all contextual equivalences [26, §4.6], but they are complete up to first-order functions, and in particular they prove all the examples of Section 2 above.

4 Equational Reasoning

Applicative equivalence is generally much simpler to demonstrate than contextual equivalence, and thus it provides a useful proof technique in itself. However, it is still quite fiddly to apply, and at higher types it involves checking that functions agree on an infinite collection of possible arguments. In this section we present an equational logic that is of similar power but much simpler to use in actual proofs.

Assertions in the logic take the form

$$s, \Gamma \vdash M_1 =_\sigma M_2$$

for open expressions $M_1, M_2 \in \mathrm{Exp}_\sigma(s, \Gamma)$. Valid assertions are derived inductively using the rules of Figure 3. To simplify the presentation we use here a notion of non-binding *univalent context* $U\langle - \rangle$, given by

$$\begin{array}{rl} U\langle - \rangle \; ::= & \langle - \rangle M \mid F\langle - \rangle \mid N = \langle - \rangle \mid \langle - \rangle = N' \\ & \mid \; \textit{if } \langle - \rangle \textit{ then } M \textit{ else } M' \\ & \mid \; \textit{if } B \textit{ then } \langle - \rangle \textit{ else } M' \mid \textit{if } B \textit{ then } M \textit{ else } \langle - \rangle . \end{array}$$

Equality:

$$\frac{}{s, \Gamma \vdash M =_\sigma M} \qquad \frac{s, \Gamma \vdash M_1 =_\sigma M_2}{s, \Gamma \vdash M_2 =_\sigma M_1} \qquad \frac{s, \Gamma \vdash M_1 =_\sigma M_2 \quad s, \Gamma \vdash M_2 =_\sigma M_3}{s, \Gamma \vdash M_1 =_\sigma M_3}$$

Congruence:

$$\frac{s, \Gamma \vdash M_1 =_\sigma M_2}{s, \Gamma \vdash U\langle M_1 \rangle =_{\sigma'} U\langle M_2 \rangle} \qquad \frac{s, \Gamma \oplus \{x : \sigma\} \vdash M_1 =_{\sigma'} M_2}{s, \Gamma \vdash \lambda x{:}\sigma.M_1 =_{\sigma \to \sigma'} \lambda x{:}\sigma.M_2}$$

Functions:

$$\beta_v \qquad \frac{}{s, \Gamma \vdash (\lambda x{:}\sigma.M)C =_{\sigma'} M[C/x]} \qquad C \text{ canonical}$$

$$\eta_v \qquad \frac{}{s, \Gamma \vdash C =_{\sigma \to \sigma'} \lambda x{:}\sigma.Cx} \qquad C \text{ canonical}$$

$$\beta_{id} \qquad \frac{}{s, \Gamma \vdash (\lambda x{:}\sigma.x)M =_\sigma M}$$

$$\beta_U \qquad \frac{}{s, \Gamma \vdash (\lambda x{:}\sigma.U\langle M \rangle)M' =_{\sigma'} U\langle (\lambda x{:}\sigma.M)M' \rangle} (x \notin fv(U\langle - \rangle))$$

Booleans:

$$\frac{}{s, \Gamma \vdash (\mathit{if\ true\ then}\ M\ \mathit{else}\ M') =_\sigma M} \qquad \frac{}{s, \Gamma \vdash (\mathit{if\ false\ then}\ M\ \mathit{else}\ M') =_\sigma M'}$$

$$\frac{s, \Gamma \vdash M_1[\mathit{true}/b] =_\sigma M_2[\mathit{true}/b] \qquad s, \Gamma \vdash M_1[\mathit{false}/b] =_\sigma M_2[\mathit{false}/b]}{s, \Gamma \oplus \{b : o\} \vdash M_1 =_\sigma M_2}$$

Names:

$$\frac{}{s, \Gamma \vdash (n = n) =_o \mathit{true}} (n \in s) \qquad \frac{}{s, \Gamma \vdash (n = n') =_o \mathit{false}} (n, n' \in s \text{ distinct})$$

$$\frac{\begin{array}{ll} s, \Gamma \vdash M_1[n/x] =_\sigma M_2[n/x] & \text{each } n \in s \\ s \oplus \{n'\}, \Gamma \vdash M_1[n'/x] =_\sigma M_2[n'/x] & \text{some fresh } n' \end{array}}{s, \Gamma \oplus \{x : \nu\} \vdash M_1 =_\sigma M_2}$$

New names:

$$\frac{}{s, \Gamma \vdash M =_\sigma \nu n.M} \qquad \frac{}{s, \Gamma \vdash \nu n.\nu n'.M =_\sigma \nu n'.\nu n.M}$$

$$\frac{s \oplus \{n\}, \Gamma \vdash M_1 =_\sigma M_2}{s, \Gamma \vdash \nu n.M_1 =_\sigma \nu n.M_2} \qquad \frac{}{s, \Gamma \vdash U\langle \nu n.M \rangle =_\sigma \nu n.U\langle M \rangle} (n \notin fn(U\langle - \rangle))$$

Fig. 3. Rules for deriving equational assertions.

Thus M is always an immediate subterm of $U\langle M\rangle$, though it may not be the first to be evaluated. This abbreviation appears in the rules for congruence, functions and new names.

General β and η-equivalences do not hold for a call-by-value system such as the nu-calculus; even so, the four rules β_v, η_v, β_{id} and β_U given here still allow considerable scope for function manipulation. In particular the β_U-rule lifts $U\langle - \rangle$ contexts through function application; this is a generalisation of Sabry and Felleisen's β_{lift} [25, Fig. 1].

The most interesting rules of the logic are those concerned with names and name creation. Two expressions with a free variable of type ν are equal if they are equal after instantiation with any existing name, and with a single representative fresh one. Name abstractions $\nu n.(-)$ can be moved past each other, and through contexts $U\langle - \rangle$, providing that name capture is avoided.

Proposition 7. *This equational theory respects evaluation:*

$$s \vdash M \Downarrow_\sigma (s')C \quad\Longrightarrow\quad s \vdash M =_\sigma \nu s'.C\,.$$

Proof. It is not hard to demonstrate, using the equational theory, that every rule for evaluation in Figure 2 preserves the property given. □

Theorem 8 (Soundness and Completeness). *Equational reasoning can be used to prove applicative equivalence, and hence also contextual equivalence:*

$$\begin{aligned} s,\Gamma \vdash M_1 =_\sigma M_2 \quad&\Longrightarrow\quad s,\Gamma \vdash M_1 \sim_\sigma M_2 \\ s,\Gamma \vdash M_1 =_\sigma M_2 \quad&\Longrightarrow\quad s,\Gamma \vdash M_1 \approx_\sigma M_2. \end{aligned}$$

Moreover, it corresponds exactly to applicative equivalence at first-order types, and to contextual equivalence at ground types:

$$\begin{aligned} s,\Gamma \vdash M_1 \sim_\sigma M_2 \quad&\Longrightarrow\quad s,\Gamma \vdash M_1 =_\sigma M_2 \qquad \sigma \textit{ first-order, ground } \Gamma \\ s \vdash M_1 \approx_\sigma M_2 \quad&\Longrightarrow\quad s \vdash M_1 =_\sigma M_2 \qquad \sigma \in \{o,\nu\}. \end{aligned}$$

Proof. Soundness follows from the fact that every rule of Figure 3 for $=_\sigma$ also holds for $\sim_\sigma$. The converse results on completeness are tedious to prove but not especially difficult; however correct use of the rules for introducing free variables of ground type is important for handling first-order functions. □

At higher types applicative equivalence is in principle more powerful than our equational reasoning. However this advantage is illusory: the only way to demonstrate it is to use some more sophisticated technique (such as logical relations) to show that particular functions can never be expressed in the nu-calculus. In practice, the equational logic is much more direct and convenient for reasoning about higher-order functions.

The sample contextual equivalences (1)–(4) from Section 2 are all confirmed immediately by the equational theory. We expand here on two further examples. First, that full β-reduction can be applied to functions with univalent bodies:

$$s,\Gamma \vdash (\lambda x{:}\sigma.U\langle x\rangle)M \approx_{\sigma'} U\langle M\rangle, \tag{7}$$

which we deduce from

$$\begin{aligned} s,\Gamma \vdash (\lambda x{:}\sigma.U\langle x\rangle)M &= U\langle(\lambda x{:}\sigma.x)M\rangle && \text{by } \beta_U \\ &= U\langle M\rangle && \text{by } \beta_{id} \text{ and congruence.} \end{aligned}$$

This extends easily to nested $U\langle-\rangle$ contexts, showing that β-reduction is valid for any function whose bound variable appears just once.

Furthermore, if a function makes no use of its argument at all, then it need not be evaluated:

$$s,\Gamma \vdash (\lambda x{:}\sigma.M)M' \approx_{\sigma'} M \qquad \text{if } x \notin fv(M). \tag{8}$$

In a certain sense then the nu-calculus is free of side-effects. To prove this, we use the univalent context *if true then M else* $\langle-\rangle$, which is certain to ignore the contents of its hole. Thus:

$$\begin{aligned} s,\Gamma \vdash (\lambda x{:}\sigma.M)M' &= (\lambda x{:}\sigma.\mathit{if\ true\ then\ } M \mathit{\ else\ } M)M' \\ &= \mathit{if\ true\ then\ } M \mathit{\ else\ } ((\lambda x{:}\sigma.M)M') && \text{by } \beta_U \\ &= M. \end{aligned}$$

Note that both (7) and (8) may include expressions with free variables, and are truly higher-order: it matters not at all what is the order of the final type σ'.

5 Relational Reasoning

The equational logic presented above is fairly simple, and powerful in that it allows correct reasoning in the presence of an unusual language feature. However it is unable to distinguish between private and public names, and thus cannot prove example (6) of Section 2. The same limitation in the operational technique of applicative equivalence is addressed by a move to logical relations; in this section we introduce a correspondingly refined scheme for relational reasoning about the nu-calculus. As with the equational theory, the aim is to provide all the useful power of operational logical relations in a more accessible form.

Assertions now take the form

$$\Gamma \vdash M_1 \; R_\sigma \; M_2$$

where $R : s_1 \rightleftharpoons s_2$ is a span such that $M_1 \in \mathrm{Exp}_\sigma(s_1,\Gamma)$ and $M_2 \in \mathrm{Exp}_\sigma(s_2,\Gamma)$. To write such assertions, we first need an explicit language to describe spans between sets of names. We build this up using disjoint sum $R \oplus R' : s_1 \oplus s_1' \rightleftharpoons s_2 \oplus s_2'$ over the following basic spans:

$$\begin{aligned} &\overrightarrow{n} : \emptyset \rightleftharpoons \{n\} && \overleftarrow{n} : \{n\} \rightleftharpoons \emptyset \\ &\emptyset : \emptyset \rightleftharpoons \emptyset && n_1 \frown n_2 : \{n_1\} \rightleftharpoons \{n_2\} \quad \text{nonempty.} \end{aligned}$$

In particular, we shall use the derived span:

$$\widehat{n} = n \frown n : \{n\} \rightleftharpoons \{n\} \quad \text{nonempty.}$$

Equational Reasoning:

$$\frac{s, \Gamma \vdash M_1 =_\sigma M_2 \quad \Gamma \vdash M_2 \; R_\sigma \; M_3 \quad s', \Gamma \vdash M_3 =_\sigma M_4}{\Gamma \vdash M_1 \; R_\sigma \; M_4} \; (R : s \rightleftharpoons s')$$

Congruence:

$$\frac{}{\Gamma \vdash x \; R_\sigma \; x} \; (x : \sigma \in \Gamma) \qquad \frac{}{\Gamma \vdash \mathit{true} \; R_o \; \mathit{true}}$$

$$\frac{\Gamma \vdash F_1 \; R_{\sigma \to \sigma'} \; F_2 \quad \Gamma \vdash M_1 \; R_\sigma \; M_2}{\Gamma \vdash (F_1 M_1) \; R_{\sigma'} \; (F_2 M_2)} \qquad \frac{}{\Gamma \vdash \mathit{false} \; R_o \; \mathit{false}}$$

$$\frac{\Gamma \oplus \{x : \sigma\} \vdash M_1 \; R_{\sigma'} \; M_2}{\Gamma \vdash (\lambda x{:}\sigma.M_1) \; R_{\sigma \to \sigma'} \; (\lambda x{:}\sigma.M_2)} \qquad \frac{\Gamma \vdash N_1 \; R_\nu \; N_2 \quad \Gamma \vdash N_1' \; R_\nu \; N_2'}{\Gamma \vdash (N_1 = N_1') \; R_o \; (N_2 = N_2')}$$

$$\frac{\Gamma \vdash B_1 \; R_o \; B_2 \quad \Gamma \vdash M_1 \; R_\sigma \; M_2 \quad \Gamma \vdash M_1' \; R_\sigma \; M_2'}{\Gamma \vdash (\mathit{if}\; B_1 \;\mathit{then}\; M_1 \;\mathit{else}\; M_1') \; R_\sigma \; (\mathit{if}\; B_2 \;\mathit{then}\; M_2 \;\mathit{else}\; M_2')}$$

Booleans:

$$\frac{\Gamma \vdash (M_1[\mathit{true}/b]) \; R_\sigma \; (M_2[\mathit{true}/b]) \quad \Gamma \vdash (M_1[\mathit{false}/b]) \; R_\sigma \; (M_2[\mathit{false}/b])}{\Gamma \oplus \{b : o\} \vdash M_1 \; R_\sigma \; M_2}$$

Names:

$$\frac{}{\Gamma \vdash n_1 \; R_\nu \; n_2} \; ((n_1, n_2) \in R)$$

$$\frac{\begin{array}{ll} \Gamma \vdash (M_1[n/x]) \; (R \oplus \widehat{n})_\sigma \; (M_2[n/x]) & \text{some fresh } n \\ \Gamma \vdash (M_1[n_1/x]) \; R_\sigma \; (M_2[n_2/x]) & \text{each } (n_1, n_2) \in R \end{array}}{\Gamma \oplus \{x : \nu\} \vdash M_1 \; R_\sigma \; M_2}$$

Name creation:

$$\frac{\Gamma \vdash M_1 \; (R \oplus \overleftarrow{n_1})_\sigma \; M_2}{\Gamma \vdash (\nu n_1.M_1) \; R_\sigma \; M_2} \qquad \frac{\Gamma \vdash M_1 \; (R \oplus \overrightarrow{n_2})_\sigma \; M_2}{\Gamma \vdash M_1 \; R_\sigma \; (\nu n_2.M_2)} \qquad \frac{\Gamma \vdash M_1 \; (R \oplus n_1 \hat{\ } n_2)_\sigma \; M_2}{\Gamma \vdash (\nu n_1.M_1) \; R_\sigma \; (\nu n_2.M_2)}$$

Fig. 4. Rules for deriving relational assertions.

It is clear that this language is enough to express all finite spans.

The rules for deriving relational assertions are given in Figure 4. The first of these integrates equational results into the logic, so that existing equational reasoning can be reused and we need only consider spans when absolutely necessary. This is followed by straightforward rules for congruence and booleans. Note that a trace of logical relations comes through in the congruence rule for application: related functions applied to related arguments give related results. As usual the most interesting rules are those concerning names.

To introduce a free variable of type ν requires checking its instantiation with all related pairs of names, and one representative fresh name. This is a weaker constraint than the corresponding rule in the equational logic, where every current name had to be considered; and it is precisely this difference that makes relational reasoning more powerful.

The final three rules handle the name creation operator $\nu n.(-)$, and capture the notion that local names may be private or public. In combination with the equational rules for new names, they are equivalent to the following general rule:

$$\frac{\Gamma \vdash M_1 \; (R \oplus S)_\sigma \; M_2}{\Gamma \vdash (\nu s_1.M_1) \; R_\sigma \; (\nu s_2.M_2)} \qquad S : s_1 \rightleftharpoons s_2 .$$

To apply such rules successfully requires some insight into how an expression uses its local names; which if any are ever revealed to a surrounding program.

Theorem 9 (Soundness). *Relational reasoning can be used to prove the corresponding operational relations:*

$$\Gamma \vdash M_1 \; R_\sigma \; M_2 \quad \Longrightarrow \quad \Gamma \vdash M_1 \; R_\sigma^{opn} \; M_2 \; .$$

By Theorem 5, these can then be used to demonstrate contextual equivalence:

$$\Gamma \vdash M_1 \; (id_s)_\sigma \; M_2 \quad \Longrightarrow \quad s, \Gamma \vdash M_1 \approx_\sigma M_2 \; .$$

Proof. We can show that the operational logical relations R_σ^{opn} satisfy all the rules of Figure 4; this in turn depends on Theorem 8, that provable equality $=_\sigma$ implies applicative equivalence $\sim_\sigma$. □

Theorem 10 (Completeness). *Relational reasoning corresponds exactly to operational logical relations up to first-order types:*

$$\Gamma \vdash M_1 \; R_\sigma^{opn} \; M_2 \quad \Longrightarrow \quad \Gamma \vdash M_1 \; R_\sigma \; M_2 \qquad \sigma \text{ first-order, ground } \Gamma .$$

By Theorem 5, the same result holds for contextual equivalence:

$$s, \Gamma \vdash M_1 \approx_\sigma M_2 \quad \Longrightarrow \quad \Gamma \vdash M_1 \; (id_s)_\sigma \; M_2 \qquad \sigma \text{ first-order, ground } \Gamma .$$

Proof. By induction on the size of Γ and structure of σ; essentially, we work through the defining clause for logical relations (Definition 4). It is significant here that evaluation is respected by the equational logic (Proposition 7), which is in turn incorporated into the relational theory. □

Thus relational reasoning provides a further practical method for reasoning about contextual equivalence. We even have that it can prove all contextual equivalences between expressions of first-order type, thanks to the corresponding (hard) result for operational logical relations. In particular we obtain a demonstration of the final example (6) from Section 2: the crucial closing steps are

$$\frac{\dfrac{x : \nu \vdash (x = n)\ (\overleftarrow{n})_o\ \mathit{false}}{\vdash (\lambda x{:}\nu.(x = n))\ (\overleftarrow{n})_{\nu\to o}\ (\lambda x{:}\nu.\mathit{false})}}{\vdash (\nu n.\lambda x{:}\nu.(x = n))\ \emptyset_{\nu\to o}\ (\lambda x{:}\nu.\mathit{false})}$$

from which we deduce

$$\nu n.\lambda x{:}\nu.(x = n) \approx_{\nu\to o} \lambda x{:}\nu.\mathit{false}$$

as required. The span $(\overleftarrow{n}) : \{n\} \rightleftharpoons \emptyset$ used here captures our intuition that the name bound to n on the left hand side is private, never revealed, and need not be matched in the right hand expression.

6 Conclusions and Further Work

We have looked at the nu-calculus, a language of names and higher-order functions, designed to expose the effect of generativity on program behaviour. Building on operational techniques of applicative equivalence and logical relations, we have derived schemes for equational and relational reasoning; where a collection of inductive rules allow for straightforward proofs of contextual equivalence. We have proved that this approach successfully captures the distinction between private and public names, and is complete up to first-order function types.

Figure 5 summarises the inclusions between the five equivalences that we have considered. For general higher types they are all distinct; at first-order function types the three right-hand equivalences are identified; and at ground types all five are the same. Furthermore, as explained after the proof of Theorem 8, the reasoning schemes of this paper in the bottom row are in practice just as powerful as the operational methods above them.

One direction for future work is to extend the language from names to the dynamically allocated *references* of Standard ML, storage cells that allow imperative update and retrieval. For integer references, appropriate denotational and operational techniques are already available [20,26]. These use relations between sets of states to indicate how equivalent expressions may make different use of local storage cells. The idea then would be to make a similar step in the logic, from name relations to these state relations.

The question of completeness remains open: can these methods be enhanced to prove all contextual equivalences? The operational method of 'predicated logical relations' [26, §4.6] does take things a little further, with an even finer analysis of name use; however the theoretical effort involved seems at present to outweigh the practical returns.

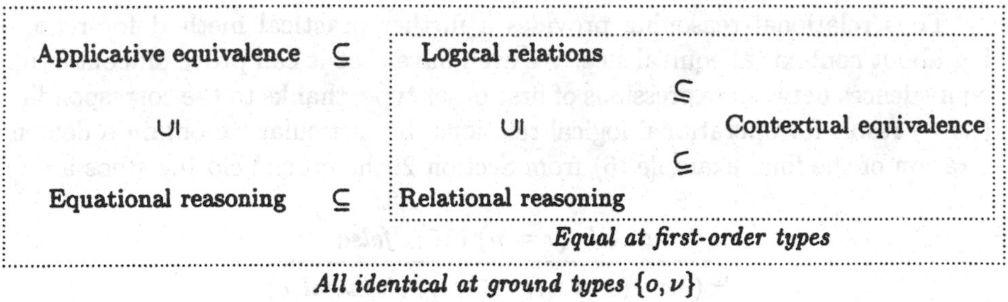

Fig. 5. Various equivalences between expressions of the nu-calculus

Separately from this, it should not be hard to implement the existing relational logic within a generic theorem prover such as Isabelle [19], as Frost and Mason have begun to do for a fragment of VTLoE [3]. The only difficulty for proof search lies in the choice of relation between local name sets. Human guidance is one solution here, but even a brute force approach would work as there are only finitely many spans $R : s_1 \rightleftharpoons s_2$ between any two given name sets. Note that we are not concerned here with an implementation of the proof that the reasoning system itself is correct (Theorem 9); what might benefit from machine assistance is the demonstration that two particular expressions are id_s-related, and hence contextually equivalent.

References

1. S. Abramsky. The lazy lambda calculus. In *Research Topics in Functional Programming*, pages 65–117. Addison Wesley, 1990.
2. M. Felleisen and R. Hieb. The revised report on the syntactic theories of sequential control and state. *Theoretical Computer Science*, 103:235–271, 1992. Also published as Technical Report 100-89, Rice University.
3. J. Frost and I. A. Mason. An operational logic of effects. In *Computing: The Australasian Theory Symposium, Proceedings of CATS '96*, pages 147–156, January 1996.
4. A. Gordon and A. Pitts, editors. *Higher Order Operational Techniques in Semantics.* Cambridge University Press, 1997. To appear.
5. A. D. Gordon. *Functional Programming and Input/Output.* Cambridge University Press, September 1994.
6. A. D. Gordon. Bisimilarity as a theory of functional programming. In *Mathematical Foundations of Programming Semantics: Proceedings of the 11th International Conference*, Electronic Notes in Theoretical Computer Science 1. Elsevier, 1995.
7. F. Honsell, I. A. Mason, S. Smith, and C. Talcott. A variable typed logic of effects. *Information and Computation*, 119(1):55–90, May 1995.
8. D. Howe. Proving congruence of bisimulation in functional programming languages. *Information and Computation*, 124(2):103–112, February 1996.
9. I. A. Mason. *The Semantics of Destructive Lisp.* PhD thesis, Stanford University, 1986. Also published as CSLI Lecture Notes Number 5, Center for the Study of Language and Information, Stanford University.

10. I. A. Mason and C. Talcott. Equivalence in functional languages with effects. *Journal of Functional Programming*, 1(3):297–327, July 1991.
11. I. A. Mason and C. Talcott. Inferring the equivalence of functional programs that mutate data. *Theoretical Computer Science*, 105:167–215, 1992.
12. I. A. Mason and C. L. Talcott. References, local variables and operational reasoning. In *Proceedings of the Seventh Annual IEEE Symposium on Logic in Computer Science*, pages 186–197. IEEE Computer Society Press, 1992.
13. R. Milner. Fully abstract models of typed λ-calculi. *Theoretical Computer Science*, 4:1–22, 1977.
14. R. Milner, J. Parrow, and D. Walker. A calculus of mobile processes, parts I and II. *Information and Computation*, 100:1–77, 1992.
15. R. Milner, M. Tofte, and R. Harper. *The Definition of Standard ML.* MIT Press, 1990.
16. E. Moggi. Notions of computation and monads. *Information and Computation*, 93(1):55–92, July 1991.
17. E. Moggi. A general semantics for evaluation logic. In *Proceedings of the Ninth Annual IEEE Symposium on Logic in Computer Science.* IEEE Computer Society Press, 1994.
18. E. Moggi. Representing program logics in evaluation logic. Unpublished manuscript, available electronically, 1994.
19. L. C. Paulson. *Isabelle: A Generic Theorem Prover.* Lecture Notes in Computer Science 828. Springer-Verlag, 1994.
20. A. Pitts and I. Stark. Operational reasoning for functions with local state. In Gordon and Pitts [4]. To appear.
21. A. M. Pitts. Evaluation logic. In *IVth Higher Order Workshop, Banff 1990*, pages 162–189. Springer-Verlag, 1991. Also published as Technical Report 198, University of Cambridge Computer Laboratory.
22. A. M. Pitts and I. Stark. Observable properties of higher order functions that dynamically create local names, or: What's *new*? In *Mathematical Foundations of Computer Science: Proceedings of the 18th International Symposium*, Lecture Notes in Computer Science 711, pages 122–141. Springer-Verlag, 1993.
23. A. M. Pitts and I. Stark. On the observable properties of higher order functions that dynamically create local names (preliminary report). In *Proceedings of the 1993 ACM SIGPLAN Workshop on State in Programming Languages*, Yale University Department of Computer Science, Research Report YALEU/DCS/RR-968, pages 31–45, 1993.
24. G. Plotkin. Call-by-name, call-by-value and the λ-calculus. *Theoretical Computer Science*, 1:125–159, 1975.
25. A. Sabry and M. Felleisen. Reasoning about programs in continuation-passing style. *Lisp and Symbolic Computation*, 6(3/4):287–358, 1993.
26. I. Stark. *Names and Higher-Order Functions.* PhD thesis, University of Cambridge, December 1994. Also published as Technical Report 363, University of Cambridge Computer Laboratory.
27. I. Stark. Categorical models for local names. *Lisp and Symbolic Computation*, 9(1):77–107, February 1996.
28. C. Talcott. Reasoning about functions with effects. In Gordon and Pitts [4]. To appear.

An Axiomatic System of Parametricity

Takeuti, Izumi

Faculty of Science, Tokyo Metropolitan University,
Minami-Osawa, Hatiozi, Tokyo 192-03 Japan

Abstract. Plotkin and Abadi have proposed a syntactic system for parametricity on a second order predicate logic. This paper shows three theorems about that system. The first is consistency of the system, which is proved by the method of relativization. The second is that polyadic parametricities of recursive types are equivalent to each other. The third is that the theory of parametricity for recursive types is self-realizable. As a corollary of the third theorem, the theory of parametricity for recursive types satisfies the term extraction property.

1 Introduction

1.1 Related Works

There are many works on parametric polymorphism. Interest in parametricity arose in two areas, model theory and formal logic. The original meaning of parametricity was as a notion for models of polymorphic calculi, such as System F, the second order lambda calculus studied by Girard [4]. More recently, many researchers have shown interest in parametricity in formal logic. This paper discusses parametricity in formal logic.

In the sense of Reynolds [8], parametricity is the property that if X is of universal type $\Pi\tau.F[\tau]$ then $XA(F[R])XB$ holds for all types A and B and for all relations R between A and B. In the original meaning, parametricity was a notion for universal types. This paper regards the parametricity notion as extended into the notion for all types, just as Plotkin and Abadi [7] do. The parametricity notion for functional types is called naturality by Hasegawa [5, 6].

Abadi, Cardelli and Curien [1] propose System R, a formal system for parametricity. It is a logical system for binary relations between terms of System F. Because it deals only with binary predicates, its expressive power is rather weak.

Hasegawa [5, 6] makes a category-like model for parametricity, which is called a relation-frame. It is a suitable model for System R. The informal logic which is used by Hasegawa [5, 6] is second order predicate logic, and formalizing his informal logic, we can obtain a formal treatment of parametricity in formal second order predicate logic. Hasegawa [5, 6] does not show the existence of the parametric relation-frame for the whole types of System F, although he shows the existence of the parametric relation-frame for some particular types of System F, such as pair types, sum types, natural numbers, and so forth.

Bellucci, Abadi, and Curien [3] construct a model of System R from a partial equivalence relation model of System F. The existence of the model gives a model theoretic proof of the consistency of the theory of parametricity.

Plotkin and Abadi [7] formalize parametricity on a second order predicate logic. In their system, the basic logic is not specialized to the traitment for parametricity, and the axioms implement the parametricity. They also discuss polyadic parametricities. The original notion of parametricity was a notion for binary predicates. Tertiary parametricity for a universal type is the property that if X is of universal type $\Pi\tau.F[\tau]$ then it holds that $F[R](XA)(XB)(XC)$ for all types A, B and C and for all predicates R for terms of type A, B and C. Quaternary parametricity and n-ary parametricity for general n are defined in the similar ways. Plotkin and Abadi propose a problem: Are the theory of binary parametricity and the theory of tertiary parametricity equivalent? How about for quaternary, or n-ary for general n? We solve a part of this problem in this paper, which is an extension of their work.

1.2 Outline

We have three main results in this paper. The first is the consistency of the theory for parametricity. This theory is given by an axiomatic system of parametricity on a basic logic **L**. The basic logic **L** is second order predicate logic over terms of System F. The consistency has been already shown by the model theoretic proof in [3], but there has been no syntactic proof of consistency to the best of the author's knowledge. Our proof is a direct proof, which uses relativization. It is given in Section 3.

The second result is that the theories of polyadic parametricities are equivalent to each others as long as types are restricted to recursive types. Recursive types are defined as

$$\mathbf{T} ::= \mathbf{1} \mid \tau \mid \mathbf{T} \to \mathbf{T} \mid \mathbf{T} \times \mathbf{T} \mid \mathbf{T} + \mathbf{T} \mid \mu\tau.\mathbf{T} \mid \nu\tau.\mathbf{T}.$$

In the definition, the constructors are encoded as the types of the system F. In $\mu\tau.T$ and $\nu\tau.T$, τ occurs only positively. We define an axiom $\mathbf{Par}_n(T)$ which asserts the n-ary parametricity of a type T, and an axiomatic system $\mathbf{Par}_n(\mathbf{T})$ as the set of axioms $\mathbf{Par}_n(T)$ for all the types T in $\mathbf{T}$. We prove that the theories of $\mathbf{Par}_m(\mathbf{T})$ and of $\mathbf{Par}_n(\mathbf{T})$ are equal to each others for arbitrary numbers $m, n \geq 2$. It is a partial answer to the problem of Plotkin and Abadi [7]. This is given in Section 4.

The third result is the realizability interpretation of the theory of the axiomatic system $\mathbf{Par}(\mathbf{T})$, where we write simply $\mathbf{Par}(\mathbf{T})$ for $\mathbf{Par}_2(\mathbf{T})$. We also write $\mathbf{Par}(T)$ for $\mathbf{Par}_2(T)$. We define a realizability interpretation of the basic logic **L** into **L** itself. and we prove that $\mathbf{Par}(\mathbf{T})$ is self-realisable, that is, for each $\mathbf{Par}(T)$ in $\mathbf{Par}(\mathbf{T})$, there is a term R such that $\mathbf{Par}(\mathbf{T})$ proves that R realizes $\mathbf{Par}(T)$. As the consequence of it, for each theorem P of $\mathbf{Par}(\mathbf{T})$, there is a term R such that $\mathbf{Par}(\mathbf{T})$ proves that R realizes P. This result infers the term extraction property for the axiomatic system $\mathbf{Par}(\mathbf{T})$. It is given in Section 5.

In Section 2 of this paper, we first define a logical system **L**, and then use it as the basic logic for the axiomatic system of parametricity. **L** is a second order predicate logic on terms of the system F. It is slightly modified from the system of the basic logic of Plotkin and Abadi [7]. Their system and **L** are logically equivalent to each other, that is, there are interpretations of their system into **L**, and of **L** into their system, which preserve provabilities and logical meanings. But there are two main differences between our logic and their logic. One is that **L** has n-ary predicate variables for an arbitrary number n, while their system restricts predicate variables to be binary. This is not an essential difference, because both systems have pairing, so polyadic predicates can be encoded into binary predicates. The other difference is that our logic does not have equality as a primitive symbol; it is defined as Leibniz equality in our logic, while their logic has equality as a primitive symbol. This difference is used essentially in relativizing predicates in the proof of the consistency. From another point of view, **L** is a subsystem of λP2 in the λ-cube which is studied by Barendregt [2], though **L** distinguishes types and formulae but λP2 does not distinguish them. The terms and types of **L** are those of System F. For example, $\forall x.\phi x \to \psi x$ is a formula, but not a type of **L**. But λP2 has $\Pi x.\phi x \to \psi x$ as a type, and $y^{\Pi x.\phi x \to \psi x}$ as a variable.

Continuing Section 2, we next define several notions in order to define the axiomatic system **Par** for parametricity. First, conformity for a type T is defined by induction on the construction of T. We use the notation $\langle T\rangle\{\Phi_1/\tau_1, \ldots, \Phi_i/\tau_i\}$ to denote the notion of conformity for the type T. An outline of the definition of conformity is the following:

- $\langle\tau\rangle\{\ldots, \Phi/\tau, \ldots\} :\equiv \Phi$ for a type variable τ.
- $\langle T \to U\rangle\{\ldots\}xy$ iff for all v and w, if $\langle T\rangle\{\ldots\}vw$ then $\langle U\rangle\{\ldots\}(xv)(yw)$.
- $\langle \Pi\tau.T\rangle\{\ldots\}xy$ iff $\langle T\rangle\{\phi/\tau, \ldots\}(x\alpha)(y\beta)$ for all types α and β, and for all predicates ϕ.

If the assignment $\{\ldots\}$ does not assign a predicate to a type variable τ, then the assignment is regarded as assigning equality to τ. We write $\langle T\rangle^=$ for $\langle T\rangle\{\}$. We say that x^T and y^T conform to each other when $\langle T\rangle^= xy$ holds. x^T is self-conforming when x conforms to x itself. $\phi^{T\to *}$ is extensional when ϕx is equivalent to ϕy if x and y conform to each other. Conformity of a type T is a property that all the terms of the type T are self-conforming. Extensionality of a type T is the property that all the predicates of predicate type $T \to *$ are extensional. Parametricity of T is the conjunction of conformity and extensionality of T. We write **Par**(T) for parametricity of T. **Par** is an axiomatic system. Its intuitive meaning is the conjunction of **Par**(T) for all types T. The intuitive meaning of parametricity is that x^T and y^T are equal to each other if and only if they conforms to each other. The axiomatic system **Par** is logically equivalent to the axiomatic system in [7]. As compared to [7] however, we divide the notion of parametricity into two notions: conformity and extensionality. The axioms in [7] only say that conformity and equality are equivalent to each other. The two notions, conformity and extensionality, are the essential points in relativizing predicates as well as Leibniz equality.

In Section 3, we show that the axiomatic system **Par** is consistent. That is the first main result of this paper. We prove consistency by the method of relativization. Because equality is defined as Leibniz equality, an equation works as a statement for predicates as well as a statement for terms. By using this point, we define the relativization. It relativizes terms by self-conformity, and relativizes predicates by extensionality. The relativization is sound and complete for the theory of **Par**, that is, a formula P is provable in the axiomatic system **Par** if and only if the relativized formula $rel(P)$ is provable in **L**.

In Section 4, we discuss the polyadic parametricity, and give a partial answer to the problem of Plotkin and Abadi [7]. Our answer is that the polyadic parametricities are equivalent to each other as long as the types are restricted to recursive types. That is the second main result of this paper. They [7] define n-ary parametricity: n-ary parametricity of a type T is that $\langle T\rangle_n^= x_1 \ldots x_n$ is equivalent to $x_1 = x_2 = \cdots = x_n$, where $\langle T\rangle_n^=$ is an n-ary conformity notation defined in a similar way to the binary. We write $\mathbf{Par}_n(T)$ for n-ary parametricity of a type T, and $\mathbf{Par}_n(\mathbf{T})$ for the set of axioms $\mathbf{Par}_n(T)$ for all the types T in $\mathbf{T}$. The result is that the theories of $\mathbf{Par}_m(\mathbf{T})$ and of $\mathbf{Par}_n(\mathbf{T})$ are equal to each other for arbitrary numbers $m, n{\ge}2$. In other words, for any T in $\mathbf{T}$, $\mathbf{Par}_n(T)$ is provable by $\mathbf{Par}_m(\mathbf{T})$.

In order to prove the result, we show that $\langle T\rangle_m^=$ and $\langle T\rangle_n^=$ are equivalent under the theory of $\mathbf{Par}_m(\mathbf{T})$. From this fact, we obtain $\langle T\rangle_n^=$ is equivalent to equality, because $\mathbf{Par}_m(T)$ implies that $\langle T\rangle_m\{\}$ is equivalent to equality. But we cannot compare $\langle T\rangle_m^=$ and $\langle T\rangle_n^=$ directly, because $\langle T\rangle_m^=$ is an m-ary predicate and $\langle T\rangle_n^=$ is n-ary. Thus we define filters $\mathbf{C}_n^m[\,]$ and $\mathbf{D}_n^m[\,]$ which translate an n-ary predicate into an m-ary predicate. The intuitive meaning of $\mathbf{C}_n^m[\Phi]$ is an internal partial equivalence relation in the domain of Φ, and the intuitive meaning of $\mathbf{D}_n^m[\Phi]$ is the symmetric transitive closure of Φ. The crucial lemma is that $\mathbf{Par}_l(\mathbf{T})$ proves

$$\forall x_1 \ldots x_k . \mathbf{D}_m^k[\langle T\rangle_m\{\ldots\}]x_1 \ldots x_k \supset \mathbf{C}_n^k[\langle T\rangle_n\{\ldots\}]x_1 \ldots x_k$$

for arbitrary numbers $k, l, m, n{\ge}2$. We have also that $\langle T\rangle_m^=$ implies $\mathbf{D}_m^m[\langle T\rangle_m^=]$ and $\mathbf{C}_n^n[\langle T\rangle_n^=]$ implies $\langle T\rangle_n^=$. Thus from this crucial lemma, we obtain the equivalence of $\langle T\rangle_m^=$ and $\langle T\rangle_n^=$.

The proof of the lemma is done by induction on the construction of the type T in $\mathbf{T}$. In the step for functional types $\rightarrow$, the symmetricity and the transitivity of $\mathbf{C}[\,]$ and $\mathbf{D}[\,]$ do essential work. In the steps for $\mu\tau.T$ and $\nu\tau.T$ of the proof, we use the following fixed point lemma. The fixed point lemma involves fixed point property, whose intuitive meaning is that the conformity of fixed point types is equivalent to the fixed point predicate of conformity. The fixed point lemma says that $\mathbf{Par}_m(\mathbf{T})$ proves the fixed point property for n-ary predicates for any arities m and n independent of each other. This lemma is crucial in the induction steps for $\mu\tau.T$ and $\nu\tau.T$. Only this lemma for $m = n$ appears in [1]. This paper extends their lemma.

In Section 5, we give a realizability interpretation to the theory of $\mathbf{Par}(\mathbf{T})$. First we define a realizability interpretation of the basic logic **L** into **L** itself. This interpretation is sound and complete in the following sense. Soundness means that for each theorem P of **L**, there is a term R such that **L** derives that R realizes P; completeness means that **L** proves that if R realizes P, then P holds.

Next we show that the axiomatic system **Par(T)** is self-realizable, that is, there is a term R_T for each axiom $\mathbf{Par}(T) \in \mathbf{Par(T)}$ such that **Par(T)** proves that R_T realizes $\mathbf{Par}(T)$. In order to prove the self-realizability of **Par(T)**, we define a notion $\langle T\rangle^{\models}$. The intuitive meaning of $\langle T\rangle^{\models}rxy$ is that r realizes $\langle T\rangle^{=}xy$. We show the crucial lemma which says that **Par(T)** proves that $\langle T\rangle^{=}_{3}xyz$ implies $\langle T\rangle^{\models}(K_T x)yz$ for some term K_T, for each type $T \in \mathbf{T}$. The proof of this crucial lemma is done by induction on construction of $T \in \mathbf{T}$.

2 Logic for Parametricity

In this section we define a logical system **L**, and a set of axioms for parametricity **Par** on **L**. In order to define **Par**, we introduce the notion of conformity and extensionality in advance.

2.1 The Basic Logic

First we define the basic logic **L**, on which the theory of parametricity is founded. It is the second order logic on the second order typed lambda calculus, which is studied as System F by Girard [4].

Definition 1. Here we define the second order logic on the second order typed lambda calculus, and we call it **L**.

The language of **L** is defined as follows:

- *Types* and *terms* are those of System F, that is:

$$T ::= \tau \mid T \to T \mid \Pi\tau.T \quad \text{and} \quad X ::= x \mid \lambda x^T.X \mid XX \mid \Lambda\tau.X \mid XT$$

where the application is done only when the types of the two terms involved are suitable. All the variables are explicitly typed, thus all the subterms are typed. But we often omit the superfixes which indicate types.

- *Predicate types*, *predicates*, and *formulae* are constructed as:

$$\begin{array}{l}\text{Predicate types: } A ::= * \mid T \to A \\ \text{Predicates: } \Phi ::= \phi^A \mid (\lambda x^T.\Phi^A)^{T\to A} \mid (\Phi^{T\to A}X^T)^A \mid P^* \\ \text{Formulae: } P ::= \Phi^* \mid P \supset P \mid \forall\tau.P \mid \forall x.P \mid \forall\phi.P\end{array}$$

All the variables and the predicate variables are explicitly typed, thus all the subexpressions are typed. But we often omit the superfixes which indicate types. Formulae are the predicates of predicate type $*$. We use the letters $x, y, \ldots$ for variables, $X, Y, \ldots$ for terms, $\tau, \upsilon, \alpha, \beta, \ldots$ for type variables, and $T, U, A, B, \ldots$ for types. A and B are also used for predicate types. $P, Q, \ldots$ are for formulae, $\phi, \psi, \ldots$ for predicate variables, and Φ, Ψ for predicates.

The inference rules are given below. In the expression "$E, \Gamma \vdash P$", E is an environment which declares types of variables such as $\{x : T, y : U, \ldots\}$, Γ is an assumption, which is a finite set of formulae, and P is a formula. An expression Γ, Δ stands for $\Gamma \cup \Delta$, and Γ, P stands for $\Gamma \cup \{P\}$, and so forth.

When X is a term or a predicate, $Env(X)$ is the least environment which is needed for constructing X. When Γ is a set $\{P_1, P_2, \ldots\}$, $Env(\Gamma) :\equiv \bigcup_i Env(P_i)$.

– Structural Rules:

$$\overline{E, \Gamma, P \vdash P}, \qquad \frac{E, \Gamma \vdash P}{E, \Gamma, Q \vdash P}, \qquad \frac{E, \Gamma \vdash P}{E, x : T, \Gamma \vdash P},$$

where $Env(P) \subset E$, $Env(Q) \subset E$ and $Env(\Gamma) \subset E$.

– Implication Introduction:

$$\frac{E, \Gamma, P \vdash Q}{E, \Gamma \vdash P \supset Q}$$

– Implication Elimination:

$$\frac{E, \Gamma \vdash P \supset Q \quad E, \Gamma \vdash P}{E, \Gamma, \vdash Q}$$

– Term Quantification Introduction:

$$\frac{E, x : T, \Gamma, \vdash P}{\Gamma \vdash \forall x.P}$$

where x does not occur free in Γ.

– Term Quantification Elimination:

$$\frac{E, \Gamma \vdash \forall x^T.P}{E, \Gamma \vdash P[X^T/x]}$$

where $Env(X) \subset E$.

– Predicate Quantification Introduction:

$$\frac{E, \Gamma \vdash P}{E, \Gamma \vdash \forall \phi.P}$$

where ϕ does not occur free in Γ.

– Predicate Quantification Elimination:

$$\frac{E, \Gamma \vdash \forall \phi^A.P}{E, \Gamma \vdash P[\Phi^A/\phi]}$$

where $Env(\Phi) \subset E$.

– Type Quantification Introduction:

$$\frac{E, \Gamma \vdash P}{E, \Gamma \vdash \forall \tau.P}$$

where τ does not occur free in Γ, and moreover, τ does not occur free in the types of free variables, nor the predicate types of free predicate variables in P.

– Type Quantification Elimination:

$$\frac{E, \Gamma \vdash \forall \tau.P}{E, \Gamma \vdash P[T/\tau]}$$

– β-Conversion:

$$\frac{E, \Gamma \vdash P}{E, \Gamma \vdash Q}$$

where there is an one-step β-reduction from P to Q, or from Q to P; and $Env(Q) \subset E$. Both $(\lambda x.X)Y$ and $(\lambda x.\Phi)X$ are regarded to be β-redexes.

If E or Γ is a infinite set, then $E, \Gamma \vdash P$ means that there are finite subsets $E' \subset$ and $\Gamma' \subset \Gamma$ such that $E', \Gamma' \vdash P$. If $\vdash P$ is derivable, then P is called a *theorem* of **L**, and we write $\mathbf{L} \vdash P$. The set of all theorems of **L** is called the *theory* of **L**. Let **A** be a set of formulae regarded as axioms. If P is derivable from **A**, then P is called a *theorem* of **A**, and the set of all formulae P such that $\mathbf{A} \vdash P$ is called the *theory* of **A**.

2.2 Preliminaries

Here are some important propositions and definitions for **L**. The propositions say that **L** is situated between $\lambda 2$ and λP2 in Lambda-Cube, which is studied by Barendregt [2].

Proposition 2. **L** *is consistent.*

Proof. By forgetting all the terms and all the inferences of term quantification, a derivation of **L** is translated into a derivation of second order propositional logic. Hence, the consistency of **L** is reduced to the consistency of second order propositional logic. □

This proof mentions a projection which translates derivations of **L** onto those of second order propositional logic.

Proposition 3. *There is a sound interpretation of* **L** *into* $\lambda P2$.

This proof uses an embedding which translates derivations of **L** into terms of λP2. By regarding a derivation of **L** as a term of λP2, and normalizing the term, we get the disjunction property and the term extracting property. Later we shall prove them again by using the realizability interpretation.

Note that the converse does not hold. **L** does not derive $x : \Pi\tau.\tau \vdash \forall\phi.\phi$, although λP2 derives its interpretation, which is $x : \Pi\tau : *.\tau \vdash x : \Pi\tau : *.\tau$.

Now we define some notations.

Definition 4.

$\bot :\equiv \forall\phi^*.\phi,\quad P{\wedge}Q :\equiv \forall\phi^*.(P{\supset}Q{\supset}\phi){\supset}\phi,\quad P{\vee}Q :\equiv \forall\phi^*.(P{\supset}\phi){\supset}(Q{\supset}\phi){\supset}\phi,$

$\exists x.P :\equiv \forall\phi^*.(\forall x.P{\supset}\phi){\supset}\phi,\quad P \supset\subset Q :\equiv (P{\supset}Q) \wedge (Q{\supset}P),$

$X =_T Y :\equiv \forall\phi^{T\to *}.\phi X \supset \phi Y,\quad \mathcal{E}q_T :\equiv \lambda xy.(x =_T y),$

We often omit the suffix T in $=_T$ and $\mathcal{E}q_T$.

In this paper, $\equiv$ denotes identity of expressions and $=_\beta$ denotes β-equality of expressions. We write $\{\dots\}$ for assignments and $[\dots]$ for substitutions.

2.3 Conformity, Extensionality and Parametricity

Now we define the conformity notation for **L**.

Definition 5. Let T be a type and $\tau_1, \dots, \tau_n$ be type variables. Let $\Phi_1, \dots, \Phi_n$ be binary predicates whose predicate types are $A_1 \to B_1 \to *, \dots, A_n \to B_n \to *$ respectively. Let T_1, T_2 be types such that $T_1 \equiv T[A_1/\tau_1, \dots, A_n/\tau_n]$ and $T_2 \equiv T[B_1/\tau_1, \dots, B_n/\tau_n]$.

Then $\langle T\rangle\{\Phi_1/\tau_1, \dots, \Phi_n/\tau_n\}$ is a binary predicate, whose predicate type is $T_1 \to T_2 \to *$. It is called the *conformity notation.* It is defined by the induction on T as below. In the definition, $\Theta :\equiv \{\Phi_1/\tau_1, \dots, \Phi_n/\tau_n\}$.

- For a type variable τ:
  $\langle\tau\rangle\{\Theta\} :\equiv \Phi$ where Θ assigns Φ to τ.

$\langle\tau\rangle\{\Theta\} :\equiv \mathcal{E}q_\tau$ where Θ does not assign a predicate to τ.
- $\langle T \to U\rangle\{\Theta\} :\equiv \lambda xy.\forall vw.\langle T\rangle\{\Theta\}vw \supset \langle U\rangle\{\Theta\}(xv)(yw)$.
- $\langle \Pi\tau.T\rangle\{\Theta\} :\equiv \lambda xy.\forall\alpha\beta.\forall\phi^{\alpha\to\beta\to *}.\langle T\rangle\{\phi/\tau,\Theta\}(x\alpha)(y\beta)$.

For example:

$$\langle \Pi\tau.\tau \to \tau\rangle\{\}XY =_\beta \forall\alpha\beta.\forall\phi^{\alpha\to\beta\to *}.\forall v^\alpha w^\beta.\phi xy \supset \phi(X\alpha v)(Y\beta w).$$

Definition 6. The notations $\langle\ \rangle^=$, **Conf**(), **Ext**() and **Par**() are defined as below:
- $\langle T\rangle^= :\equiv \langle T\rangle\{\}$.
- For a type T and a term X^T: $\mathbf{Conf}_T(X) :\equiv \langle T\rangle^= XX$.

We call this the *self-conformity* of X. When it holds, we say that X is *self-conforming*.
- $\mathbf{Conf}(T) :\equiv \forall x^T.\mathbf{Conf}_T(x)$. We call this the *conformity* of T.
- For predicate type $A \equiv T_1 \to \cdots \to T_i \to *$ and a predicate Φ^A:
  $\mathbf{Ext}_A(\Phi) :\equiv \forall x_1^{T_1}\ldots x_n^{T_n} y_1^{T_1}\ldots y_n^{T_n}.$
  $$\langle T_1\rangle^= x_1y_1 \supset \cdots \supset \langle T_n\rangle^= x_ny_n \supset \Phi x_1\ldots x_i \supset \Phi y_1\ldots y_n.$$

We call this the *extensionality* of Φ. When it holds, we say that Φ is *extensional*.
- $\mathbf{Ext}(T) :\equiv \forall\phi^{T\to *}.\mathbf{Ext}_{T\to *}(\phi)$. We call this the *extensionality* of T.
- $\mathbf{Par}(T) :\equiv \mathbf{Conf}(T) \wedge \mathbf{Ext}(T)$, which is called the *parametricity* of T.

In **L**, $\mathbf{Conf}(T)$ is equivalent to: $\forall x^T y^T.x = y \supset \langle T\rangle^= xy$.
$\mathbf{Ext}(T)$ is equivalent to: $\forall x^T y^T.\langle T\rangle^= xy \supset x = y$.
And $\mathbf{Par}(T)$ is equivalent to: $\forall x^T y^T.x = y \supset\subset \langle T\rangle^= xy$.

Definition 7. **Par** is the set consisting of the axioms $\forall\tau_1\ldots\tau_n.\mathbf{Par}(T)$ where $\tau_1,\ldots,\tau_n$ are all the free type variables of T, for all types T.

The theory of **Par** is logically equivalent to the system of [7]. In the next subsection we will mention that the parametricities infer the universalities of universal types, where the universalities are in a sense of categorical theory.

2.4 Recursive Types

Here are some data structures which are very often used. The theory of parametricity proves the universalities of these structures.

Definition 8. (Unit Type, Pair Type, Sum Type)
Unit type: $\mathbf{1} :\equiv \Pi\tau.\tau \to \tau$.
Pair type: $T \times U :\equiv \Pi\tau.(T \to U \to \tau) \to \tau$.
Sum type: $T + U :\equiv \Pi\tau.(T \to \tau) \to (U \to \tau) \to \tau$.

We call $T \to U$ a functional type. It is well-known that the theory of parametricity proves the universalities of these structures. For example:

Proposition 9. $\mathbf{Par}(T \times U), \mathbf{Par}(T), \mathbf{Par}(U) \vdash \forall x^{T\times U}.x = \boldsymbol{\pi}(\mathbf{pl}x)(\mathbf{pr}x)$, *where $\boldsymbol{\pi}$ is the standard pairing, and* **pl** *and* **pr** *are the standard projections.*

Definition 10. (Fixed point type)

$$\mu\tau.T :\equiv \Pi\tau.T \to \tau \to \tau, \quad \nu\tau.T :\equiv \Pi v.(\Pi\tau.(\tau \to T) \to \tau \to v) \to v,$$

where τ occurs only positively in T. $\mu\tau.T$ is called the *least fixed point type*, and $\nu\tau.T$ is called the *greatest fixed point type.*

We discuss the properties of fixed point types in Section 4.3.

Definition 11. $\mathbf{T}$ is the set of all the types which are constructed from $\mathbf{1}$ and type variables by the pair type $\times$, the sum type $+$, the functional type $\to$ and the fixed point types μ and ν. Types in $\mathbf{T}$ are called *recursive types.* $\mathbf{Par}(\mathbf{T})$ is the set consisting of axioms $\mathbf{Par}(T)$ for all types $T \in \mathbf{T}$.

3 Consistency of the Parametricity Axioms

In this section and the next section we prove the consistency of **Par**. It is done by the method of relativization, and is one of main results in this paper.

3.1 Relativized Parametricity

Definition 12. Here we define the notations $\langle T\rangle_{rel}\{\ldots\}$ and $\langle T\rangle^{=}_{rel}$, which are called *relativized conformity notations.* They are defined by mutual induction on T. The definition of $\langle\ \rangle_{rel}\{\ \}$ is as below:

- For a type variable τ:
  $\langle\tau\rangle_{rel}\{\Theta\} :\equiv \Phi$ where Θ assigns Φ to τ.
  $\langle\tau\rangle_{rel}\{\Theta\} :\equiv \mathcal{E}q_\tau$ where Θ does not assign a predicate to τ.
- $\langle T \to U\rangle_{rel}\{\Theta\} :\equiv$
  $\lambda xy.\forall vw.\langle T\rangle^{=}_{rel}vv \supset \langle T\rangle^{=}_{rel}ww \supset \langle T\rangle_{rel}\{\Theta\}vw \supset \langle U\rangle_{rel}\{\Theta\}(xv)(yw).$
- $\langle\Pi\tau.T\rangle_{rel}\{\Theta\} :\equiv \lambda xy.\forall\alpha\beta\forall\phi.\langle T\rangle_{rel}\{\phi/\tau,\Theta\}(x\alpha)(y\beta)$.

The definition of $\langle T\rangle^{=}_{rel}$ is: $\langle T\rangle^{=}_{rel} :\equiv \langle T\rangle_{rel}\{\}$.

$\langle -\rangle_{rel}\{\ldots\}$ is defined in the similar way to the definition of $\langle -\rangle\{\ldots\}$. The difference between them is that the quantifier is relativized in the definition of $\langle T \to U\rangle_{rel}\{\ldots\}$.

Definition 13. Here are the definitions of relativized self-conformity and relativized extensionality.

$$\mathbf{Conf}^{rel}_T(X) :\equiv \langle T\rangle^{=}_{rel}XX,$$

$$\begin{aligned}\mathbf{Ext}^{rel}_{T_1\to\cdots\to T_n\to *}(\Phi) :\equiv\ & \forall x_1^{T_1}\ldots x_n^{T_n} y_1^{T_1}\ldots y_n^{T_n}.\\ & \mathbf{Conf}^{rel}(x_1) \supset \cdots \supset \mathbf{Conf}^{rel}(x_n) \supset\\ & \mathbf{Conf}^{rel}(y_1) \supset \cdots \supset \mathbf{Conf}^{rel}(y_n) \supset\\ & \langle T_1\rangle^{=}_{rel}x_1y_1 \supset \cdots \supset \langle T_n\rangle^{=}_{rel}x_ny_n \supset \Phi x_1\ldots x_n \supset \Phi y_1..y_n.\end{aligned}$$

Now we define parametricity relativization. This relativization is a translation from formulae into formulae which rewrites quantifiers. We define the translation for predicates for convenience of definition.

Definition 14. For a predicate P, $rel(P)$ is defined by induction on P as follows:
$rel(\phi) :\equiv \phi$, for a predicate variable ϕ.
$rel(\Phi X) :\equiv rel(\Phi)X$, for a predicate ΦX.
$rel(\lambda x.\Phi x) :\equiv \lambda x.rel(\Phi x)$, for a predicate $\lambda x.\Phi x$.
$rel(P \supset Q) :\equiv rel(P) \supset rel(Q)$.
$rel(\forall x^T.P) :\equiv \forall x^T.\mathbf{Conf}_T^{rel}(x) \supset rel(P)$.
$rel(\forall \phi^A.P) :\equiv \forall \phi^A.\mathbf{Ext}_A^{rel}(\phi) \supset rel(P)$.
$rel(\forall \tau.P) :\equiv \forall \tau.rel(P)$.

The intuitive meaning of this relativization is that terms are relativized by self-conformity, and predicates are relativized by extensionality. Note that Leibniz equality and extensionality play essential roles in relativizing a formula $x = y$.

3.2 Soundness and Completeness

This subsection is aimed to prove the following theorem.

Theorem 15. *The above parametricity relativization is sound and complete for the theory of* **Par**, *that is,* **Par** $\vdash P$ *iff* **L** $\vdash rel(P)$ *for every closed formula* P.

To prove this, we need to prove some lemmata. Theorem 15 is proved from the soundness and the completeness of the relativization. We shall obtain the consistency of **Par** from the soundness. First we show the soundness.

Lemma 16. *Let* X *be a term of type* T, *and let* $x_1^{T_1} \ldots x_n^{T_n}$ *be all the free variables in* X. *Then* **L** *derives:*

$$Env(X), Env(Y), \langle T_1\rangle^{=}_{rel} x_1 y_1, \ldots, \langle T_n\rangle^{=}_{rel} x_n y_n \vdash \langle T\rangle^{=}_{rel} XY,$$

where $Y :\equiv X[y_1/x_1, \ldots, y_n/x_n]$.

Proof. By induction on the construction of X. □

Lemma 17. *Let* Φ *be a predicate, and let* $x_1, \ldots, x_m, \phi_1, \ldots, \phi_n$ *be all the variables and the predicate variables in* Φ. *Then* **L** *derives:*

$$Env(\Phi), \mathbf{Conf}^{rel}(x_1), \ldots, \mathbf{Conf}^{rel}(x_m), \mathbf{Ext}^{rel}(\phi_1), \ldots, \mathbf{Ext}^{rel}(\phi_n) \vdash \mathbf{Ext}^{rel}(rel(\Phi)).$$

Proof. By induction on the construction of Φ. □

Lemma 18. **(Soundness of Relativization)** *Let $P, Q_1, \ldots, Q_n$ be formulae, and let $x_1 \ldots x_l, \phi_1, \ldots, \phi_m$ be all the variables and predicate variables in $P, Q_1, \ldots, Q_n$. If $Env(P, Q_1, \ldots, Q_n)$, **Par**, $Q_1, \ldots, Q_n, \vdash P$, then **L** derives:*

$$Env(P), \mathbf{Conf}^{rel}(x_1), \ldots, \mathbf{Conf}^{rel}(x_l), \\ \mathbf{Ext}^{rel}(\phi_1), \ldots, \mathbf{Ext}^{rel}(\phi_m), rel(Q_1), \ldots, rel(Q_n) \vdash rel(P).$$

Proof. By induction on the derivation of $Env(P)$, **Par**, $Q_1, \ldots, Q_n, \vdash P$. □

Next we show completeness.

Proposition 19. **Par**, $Env(P) \vdash P \supset\subset rel(P)$ *for any formula* P.

Lemma 20. **(Completeness of Relativization)** *Let P be a formula, and let $x_1, \ldots, x_m, \phi_1 \ldots \phi_n$ be all the variables and the predicate variables in P. If **L** derives:*

$$Env(P) \vdash \mathbf{Conf}^{rel}(x_1) \supset \cdots \supset \mathbf{Conf}^{rel}(x_m) \supset \\ \mathbf{Ext}^{rel}(\phi_1) \supset \cdots \supset \mathbf{Ext}^{rel}(\phi_n) \supset rel(P),$$

then **Par** $\vdash P$.

Proof. By induction on P. □

Corollary 21. **(Consistency of Par)** **Par** *is consistent.*

Proof. From soundness, if **Par** $\vdash \bot$, then **L** $\vdash rel(\bot)$. But $rel(\bot) \equiv \bot$. Thus, if **Par** is inconsistent, then **L** is inconsistent, so second order propositional logic is inconsistent. □

4 Polyadic Parametricity

We discussed parametricity only for binary relations until the section before. Plotkin and Abadi [7] give the definition of polyadic parametricity. In this section we discuss the equivalence among the theories of polyadic parametricities.

4.1 Definitions, Conjecture and Theorem

The definition of polyadic parametricity is quite similar to the binary one. It is essentially equivalent to that in [7] just the same as for binary relations.

Definition 22. $\mathcal{E}q^n_T :\equiv \lambda x_1 \ldots x_n.(x_1 =_T x_2 \wedge x_1 =_T x_3 \wedge \cdots \wedge x_1 =_T x_n)$.

Definition 23. Let T be a type, n be a number ≥ 2 and $\Phi_1, \ldots, \Phi_m$ be n-ary predicate variables, where Φ_i is of type $U_{i,1} \to \cdots \to U_{i,n} \to *$.

Then $\langle T \rangle_n \{\Phi_1/\tau_1, \ldots, \Phi_m/\tau_m\}$ is a predicate of type $T_1 \to \cdots \to T_n \to *$, where $T_i :\equiv T[U_{1,i}/\tau_1, \ldots, U_{m,i}/\tau_m]$. It is defined by induction on T as below.

- For a type variable: $\langle \tau \rangle_n \{\Theta\} :\equiv \Phi$ where Θ assigns Φ to τ.

$\langle\tau\rangle_n\{\Theta\} :\equiv \mathcal{E}q^n_\tau$ where Θ does not assign a predicate to τ.

- $\langle T \to U\rangle_n\{\Theta\} :\equiv$
  $\lambda x_1 \ldots x_n.\forall y_1 \ldots y_n.\langle T\rangle_n\{\Theta\}y_1 \ldots y_n \supset \langle U\rangle_n\{\Theta\}(x_1y_1)\ldots(x_ny_n)$.
- $\langle \Pi\tau.T\rangle_n\{\Theta\} :\equiv$
  $\lambda x_1 \ldots x_n.\forall \alpha_1 \ldots \alpha_n.\forall \phi^{\alpha_1 \to \cdots \to \alpha_n \to *}.\langle T\rangle_n\{\phi/\tau, \Theta\}(x_1\alpha_1)\ldots(x_n\alpha_n)$.

Definition 24. The definitions of $\langle T\rangle^=_n$, $\mathbf{Par}_n(T)$, $\mathbf{Par}_n$ and $\mathbf{Par}_n(\mathbf{T})$ are similar to those for the binary case. $\langle T\rangle^=_n$ and $\mathbf{Par}_n(T)$ are defined as:

$$\langle T\rangle^=_n :\equiv \langle T\rangle_n\{\}, \quad \mathbf{Par}_n(T) :\equiv \forall x_1^T \ldots x_n^T.\mathcal{E}q x_1 \ldots x_n \supset\subset \langle T\rangle^=_n x_1 \ldots x_n.$$

$\mathbf{Par}_n$ is the set consisting of the axioms $\forall\tau_1 \ldots \tau_n.\mathbf{Par}_n(T)$ where $\tau_1, \ldots, \tau_n$ are all the free type variables of T, for all types T. $\mathbf{Par}_n(\mathbf{T})$ is the set consisting of axioms $\mathbf{Par}_n(T)$ for all types $T \in \mathbf{T}$.

Lemma 25. *$\mathbf{Par}_n$ is consistent. Moreover, $\bigcup_{m \leq n} \mathbf{Par}_m$ is consistent for any n. Therefore, $\bigcup_n \mathbf{Par}_n$ is consistent.*

Proof. Similarly to the case for binary parametricity, there is a suitable relativization which is sound and complete for the theory of $\bigcup_{m \leq n} \mathbf{Par}_m$. □

We make this conjecture for the polyadic parametricity: $\mathbf{Par}_m \vdash \mathbf{Par}_n(T)$ for all type T and $m, n \geq 2$. This conjecture says that all of $\mathbf{Par}_n$'s are equivalent to each other for arbitrary numbers $n \geq 2$. However, we have only the following theorem yet.

Theorem 26. $\mathbf{Par}_m(\mathbf{T}) \vdash \mathbf{Par}_n(T)$ *for all types T in $\mathbf{T}$ and all $m, n \geq 2$.*

This theorem says that all of $\mathbf{Par}_n(\mathbf{T})$'s are equivalent to each other for arbitrary numbers $n \geq 2$. We will spend the remaining part of this section in showing the proof of it.

4.2 Functor

We have to define the notion of functor and some terms in order to prove Theorem 26.

Definition 27. When a type variable τ occurs only positively in a type T, then the operation which makes $T[U/\tau]$ from U is called a *functor*. When a type variable τ occurs only negatively in a type T, then the operation which makes $T[U/\tau]$ from U is called a *contravariant functor*.

Suppose that $C[\,]$ is a context which makes a type $C[T]$ from a type T, and that a functor or contravariant functor F makes $C[T]$ from T. Then $C[-]$ denotes the functor or contravariant functor F.

Definition 28. A functor or a contravariant functor F is said to be in $\mathbf{T}$ if $F\tau \in \mathbf{T}$.

Definition 29. $I_T :\equiv \lambda x^T.x, \quad X \circ Y :\equiv \lambda x.X(Yx)$.

Definition 30. We define the operation of functors and contravariant functors for some terms. Let X be a term of type $T \to U$, F be a functor, or a contravariant functor, then $\langle X \rangle_F$ is a term of type $FT \to FU$ if F is a functor, or of type $FU \to FT$ if F is a contravariant functor, and is defined as follows.

If τ does not occur in $F\tau$, then $\langle X \rangle_F :\equiv I_{F\tau}$.

Otherwise:

- When $F\tau \equiv \tau$, then $\langle X \rangle_F :\equiv X$.
- When $F\tau \equiv G\tau \to H\tau$, then $\langle X \rangle_F :\equiv \lambda x.\langle X \rangle_H \circ x \circ \langle X \rangle_G$.
- When $F\tau \equiv \Pi v.C[\tau, v]$, then $\langle X \rangle_F :\equiv \Lambda v.\langle X \rangle_{C[-,v]}$.

Definition 31. We define substitution of types which occur in functors. When F is a functor and θ is a substitution of types, then $F\theta$ is a functor which makes $F\theta\tau \equiv (F\tau)\theta$ from a fresh variable τ. Substitution of types for contravariant functors is defined similarly.

These identities hold: $FT\theta \equiv F\theta(T\theta)$, $(\langle X \rangle_F)\theta \equiv \langle X\theta \rangle_{F\theta}$, and $(\mu\tau.F\tau)\theta \equiv \mu\tau.F\theta\tau$.

Definition 32. Let F be a functor or a contravariant functor, Θ be an assignment of n-ary predicates for type variables, and Φ be an n-ary predicate. Then: $F^\Theta\Phi :\equiv \langle F\tau \rangle_n\{\Phi/\tau, \Theta\}$.

4.3 Fixed Point Type and Fixed Point Predicate

In this subsection, we define fixed point predicates, and show some properties of fixed point types and fixed point predicates.

Definition 33. Let Φ, Ψ be predicates with the same predicate type. Then:

$$\Phi \sqsubseteq \Psi :\equiv \forall x_1 \dots x_n.\Phi x_1 \dots x_n \supset \Psi x_1 \dots x_n, \quad \text{and} \quad \Phi \cong \Psi :\equiv \Phi \sqsubseteq \Psi \wedge \Psi \sqsubseteq \Phi.$$

Definition 34. Let $\Phi[\phi^A]$ be a predicate of predicate type $A \equiv T_1 \to \cdots T_n \to *$, where ϕ occurs only positively. Then:

$$\mu\phi.\Phi[\phi] :\equiv \lambda x_1 \dots x_n.\forall\phi.(\Phi[\phi] \sqsubseteq \phi) \supset \phi x_1 \dots x_n,$$

$$\nu\phi.\Phi[\phi] :\equiv \lambda x_1 \dots x_n.\exists\phi.(\phi \sqsubseteq \Phi[\phi]) \wedge \phi x_1 \dots x_n.$$

$\mu\phi.\Phi[\phi]$ is called the *least fixed point predicate*, and $\nu\phi.\Phi[\phi]$ is called the *greatest fixed point predicate*.

Proposition 35. (Inferences of Fixed Point Predicate)

$$Env(\Phi), Env(\Psi), \Phi[\Psi] \sqsubseteq \Psi \vdash \mu\phi.\Phi[\phi] \sqsubseteq \Psi, \quad Env(\Phi) \vdash \Phi \cong \mu\phi.\Phi[\phi],$$

$$Env(\Phi), Env(\Psi), \Psi \sqsubseteq \Phi[\Psi] \vdash \Psi \sqsubseteq \nu\phi.\Phi[\phi], \quad Env(\Phi) \vdash \Phi \cong \nu\phi.\Phi[\phi].$$

For a functor F, we write μF for $\mu\tau.F\tau$, and νF for $\nu\tau.F\tau$. It is well-known that μF can encode recursive structures, and νF can encode lazy structures.

Definition 36. For a functor F and a type T, The terms $\mathbf{c}_{\mu F,T}$ and $\mathbf{c}_{\nu F,T}$ are the standard encodings of the constructors of type $(FT \to T) \to \mu F \to T$ and of $(T \to FT) \to T \to \nu F$. $\boldsymbol{f}_{\mu F}$ and $\boldsymbol{f}_{\nu}$ are the standard encodings of the unfold operators of type $\mu F \to F\mu F$ and of type $\nu F \to F\nu F$. $\boldsymbol{g}_{\mu F}$ and $\boldsymbol{g}_{\nu}$ are the standard encodings of the unfold operators of type $F\mu F \to \mu F$ and of type $F\nu F \to \nu F$.

We have by an easy calculation that $\mathbf{c}_{\mu F,T}X \circ \boldsymbol{f}_{\mu F} =_\beta X \circ \langle \mathbf{c}_{\mu F}X\rangle_F$ for each term $X^{FT\to T}$, and also $\boldsymbol{g}_{\nu F} \circ \mathbf{c}_{\mu F,T}X =_\beta \langle \mathbf{c}_{\mu F}X\rangle_F \circ X$ for each term $X^{T\to FT}$. These are a standard encoding of recursive structure and lazy structure. It is well-known that the axioms of parametricities prove that μF is a initial algebra and νF is a final coalgebra.

We will prove the fixed point lemma 40. We define some notations in advance.

Definition 37. $T^0 \to * :\equiv *,\quad T^{n+1} \to * :\equiv T \to T^n \to *.$

Definition 38. Let Φ be a predicate of predicate type $T^n \to *$, and X be a term of type $U \to T$. Then, $\Phi \circ X :\equiv \lambda x_1^U \ldots x_n^U.\Phi(Xx_1)\ldots(Xx_n)$.

Definition 39. Let F be a functor, $\phi_1, \ldots, \phi_m$ be predicate variables where ϕ_i is of predicate type $U_m^n \to *$, θ be a substitution of types $[U_1/\tau_1, \ldots, U_m/\tau_m]$, and Θ be a assignment $\{\phi/\tau_1, \ldots, \phi/\tau_m\}$. Then

$$\mathbf{Fix}_n(\mu F, \theta) :\equiv \forall\phi_1 \ldots \phi_m.(\langle\mu F\rangle_n\{\Theta\} \cong \mu\psi^{\mu F\theta^n \to *}.(F^\Theta\psi) \circ \boldsymbol{g}_{\mu F\theta})$$

$$\mathbf{Fix}_n(\nu F, \theta) :\equiv \forall\phi_1 \ldots \phi_m.(\langle\nu F\rangle_n\{\Theta\} \cong \mu\psi^{\nu F\theta^n \to *}.(F^\Theta\psi) \circ \boldsymbol{g}_{\nu F\theta})$$

We call $\mathbf{Fix}_n(T, \theta)$ *fixed point property* of T for n-ary predicates.

The intuitive meaning of fixed point property is that the conformity of fixed point type is equivalent to the fixed point of conformity.

Lemma 40. (Fixed Point Lemma) *Let F be a functor in* $\mathbf{T}$*, and $F\theta$ also be a functor in* $\mathbf{T}$*. Then* $\mathbf{Par}_m(\mathbf{T}) \vdash \mathbf{Fix}_n(\mu F, \theta)$ *and* $\mathbf{Par}_m(\mathbf{T}) \vdash \mathbf{Fix}_n(\nu F, \theta)$ *for any* $m, n \geq 2$.

Note that we can choose the suffix numbers m and n independently. In [7], this lemma is mentioned, but it is restricted for the same suffix n.

4.4 Crucial Lemma

This subsection is spent on proving Lemma 45. It is crucial for the proof of our main theorem 26.

Definition 41. Let Φ be a predicate of predicate type $T^n \to *$. Then:

$$\hat{\mathfrak{b}}[\Phi] \equiv \lambda x^T y^T.(\Phi x \ldots x \wedge \bigwedge_{i=1,\ldots,n} \forall z_1 \ldots z_{i-1} z_{i+1} \ldots z_n. \Phi z_1 \ldots z_{i-1} x z_{i+1} \ldots z_n \supset\subset \Phi z_1 \ldots z_{i-1} y z_{i+1} \ldots z_n),$$

$$\bar{\mathfrak{b}}[\Phi] :\equiv \lambda x^T y^T.\exists z_1 \ldots z_n.(\Phi z_1 \ldots z_n \wedge (\bigvee_{i=1,\ldots,n} x = z_i) \wedge (\bigvee_{i=1,\ldots,n} y = z_i)),$$

$$\check{\mathfrak{b}}[\Phi] :\equiv \mu\phi^{T \to T \to *}.\lambda xy.\exists z.\bar{\mathfrak{b}}[\Phi]xz \wedge (z = y \vee \phi zy).$$

The binary relation $\hat{\mathfrak{b}}[\Phi]$ denotes a partial equivalence relation in a domain in which Φ holds. If $\hat{\mathfrak{b}}[\Phi]xy$, then x and y behave in the same way for Φ. The binary relation $\bar{\mathfrak{b}}[\Phi]$ is the symmetric closure of Φ, and $\check{\mathfrak{b}}[\Phi]$ is the transitive closure of $\bar{\mathfrak{b}}[\Phi]$.

Definition 42. Let Φ be a predicate of predicate type $T^n \to *$. Then:

$$\mathbf{C}^m_n[\Phi] :\equiv \lambda x_1 \ldots x_m.\hat{\mathfrak{b}}[\Phi]x_1x_2 \wedge \hat{\mathfrak{b}}[\Phi]x_1x_3 \wedge \ldots \wedge \hat{\mathfrak{b}}[\Phi]x_1x_m,$$

$$\mathbf{D}^m_n[\Phi] :\equiv \lambda x_1 \ldots x_m.\check{\mathfrak{b}}[\Phi]x_1x_2 \wedge \check{\mathfrak{b}}[\Phi]x_1x_3 \wedge \ldots \wedge \check{\mathfrak{b}}[\Phi]x_1x_m.$$

$\mathbf{C}^m_n[\,]$ and $\mathbf{D}^m_n[\,]$ are partial equivalence relations of arity m. $\mathbf{C}^m_n[\Phi]$ is an internal relation over the domain of Φ, and $\mathbf{D}^m_n[\Phi]$ is the closure of Φ.

Proposition 43. $\mathbf{D}[\,]$ *is a monotone context, that is,* ϕ *occurs only positively in* $\mathbf{D}^m_n[\phi]$. *Thus* $\mathbf{L}$ *derives:*

$$Env(\Phi), Env(\Psi), \Phi \sqsubseteq \Psi \vdash \mathbf{D}[\Phi] \sqsubseteq \mathbf{D}[\Psi].$$

Definition 44. Let Θ be an assignment $\{\Phi_1^{T^n_1 \to *}/\tau_1, \ldots, \Phi_l^{T^n_l \to *}/\tau_l\}$. Then:

$$\mathbf{C}^m_n[\Theta] :\equiv \{\mathbf{C}^m_n[\Phi_1]/\tau_1, \ldots, \mathbf{C}^m_n[\Phi_l]/\tau_l\}, \mathbf{D}^m_n[\Theta] :\equiv \{\mathbf{D}^m_n[\Phi_1]/\tau_1, \ldots, \mathbf{D}^m_n[\Phi_l]/\tau_l\}.$$

Lemma 45. *Let* $T, T_1, \ldots, T_i, T'_1, \ldots, T'_{i'}, U_1, \ldots, U_j, U'_1, \ldots, U'_{j'}$ *be types in* $\mathbf{T}$. *Suppose that type variables* $\tau_1, \ldots, \tau_i, \tau'_1, \ldots, \tau'_{i'}$ *occur only positively in* T, *and* $\upsilon_1, \ldots, \upsilon_j, \upsilon'_1, \ldots, \upsilon'_{j'}$ *occur only negatively in* T. *Let*

$$\Phi_1{}^{T^m_1 \to *}, \ldots, \Phi_i{}^{T^m_i \to *}, \Phi'_1{}^{T'^n_1 \to *}, \ldots, \Phi'_{i'}{}^{T'^n_{i'} \to *},$$

$$\Psi_1{}^{U^n_1 \to *}, \ldots, \Psi_j{}^{U^n_j \to *}, \Psi'_1{}^{U'^m_1 \to *}, \ldots, \Psi'_{j'}{}^{U'^m_{j'} \to *}$$

be predicates. Let assignments Θ, Θ', Ξ *and* Ξ' *be defined as:*

$$\Theta :\equiv \{\Phi_1/\tau_1, \ldots, \Phi_i/\tau_i\}, \quad \Theta' :\equiv \{\Phi'_1/\tau'_1, \ldots, \Phi'_{i'}/\tau'_{i'}\},$$

$$\Xi :\equiv \{\Psi_1/\upsilon_1, \ldots, \Psi_j/\upsilon_j\}, \quad \Xi' :\equiv \{\Psi'_1/\upsilon'_1, \ldots, \Psi'_{j'}/\upsilon'_{j'}\}.$$

Then for arbitrary numbers $k, l, m, n \geq 2$, *it holds that:*

$$\mathbf{Par}_l(\mathbf{T}) \vdash \mathbf{D}^k_m[\langle T\rangle_m\{\Theta, \mathbf{C}^m_n[\Theta'], \mathbf{D}^m_n[\Xi], \Xi'\}] \sqsubseteq \mathbf{C}^k_n[\langle T\rangle_n\{\mathbf{D}^n_m[\Theta], \Theta', \Xi, \mathbf{C}^n_m[\Xi']\}].$$

Proof. The proof is by induction on the construction of T. We use the following notations:

$$M :\equiv \{\Theta, \mathbf{C}^m_n[\Theta'], \mathbf{D}^m_n[\Xi], \Xi'\}, \quad \text{and,} \quad N :\equiv \{\mathbf{D}^n_m[\Theta], \Theta', \Xi, \mathbf{C}^n_m[\Xi']\}.$$

M is an assignment of predicates with m arguments, and N is an assignment of predicates with n arguments. The substitution θ is defined as:

$$\theta :\equiv [T_1/\tau_1, \ldots, T_i/\tau_i, T'_1/\tau'_1, \ldots, T'_{i'}/\tau_{i'}, U_1/\upsilon_1, \ldots, U_j/\upsilon_j, U'_1/\upsilon'_1, \ldots, U'_{j'}/\upsilon_{j'}]$$

For an arbitrary type T, $\langle T\rangle_m\{M\}$ is a predicate of predicate type $T\theta^m \to *$, and $\langle T\rangle_n\{N\}$ is a predicate of predicate type $T\theta^n \to *$.

We show the proof step only for functional types $\to$ and for least fixed types μ. The steps for unit type $\mathbf{1}$, pair type $\times$, and sum type $+$ are easy, and the step for greatest fixed point types ν are similar to that for μ.

The step for least fixed point types μ is shown by using the fixed point lemma 40. The outline of the step for the least fixed point types is as below:

By the fixed point lemma, we have

$$\langle \mu F\rangle_m\{M\} \cong \mu\phi.F^M\phi \circ \mathbf{g}_{\mu F\theta}, \quad \text{and} \quad \langle \mu F\rangle_n\{N\} \cong \mu\phi.F^N\phi \circ \mathbf{g}_{\mu F\theta}.$$

The induction hypothesis is $F^M(\mathbf{C}^m_n[\langle \mu F\rangle_n\{N\}]) \sqsubseteq \mathbf{C}^m_n[F^N(\langle \mu F\rangle_n\{N\})]$, thus

$$\mu\phi.F^M\phi \sqsubseteq \mathbf{C}^m_n[F^N(\langle \mu F\rangle_n\{N\})] \circ \mathbf{g}_{\mu F\theta}.$$

Therefore we obtain $\langle \mu F\rangle_m\{M\} \sqsubseteq \mathbf{C}^n_n[\langle \mu F\rangle_n\{N\}]$.

The proof of the step for a functional type $T \to U$ is as below:

Our goal in this step is to derive

$$\mathbf{D}^k_m[\langle T \to U\rangle_m\{M\}] \sqsubseteq \mathbf{C}^k_n[\langle T \to U\rangle_n\{N\}].$$

It is enough to show

$$\bar{\mathfrak{b}}[\langle T \to U\rangle_m\{M\}] \sqsubseteq \hat{\mathfrak{b}}[\langle T \to U\rangle_n\{N\}],$$

because then we get the target formula by operating with $\mathbf{D}^k_2[\;]$ on both sides. The induction hypotheses of the case $k = 2$ are:

$$\check{\mathfrak{b}}[\langle T\rangle_m\{N\}] \sqsubseteq \hat{\mathfrak{b}}[\langle T\rangle_m\{M\}], \quad \bar{\mathfrak{b}}[\langle U\rangle_m\{M\}] \sqsubseteq \hat{\mathfrak{b}}[\langle U\rangle_n\{N\}].$$

We can derive $\hat{\mathfrak{b}}[\langle T \to U\rangle_m\{N\}]xy$ from $\bar{\mathfrak{b}}[\langle T \to U\rangle_m\{M\}]xy$. In the derivation, the symmetricity of $\bar{\mathfrak{b}}[\;]$, and the transitivity of $\hat{\mathfrak{b}}[\;]$ and $\check{\mathfrak{b}}[\;]$ do essential work. □

4.5 Proof of the Theorem

Proof of Theorem 26. For each type $T \in \mathbf{T}$, we have

$$\mathbf{Par}_m(\mathbf{T}) \vdash \mathbf{D}^n_m[\langle T\rangle^=_m] \sqsubseteq \mathbf{C}^n_n[\langle T\rangle^=_n], \quad \mathbf{Par}_m(\mathbf{T}) \vdash \mathbf{D}^n_n[\langle T\rangle^=_n] \sqsubseteq \mathbf{C}^n_m[\langle T\rangle^=_m]$$

by Lemma 45. $\mathbf{Par}_m(\mathbf{T})$ also proves $\mathcal{E}q^m_T \cong \langle T\rangle^=_m$. $\mathbf{L}$ derives $\mathcal{E}q^n_T \cong \mathbf{D}^n_m[\mathcal{E}q^m_T]$, and $\mathcal{E}q^n_T \cong \mathbf{C}^n_m[\mathcal{E}q^m_T]$. Thus we obtain the theorem.

5 Realizability Interpretation

In this section we discuss the realizability interpretation for **L**, and we show the self-realizability of **Par(T)**. By using this self-realizability, we obtain the term extraction property of **Par(T)**.

5.1 Realizability of the Basic Logic

First we define the realizability interpretation of **L** into **L** itself.

Definition 46. A figure $\{(\Phi_1, \Psi_1)/\phi_1, \ldots, (\Phi_n, \Psi_n)/\phi_n\}$ denotes the assignment which assigns predicates (Φ_i, Ψ_i) to ϕ_i, where the predicate type of Φ_i is A, which is the same as the predicate type of ϕ_i, and the predicate type of Ψ_i is $T \to A$ for some type T. It is called a *double assignment*.

Definition 47. For a double assignment $\Xi \equiv \{(\Phi_1, \Psi_1)/\phi_1, \ldots, (\Phi_n, \Psi_n)/\phi_n\}$, $[\Xi]$ stands for the substitution such that $[\Xi] :\equiv [\Phi_1/\phi_1 \ldots \Phi_n/\phi_n]$.

Definition 48. We define a type $\mathbf{r}^{\Xi}(\Phi)$ for a predicate Φ and a double assignment Ξ. It will be the type of a realizer of Φ with the double assignment Ξ.

$\mathbf{r}^{\Xi}(\phi) :\equiv T$, where Ξ assigns $(\Phi^A, \Psi^{T\to A})$ to ϕ,
$\mathbf{r}^{\Xi}(\phi) :\equiv \mathbf{1}$, where Ξ does not assign predicates to ϕ,
$\mathbf{r}^{\Xi}(\lambda x.\Phi) :\equiv \mathbf{r}^{\Xi}(\Phi)$,
$\mathbf{r}^{\Xi}(\Phi X) :\equiv \mathbf{r}^{\Xi}(\Phi)$,
$\mathbf{r}^{\Xi}(P \supset Q) :\equiv \mathbf{r}^{\Xi}(P) \to \mathbf{r}^{\Xi}(Q)$,
$\mathbf{r}^{\Xi}(\forall x^T.P) :\equiv T \to \mathbf{r}^{\Xi}(P)$,
$\mathbf{r}^{\Xi}(\forall \tau.P) :\equiv \Pi\tau.\mathbf{r}^{\Xi}(P)$,
$\mathbf{r}^{\Xi}(\forall \phi^A.P) :\equiv \Pi\tau.\mathbf{r}^{\Xi'}(P)$, where $\Xi' \equiv \{(\phi, \psi)/\phi, \Xi\}$ for a fresh $\psi^{\tau\to A}$.

Definition 49. For a predicate Φ, a double assignment Ξ and a term R of type $\mathbf{r}^{\Xi}(\Phi)$, we define the realizability interpretation $R\{\Xi\} \models \Phi$. It is defined by induction on Φ.

$R\{\Xi\} \models \phi :\equiv \lambda x_1 \ldots x_n.(\Phi x_1 \ldots x_n \wedge \Psi R x_1 \ldots x_n,)$
where Ξ assigns (Φ, Ψ) to ϕ,
$R\{\Xi\} \models \phi :\equiv \phi$, where Ξ does not assign predicates to ϕ,
$R\{\Xi\} \models \lambda x.\Phi :\equiv \lambda x.(R\{\Xi\} \models \Phi)$,
$R\{\Xi\} \models \Phi X :\equiv (R\{\Xi\} \models \Phi)X$,
$R\{\Xi\} \models P \supset Q :\equiv (\forall r.(r\{\Xi\} \models P) \supset (Rr\{\Xi\} \models Q)) \wedge (P \supset Q)[\Xi]$,
$R\{\Xi\} \models \forall x.P :\equiv \forall x.Rx\{\Xi\} \models P$,
$R\{\Xi\} \models \forall \tau.P :\equiv \forall \tau.R\tau\{\Xi\} \models P$,
$R\{\Xi\} \models \forall \phi^A.P :\equiv \forall \tau.\forall \phi^A \psi^{\tau\to A}.R\tau\{(\phi, \psi)/\phi, \Xi\} \models P$.

$\mathbf{r}(\Phi)$ stands for $\mathbf{r}^{\{\ \}}(\Phi)$ and $R \models P$ stands for $R\{\} \models P$.

Proposition 50. (Completeness) *For each formula P and each assignment Ξ,* **L** *derives $E, (R\{\Xi\} \models P) \vdash P[\Xi]$ for a suitable environment E.*

Proof. By induction on P. □

Theorem 51. (Soundness) *If* $\vdash P$ *then there is a closed term* R *such that* $\mathbf{L}$ *derives* $\vdash (R \models P)$.

Proof. By induction on the proof of $\vdash P$. □

5.2 Self-Realizability of Parametricity

Lemma 52. *Let* $\mathbf{A}$ *be a set of axioms. If* $\mathbf{A} \vdash (R \models \forall x.\exists y.P)$, *then there is a term* X *such that* $\mathbf{A} \vdash \forall x.P[Xx/y]$.

Lemma 53. *Let* $\mathbf{A}$ *be a set of axioms. Suppose that for each* $A \in \mathbf{A}$ *there is a term* R *such that* $\mathbf{A} \vdash (R \models A)$. *If* $\mathbf{A} \vdash P$, *then there is a term* R *such that* $\mathbf{A} \vdash (R \models P)$.

Corollary 54. *Let* $\mathbf{A}$ *be a set of axioms. Suppose that for each* $A \in \mathbf{A}$ *there is a term* R *such that* $\mathbf{A} \vdash (R \models A)$. *If* $\mathbf{A} \vdash \forall x.\exists y.P$, *then there is a term* X *such that* $\mathbf{A} \vdash \forall x.P[Xx/y]$.

We call $\mathbf{A}$ *self-realizable* when there is a term R_A such that $\mathbf{A} \vdash (R_A \models A)$ for each $A \in \mathbf{A}$. If $\mathbf{A}$ is self-realizable, the theory of $\mathbf{A}$ satisfies term extraction.

We shall show that $\mathbf{Par(T)}$ is self-realizable. We want to show that for each T in $\mathbf{T}$ there is a term R_T such that $\mathbf{Par(T)} \vdash R_T \models \mathbf{Conf}(T)$, which is equivalent to $\mathbf{Par(T)}, x^T \vdash R_T x \models \langle T \rangle^{=} xx$. For the purpose of showing that, we first will define a predicate $\langle T \rangle^{\models}$ for each T, and show that $\mathbf{Par(T)} \vdash \forall x^T.\langle T \rangle^{\models}(K_T x)xx$ for some term K_T.

Definition 55. Let $\tau_1, \ldots, \tau_n$ be all the free variables in T. Then define a predicate

$$\langle T \rangle^{\models} :\equiv \lambda r.(r\{\Xi\} \models \langle T \rangle\{\phi_1/\tau_1, \ldots, \phi_n/\tau_n\})$$

where $\phi_1^{\tau_1 \to \tau_1 \to *}, \ldots, \phi_n^{\tau_n \to \tau_n \to *}$ are fresh predicate variables, and Ξ is a double assignment such that $\Xi \equiv \{(\mathcal{E}q^2_{\tau_1}, \mathcal{E}q^3_{\tau_1})/\phi_1, \ldots, (\mathcal{E}q^2_{\tau_n}, \mathcal{E}q^3_{\tau_n})/\phi_n\}$.

Lemma 56. *Let* T *be a type in* $\mathbf{T}$. *Then there are terms* J_T *and* K_T *such that*

$$\mathbf{Par(T)} \vdash \forall xyz.\langle T \rangle^{=}_{3} xyz \supset \langle T \rangle^{\models}(K_T x)yz,$$

$$\mathbf{Par(T)} \vdash \forall rxy.\langle T \rangle^{\models} rxy \supset \langle T \rangle^{=}_{3}(J_T r)xy$$

Proof. By induction of construction on T. □

Lemma 57. (Realizability of Conformity) *For a type* $T \in \mathbf{T}$, *there is a term* R *such that* $\mathbf{Par(T)} \vdash R \models \mathbf{Conf}(T)$.

Proof. By Lemma 56. □

Lemma 58. (Self-Realizability of Extensionality) *For each type* T, *there is a term* R *such that* $\mathbf{Ext}(T) \vdash (R \models \mathbf{Ext}(T))$.

By using these lemmata, we obtain the next theorem.

Theorem 59. *For each type $T \in \mathbf{T}$, there is a term R such that* $\mathbf{Par}(\mathbf{T}) \vdash R \models \mathbf{Par}(T)$.

Corollary 60. $\mathbf{Par}(\mathbf{T})$ *satisfies the term extraction property.*

6 Conclusion

This paper has discussed the system of parametricity of Plotkin and Abadi [7]. We have pointed out three new results.

The first is consistency of the parametricity. The second is that the polyadic parametricities for recursive types are equivalent to each other. The third is a sound realizability interpretation of the theory of parametricity for recursive types.

Acknowledgement
I am greatly indebted to M. Sato for his total support for this work. I am deeply indebted to J. R. Hindley for carefully reading the draft of this paper and checking my English expression. I am very indebted to G. D. Plotkin, and P. Dybjer for many previous discussions and comments. I would like to thank very much T. Ito, and M. H. Takahashi for their encouragement and comments. I would like to thank R. Hasegawa, P. O'Hearn, Y. Akama, D. Suzuki, M. Hasegawa, and the referees for discussions and comments.

References

1. M. Abadi, L. Cardelli and P.-L. Curien. Formal parametric polymorphism. *Theoretical Computer Science,* 121, 9–58, 1993.
2. H. Barendregt. Lambda Calculi with Types. In *Handbook of Logic in Computer Science,* Vol. 2, 117–309, Oxford Univ. Press, 1992.
3. R. Bellucci, Martin Abadi, P.-L. Curien. A Model for Formal Parametric Polymorphism: A PER Interpretation for system R. In *Typed Lambda Calculi and Applications,* Lecture Notes in Computer Science, 902, Springer-Verlag, pp.32–46, 1995.
4. J.-Y. Girard, et al., *Proofs and Types* Cambridge Univ. Press, 1989.
5. R. Hasegawa. Parametricity of extensionally collapsed term models of polymorphism and their categorical properties. In *Proceeding of Theoretical Aspects of Computer Science,* Lecture Notes In Computer Science 526, pp. 495–512, Springer-Verlag 1991.
6. R. Hasegawa. Categorical data types in parametric polymorphism. In *Mathematical Structures in Computer Science,* 4, 71–109, 1994.
7. G. Plotkin and M. Abadi. A Logic for parametric polymorphism. In *Typed Lambda Calculi and Applications,* Lecture Notes in Computer Science, 664, Springer-Verlag, pp.361–375, 1993.
8. J. C. Reynolds. Types, abstraction and parametric polymorphism. In R. E. A. Mason, editor, *Information Processing 83,* pp. 513–523, Elsevier, 1983.

Inhabitation in Typed Lambda-Calculi (A Syntactic Approach)*

Paweł Urzyczyn

Institute of Informatics
University of Warsaw
ul. Banacha 2, 02-097 Warszawa, Poland
urzy@mimuw.edu.pl

Abstract. A type is inhabited (non-empty) in a typed calculus iff there is a closed term of this type. The inhabitation (emptiness) problem is to determine if a given type is inhabited. This paper provides direct, purely syntactic proofs of the following results: the inhabitation problem is PSPACE-complete for simply typed lambda-calculus and undecidable for the polymorphic second-order and higher-order lambda calculi (systems **F** and $\mathbf{F}_\omega$).

1 Introduction

This paper is mostly of methodological and educational interest. Out of the three main theorems, two have been well-known for many years. The author does not know about a published proof of the third one, but it might have been known to some logicians already. However, the author believes that all the proofs below are much simpler than the existing ones. The motivation for this work was to provide a way to explain these results in a comprehensible way to a reader familiar with lambda-calculus but not necessarily with the proof theory and model theory of intuitionistic logic. The author hopes that the goal was achieved, at least in part.

Let us begin with explaining the problem. In a typed lambda-calculus (Curry or Church style) we normally assume a fixed set of types, and all these types may be assigned to term variables. Thus, there may be terms of arbitrary types, depending on the choice of free variables. However, it is normally the case that not all types may be assigned to closed terms. For instance, the type $\forall\alpha\alpha$ cannot be assigned to any closed term in the second-order lambda-calculus. Such an expression would be quite artificial, as it would have to have all possible types at once. It is less obvious that, for instance, the type $((\alpha \to \beta) \to \alpha) \to \alpha$ is also *empty*, i.e. it cannot be assigned to any closed term. From a programmer's point of view, an empty type means a specification that should be avoided as it cannot be fulfilled by any program phrase. From a logician's point of view, under an appropriate Curry-Howard isomorphism, an *inhabited* (non-empty) type is just a

* This work was partly supported by NSF grant CCR-9417382, KBN Grant 8 T11C 034 10 and by ESPRIT BRA7232 "Gentzen".

provable formula. Thus, the above two examples must indeed be empty: the formula $\forall\alpha\alpha$ is a definition of falsity in the second-order propositional intuitionistic logic, and the other example is the famous Peirce's law — a purely implicational formula which is a classical, but not an intuitionistic tautology.

The inhabitation problem, which is the subject of this paper, can be stated as follows:

Given a type A, is there a closed term of type A?

But for any reasonable typed calculus, we can equivalently rephrase our problem so that it becomes more general:

Given a basis Γ and a type A, is there a term M such that $\Gamma \vdash M : A$?

Clearly, the first version is a special case of the second one. To see that there is also an effective reduction the other way, assume that Γ consists of the declarations $(x_i : A_i)$, for $i = 1, \ldots, n$. Then $\Gamma \vdash M : A$ holds iff there is a closed term of type $A_1 \to \cdots \to A_n \to A$.

The inhabitation problem was first addressed by logicians, as the provability problem for the corresponding logics. The simply-typed lambda-calculus corresponds to ordinary propositional intuitionistic logic, and it has been long well-known that the latter is decidable. This can be obtained from various finite model properties (w.r.t. Kripke models, Heyting algebras etc.) or with help of cut-elimination. (See for instance [5, 10, 6].) A precise characterization of the complexity of this logic was given by Statman [11], who first proved that it is PSPACE-complete. It follows that inhabitation for simply-typed lambda-calculus is also PSPACE-complete. We show in Section 2 how to directly prove the latter fact. As in the original paper of Statman, we use the QBF problem for PSPACE-hardness, but we reduce it directly to the implicational fragment of intuitionistic propositional logic, i.e., to ordinary simple types. Another source of simplification is that we work with lambda-calculus terms in normal form rather than with cut-free proofs.

The second-order lambda-calculus corresponds to the second-order propositional intuitionistic logic, see e.g. [9]. The best known proof that this logic is undecidable is due to Löb [8]. The undecidability result folows from a reduction of classical first-order logic into the second-order propositional intuitionistic logic. The proof is very complex and difficult to follow. In addition it is based on a semantic approach: it uses a completeness theorem for first-order intuitionistic logic extended with free second-order notions. The proof of Löb has been studied and slightly simplified by Arts [1] and Arts and Dekkers [2], but the resulting presentations are still quite complicated.

We reprove Löb's theorem using a coding suggested by Löb himself on page 716 of [8]. However, our proof is syntactic, by a reduction from a restricted fragment of intuitionistic first-order logic, which is simple but powerful enough to remain undecidable. Section 5 introduces this restricted first-order calculus, called system **P**, and Section 7 contains a purely syntactic reduction from provability in system **P** to inhabitation in system **F**. The only semantic element in our proof is the soundness theorem for ordinary first-order classical logic used in Section 5. Replacing this with a syntactic argument about normal proofs is of course possible, but probably pointless.

Another known proof is due to Gabbay [4, 5]. But the system he proves undecidable is an extension of second-order propositional intuitionistic logic with an extra axiom scheme

$$\forall\alpha(A \vee B) \to (A \vee \forall\alpha B), \quad (\alpha \text{ not free in } A),$$

which is not intuitionistically valid. The proof is via completeness theorem for this logic w.r.t. a class of Kripke models.

The proof for system **F**, which we give in Section 7 is considerably simpler than the original Löb's proof, but it is still a bit involved. To give the reader a feeling of the proof method without going into details, we first prove that inhabitation in system $\mathbf{F}_\omega$ is undecidable. This is done in Section 6. Due to the availability of free constructor variables, one can obtain the result in a simpler way, avoiding most of the details. In Section 7, we refine the approach of Section 6 to obtain a proof for system **F**. In fact, for system $\mathbf{F}_\omega$, one does not have to use the same method. It is possible to have even a simpler proof, by a direct machine coding. We give such a proof in Section 4. Whether this approach could be improved to work for system **F** or not, should be a matter of further investigation.

Concluding, let us point out some related results. Inhabitation for intersection types is undecidable, see [13]. The dual problem to type inhabitation, the type reconstruction problem (given a term, can it be assigned a type?) is PTIME-complete for the simply-typed lambda-calculus, and undecidable for both systems **F** and $\mathbf{F}_\omega$, as well as for intersection types, see [3, 14, 12, 7].

In what follows we assume the reader to be familiar with typed lambda calculi, including systems **F** and $\mathbf{F}_\omega$. A reader who does not feel comfortable with $\mathbf{F}_\omega$ may skip Sections 4 and 6, and disregard any mention of "constructors" in Section 3. With the Curry-Howard isomorphism always in mind, we take the liberty to identify lambda-terms with proofs, and we use simplified notation of the form e.g. $\Gamma, A \vdash B$ instead of $\Gamma, x{:}A \vdash M : B$, whenever it is not essential which variable x and term M are considered.

2 Simply-typed lambda calculus

In this section we reprove a result of Statman [11], that the inhabitation problem for the finitely-typed lambda calculus is PSPACE-complete. As we are only interested in the implicational fragment of intuitionistic propositional logic, we can make this proof shorter and perhaps easier to follow than the original one. First we prove that our problem is in PSPACE:

Lemma 1. *There is an alternating polynomial time algorithm (and thus also a deterministic polynomial space algorithm) to determine whether a given type A is inhabited in a given basis Γ in the simply-typed lambda calculus.*

Proof. If a type is inhabited, it is inhabited by a term in an long normal form. To determine if there exists a term M in an long normal form, satisfying $\Gamma \vdash M : A$, we proceed as follows:

- If $A = A_1 \to A_2$, then M must be an abstraction, $M \equiv \lambda x{:}A_1.M'$. Thus, we look for an M' satisfying $\Gamma, x{:}A_1 \vdash M' : A_2$.
- If A is a type variable, then M is an application of a variable to a sequence of terms. We nondeterministically choose a variable z, declared in Γ to be of type $A_1 \to \cdots \to A_n \to A$. If there is no such variable, we reject. If $n = 0$ then we accept. If $n > 0$, we answer in parallel the questions if A_i are inhabited in Γ.

This alternating (or recursive) procedure is repeated as long as there are new questions of the form $\Gamma \vdash ? : A$. Note that if there are two variables in Γ, say x and y, declared to be of the same type B, then each term M can be replaced with $M[x/y]$ with no change of type. This means that a type A is inhabited in Γ iff it is inhabited in $\Gamma - \{y : B\}$, and we can only consider bases with all declared types being different. At each step of our procedure, the basis Γ either stays the same or it expands. Thus the number of steps (depth of recursion) does not exceed the squared number of subformulas of types in Γ, A, where $\Gamma \vdash ? : A$ is the initially posed question. ∎

To show PSPACE-hardness, we define a reduction from the satisfiability problem for classical second-order propositional formulas (QBF). Assume that a second-order propositional formula Φ is given. Without loss of generality we may assume that the negation symbol $\neg$ does not occur in Φ, but in the context $\neg p$, where p is a propositional variable.

Assume that all bound variables of Φ are different and that no variable occurs both free and bound. For each propositional variable p, occurring in Φ (free or bound), let α_p and $\alpha_{\neg p}$ be fresh type variables. Also, for each subformula φ of Φ, let α_φ be a fresh type variable. We construct a basis Γ_Φ from the following types:

- $(\alpha_p \to \alpha_\psi) \to (\alpha_{\neg p} \to \alpha_\psi) \to \alpha_\varphi$, for each subformula φ of the form $\forall p \psi$;
- $(\alpha_p \to \alpha_\psi) \to \alpha_\varphi$ and $(\alpha_{\neg p} \to \alpha_\psi) \to \alpha_\varphi$, for each φ of the form $\exists p \psi$;
- $\alpha_\psi \to \alpha_\vartheta \to \alpha_\varphi$, for each subformula φ of the form $\psi \wedge \vartheta$;
- $\alpha_\psi \to \alpha_\varphi$ and $\alpha_\vartheta \to \alpha_\varphi$, for each subformula φ of the form $\psi \vee \vartheta$.

If v is a zero-one valuation of propositional variables, then Γ_v is defined as Γ extended with the type variables

- α_p, when $v(p) = 1$;
- $\alpha_{\neg p}$, when $v(p) = 0$.

The following lemma is proven by a routine induction w.r.t. the length of formulas. Details are left to the reader.

Lemma 2. *For every subformula φ of Φ, and every valuation v, defined on the free variables of φ, the type α_φ is inhabited in Γ_v iff $v(\varphi) = 1$.*

From Lemma 2 we obtain PSPACE-hardness, since the reduction can be performed in logarithmic space. This, together with Lemma 1 implies the main result of this Section.

Theorem 3. *The inhabitation problem for simply-typed lambda-calculus is complete for polynomial space.*

3 Fully applied polymorphic terms

Unless stated otherwise, all lambda terms considered below are Church terms of system $\mathbf{F}_\omega$. The definitions and lemmas of the present section apply however as well to system **F**, under the restriction that the only constructors are types. Each time we consider a typed term without specifying a basis, we assume that a basis is implicitely given. We use the notation M^A to stress that $M : A$ holds in an appropriately defined basis. An *atom* type is either a type variable or a type of the form $\alpha\vec{A}$, where α is a constructor variable and $\vec{A}$ are constructors of appropriate kinds.

It is convenient to generalize the notion of an long normal form to polymorphic terms. A term M in normal form is said to be *fully applied* iff each subterm of M of a functional or universal type is either an abstraction or occurs in an applied position. More formally, we define fully applied terms of system $\mathbf{F}_\omega$ by induction as follows:

1. If N^B is fully applied then so is $(\lambda x{:}A.N)^{A\to B}$;
2. If N^B is fully applied then so is $(\Lambda\alpha.N)^{\forall\alpha B}$;
3. If x^A is a term variable, B is an atom type, and for some sequence $\vec{\Delta}$ of types (constructors) and fully applied terms we have $x\vec{\Delta} : B$, then $x\vec{\Delta}$ is fully applied.

In clause (2) above, the variable α may be a constructor variable of an arbitrary kind, and in clause (3) the sequence $\vec{\Delta}$ may also involve arbitrary constructors.

It should be obvious from the definition that fully applied terms have the following properties:

- A fully applied term of a functional type is an abstraction;
- A fully applied term of a universal type is a type (constructor) abstraction;
- A fully applied term of an atom type is an application or a variable.

The following lemma holds for both systems, **F** and $\mathbf{F}_\omega$.

Lemma 4. *If* $\Gamma \vdash M : A$ *then there exists a fully applied term* M' *such that* $\Gamma \vdash M' : A$.

A *target* of a type A is the type (constructor) variable that occurs in A along the rightmost path. A basis Γ is *ordinary* if targets of all types in the range of Γ are free variables. The following easy lemma holds for both systems: **F** and $\mathbf{F}_\omega$.

Lemma 5. *Let* Γ *be an ordinary basis and let* $\Gamma \vdash M : \alpha$ *or* $\Gamma \vdash M : \alpha\vec{A}$, *where* α *is a type (constructor) variable and* M *is a fully applied normal form. Then* $M \equiv x\vec{\Delta}$, *where* α *is the target of* $\Gamma(x)$.

4 Inhabitation in $\mathbf{F}_\omega$ — first proof

In this section we show that the inhabitation problem for $\mathbf{F}_\omega$ is undecidable. We use the simplest method: a direct coding of a Turing Machine. We assume that a deterministic TM must have the following properties:

- it never attempts to move left, when reading the leftmost tape cell;
- it never writes a blank;
- there is only one final state.

Tape symbols and internal states are coded with help of (different) type variables and are identified with these variables. In addition we fix a type variable $\#$, which will denote an empty sequence, and one constructor variable $\kappa : \mathbf{Prop} \Rightarrow \mathbf{Prop} \Rightarrow \mathbf{Prop}$. All the above variables are free everywhere in the construction. All bound variables are type variables.

A finite sequence $w = a \cdot w'$ is coded as the type $[w] = \kappa a[w']$, where $[w']$ codes w' (the variable $\#$ codes the empty sequence). The notation $\langle A, B, C\rangle$ abbreviates $\kappa A(\kappa BC)$. A Turing Machine ID of the form (w, q, v) is understood so that wv is the tape contents, q is the current state, and the first symbol of v is being scanned. The word v extends to the right up to (and including) the first blank. Such an ID is coded as $\langle [w^R], q, [v]\rangle$ (where w^R stands for the reverse of w), i.e. as $\kappa[w^R](\kappa q[v])$. That is, the initial ID (with empty input word) is coded by $D_0 = \kappa\#(\kappa q_0(\kappa(\text{blank})\#))$.

The transition relation of a TM is coded as a set of types, each representing one possible clause of the form $(q, a) \vdash (b, p, \varepsilon)$, where ε is either -1 or 0 or $+1$, and denotes "left", "stay" or "right" respectively. The coding depends on v as follows: A "stay" clause $(q, a) \vdash (b, p, 0)$ is coded by the following type:

$$\forall\alpha\beta(\langle\alpha, p, \kappa b\beta\rangle \to \langle\alpha, q, \kappa a\beta\rangle).$$

A "left" clause $(q, a) \vdash (b, p, -1)$ is coded by a set of types, one for each tape symbol c, of the following form:

$$\forall\alpha\beta(\langle\alpha, p, \kappa c(\kappa b\beta)\rangle \to \langle\kappa c\alpha, q, \kappa a\beta\rangle).$$

A "right" clause $(q, a) \vdash (b, p, +1)$, where $a \neq \text{blank}$, is coded by a single type:

$$\forall\alpha\beta(\langle\kappa b\alpha, p, \beta\rangle \to \langle\alpha, q, \kappa a\beta\rangle).$$

Finally, a "right" clause $(q, \text{blank}) \vdash (b, p, +1)$, is represented by

$$\forall\alpha\beta(\langle\kappa b\alpha, p, \kappa(\text{blank})\#\rangle \to \langle\alpha, q, \kappa(\text{blank})\#\rangle).$$

Assuming f is the final state, the final ID's are represented by the type $A_0 = \forall\alpha\beta\langle\alpha, f, \beta\rangle$.

Let A_i, for $i = 1, \ldots, n$ be all types used to code the transition function. Let Γ be the context consisting of type assumptions $x_i : A_i$, for $i = 0, \ldots, n$, thus including the type A_0. The next lemma states the correctness of our representation (recall that D_0 is the type of the initial ID).

Lemma 6. *The machine converges on empty input word iff there exists M such that $\Gamma \vdash M : D_0$.*

Proof. For the "only if" part, let $\mathcal{D}_0, \ldots, \mathcal{D}_p$ be the sequence of ID's that forms a converging computation on empty input ($\mathcal{D}_p$ is a final ID), and let $D_0, \ldots, D_p$ be types representing these ID's. By a "backward" induction we prove that for all i there are M_i with $\Gamma \vdash M_i : D_i$. The "if" part is based on the following claim:

Suppose that $\mathcal{D}$ is an ID coded by a type D and that $\mathcal{D}'$, coded by D', is obtained from $\mathcal{D}$ in one machine step. Let $\Gamma \vdash M : D$ for some fully applied M. Then either $\mathcal{D}$ is final or $M = x_iBCM'$, for some B, C and some M' such that

$$\Gamma \vdash M' : D'.$$

Note that M' must be shorter than M, thus the above may only be applied a finite number of times. Details are left to the reader. ∎

The above lemma gives an effective reduction of the halting problem for TM's to inhabitation for $\mathbf{F}_\omega$. Therefore we obtain:

Theorem 7. *The inhabitation problem for system $\mathbf{F}_\omega$ is undecidable.*

Let us stress here that this construction works only because we have a free constructor variable κ. A type that begins with κ must necessarily be an application and thus κ may be used to build sequences of rigid "cells" able to store data, the contents of such cells never being accessed from outside and decomposed. This approach does not seem to be easily applicable to to system $\mathbf{F}$. For instance, a naive way of coding $a_1a_2\ldots a_n$ into something like $a_1 \to a_2 \to \cdots a_n \to \#$ does not work, because e.g. codes of ab, ba and abb are equivalent. The author does not know how to replace κ by a fixed type constructor definable in $\mathbf{F}$, so that the proof above remains correct. In fact, it seems that such a fixed constructor may not exist at all, for the following reason. An application κAB may neither bind type variables free in A or B, nor it can introduce new variable occurrences into A or B. Thus, even if many applications of κ can possibly introduce deep nesting of quantifier bindings, the scopes of quantifiers cannot "interleave", and may be considered disjoint. On the other hand, it is necessary in Section 5 to use interleaving quantifier scopes.

5 Provability in first-order logic

It is well-known that both classical and intuitionistic first-order logic are undecidable. However, the undecidabilty result holds also for quite restricted first-order systems. In this section we consider one such system, which we call system $\mathbf{P}$, and which is a fragment of intuitionistic first-order calculus. The language of system $\mathbf{P}$ consists of the following symbols:

- binary predicate letters (relation symbols);
- individual variables;

- the symbol **false**;
- the implication sign "$\rightarrow$" and the universal quantifier "$\forall$".

The restriction to binary relations is not essential, and is introduced to uniformize further consideration. The symbol **false** is introduced here for presentation reasons. In fact, we do not formally associate any particular logical meaning to this symbol, so it may be considered just to be a nullary relation symbol (this in turn simulated by a binary one). Later in Sections 6 and 7, we indentify individual variables with type variables, so we now adopt the convention that individual variables of system **P** are written with Greek letters.

The set of formulas of system **P** is a proper subset of first-order formulas over the above alphabet, and is divided into the following three categories:

Atomic formulas: the constant **false**, and expressions $\mathcal{P}(\alpha, \beta)$, where $\mathcal{P}$ is a predicate letter and α, β are individual variables;

Universal formulas: expressions of the form $\forall\vec{\alpha}(A_1 \rightarrow A_2 \rightarrow \cdots \rightarrow A_n)$, where all A_i's are atomic formulas, $A_i \neq$ **false** for $i \neq n$, and each free individual variable of A_n must also occur free in some A_i with $i < n$;

Existential formulas: formulas of the form $\forall\vec{\alpha}(A_1 \rightarrow A_2 \rightarrow \cdots \rightarrow A_{n-1} \rightarrow \forall\beta(A_n \rightarrow \mathbf{false}) \rightarrow \mathbf{false})$, where all A_i's are atomic formulas, and all $A_i \neq$ **false**.

In addition we require that formulas of **P** contain no dummy quantifiers i.e., each quantified variable must actually occur in its scope. (This assumption, as well as the restriction on variable occurrences in universal formulas, is not essential, but simplifies the proofs.)

The name "existential formula" is of course used only for simplicity as it should rather be "universally-existential". Indeed, a classical equivalent of an "existential" formula as above is $\forall\vec{\alpha}(A_1 \rightarrow A_2 \rightarrow \cdots \rightarrow A_{n-1} \rightarrow \exists\beta A_n)$, assuming the ordinary meaning of **false**. Intutionistically, these two formulas become equivalent if **false** is replaced by a universally quantified propositional variable.

The proof rules of system **P** are ordinary natural deduction rules:

(Axiom) $\quad \Gamma, A \vdash A;$

(E$\rightarrow$) $$\frac{\Gamma \vdash A \rightarrow B, \ \Gamma \vdash A}{\Gamma \vdash B}$$

(I$\rightarrow$) $$\frac{\Gamma, A \vdash B}{\Gamma \vdash A \rightarrow B}$$

(I$\forall$) $$\frac{\Gamma \vdash B}{\Gamma \vdash \forall\alpha\, B}$$ (where α is not free in Γ)

(E$\forall$) $$\frac{\Gamma \vdash \forall\alpha B}{\Gamma \vdash B\{\beta/\alpha\}}$$

In the quantifier elimination rule above, β is an individual variable (there are no composite terms). Note that there is no rule for **false**, which means that **false** behaves like a propositional variable rather than a logical constant. However, it does not make difference for us, as long as we only ask about consistency (non-provability of **false**) of sets of **P**-formulas. Indeed, suppose that a set of assumptions Γ is consistent in **P**. By induction w.r.t. the length of normal proofs, one can show that all atomic formulas provable in Γ with help of **false**-elimination are also provable in **P**. (Consider the shortest proof with a **P**-consistent set of assumptions which uses **false**-elimination to prove something that cannot be proven in **P**.) It follows that a **P**-consistent set of assumptions remains consistent after allowing **false**-elimination.

Lemma 8. *For every deterministic two-counter automaton* **M**, *one can effectively construct a finite set* $\Gamma_{\mathbf{M}}$ *of* **P**-*formulas, such that*

M *terminates on input* $(0,0)$ *iff* $\Gamma_{\mathbf{M}} \vdash$ **false** *holds in system* **P**.

Proof. We leave the proof to the full version of the paper. Here we would like to give only an overview of the idea.

An ID of our automaton of the form $C = \langle Q, m, n\rangle$ (the current state is Q, the value of the first counter is m, and the value of the second counter is n) is represented by the following set Γ_C of assumptions:

- $Q(\alpha)$;
- $P_1(\alpha, \alpha_0)$ and $P_1(\alpha_{i-1}, \alpha_i)$, for $i = 1, \ldots, m$;
- $P_2(\alpha, \beta_0)$ and $P_1(\beta_{i-1}, \beta_i)$, for $i = 1, \ldots, n$;
- $D(\alpha)$, $D(\alpha_i)$ and $D(\beta_j)$, for $i < m$ and $j < n$;
- $E(\alpha_m)$ and $E(\beta_n)$.

In the above, Q is a predicate letter representing the state Q, and the α's and β's form two "chains" representing the values of the counters. The role of D and E is to distinguish the end of a chain from an "internal element".

To guarantee the correctness of this representation, we need the following set of formulas Γ_1:

- $\forall\alpha\beta(S(\alpha,\beta) \to D(\beta))$;
- $\forall\alpha(D(\alpha) \to \forall\beta(P_1(\alpha,\beta) \to \mathbf{false}) \to \mathbf{false})$;
- $\forall\alpha(D(\alpha) \to \forall\beta(P_2(\alpha,\beta) \to \mathbf{false}) \to \mathbf{false})$;
- $\forall\alpha(\forall\beta(S(\alpha,\beta) \to \mathbf{false}) \to \mathbf{false})$.

If we think of these formulas as of existential ones, their meaning is to guarantee that a chain representing a counter can be extended as long as the end is not reached, and that there is always a next moment. The transition function of our automaton is represented by the set of formulas Γ_2, which contains 3 or 4 formulas for each internal state of our automaton.

For instance, assume that the machine instruction for state Q be to increase the first counter by 1 and go to state Q'. Then Γ_2 contains the formulas:

- $\forall\alpha\beta(Q(\alpha) \to S(\alpha,\beta) \to Q'(\beta))$, to represent the change of state;
- $\forall\alpha\beta\gamma\delta(Q(\alpha) \to S(\alpha,\beta) \to P_1(\alpha,\gamma) \to P_1(\beta,\delta) \to P_1(\delta,\gamma))$, to lenghten up the chain;
- $\forall\alpha\beta\gamma\delta(Q(\alpha) \to S(\alpha,\beta) \to P_1(\beta,\delta) \to D(\delta))$; to prevent an "artificial" zero;
- $\forall\alpha\beta\gamma(Q(\alpha) \to S(\alpha,\beta) \to P_2(\alpha,\gamma) \to P_2(\beta,\gamma))$, to leave the other counter unmodified.

In addition, we put into Γ_2 the formula

$$\forall\alpha(Q_f(\alpha) \to \mathbf{false}),$$

where Q_f is the final state. Let $\Gamma = \Gamma_1 \cup \Gamma_2$. We define Γ_M to be $\Gamma \cup \Gamma_{C_0}$, where C_0 is the initial ID. The "only if" part of our hypothesis follows from the following claim that is proven by induction on the length of computations:

*Let C be an arbitrary ID. If a final ID is reachable from C then $\Gamma_C \cup \Gamma \vdash$ **false**.*

To prove the "if" part, we assume that the machine does not terminate, and we construct a first-order classical model for the set Γ_{M} in which the nullary predicate letter **false** is interpreted as falsity (and thus deserves its name). It follows that **false** cannot be proved from Γ_{M}, even classically. ∎

Corollary 9. *Provability in system* **P** *is undecidable.*

6 Inhabitation in $\mathbf{F}_\omega$ — second proof

This section is not essential for the technical contents of rest of the paper. Its aim is to help the reader understand Section 7, where we apply the same approach, but with more technical difficulties. It is much easier to do the same with help of higher-order constructors.

Below we define a fragment of system $\mathbf{F}_\omega$, denoted $\mathbf{F}_\omega^{\mathbf{P}}$, which represents the system **P** of Section 5. The system $\mathbf{F}_\omega^{\mathbf{P}}$ is obtained from $\mathbf{F}_\omega$ by imposing appropriate syntactic restrictions on types. We begin with type and constructor variables:

- All constructor variables are either type variables or are of kind **Prop** $\Rightarrow$ **Prop** $\Rightarrow$ **Prop**. There is one distinguished type variable **false**.
- Type variables, except **false**, are identified with the individual variables of **P**, while the first-order constructor variables are identified with the predicate letters of **P**.

In order to simulate **P** within $\mathbf{F}_\omega^{\mathbf{P}}$, we require that constructors may be applied only to individual variables, and only individual variables, except **false**, may be quantified. This restricts the use of constructor variables in types so that they occur as binary relation symbols applied to individual variables. Under this restriction, we may think of types as exactly corresponding to first-order formulas.

In order to obtain an exact correspondence between $\mathbf{F}_\omega^\mathrm{P}$ and $\mathbf{P}$, we define the set of types of $\mathbf{F}_\omega^\mathrm{P}$ (called also $\mathbf{F}_\omega^\mathrm{P}$-formulas) to consists exclusively of types representing formulas of $\mathbf{P}$. We identify these types with the underlying formulas. That is, we have again three kinds of expressions:

Atomic formulas: the variable **false**, and constructor applications of the form $\mathcal{P}\alpha\beta$, where $\mathcal{P}$ is a predicate letter and α, β are individual variables;

Universal formulas: types of the form $\forall\vec{\alpha}(A_1 \to A_2 \to \cdots \to A_n)$, where all A_i's are atomic formulas, $A_i \neq$ **false** for $i \neq n$, and each free individual variable of A_n must also occur free in some A_i with $i < n$;

Existential formulas: types of the form $\forall\vec{\alpha}(A_1 \to A_2 \to \cdots \to A_{n-1} \to \forall\beta(A_n \to \mathbf{false}) \to \mathbf{false})$, where all A_i's are atomic formulas, and all $A_i \neq$ **false**.

Recall that we also assume that formulas contain no dummy quantifiers.

An $\mathbf{F}_\omega$-basis is called an $\mathbf{F}_\omega^\mathrm{P}$-*basis* iff it consists solely of $\mathbf{F}_\omega^\mathrm{P}$-types. We say that the $\mathbf{F}_\omega$-judgement $\Gamma \vdash M : A$ holds in $\mathbf{F}_\omega^\mathrm{P}$ iff all types occurring in it (including all types in M) are $\mathbf{F}_\omega^\mathrm{P}$-formulas. It should be obvious that

$$\Gamma \vdash M : A \text{ holds in } \mathbf{F}_\omega^\mathrm{P}, \text{ for some } M \quad \text{iff} \quad \Gamma \vdash A \text{ in } \mathbf{P}.$$

However, a valid $\mathbf{F}_\omega$-judgement $\Gamma \vdash M : A$, with an $\mathbf{F}_\omega^\mathrm{P}$-basis Γ and an atomic $\mathbf{F}_\omega^\mathrm{P}$-formula A, does not have to hold in $\mathbf{F}_\omega^\mathrm{P}$. This is because the term M may contain occurrences of types of the form e.g., $\mathcal{P}(B, C)$, where the predicate letter $\mathcal{P}$ is applied to some complex types B, C, and not to type variables. But such a term can always be replaced by another one, which makes a valid $\mathbf{F}_\omega^\mathrm{P}$ judgement.

Lemma 10. *Let Γ be an $\mathbf{F}_\omega^\mathrm{P}$-basis, and let an atomic formula A be inhabited in Γ in the system $\mathbf{F}_\omega$. Then A is inhabited in Γ in the system $\mathbf{F}_\omega^\mathrm{P}$, and thus it is provable in $\mathbf{P}$ from Γ.*

Proof. We first prove the following claim. Let $\mathcal{P}$ be of kind **Prop** $\Rightarrow$ **Prop** $\Rightarrow$ **Prop**. If Γ is an $\mathbf{F}_\omega^\mathrm{P}$-basis and $\Gamma \vdash M : \mathcal{P}AB$ holds in $\mathbf{F}_\omega$, for some types A and B, then A and B are type variables. To prove the claim, suppose the contrary, and assume that M is the shortest possible fully applied counterexample. By Lemma 5, we must have a variable z declared in Γ with target $\mathcal{P}$, such that $M \equiv z\vec{\Delta}$. If the type of z is an atomic formula, then A and B must be variables (and $\vec{\Delta}$ is empty). Otherwise, the type of z is an universal formula, say $\forall\vec{\alpha}(A_1 \to A_2 \to \cdots \to A_n \to \mathcal{P}\alpha_1\alpha_2)$. Then $M \equiv z\vec{C}\vec{N}$, for some types $\vec{C}$ and some terms $\vec{N}$. Types A and B occur in $\vec{C}$ at places corresponding to occurrences of the variables α_1 and α_2 in $\vec{\alpha}$. But α_1 and α_2 must occur in some of the A_i's. Thus, one of the terms in N is actually of type $\mathcal{Q}CD$, where $\mathcal{Q}$ is a predicate letter and either C or D is a non-variable type (equal to either A or B). This gives a counterexample to our claim, shorter than M.

Now suppose that an atomic formula A is inhabited in Γ in the system $\mathbf{F}_\omega$, say $\Gamma \vdash M : A$ with fully applied M. We proceed by induction on M. If A has the form $\mathcal{P}\alpha\beta$ then either M is a variable (and the hypothesis is obvious) or

$M \equiv z\vec{C}\vec{N}$, for some types $\vec{C}$ and some terms $\vec{N}$. It follows from the previous claim that the types $\vec{C}$ must be just type variables, as otherwise one of the terms in $\vec{N}$ would contradict the claim. Thus, the induction hypothesis applies for $\vec{N}$, i.e., types of these terms are inhabited in $\mathbf{F}_\omega^\mathrm{P}$. Let $\vec{N}'$ be a vector of inhabitants. Thus A is also inhabited, since the application $z\vec{C}\vec{N}'$ does not introduce unwanted types.

It remains to consider the case when $A \equiv \mathbf{false}$. One possibility is that $M \equiv z\vec{C}\vec{N}$, where the type of z is an universal formula (with target **false**). Then we proceed as in the previous case. Otherwise, M must be of the form $z\vec{C}\vec{N}L$, where $\vec{C}$ must be variables and $\vec{N}$ may be replaced by $\vec{N}'$, just as in the previous case. The term L is of type $\forall\beta(B \to \mathbf{false})$, for some atomic B, and has the form $\Lambda\beta.\lambda y{:}B.L_1^{\mathbf{false}}$. Thus, $\Gamma, y : B \vdash L_1 : \mathbf{false}$ and $\Gamma, y : B$ is an $\mathbf{F}_\omega^\mathrm{P}$ basis. This allows us to apply the induction hypothesis to L_1. ∎

From the above Lemma, and from Lemma 8, we conclude that a given two-counter automaton $\mathbf{M}$ terminates on input $(0,0)$ iff $\Gamma_\mathbf{M} \vdash \mathbf{false}$ is a valid $\mathbf{F}_\omega$-judgement. Thus, the second proof of Theorem 7 is completed.

The concluding observation is that although it is obvious that the above property holds for $\mathbf{F}_\omega^\mathrm{P}$, it is not obvious at all for the full system $\mathbf{F}_\omega$. Luckily, the proof is still easy, due to the fact that relation symbols are represented by free constructors. In the next Section, we reprove this once again, but this time for system $\mathbf{F}$. The main technical problem that must be solved for this proof is how to replace the free constructor variables with type expressions, so that an analogue of Lemma 10 is true.

7 Inhabitation in F

The departure point for this section is a given set $\Gamma_\mathbf{M}$ of closed formulas of $\mathbf{P}$, as in Section 5 (equivalently seen as a set of $\mathbf{F}_\omega^\mathrm{P}$ formulas, with no free individual variables). The goal is to effectively define a coding $\overline{\Gamma}_\mathbf{M}$ of the formulas (types) in $\Gamma_\mathbf{M}$, so that $\Gamma_\mathbf{M} \vdash \mathbf{false}$ holds in $\mathbf{P}$ if and only if $\overline{\Gamma}_\mathbf{M} \vdash \mathbf{false}$ holds in system $\mathbf{F}$.

It is technically useful to assume that we have three disjoint sorts of type variables in system $\mathbf{F}$: the individual variables, the auxiliary variables used for the coding, and one distinguished type variable **false**. Of course, the variable **false** is used to code the symbol **false**. The coding of predicate letters uses a number of fresh auxiliary type variables, which remain free in all types. For each predicate letter $\mathcal{P}$ we need three such type variables denoted p, p_1 and p_2 (the convention is to use p for $\mathcal{P}$, and q for $\mathcal{Q}$ etc.) and we will use three more type variables ξ, η_1 and η_2. We code an atomic formula $\mathcal{P}(\alpha,\beta)$ as the type

$$\overline{\mathcal{P}(\alpha,\beta)} \equiv [(\alpha \to p_1) \to (\beta \to p_2) \to p] \to \xi.$$

To avoid rewriting this long type each time, we will abbreviate the argument part to $\mathcal{P}_{\alpha\beta}$, so that we write

$$\overline{\mathcal{P}(\alpha,\beta)} \equiv \mathcal{P}_{\alpha\beta} \to \xi.$$

We use the abbreviations $\mathcal{P}(A,B)$ and $\mathcal{P}_{AB}$ to denote the types $\mathcal{P}(\alpha,\beta)[A/\alpha,B/\beta]$ and $\mathcal{P}_{\alpha\beta}[A/\alpha,B/\beta]$, respectively.

To define codes of complex formulas, we need an additional set of types $\mathcal{U}(\alpha)$, for each individual variable α. This set consists of all types of the form

$$(\alpha \to p_i) \to \eta_1$$

for all $\mathcal{P}$ and all $i = 1,2$, and one more type

$$\alpha \to \eta_2.$$

To simplify notation let us use the following conventions. We write $\Gamma \vdash \mathcal{U}(\alpha)$ to mean that $\Gamma \vdash A$, for all $A \in \mathcal{U}(\alpha)$. An expression of the form "$\mathcal{U}(\alpha) \to B$" stands for "$A_1 \to \cdots A_n \to B$", where $A_1,\ldots,A_n$ are all members of $\mathcal{U}(\alpha)$. We write $\mathcal{U}(\vec{\alpha})$ for the union of all $\mathcal{U}(\alpha)$, where α occurs in $\vec{\alpha}$, and we also use the abbreviation $\mathcal{U}(A)$ for $\mathcal{U}(\alpha)[A/\alpha]$.

The existential and universal formulas are now coded as follows: each atomic formula $\mathcal{P}\alpha\beta$ is replaced by its code $\mathcal{P}(\alpha,\beta)$, and each quantifier $\forall\alpha$ is relativized to $\mathcal{U}(\alpha)$, e.g., a formula of the form $\forall\alpha\beta(\mathcal{P}\alpha\beta \to \cdots)$ is coded as $\forall\alpha\beta(\mathcal{U}(\alpha) \to \mathcal{U}(\beta) \to \mathcal{P}(\alpha,\beta) \to \cdots)$.

We will use the notation $\overline{\varphi}$ for a code of a formula φ. A set Γ of **P**-formulas is represented by an **F**-basis, denoted $\overline{\Gamma}$, consisting of all codes of formulas in Γ, and in addition, of all types $\mathcal{U}(\alpha)$, for each individual variable α that is free in Γ. Our goal is to show the equivalence,

$$\Gamma \vdash_{\mathbf{P}} \textbf{false} \quad \text{iff} \quad \overline{\Gamma} \vdash_{\mathbf{F}} \textbf{false},$$

for each finite set Γ of **P**-formulas. The difficult part is the "if" part: if a code of a first-order judgement is derivable in **F**, then such a derivation may be translated back to a valid derivation in **P** of the original judgement (cf. Lemma 10). In other words, no "ad hoc" proof is possible.

From now on, all terms, types, judgements etc. are terms, types, judgements etc. of system **F**, unless stated otherwise. If we refer to formulas of **P**, we mean their codes. If not stated otherwise, all proofs are assumed to be normal and fully applied.

A crucial role in eliminating ad hoc proofs is played by the relativizer $\mathcal{U}(\alpha)$. Lemma 12 below states that each type A satisfying $\mathcal{U}(A)$ can be replaced by an individual variable. To prove this we need the following definitions.

An **F** basis Γ is *good* iff it satisfies the following restrictions:

- it is ordinary, i.e., targets of all types in Γ are free variables;
- no individual variable is a target in Γ;
- p_i is not a target in Γ, for any $\mathcal{P}$ and $i \in \{0,1,2\}$;
- the variables η_1,η_2 may occur as targets only in types of the form $\mathcal{U}(\alpha)$;

The last statement should be understood as follows: each occurrence of an η_i is a part of a full set of assumptions of the form $\mathcal{U}(\alpha)$, for some individual variable α.

We say that an individual variable α is an *upper bound* for a type A in a basis Γ iff

$$\Gamma, A \vdash \alpha,$$

A variable α is a *lower bound* for A in Γ iff for every predicate letter $\mathcal{P}$ and every $i \in \{1, 2\}$ we have:

$$\Gamma, A \to p_i \ \vdash \alpha \to p_i. \tag{1}$$

Finally, we say that α *represents* A in Γ iff α is both an upper and lower bound.

Lemma 11. *Let β be an upper bound for A in an ordinary basis Γ, where β is not a target. Let α be an individual variable such that the condition (1) above holds for some $\mathcal{P}$ and $i \in \{1, 2\}$. In addition, assume that p_i is not a target in Γ. Then $\alpha = \beta$.*

Proof. Assume $\Gamma, A \to p_i \vdash \alpha \to p_i$. Since $\Gamma, A \vdash \beta$, we also have $\Gamma, \beta \to p_i \vdash A \to p_i$, and thus $\Gamma, \beta \to p_i \vdash N : \alpha \to p_i$, for some N. It means that $\Gamma, \beta \to p_i, \alpha \vdash N' : p_i$, where N' may be assumed to be fully applied. There is only one declaration with target p_i, thus we must have $\Gamma, \beta \to p_i, \alpha \vdash \beta$, which is impossible (unless $\beta = \alpha$) as there is no declaration with target β. ∎

Lemma 12. *Let Γ be a good basis, and let $\Gamma \vdash \mathcal{U}(A)$, for some type A. Then there is a (unique) individual variable β representing A in Γ. In addition, Γ must contain the assumptions $\mathcal{U}(\beta)$.*

Proof. It follows immediately from Lemma 11, that there is at most one variable representing A. Because $\Gamma \vdash \mathcal{U}(A)$, we have in particular $\Gamma \vdash A \to \eta_2$, that is, $\Gamma, A \vdash \eta_2$. Since η_2 may occur in Γ only as target of some $\mathcal{U}(\beta)$, we must have $\Gamma, A \vdash N : \beta$, for some term N. We conclude that β is an upper bound. On the other hand, we also have $\Gamma \vdash (A \to p_i) \to \eta_1$, that is, $\Gamma, (A \to p_i) \vdash \eta_1$.

Again, η_1 may occur in Γ only as target of some $\mathcal{U}(\alpha)$, and we get $\Gamma, (A \to p_i) \vdash \alpha \to q_j$. Since q_j is not a target in Γ, it must be that $q_j = p_i$, that is, we actually have $\Gamma, (A \to p_i) \vdash \alpha \to p_i$. From Lemma 11, we conclude that $\alpha = \beta$. Repeating this argument for all $\mathcal{P}$ and i, we get that β is also a lower bound. ∎

A basis Γ is called a **P**-instance iff it has the form $\overline{\Delta} \cup \mathrm{U} \cup \Pi$, where:

- Δ is a set of **P**-formulas;
- U consists of formulas of the form $\mathcal{U}(\alpha)$, where α is an individual variable (not free in Δ);
- Π consists of instances of atomic formulas, i.e., types of the form $\mathcal{P}(A, B)$, such that $\Gamma \vdash \mathcal{U}(A)$ and $\Gamma \vdash \mathcal{U}(B)$.

Clearly, every **P**-instance is a good basis. Thus, by Lemma 12, each of the types substituted into atomic formulas is represented in Γ by a (unique) individual variable.

Let $\Gamma = \overline{\Delta} \cup \mathrm{U} \cup \Pi$ be a **P**-instance as above. By $|\Pi|$ we denote the set of atomic formulas obtained from Π, by replacing each type $\mathcal{P}(A, B)$ with the atomic formula $\mathcal{P}(\alpha, \beta)$, where α and β represent A and B, respectively. Then by $|\Gamma|$ we denote the set $\Delta \cup |\Pi|$. In the particular case when the last component is empty, we have of course $|\Gamma| = \Delta$.

A crucial step in the proof of our main Lemma 14 is that whenever Γ is a **P**-instance and $\Gamma \vdash \mathcal{P}(A, B)$, with some types A, B, satisfying $\Gamma \vdash \mathcal{U}(A)$ and $\Gamma \vdash \mathcal{U}(B)$, then also $|\Gamma| \vdash_{\mathbf{P}} \mathcal{P}(\alpha, \beta)$, for α, β representing A, B. For the induction to work, we need a slightly more general statement.

Lemma 13. *Let Γ be a **P**-instance, and let $\Sigma = \{(y_i : \mathcal{P}^i_{A_iB_i}) \mid i = 1, \ldots, n\}$. Assume that $\Gamma, \Sigma \vdash \xi$, and $\Gamma, \Sigma \vdash \mathcal{U}(A_i)$ and $\Gamma, \Sigma \vdash \mathcal{U}(B_i)$, for all $i \in \{1, \ldots, n\}$. Then there exists i such that $|\Gamma| \vdash_{\mathbf{P}} \mathcal{P}^i(\alpha, \beta)$, for some variables α, β, representing A_i and B_i, respectively, in the basis Γ, Σ.*

Proof. By Lemma 12, for all $i = 1, \ldots, n$, there are variables α_i and β_i representing A_i and B_i in Γ, Σ. We begin with showing the following claim:

Suppose that, for some types A and B, we have $\Gamma, \Sigma \vdash N : \mathcal{P}_{AB}$ with $\Gamma, \Sigma \vdash \mathcal{U}(A)$ and $\Gamma, \Sigma \vdash \mathcal{U}(B)$. Then $P = \mathcal{P}^i$, for some i, and the variables α_i and β_i represent A and B, respectively, in the basis Γ, Σ.

To prove the claim, assume that N is fully applied. Thus it must be an abstraction, and we have:

$$\Gamma, \Sigma, (p_0 \to A) \to p_1, (p_0 \to B) \to p_2 \vdash N' : p.$$

Since p is not a target in Γ, the term N' has the form $y_i N_1 N_2$, where $y_i : \mathcal{P}^i_{A_iB_i}$ is in Σ (in particular, $\mathcal{P}^i = \mathcal{P}$). This means that

$$\Gamma, \Sigma, (p_0 \to A) \to p_1, (p_0 \to B) \to p_2 \vdash N_1 : (p_0 \to A_i) \to p_1, \text{ and}$$
$$\Gamma, \Sigma, (p_0 \to A) \to p_1, (p_0 \to B) \to p_2 \vdash N_2 : (p_0 \to B_i) \to p_2,$$

for some $i \in \{1, \ldots, n\}$. Now observe that there are variables α and β, representing A and B in Γ. We have in fact $\alpha_i = \alpha$ and $\beta_i = \beta$. Indeed, $\Gamma, \Sigma, (p_0 \to A_i) \to p_1 \vdash (p_0 \to \alpha_i) \to p_1$ must hold, and it follows that $\Gamma, \Sigma, (p_0 \to A) \to p_1, (p_0 \to B) \to p_2 \vdash (p_0 \to \alpha_i) \to p_1$. By Lemma 11, this means that $\alpha_i = \alpha$. Similar argument yields $\beta_i = \beta$.

The main proof is by induction w.r.t. the size of fully applied proofs (terms). First observe that if $\Gamma, \Sigma \vdash M : \xi$ then M must be an application of a variable z to a sequence of terms and types. We consider two cases, depending on the type of z (the target of which must be ξ).

Case 1. The type of z is an instance of an atomic formula, say $\mathcal{P}(A, B)$. Of course then $M \equiv zN$, where $\Gamma, \Sigma \vdash N : \mathcal{P}_{AB}$. By the above claim, we have $P = \mathcal{P}^i$ and α_i and β_i must represent A and B.

But this means that the formula $P(\alpha_i, \beta_i)$ is in $|\Gamma|$, and thus it is provable.

Case 2. The type of z is a universal formula of the form $\forall\vec{\alpha}(\mathcal{U}(\vec{\alpha}) \to A_1 \to \cdots \to A_n \to \mathcal{P}(\gamma, \delta))$. The term M has the form $M \equiv z\vec{D}\vec{U}\vec{N}L$, where $\vec{D}$ is a sequence of types substituted for $\vec{\alpha}$, the symbol $\vec{U}$ denotes a sequence of proofs for the components of $\mathcal{U}(\vec{D})$, and $\vec{N}$ is a sequence of proofs for types $A_j[\vec{D}/\vec{\alpha}]$. Finally,

L is a proof of $\mathcal{P}_{AB}$, where A and B are the components of $\vec{D}$ corresponding to δ and γ. By Lemma 12, there are (unique) variables $\vec{\beta}$, representing the types $\vec{D}$ in Γ, Σ.

Assume that $N_j \equiv \lambda x : Q_{GH}.N_j'$ is a proof of $A_j[\vec{D}/\vec{\alpha}] = Q(G,H) = Q_{GH} \to \xi$. We apply the induction hypothesis to N_j' in the basis Γ, Σ, Q_{GH}. There are two cases. First suppose that $|\Gamma| \vdash_{\mathbf{P}} \mathcal{P}^i(\alpha,\beta)$, for α,β representing A_i and B_i for some i. Then we are done. Otherwise, $|\Gamma| \vdash_{\mathbf{P}} Q(\alpha,\beta)$ and α,β represent G and H, and this case may be assumed to hold for all j. Clearly, α,β occur in $\vec{\beta}$ at the positions corresponding to occurrences of G and H in $\vec{D}$. Thus, we have $|\Gamma| \vdash_{\mathbf{P}} A_j[\vec{\beta}/\vec{\alpha}]$, for all j.

We conclude that $|\Gamma| \vdash_{\mathbf{P}} \mathcal{P}(\alpha,\beta)$, where α,β are variables occurring in $\vec{\beta}$ at the same positions as γ,δ occur in $\vec{\alpha}$. It suffices to show that $\mathcal{P} = \mathcal{P}^i$, for some i with α,β representing A_i, B_i. This follows from the claim we made at the beginning, applied to the term $L : \mathcal{P}_{AB}$. ∎

Lemma 14. *Let* Γ *be a* **P***-instance. If* $\Gamma \vdash$ **false** *then* $|\Gamma| \vdash_{\mathbf{P}}$ **false**.

Proof. We proceed by induction w.r.t. length of proofs in **F**. Suppose that $\Gamma \vdash M :$ **false**. As in the previous proof, M must be an application of a variable z to a sequence of terms and types, and we have two cases.
Case 1. The type of z is a universal formula of the form $\forall\vec{\alpha}(\mathcal{U}(\vec{\alpha}) \to A_1 \to \cdots \to A_n \to \mathbf{false})$. Then $M \equiv z\vec{D}\vec{U}\vec{N}$, where $\vec{D}$ is a sequence of types substituted for $\vec{\alpha}$, the symbol $\vec{U}$ denotes a sequence of proofs for the components of $\mathcal{U}(\vec{D})$, and $\vec{N}$ is a sequence of proofs for types $A_j[\vec{D}/\vec{\alpha}]$. For each j there are Q and G, H with $A_j[\vec{D}/\vec{\alpha}] = Q(G,H) = Q_{GH} \to \xi$. The term N_j must be an abstraction, say $N_j \equiv \lambda x{:}Q_{GH}.N_j'$, and we have $\Gamma, x{:}Q_{GH} \vdash N_j' : \xi$. From Lemma 13 (applied to a one-element set $\Sigma = \{(x{:}Q_{GH})\}$) we obtain $|\Gamma| \vdash_{\mathbf{P}} Q(\gamma,\delta)$, where γ,δ represent G and H. In other words, if $\vec{\beta}$ is the sequence of variables representing $\vec{D}$, then for all j we have $|\Gamma| \vdash_{\mathbf{P}} A_j[\vec{\beta}/\vec{\alpha}]$. It follows that $|\Gamma| \vdash_{\mathbf{P}}$ **false**.
Case 2. The type of z is an existential formula $\forall\vec{\alpha}(\mathcal{U}(\vec{\alpha}) \to A_1 \to \cdots \to A_n \to \forall\gamma(\mathcal{U}(\gamma) \to \mathcal{P}(\delta,\gamma) \to \mathbf{false}) \to \mathbf{false})$. The term M has the form $z\vec{D}\vec{U}\vec{N}L$, where the meaning of $\vec{D}$, $\vec{U}$ and $\vec{N}$ is similar as in Case 1. The term L has type $\forall\gamma(\mathcal{U}(\gamma) \to \mathcal{P}(A,\gamma) \to \mathbf{false})$, where A is the component of $\vec{D}$ corresponding to δ. Using Lemma 13, we obtain as in Case 1, that $|\Gamma| \vdash_{\mathbf{P}} A_i[\vec{\beta}/\vec{\alpha}]$ for all $i \leq n$. Thus, it suffices to show that $|\Gamma| \vdash_{\mathbf{P}} \forall\gamma(\mathcal{P}(\alpha,\gamma) \to \mathbf{false})$, where α is the component of $\vec{\beta}$ corresponding to δ, i.e., representing the type A.

As we observed, L has type $\forall\gamma(\mathcal{U}(\gamma) \to \mathcal{P}(A,\gamma) \to \mathbf{false})$. This means that we have $\Gamma, \mathcal{U}(\gamma), \mathcal{P}(A,\gamma) \vdash$ **false**, with a shorter proof. By the induction hypothesis, $|\Gamma|, \mathcal{P}(\alpha,\gamma) \vdash_{\mathbf{P}}$ **false**, where α represents A. It follows that $|\Gamma| \vdash \forall\gamma(\mathcal{P}(\alpha,\gamma) \to \mathbf{false})$, as desired. ∎

Corollary 15. *Let* Δ *be a set of* **P***-formulas. Then* $\overline{\Delta} \vdash$ **false** *iff* $\Delta \vdash_{\mathbf{P}}$ **false**.

It follows that there is an effective reduction from **P** to **F**. Together with Lemma 8, this gives a proof of Löb's theorem.

Theorem 16. *The inhabitation problem for system* **F** *is undecidable.*

Let us note here that the above proof is based on the possibility of nested quantification. The author conjectures that the universal fragment of the second-order propositional intuitionistic logic is decidable.

Acknowledgments

This work originated from e-mail discussions with Daniel Leivant. Morten Heine Sørensen made me aware of the work of Thomas Arts, who provided copies of relevant papers. Aleksy Schubert proofread an initial draft version.

References

1. Arts, T., Embedding first order predicate logic in second order propositional logic, Master Thesis, Katholieke Universiteit Nijmegen, 1992.
2. Arts, T., Dekkers, W., Embedding first order predicate logic in second order propositional logic, Technical report 93-02, Katholieke Universiteit Nijmegen, 1993.
3. Barendregt, H.P., Lambda calculi with types, chapter in: *Handbook of Logic in Computer Science* (S. Abramsky, D.M. Gabbay, T. Maibaum eds.), Oxford University Press, 1990, pp. 117–309.
4. Gabbay, D.M., On 2nd order intuitionistic propositional calculus with full comprehension, *Arch. math. Logik,* **16** (1974), 177–186.
5. Gabbay, D.M., *Semantical Investigations in Heyting's Intuitionistic Logic*, D. Reidel Publ. Co., 1981.
6. Girard, J.-Y., *Proofs and Types,* Cambridge Tracts in Theoretical Computer Science **7**, Cambridge University Press, Cambridge, 1989.
7. Leivant, D., Polymorphic type inference, *Proc. 10-th ACM Symposium on Principles of Programming Languages*, ACM, 1983.
8. Löb, M. H., Embedding first order predicate logic in fragments of intuitionistic logic, *Journal Symb. Logic,* **41**, No. 4 (1976), 705–718.
9. Prawitz, D., Some results for intuitionistic logic with second order quantification rules, in: *Intuitionism and Proof Theory* (A. Kino, J. Myhill, R.E. Vesley eds.), North-Holland, Amsterdam, 1970, pp. 259–270.
10. Rasiowa, H., Sikorski, R., *The Mathematics of Metamathematics*, PWN, Warsaw, 1963.
11. Statman, R., Intuitionistic propositional logic is polynomial-space complete, *Theoretical Computer Science* **9**, 67–72 (1979)
12. Urzyczyn, P., Type reconstruction in $\mathbf{F}_\omega$, to appear in *Mathematical Structures in Computer Science*. Preliminary version in *Proc. Typed Lambda Calculus and Applications* (M. Bezem and J.F. Groote, eds.), LNCS 664, Springer-Verlag, Berlin, 1993, pp. 418–432.
13. Urzyczyn, P., The emptiness problem for intersection types, *Proc. 9th IEEE Symposium on Logic in Computer Science*, 1994, pp. 300–309. (To appear in *Journal of Symbolic Logic.*)
14. Wells, J.B., Typability and type checking in the second-order λ-calculus calculus are equivalent and undecidable, *Proc. 9th IEEE Symposium on Logic in Computer Science*, 1994, pp. 176–185. (To appear in *Annals of Pure and Applied Logic.*)

Weak and Strong Beta Normalisations in Typed λ-Calculi

Hongwei Xi

Department of Mathematical Sciences
Carnegie Mellon University
Pittsburgh, PA 15213, USA

Abstract. We present translation techniques to study relations between weak and strong beta-normalisations in various typed lambda-calculi. We first introduce a translation which translates lambda-terms into lambda-I-terms, and show that every lambda-term is strongly beta-normalisable if and only if its translation is weakly beta-normalisable. We then prove that the translation preserves typability of lambda-terms in various typed lambda-calculi. This enables us to establish the equivalence between weak and strong beta-normalisations in these typed lambda-calculi. This translation can handle Curry typing as well as Church typing, strengthening some recent closely related results. This may bring some insights into answering whether weak and strong beta-normalisations in all pure type systems are equivalent.

1 Introduction

In various typed λ-calculi, one of the most interesting and important properties of λ-terms is *how* they can be β-reduced to β-normal forms. A λ-term M is said to be *weakly β-normalisable* ($\mathcal{WN}_\beta(M)$) if it can be β-reduced to a β-normal form in *some* way while M is *strongly β-normalisable* ($\mathcal{SN}_\beta(M)$) if *every* β-reduction sequence starting from M terminates with a β-normal form. If every λ-term in a λ-calculus λS is weakly/strongly β-normalisable, we say that λS enjoys *weak/strong β-normalisation* ($\lambda S \models \mathcal{WN}_\beta / \lambda S \models \mathcal{SN}_\beta$).

A conjecture presented by Barendregt at *The Second International Conference on Typed Lambda Calculi and Applications* states that $\lambda S \models \mathcal{WN}_\beta$ implies $\lambda S \models \mathcal{SN}_\beta$ for every pure type system λS [2]. One can certainly strengthen the conjecture by asking whether it is provable in the first-order Peano arithmetic. We will argue that this is the case for various typed λ-calculi including all systems in the λ-cube [2].

We achieve this via introducing a translation $\|\cdot\|$ which translates a λ-term M into a λI-term $\|M\|$. We then show that $\mathcal{WN}_\beta(\|M\|)$ implies $\mathcal{SN}_\beta(M)$ for every λ-term M and $\|\cdot\|$ preserves the typability of λ-terms in various typed λ-calculi.

An immediate consequence of our technique is that it can also be used to infer $\lambda S \models \mathcal{SN}_\beta$ from $\lambda S \models \mathcal{WN}_\beta$ for many typed λ-calculi λS. In a retrospection, $\lambda S \models \mathcal{SN}_\beta$ were usually proven later than $\lambda S \models \mathcal{WN}_\beta$ for many λS such as

$\lambda^{\rightarrow}$, $\lambda\cap^{-}$, $\lambda 2$ and λC, involving more complicated methods. In addition, our technique can help estimate bounds on $\mu(M)$ for certain typable λ-terms M, where $\mu(M)$ is the number of β-reduction steps in a longest β-reduction sequence from M. We will demonstrate this for the simply typed λ-terms.

The next section presents some preliminaries and basic known results. We define translations $\|\cdot\|$ and $|\cdot|$ recursively in Section 3, and prove some properties on them. We present several applications of $\|\cdot\|$ in Section 4 involving both Curry typing and Church typing, and point out some available methods for removing type dependencies on terms. Finally, we mention some related work and draw some conclusions on our technique.

2 Preliminaries

We give a brief explanation on the notions and terminology used in this paper. Most details can be found in [1], [20] and [2].

Definition 1. The set Λ of λ-terms is defined inductively as follows.

- (variable) There are infinitely many variables $x, y, z, \ldots$ in Λ; variables are subterms of themselves.
- (abstraction) If $M \in \Lambda$ then $(\lambda x.M) \in \Lambda$; N is a subterm of $(\lambda x.M)$ if N is $(\lambda x.M)$ or a subterm of M.
- (application) If $M_1, M_2 \in \Lambda$ then $M_1(M_2) \in \Lambda$; N is a subterm of $M_1(M_2)$ if N is $M_1(M_2)$ or a subterm of M_i for some $i \in \{1, 2\}$.

Note that an application $M(N)$ is written in a form which is different from that in [1]. The main advantage of this notation is that it copes nicely with the translations we shall define later.

Let Var be the set of all variables. The set $\mathbf{FV}(M)$ of free variables in a λ-term M is defined as follows.

$$\mathbf{FV}(M) = \begin{cases} \{M\} & \text{if } M \in \text{Var}; \\ \mathbf{FV}(M_1)\backslash\{x\} & \text{if } M = (\lambda x.M_1); \\ \mathbf{FV}(M_1) \cup \mathbf{FV}(M_2) & \text{if } M = M_1(M_2). \end{cases}$$

An occurrence of a variable x in a term M is bound if it occurs in some term M_1 where $\lambda x.M_1$ is a subterm of M; an occurrence of a variable is free if it is not bound.

The set Λ_I of λI-terms is the maximal subset of Λ such that, for every term $M \in \Lambda_I$, if $(\lambda x.N)$ is a subterm of M then $x \in \mathbf{FV}(N)$.

$[N/x]M$ stands for substituting N for all free occurrences of x in M. α-conversion or renaming bound variables may have to be performed in order to avoid name collisions. Also certain substitution properties, such as the substitution lemma (Lemma 2.1.16 [Bar84]), are assumed.

Definition 2. A term of form $(\lambda x.M)(N)$ is called a β-redex, and $[N/x]M$ is called the contractum of the β-redex; $M_1 \leadsto_\beta M_2$ stands for a β-reduction step in which M_2 is obtained from replacing some β-redex in M_1 with its contractum; a β-reduction sequence is a possibly infinite sequence of form

$$M_1 \leadsto_\beta M_2 \leadsto_\beta \ldots;$$

a λ-term is in β-normal form if it contains no β-redexes; $(\lambda x.M)(N)$ is a β_I-redex if $x \in \mathbf{FV}(M)$.

We use R for β-redexes. $\leadsto_\beta^n$, $\leadsto_\beta^+$ and $\leadsto_\beta^*$ stand for β-reduction sequences consisting of n steps, a positive number of steps and a nonnegative number of steps, respectively.

Definition 3. A λ-term M is weakly β-normalisable ($\mathcal{WN}_\beta(M)$) if there exists a β-reduction sequence $M \leadsto_\beta^* N$ such that N is in β-normal form; M is strongly β-normalisable ($\mathcal{SN}_\beta(M)$) if all β-reduction sequences from M are finite.

We write $\lambda S \models \mathcal{WN}_\beta$ ($\lambda S \models \mathcal{SN}_\beta$) if all λ-terms in λS are weakly (strongly) normalisable.

Let $\mu(M) = \infty$ if there exists an infinite β-reduction sequence from M, and let $\mu(M)$ be the number of β-reduction steps in a longest β-reduction sequence from M if $\mathcal{SN}_\beta(M)$, which is guaranteed to exist by König's Lemma.

We adopt for every number n

$$n + \infty = \infty + n = \infty + \infty = \infty$$

in the following arithmetic involving ∞.

Lemma 4. *Given $M_R = R(M_1)\ldots(M_n)$ and $M_N = N(M_1)\ldots(M_n)$, where $R = (\lambda x.N_1)(N_2)$ and $N = [N_2/x]N_1$; then $\mu(M_R) \leq 1 + \mu(M_N) + \mu(N_2)$.*

Proof. It suffices to prove this when $\mu(M_N) < \infty$ and $\mu(N_2) < \infty$. Please see the proof of Lemma 4.3.3 (1) in [2]. □

Now let us state the conservation theorem for λI-calculus, upon which this paper is established.

Theorem Church *For every λI-term M, $\mathcal{WN}_\beta(M)$ if and only if $\mathcal{SN}_\beta(M)$.*

Proof. See the proof of Theorem 11.3.4 in [1] or the proof of Theorem 21 in [34]. □

3 Translations

Translations $|\cdot|$ and $\|\cdot\|$ translate λ-terms into λI-terms.

$$\|x\| = x(\bullet) \qquad \|M_1M_2\| = \|M_1\|(|M_2|)$$
$$\|(\lambda x.M)\| = (\lambda x.|M|(\langle x, \bullet\rangle)) \qquad |M| = (\lambda \bullet .\|M\|)$$

Note that $\bullet$ and $\langle , \rangle$ are two distinct fresh variables and $\langle x, \bullet\rangle$ stands for $\langle , \rangle(x)(\bullet)$.

Examples Let $K_I = (\lambda x.\lambda y.y)$, $\omega = (\lambda x.x(x))$ and $\Omega = \omega(\omega)$; then

$$\begin{array}{lll} K_I(\Omega) & \leadsto_\beta & (\lambda y.y) \\ \|(\lambda y.y)\| & \leadsto^*_\beta & (\lambda y.y(\langle y, \bullet\rangle)) \\ \|K_I\| & \leadsto^*_\beta & (\lambda x.\lambda y.y(\langle y, \langle x, \bullet\rangle\rangle)) \\ \|K_I(\Omega)\| & \leadsto^*_\beta & (\lambda y.y(\langle y, \langle|\Omega|, \bullet\rangle\rangle)) \\ \|\omega\| & \leadsto^*_\beta & (\lambda x.x(\langle x, \bullet\rangle)(\lambda \bullet .x(\bullet))) \\ \|\Omega\| & = & \|\omega\|(|\omega|) \\ & \leadsto^*_\beta & (\lambda x.x(\langle x, \langle|\omega|, \bullet\rangle\rangle)(\lambda \bullet .x(\bullet)))(|\omega|) \\ & \leadsto^*_\beta & (\lambda x.x(\langle x, \langle|\omega|, \langle|\omega|, \bullet\rangle\rangle\rangle)(\lambda \bullet .x(\bullet)))(|\omega|) \\ & \leadsto^*_\beta & \ldots\ldots \end{array}$$

Proposition 5. *$\|\cdot\|$ and $|\cdot|$ have the following properties.*

1. *Given any λ-term M; $\|M\|$ and $|M|$ are λI-terms and there exists only one free occurrence of $\bullet$ in $\|M\|$.*
2. *If M_s is a subterm of M then $\|M_s\|$ is a subterm of $\|M\|$.*
3. *$[|N|/x]\|M\| \leadsto^*_\beta \|[N/x]M\|$ for any M and N.*

Proof. (1) and (2) can be readily proven by a structural induction on λ-terms. Since $|N|(\bullet) \leadsto_\beta \|N\|$ for every λ-term N, (3) can also be proven by a structural induction on M. □

We adopt $n \leq \infty$ for every number n and $\infty \leq \infty$.

Lemma 6. *(Main Lemma) For every λ-term M, $\mu(M) \leq \mu(\|M\|)$.*

Proof. It suffices to show this when $\mu(\|M\|) < \infty$. We proceed by an induction on $\mu(\|M\|)$ and the structure of M, lexicographically ordered.

- M is of form $(\lambda x.N)$. Then $\mu(N) \leq \mu(\|N\|)$ by induction hypothesis. Hence,
$$\mu(M) = \mu(N) \leq \mu(\|N\|) < \mu(|N|(\langle x, \bullet\rangle)) = \mu(\|M\|).$$
- M is of form $x(M_1)\ldots(M_n)$ for some $n \geq 0$. Then
$$\|M\| = \|x\|(|M_1|)\ldots(|M_n|).$$
By induction hypothesis, $\mu(M_i) \leq \mu(\|M_i\|) = \mu(|M_i|)$ for $i = 1, \ldots, n$. This yields
$$\mu(M) = \mu(M_1) + \cdots + \mu(M_n) \leq \mu(\|M_1\|) + \cdots + \mu(\|M_n\|)$$
$$= \mu(|M_1|) + \cdots + \mu(|M_n|) = \mu(\|M\|).$$

- M is of form $R(M_1)\dots(M_n)$ for some β-redex $R = (\lambda x N_1)(N_2)$. By definition, $\|M\| = \|R\|(|M_1|)\dots(|M_n|)$. Let

$$N = [N_2/x]N_1 \quad \text{and} \quad M_N = N(M_1)\dots(M_n).$$

With Proposition 5,

$$\begin{aligned}\|R\| &= \|(\lambda x.N_1)\|(|N_2|) = (\lambda x.|N_1|(\langle x, \bullet\rangle))(|N_2|) \\ &\leadsto_\beta ([|N_2|/x]|N_1|)(\langle |N_2|, \bullet\rangle) \leadsto_\beta^* |[N_2/x]N_1|(\langle |N_2|, \bullet\rangle) \\ &= |N|(\langle |N_2|, \bullet\rangle) \leadsto_\beta [\langle |N_2|, \bullet\rangle/\bullet]\|N\|.\end{aligned}$$

We can take $\mu(\|N_2\|)$ β-reduction steps to reduce $\|N_2\|$ to some λ-term $\star$ in β-normal form. Since there is one free occurrence of $\bullet$ in $\|N\|$, we can take at least $2 + \mu(\|N_2\|)$ β-reduction steps to reduce $\|R\|$ to $\|N\|^+ = [\langle \star, \bullet\rangle/\bullet]\|N\|$. Let $\|M_N\|^+ = \|N\|^+(|M_1|)\dots(|M_n|)$, then

$$2 + \mu(\|N_2\|) + \mu(\|M_N\|^+) \le \mu(\|M\|).$$

Notice that $\mu(\|M_N\|) \le \mu(\|M_N\|^+)$ since we can simply treat $\langle \star, \bullet\rangle$ as if it were $\bullet$. By induction hypothesis, $\mu(N_2) \le \mu(\|N_2\|)$ and $\mu(M_N) \le \mu(\|M_N\|)$ since $\mu(\|N_2\|) \le \mu(\|M_N\|) < \mu(\|M\|)$. With Lemma 4,

$$\begin{aligned}\mu(M) &\le 1 + \mu(N_2) + \mu(M_N) < 2 + \mu(\|N_2\|) + \mu(\|M_N\|) \\ &\le 2 + \mu(\|N_2\|) + \mu(\|M_N\|^+) \le \mu(\|M\|).\end{aligned}$$

Therefore, $\mu(M) \le \mu(\|M\|)$ for every λ-term M. □

Corollary 7. *For every λ-term M, $\mathcal{WN}_\beta(\|M\|)$ implies $\mathcal{SN}_\beta(M)$.*

Proof. Note that $\|M\|$ is a λI-term. By Theorem Church, $\mathcal{WN}_\beta(\|M\|)$ implies $\mathcal{SN}_\beta(\|M\|)$, namely, $\mu(\|M\|) < \infty$. Therefore, $\mu(M) \le \mu(\|M\|) < \infty$ by Lemma 6, and this yields $\mathcal{SN}_\beta(M)$. □

Lastly, we mention a translation $(\!|\cdot|\!)$ which is a minor variant of $|\cdot|$. $(\!|\cdot|\!)$ can also be used to establish a version of Lemma 6.

$$\begin{aligned}&(\!|x|\!) = x \\ &(\!|(\lambda x.M)|\!) = (\lambda \bullet .\lambda x.(\!|M|\!)(\langle x, \bullet\rangle)) \\ &(\!|M(N)|\!) = (\lambda \bullet .(\!|M|\!)(\bullet)((\!|N|\!)))\end{aligned}$$

It becomes clear that the use of $\bullet$ is to collect λ-terms which might be thrown away in β-reductions.

4 Applications

With the help of Corollary 7, we are now ready to prove the equivalence between $\lambda S \models \mathcal{WN}_\beta$ and $\lambda S \models \mathcal{SN}_\beta$ in various typed λ-calculi. This amounts to typing $\|M\|$ in these systems when given a typed term M. Since the method that we use cannot deal with all *pure type systems* uniformly, we choose some appropriate λ-calculi to illustrate our technique. We also mention that the translation $\|\cdot\|$ enables us to easily assign a bound for the length of β-reduction sequences from simply typed λ-terms.

4.1 $\lambda^{\rightarrow}$-Curry and $\lambda^{\rightarrow}$-Church

We assume the familiarity of the reader with $\lambda^{\rightarrow}$-Curry ($\lambda^{\rightarrow}$ with Curry typing) and $\lambda^{\rightarrow}$-Church ($\lambda^{\rightarrow}$ with Church typing). Translations $\|\cdot\|$ and $|\cdot|$ on simple types are given as follows.

$$\|b\| = b \qquad \|A \rightarrow B\| = |A| \rightarrow \|B\| \qquad |A| = [] \rightarrow \|A\|$$

Note that b ranges over base types and $[]$ is some base type. Translations $\|\cdot\|$ and $|\cdot|$ on terms in $\lambda^{\rightarrow}$-Church are defined below.

$$\|x\| = x(\bullet) \qquad \|M_1(M_2)\| = \|M_1\|(|M_2|)$$
$$\|(\lambda x : A.M)\| = \lambda x : |A|.|M|(\langle x, \bullet\rangle) \quad |M| = \lambda\bullet : [].\|M\|$$

Lemma 8. *If M is a term in $\lambda^{\rightarrow}$-Curry ($\lambda^{\rightarrow}$-Church) with type A, then $\|M\|$ is a term in $\lambda^{\rightarrow}$-Curry ($\lambda^{\rightarrow}$-Church) with type $\|A\|$.*

Proof. Note that $\langle , \rangle$'s are given types of form $A \rightarrow ([] \rightarrow [])$, and they are different variables if their types are different. A structural induction on M yields the results. □

By a method invented by Turing [9] and, independently, by Prawitz [26], it can be readily proven that every term M in $\lambda^{\rightarrow}$-Curry($\lambda^{\rightarrow}$-Church) is $\mathcal{WN}_\beta$, and therefore $\mathcal{SN}_\beta(M)$ by Corollary 7 since Lemma 8 yields $\mathcal{WN}_\beta(\|M\|)$. This is clearly an arithmetisable proof. We pursue this further in the next section.

4.2 An Upper Bound for β-Reduction Sequences in $\lambda^{\rightarrow}$-Church

$\lambda^{\rightarrow}$ means $\lambda^{\rightarrow}$-Church in this section. The size of a $\lambda^{\rightarrow}$-term is defined below.

$$\mathbf{size}[x] = 1 \quad \mathbf{size}[(\lambda x : A.M)] = 1 + \mathbf{size}[M] \quad \mathbf{size}[M(N)] = \mathbf{size}[M] + \mathbf{size}[N]$$

The rank $\rho(A)$ of a simple type A is defined as follows.

$$\rho(A) = \begin{cases} 0 & \text{if } A \text{ is a base type;} \\ 1 + \max\{\rho(A_0), \rho(A_1)\} & \text{if } A = A_0 \rightarrow A_1. \end{cases}$$

The rank $\rho(R)$ of a β-redex $R = (\lambda x : A.M)(N)$ of type B is $\rho(A \rightarrow B)$, and the rank $\hat{\rho}(M)$ of a $\lambda^{\rightarrow}$-term M is

$$\hat{\rho}(M) = \max\{\rho(R) : R \text{ is a } \beta\text{-redex in } M\}.$$

We define functions $2_0(n) = n$ and $2_{k+1}(n) = 2^{2_k(n)}$ for $k = 0, 1, 2, \ldots$.

Theorem 9. *$\mu(M) < 2_{k+1}(n)$ for every $\lambda^{\rightarrow}I$-term M, where $k = \hat{\rho}(M)$ and $n = \mathbf{size}[M]$.*

Proof. Please see [34] for a proof using the standardisation theorem in the untyped λ-calculus. A similar result can also be found in [29]. □

Corollary 10. *$\mu(M) < 2_{2k+1}(5n)$ for every $\lambda^{\rightarrow}$-term M, where $k = \hat{\rho}(M)$ and $n = \mathbf{size}[M]$.*

Proof. It can be readily verified that $\rho(|A|) \leq 2\rho(A)$ for every simple type A, yielding $\hat{\rho}(\|M\|) \leq 2\hat{\rho}(M)$. Also notice $\mathbf{size}[\|M\|] \leq 5{\cdot}\mathbf{size}[M]$ for every $\lambda^{\rightarrow}$-term M. Since $\|M\|$ is a $\lambda^{\rightarrow}I$-term, $\mu(\|M\|) < 2_{2k+1}(5n)$ by Theorem 9. Therefore, $\mu(M) \leq \mu(\|M\|) < 2_{2k+1}(5n)$ by Lemma 6. □

A result of a similar form is mentioned in [10], and is proven in [29].

The following translation $[\![\cdot]\!]$ translates $\lambda^{\rightarrow}$-terms into $\lambda^{\rightarrow}I$-terms of the same types.

$$
\begin{aligned}
[\![x]\!] &= x; \\
[\![M(N)]\!] &= [\![M]\!]([\![N]\!]) \\
[\![(\lambda x : A.M)]\!] &= (\lambda x : A.\lambda x_1 : A_1.\dots\lambda x_n : A_n.\langle [\![M]\!](x_1)\dots(x_n), x\rangle), \\
&\quad \text{where } M \text{ is of type } A_1 \rightarrow \dots \rightarrow A_n \rightarrow B \\
&\quad \text{and } B \text{ is a base type;}
\end{aligned}
$$

Note that $\langle , \rangle$ is a fresh variable of type $B \rightarrow (A \rightarrow B)$ in the definition. It is interesting to know that $\mu(M) \leq \mu([\![M]\!])$ for every $\lambda^{\rightarrow}$-term M. This is intimately related to the translation used in [29]. $[\![\cdot]\!]$ is far less flexible than $\|\cdot\|$ since it can only translate $\lambda^{\rightarrow}$-terms. For the reader who is interested in this subject, we point out that our technique can also be applied to the typed λ-calculus with let-polymorphism.

4.3 $\lambda 2$ with Curry Typing

The second-order polymorphic typed λ-calculus $\lambda 2$ is originally introduced in [11] and [28], where Church typing is involved. $\lambda 2$ with Curry typing can be formulated as follows.

Syntax

$$
\begin{array}{lll}
\text{Types} & A & ::= X \mid (A_1 \rightarrow A_2) \mid \forall X.A \\
\text{Contexts} & \Gamma & ::= \cdot \mid \Gamma, x : A \\
\text{Terms} & M & ::= x \mid (\lambda x.M) \mid M_1(M_2)
\end{array}
$$

We use X, Y for type variables, A, B for types, Γ for contexts, M, N for terms and x for variables assuming that any variable can be declared at most once in a context. We write $[B/X]A$ for the type obtained by substituting B for all free occurrences of X in A.

Typing Rules

$$\frac{x : A \in \Gamma}{\Gamma \vdash x : A}(var)$$

$$\frac{\Gamma, x : A \vdash M : B}{\Gamma \vdash (\lambda x.M) : A \to B}(\to I) \qquad \frac{\Gamma \vdash M : A \to B \qquad \Gamma \vdash N : A}{\Gamma \vdash M(N) : B}(\to E)$$

$$\frac{\Gamma \vdash M : A}{\Gamma \vdash M : \forall X.A}(\forall I)^* \qquad \frac{\Gamma \vdash M : \forall X.A}{\Gamma \vdash M : [B/X]A}(\forall E)$$

Γ must contain no free occurrences of X in rule $\forall I$. $\mathcal{D} :: \Gamma \vdash M : A$ denotes a typing derivation with conclusion $\Gamma \vdash M : A$.

Definition 11. M is a $\lambda 2$-term if $\Gamma \vdash M : A$ is derivable for some Γ and A.

Let $\Lambda 2$ be the set of all $\lambda 2$-terms, and we show that $\Lambda 2$ is closed under $\|\cdot\|$ and $|\cdot|$. Translations $\|\cdot\|$ and $|\cdot|$ on types are given below, where $\bot = \forall X.X$.

$$\|X\| = X \qquad \|A \to B\| = |A| \to \|B\|$$

$$\|\forall X.A\| = \forall X.\|A\| \qquad |A| = \bot \to \|A\|$$

Given $\Gamma = x_1 : A_1, \ldots, x_n : A_n$, then

$$|\Gamma| = \langle,\rangle : \forall X.X \to (\bot \to \bot), x_1 : |A_1|, \ldots, x_n : |A_n| \text{ and } \|\Gamma\| = |\Gamma|, \bullet : \bot.$$

It can be readily verified that $|\Gamma, x : A|, \bullet : \bot \vdash \langle x, \bullet \rangle : \bot$ is derivable for any Γ and A.

Lemma 12. *Given $M \in \Lambda 2$, then $\|M\| \in \Lambda 2$.*

Proof. By definition, there exists a derivation $\mathcal{D}$ with conclusion $\Gamma \vdash M : A$ for some Γ and A. We proceed by induction on the structure of $\mathcal{D}$ to show that there exists $\|\mathcal{D}\| :: \|\Gamma\| \vdash \|M\| : \|A\|$. Since $\|\Gamma\| = |\Gamma|, \bullet : \bot$, we can define $|\mathcal{D}| :: |\Gamma| \vdash |M| : |A|$ as follows when $\|\mathcal{D}\|$ is constructed.

$$\frac{\|\mathcal{D}\| :: |\Gamma|, \bullet : \bot \vdash \|M\| : \|A\|}{|\Gamma| \vdash |M| = (\lambda \bullet.\|M\|) : |A| = \bot \to \|A\|}(\to I).$$

We do a case analysis according to the last typing rule applied in $\mathcal{D}$.

- If $\mathcal{D}$ ends with (*var*), $(\to E)$ or $(\forall I)$, the case is trivial.
- $\mathcal{D}$ is of form

$$\frac{\mathcal{D}_1 :: \Gamma_1 \vdash M_1 : A_2}{\mathcal{D} :: \Gamma \vdash (\lambda x.M_1) : A_1 \to A_2}(\to I),$$

where $\Gamma_1 = \Gamma, x : A_1$ and $M = (\lambda x.M_1)$ and $A = A_1 \to A_2$. By induction hypothesis, we have

$$\frac{\dfrac{|\mathcal{D}_1| :: |\Gamma_1|, \bullet : \bot \vdash |M_1| : |A_2| = \bot \to \|A_2\| \qquad |\Gamma_1|, \bullet : \bot \vdash \langle x, \bullet \rangle : \bot}{|\Gamma_1|, \bullet : \bot \vdash |M_1|(\langle x, \bullet \rangle) : \|A_2\|}(\to E)}{|\Gamma|, \bullet : \bot \vdash (\lambda x.|M_1|(\langle x, \bullet \rangle)) : \|A\| = |A_1| \to \|A_2\|}(\to I)$$

Since $\|\Gamma\| = |\Gamma|, \bullet : \bot$ and $\|M\| = (\lambda x.|M_1|(\langle x, \bullet \rangle))$, we have derived $\|\Gamma\| \vdash \|M\| : \|A\|$.

– $\mathcal{D}$ is of form

$$\frac{\mathcal{D}_1 :: \Gamma \vdash M : \forall X.A_1}{\mathcal{D} :: \Gamma \vdash M : [B/X]A_1}(\forall E),$$

where $A = [B/X]A_1$. By induction hypothesis, we have

$$\frac{\|\mathcal{D}_1\| :: \|\Gamma\| \vdash \|M\| : \|\forall X.A_1\| = \forall X.\|A_1\|}{\|\Gamma\| \vdash \|M\| : [\|B\|/X]\|A_1\|}(\forall E).$$

Notice $[\|B\|/X]\|A_1\| = \|[B/X]A_1\| = \|A\|$. Hence, we have derived $\|\Gamma\| \vdash \|M\| : \|A\|$.

Therefore, $\|M\|$ is a $\lambda 2$-term. □

Theorem 13. *$\lambda 2 \models \mathcal{WN}_\beta$ if and only if $\lambda 2 \models \mathcal{SN}_\beta$.*

Proof. Given any term $M \in \Lambda 2$, we have $\|M\| \in \Lambda 2$ by Lemma 12. If $\lambda 2 \models \mathcal{WN}_\beta$ then $\mathcal{WN}_\beta(\|M\|)$. This implies that $\mathcal{SN}_\beta(M)$ by Corollary 7. Therefore, if $\lambda 2 \models \mathcal{WN}_\beta$ then $\lambda 2 \models \mathcal{SN}_\beta$. The other direction is trivial. □

Note that this is a recent result, which can be formulated in the first-order Peano arithmetic. On the other hand, no proofs of $\lambda 2 \models \mathcal{SN}_\beta$ can be established in the second-order Peano arithmetic. After Girard had proven that $\lambda 2 \models \mathcal{WN}_\beta$, several people (including Girard himself) modified his definition of reducibility candidates and proved that $\lambda 2 \models \mathcal{SN}_\beta$, but none of these proofs achieve this by showing the equivalence between $\lambda 2 \models \mathcal{WN}_\beta$ and $\lambda 2 \models \mathcal{SN}_\beta$.

4.4 $\lambda\omega$ with Church Typing

We choose $\lambda\omega$, the higher order polymorphic typed λ-calculus, to show that our technique can also work with Church typing. Here we can see how to handle types which contain β-redexes. Types depending on terms are avoided intentionally in this case for some reason which will be explained later.

Syntax

Kinds	$\kappa ::= * \mid \kappa_1 \to \kappa_2$
Constructors	$\tau ::= t \mid (\lambda t : \kappa.\tau) \mid \tau_1(\tau_2) \mid \tau_1 \to \tau_2 \mid \forall t : \kappa.\tau$
Kind Contexts	$\Delta ::= \cdot \mid \Delta, t : \kappa$
Type Contexts	$\Gamma ::= \cdot \mid \Gamma, x : \tau$
Terms	$e ::= x \mid (\lambda x : \tau.e) \mid (\lambda t : \kappa.e) \mid e_1(e_2) \mid e(\tau)$

We use k for kinds, τ for constructors, t for variables over constructors, Δ for kind contexts, Γ for type contexts, e for terms and x for variables. Variable can be declared at most once in a context. τ is a type if $\Delta \vdash \tau : *$ is derivable for some Δ.

Typing Rules

$$\frac{t:\kappa\in\Delta}{\Delta\vdash t:\kappa}$$

$$\frac{\Delta,t:\kappa_1\vdash\tau:\kappa_2}{\Delta\vdash(\lambda t:\kappa_1.\tau):\kappa_1\to\kappa_2}\qquad\frac{\Delta\vdash\tau_1:\kappa_1\to\kappa_2\qquad\Delta\vdash\tau_2:\kappa_1}{\Delta\vdash\tau_1(\tau_2):\kappa_2}$$

$$\frac{\Delta,t:\kappa\vdash\tau:*}{\Delta\vdash(\forall t:k.\tau):*}\qquad\frac{\Delta\vdash\tau_1:*\qquad\Delta\vdash\tau_2:*}{\Delta\vdash\tau_1\to\tau_2:*}$$

We write $\Delta\vdash\Gamma$ if $\Delta\vdash\tau:*$ for every $x:\tau\in\Gamma$.

$$\frac{\Delta\vdash\Gamma\qquad x:\tau\in\Gamma}{\Gamma\vdash_\Delta x:\tau}(var)\qquad\frac{\Gamma\vdash_\Delta e:\tau_1\qquad\tau_1=_\beta\tau_2\qquad\Delta\vdash\tau_2:*}{\Gamma\vdash_\Delta e:\tau_2}(=_\beta)$$

$$\frac{\Gamma,x:\tau_1\vdash_\Delta e:\tau_2}{\Gamma\vdash_\Delta(\lambda x:\tau_1.e):\tau_1\to\tau_2}(\to I)\qquad\frac{\Gamma\vdash_\Delta e_1:\tau_1\to\tau_2\qquad\Gamma\vdash_\Delta e_2:\tau_1}{\Gamma\vdash_\Delta e_1(e_2):\tau_2}(\to E)$$

$$\frac{\Gamma\vdash_{\Delta,t:k}e:\tau}{\Gamma\vdash_\Delta(\lambda t:k.e):(\forall t:k.\tau)}(\forall I)^*\qquad\frac{\Gamma\vdash_\Delta e:(\forall t:k.\tau_1)\qquad\Delta\vdash\tau_2:k}{\Gamma\vdash_\Delta e(\tau_2):[\tau_2/t]\tau_1}(\forall E)$$

Now we can define transforms $\|\cdot\|$ and $|\cdot|$ on constructors.

$$\|t\|=t\qquad\|\lambda t:k.\tau\|=\lambda t:k.\|\tau\|$$
$$\|\tau_1(\tau_2)\|=\|\tau_1\|(\|\tau_2\|)\qquad\|\forall t:k.\tau\|=\forall t:k.\|\tau\|$$
$$\|\tau_1\to\tau_2\|=|\tau_1|\to\|\tau_2\|\qquad|\tau|=\bot\to\|\tau\|$$

Notice that $|\cdot|$ is only defined on types.

Proposition 14. *We have the following.*

1. $\|[\tau_2/t]\tau_1\|=[\|\tau_2\|/t]\|\tau_1\|$.
2. $\tau_1=_\beta\tau_2$ *implies* $\|\tau_1\|=_\beta\|\tau_2\|$.

Proof. (1) follows from a structural induction on τ_1, and (2) follows from (1). □

Now we define $\|\cdot\|$ and $|\cdot|$ on $\lambda\omega$-terms and type contexts.

$$\|x\|=x(\bullet)\qquad\|e_1(e_2)\|=\|e_1\|(|e_2|)$$
$$\|(\lambda x:\tau.e)\|=(\lambda x:|\tau|.|e|(\langle x,\bullet\rangle_{|\tau|}))\qquad\|(\lambda t:\kappa.e)\|=(\lambda t:\kappa.\|e\|)$$
$$\|e(\tau)\|=\|e\|(\|\tau\|)\qquad|e|=(\lambda\bullet:\bot.\|e\|)$$

$$|\cdot|=\langle,\rangle:\forall t:*.\,t\to(\bot\to\bot)\qquad|\Gamma,x:\tau|=|\Gamma|,x:|\tau|\qquad\|\Gamma\|=|\Gamma|,\bullet:\bot$$

Note that $\langle x,\bullet\rangle_{|\tau|}$ stands for $\langle,\rangle(|\tau|)(x)(\bullet)$.

Lemma 15. *If* $\Gamma\vdash e:\tau$ *is derivable, then both*

$$\|\Gamma\|\vdash\|e\|:\|\tau\|\quad and\quad|\Gamma|\vdash|e|:|\tau|$$

are also derivable.

Proof. This follows from a structural induction on the derivation of $\Gamma \vdash e : \tau$. □

Theorem 16. $\lambda\omega \models \mathcal{WN}_\beta$ *if and only if* $\lambda\omega \models \mathcal{SN}_\beta$.

Proof. Following the proof of Lemma 6, we can establish $\mu(e) \leq \mu(\|e\|)$ accordingly. It is easy to observe that $\mathcal{SN}_\beta(\tau)$ for every constructor τ since τ corresponds to a λ-term in some simply typed λ-calculus with constants. Given a $\lambda\omega$-term e and $\|e\| \leadsto_\beta^* e_*$; it is easy to note that every β-redex r in e_* is either a β_I-redex or of form $(\lambda t : k.e_0)\tau$. Hence, r can always be regarded as a perpetual β-redex [3] since one can only substitute constructors for the variables in τ. Therefore, $\mathcal{WN}_\beta(\|e\|)$ implies $\mathcal{SN}_\beta(\|e\|)$. The rest is straightforward. □

Through $\lambda 2$ with Curry typing and $\lambda\omega$ with Church typing we can see that $\|\cdot\|$ handles Curry typing and Church typing with virtually no difference. This is quite desirable. We will return to this point when we mention another translation $[\cdot]$ in Section 5.

4.5 Types Depending on Terms and λ-cube

There exists a serious problem in handling dependent types if we apply our technique directly. Given two distinct λ-terms a and b such that $a =_\beta b$; let e be a term of type $P(a)$, then e is also of type $P(b)$ since $P(a) =_\beta P(b)$; if we assign type $P(\|a\|)$ to $\|e\|$, then type $P(\|b\|)$ can also be assigned to $\|e\|$; when $\|a\| \neq_\beta \|b\|$, this may not be possible since $P(\|a\|)$ and $P(\|b\|)$ are two different types; $\|a\| \neq_\beta \|b\|$ often holds when $a \neq b$ (try $a = (\lambda x.x)$ and $b = a(a)$); this is a serious conflict.

Fortunately, there exist some methods in [13] and [12] to remove type dependencies on terms for certain typed λ-calculi. A proof in [12] shows that $\lambda\omega \models \mathcal{SN}_\beta$ implies $\lambda C \models \mathcal{SN}_\beta$, where λC stands for the calculus of construction. Following this example, we can verify that $\lambda S \models \mathcal{WN}_\beta$ if and only if $\lambda S \models \mathcal{SN}_\beta$ for every system λS in λ-cube [2]. Again we point out that this is a result which can be formulated in the first-order Peano arithmetic.

We emphasise that removing type dependencies is a highly *ad hoc* approach. The above conflict has to be resolved before translation techniques can be truly applied to λ-calculi with dependent types.

5 Related Work

The research on deriving strong normalisation (SN) from weak normalisation (WN) has lasted for at least thirty years. Nederpelt[22], Klop[18], Karr[17], de Groot[8], and Kfoury and Wells[21] have all invented techniques to infer SN from WN. Their techniques all require introducing some notions of reduction different from β-reduction, deriving strong β-normalisation from weak normalisation of these newly introduced notions of reduction. For example, Klop's technique amounts to introducing a pairing constant $[\cdot,\cdot]$, a π-reduction $\leadsto_\pi$ and a

κ-reduction $\leadsto_\kappa$ as follows [30]:

$$[M_1, M_2]N \leadsto_\pi [M_1 N, M_2] \qquad\qquad [M_1, M_2] \leadsto_\kappa M_1$$

These techniques are successful when applied to λ-calculi λS for which there exist syntactic proofs of $\lambda S \models \mathcal{WN}_\beta$. If one tries to prove $\lambda 2 \models \mathcal{SN}_\beta$, it is very doubtful that one can gain much (if anything) by arguing that $\lambda 2$ with some newly introduced reductions enjoys weak normalisation. Besides, one has to have a *proof of* $\lambda S \models \mathcal{WN}_\beta$ *at hand* in order to infer $\lambda S \models \mathcal{SN}_\beta$. This is quite different from our technique since we derive $\lambda S \models \mathcal{SN}_\beta$ from $\lambda S \models \mathcal{WN}_\beta$ with no need of a proof of $\lambda S \models \mathcal{WN}_\beta$. In other words, the above techniques do not help much in establishing the equivalence between $\mathcal{WN}_\beta$ and $\mathcal{SN}_\beta$ in various typed λ-calculi.

Gandy[10] interprets simply typed λ-terms as strictly increasing functionals. His method, now called *functional interpretations*, can yield an upper bound for the lengths of β-reduction sequences from simply typed λ-terms. In this direction further work can be found in [24] and [25] and [16]. One can somewhat view our technique as a syntactic realisation of Gandy's idea.

Schwichtenberg[29] shows an upper bound for the lengths of β-reduction sequences from simply typed λ-terms. In his proof, he uses a translation closely related to the translation $\|\cdot\|$, which translates simply typed λ-terms into simply typed λI-terms. This translation can only work with $\lambda^\rightarrow$-terms. Translating λP-terms into $\lambda^\rightarrow$-terms [13], Springintveld[31] then uses Schwichtenberg's upper bound to establish an upper bound for the lengths of β-reduction sequences from λP-terms.

Sørensen and the author independently discovered the following translation $[\cdot]$, which translates a λ-term M into a λI-term $[M]$ such that $\mathcal{WN}_\beta([M])$ implies $\mathcal{SN}_\beta(M)$.

$$\begin{aligned}
[x] &= x \\
[(\lambda x.M)] &= (\lambda k_1.k_1(\lambda x.\lambda k_2.\langle [M](k_2), x\rangle)) \\
[M_1(M_2)] &= (\lambda k_1.[M_0](\lambda k_2.k_2([M_1])(k_1)))
\end{aligned}$$

Note that $\langle , \rangle$ is a fresh variable and a term of form $\langle M, N\rangle$ stands for $\langle , \rangle(M)(N)$. $[\cdot]$ is a minor variant of Plotkin's call-by-name continuation passing style translation. Sørensen then proved the equivalence between $\mathcal{WN}_\beta$ and $\mathcal{SN}_\beta$ in various typed λ-calculi including the simply typed λ-calculus, the simply typed λ-calculus with positive recursive types, and $\lambda\omega$. The author showed the equivalence between $\mathcal{WN}_\beta$ and $\mathcal{SN}_\beta$ in $\lambda^\rightarrow$ and $\lambda 2$ with Church-typing, and mentioned that $[\cdot]$ can also be applied to $\lambda\omega$.

$[\cdot]$ cannot be applied to terms in $\lambda 2$ with Curry typing since $[M]$ may not be a well-typed $\lambda 2$-term for some $M \in \Lambda 2$. Some explanation can be found in [14] and [15]. Also $[\cdot]$ has trouble working with $\lambda\cap^-$-terms. Reasoning with $[\cdot]$ is usually more difficult than with $\|\cdot\|$ since one has to deal with a lot of administrative β-redexes introduced by CPS-transformation [23].

6 Conclusion and Future Work

We have demonstrated some applications of our technique. The reader is encouraged to verify that this technique also works for $\lambda\cap^-$ with Curry typing [6, 7, 20], $\lambda\cap^-$ with Church typing [21] and the λ-calculi with positive recursive types [32]. On the other hand, the translation $[\cdot]$, which exploits continuations, has troubles handling Curry typing [30], and is less robust than translation $\|\cdot\|$. Besides, $\|\cdot\|$ — in the author's opinion — leads to much more straightforward reasoning than $[\cdot]$ does.

All the presented proofs of equivalence between $\mathcal{WN}_\beta$ and $\mathcal{SN}_\beta$ in various typed λ-calculi can be formulated in the first-order Peano arithmetic. Therefore, the following conjecture seems to have some grounds.

Conjecture 17. *For every pure type system λS, the equivalence between $\lambda S \models \mathcal{WN}_\beta$ and $\lambda S \models \mathcal{SN}_\beta$ can be established in first-order Peano arithmetic.*

Our translation technique can only work in typed settings. On the other hand, Conjecture 17 has a counterpart in untyped settings. We present it here for the sake of completeness.

Problem *Let Λ be the set of all λ-terms; given $\mathcal{S} \subseteq \Lambda$ which satisfies the following conditions:*

1. *$\mathcal{S}$ is closed under β-reduction, and*
2. *$C[x(x_1)\dots(x_n)] \in \mathcal{S}$ for every $C[M] \in \mathcal{S}$, where $C[]$ is a context and $\mathbf{FV}(M) = \{x_1, \dots, x_n\}$ and x is fresh;*

if all λ-terms in $\mathcal{S}$ are weakly normalisable, are all λ-terms in $\mathcal{S}$ strongly normalisable?

Like $[\cdot]$, $\|\cdot\|$ has a serious problem handling types depending on terms since it does not preserve the β-equivalence between terms (it would be useless for our purpose if it did). We are exploring the possibility of generalising $\|\cdot\|$ to a family of translations, making the technique applicable to dependent types. Another direction is to adapt the technique into systems such as Gödel's T, where constants and δ-reduction rules are allowed. Although it is shown in [18] that $\mathcal{T} \models \mathcal{WN}_\beta$ implies $\mathcal{T} \models \mathcal{SN}_\beta$, Klop's proof (Chapter II. 6.1.7) needs Schütte's proof of $\mathcal{T} \models \mathcal{WN}_\beta$, which cannot be formulated in HA. Therefore, it is still interesting to verify whether it can be proven in HA that $\mathcal{T} \models \mathcal{WN}_\beta$ implies $\mathcal{T} \models \mathcal{SN}_\beta$.

7 Acknowledgment

I thank Frank Pfenning, Peter Andrews and Richard Statman for their support and for providing me a nice work-environment. I also thank some anonymous referees for their constructive comments on this paper.

References

1. H.P. Barendregt (1984), *The Lambda Calculus: Its Syntax and Semantics*, North-Holland Publishing Company, Amsterdam.
2. H.P. Barendregt (1992), Lambda calculi with types, in *Handbook of Logic in Computer Science Vol. 2*, edited by S. Abramsky, D. M. Gabbay and T.S.E. Maibaum, Clarendon Press, Oxford, pp. 117-309.
3. J.A. Bergstra and J.W. Klop (1982), Strong normalization and perpetual reductions in the lambda calculus, *J. Inform. Process. Cybernet.* 18, pp. 403-417.
4. A. Church, (1941), *The Calculi of Lambda Conversion*, Princeton University Press, Princeton.
5. Th. Coquand and G. Huét (1988), The calculus of constructions, *Information and Computation* vol. 76, pp. 95-120.
6. M.Coppo and M.Dezani-Ciancaglini (1980), An extension of basic functionality theory for the lambda-calculus, *Notre Dame Journal of Formal Logic* 21 (4), pp. 685-693.
7. M.Coppo, M.Dezani-Ciancaglini and B.Venneri (1981), Functional characters of solvable terms, *Zeit. Math. Logik und Grundlagen der Math.* 27 (1), pp. 45-58.
8. P. de Groote (1993), The conservation theorem revisited, *Typed Lambda Calculi and Applications: TLCA'93*, LNCS vol. 664, Springer, pp. 163-178.
9. R.O. Gandy (1980), An early proof of normalisation by A.M. Turing, in *To H.B. Curry: Essays on Combinatory Logic, Lambda Calculus and Formalism*, Academic Press, pp. 453-456.
10. R.O. Gandy (1980), Proofs of strong normalisation, in *To H.B. Curry: Essays on Combinatory Logic, Lambda Calculus and Formalism*, Academic Press, pp. 457-478.
11. J.-Y. Girard (1972), *Interprétation Fonctionnelle et Élimination des Coupures de l'Arithmétique d'Ordre Supérieur*, Thèse de Doctorat d'Etat, Université Paris VII.
12. H. Geuvers and M.J. Nederhof (1991), A modular proof of strong normalisation for the calculus of constructions, *J. Functional Programming* 1, 155-189.
13. R. Harper, F. Honsell and G. Plotkin (1987), A framework for defining logics. In *Proc. Second Annual Symposium on Logic in Computer Science*, I.E.E.E., Ithaca, N.Y., pages 194-204.
14. R. Harper and M. Lillibridge (1993), Explicit polymorphism and CPS conversion, In *Proceedings of the Twentieth ACM Symposium on Principles of Programming Languages*, pp. 206-209
15. R. Harper and M. Lillibridge (1993), Polymorphic type assignment and CPS conversion, *LISP and Symbolic Computation*, vol. 6, 3(4), pp. 361-380.
16. Stefan Kahrs (1995), *Towards a Domain Theory for Termination Proofs*, Laboratory for Foundation of Computer Science, Report 95-314, Department of Computer Science, The University of Edinburgh.
17. M. Karr (1985), "Delayability" in proofs of strong normalisabilities in the typed lambda-calculus, in *Mathematical Foundation of Computer Software*, edited by H. Ehrig, C. Floyd, M. Nïvat and J. Thatcher, LNCS vol. 185, Springer, pp. 208-222.
18. J.W. Klop (1980), *Combinatory Reduction Systems*, Ph.D. thesis, CWI, Amsterdam, Mathematical Center Tracts, No. 127.

19. J.W. Klop, (1992), Term rewriting systems, in *Handbook of Logic in Computer Science Vol.2*, edited by S. Abramsky, D. M. Gabbay and T.S.E. Maibaum, Clarendon Press, Oxford, pp. 1-116.
20. J.L. Krivine (1990), *Lambda-calcul, Types et Modèles*, Masson, Paris.
21. A.J. Kfoury and J.B. Wells (1994), *New notions of reduction and non-semantic proofs of β-strong normalisation in typed λ-calculi*, Tech. Rep. 94-104, Computer Science Department, Boston University.
22. R.P. Nederpelt (1973), *Strong normalization in a typed lambda calculus with lambda structured types*, Ph.D. thesis, Technische Hogeschool Eindhoven.
23. G. Plotkin (1975), Call-by-name, call-by-value, and the lambda calculus, *Theoretical Computer Science*, vol 1, pp. 125-159.
24. J. van de Pol (1994), Termination proofs for higher-order rewrite systems, in *Higher Order Algebra, Logic and Term Rewriting*, edited by J. Heering, K. Meinke, B. Möller and T. Nipkow, LNCS vol. 816, Springer, pp. 305-325.
25. J. van de Pol and H. Schwichtenberg (1995), Strict functionals for termination proofs, *Typed lambda calculi and applications: TLCA '95*, LNCS vol. 902, Springer, pp. 350-364.
26. D. Prawitz (1965), *Natural Deduction: A proof theoretical study*, Almquist & Wiksell publishing company, Stockholm.
27. D. Prawitz (1971), Ideas and results of proof theory, in *Proceedings of the 2nd Scandinavian Logic Symposium*, edited by J.E. Fenstad, North-Holland Publishing Company, Amsterdam.
28. J. Reynolds (1974), Towards a theory of type structure, *Colloquium sur la Programmation*, LNCS vol. 19, Springer, pp. 408-423.
29. H. Schwichtenberg (1991), An upper bound for reduction sequences in the typed lambda-calculus, *Archive for Mathematical Logic*, 30, 405-408.
30. M.H. Sørensen (1996), *Strong Normalization from Weak Normalization by A-Translation in Typed lambda-Calculi*, Manuscript announced on the types mailing list, February.
31. J. Springintveld (1993), Lower and upper bounds for reductions of types in $\lambda\underline{\omega}$ and λP, in *Typed lambda calculi and applications: TLCA'93*, LNCS vol. 664, Springer, pp. 391-405.
32. P. Urzyczyn (1995), Positive recursive type assignment, in *Mathematical Foundations of Computer Science*, LNCS vol. 969, Springer, pp. 382-391.
33. H. Xi (1996), *On weak and strong normalisations*, Research Report 96-187, Department of Mathematical Sciences, Carnegie Mellon University, Pittsburgh.
34. H. Xi (1996), *Upper bounds for standardisations and an application*, Research Report 96-189, Department of Mathematical Sciences, Carnegie Mellon University, Pittsburgh.
35. H. Xi (1996), *An induction measure on λ-terms and its applications*, Research Report 96-192, Department of Mathematical Sciences, Carnegie Mellon University, Pittsburgh.

Author Index

Lecture Notes in Computer Science

For information about Vols. 1–1129

please contact your bookseller or Springer-Verlag

Vol. 1130: M. Haveraaen, O. Owe, O.-J. Dahl (Eds.), Recent Trends in Data Type Specification. Proceedings, 1995. VIII, 551 pages. 1996.

Vol. 1131: K.H. Höhne, R. Kikinis (Eds.), Visualization in Biomedical Computing. Proceedings, 1996. XII, 610 pages. 1996.

Vol. 1132: G.-R. Perrin, A. Darte (Eds.), The Data Parallel Programming Model. XV, 284 pages. 1996.

Vol. 1133: J.-Y. Chouinard, P. Fortier, T.A. Gulliver (Eds.), Information Theory and Applications II. Proceedings, 1995. XII, 309 pages. 1996.

Vol. 1134: R. Wagner, H. Thoma (Eds.), Database and Expert Systems Applications. Proceedings, 1996. XV, 921 pages. 1996.

Vol. 1135: B. Jonsson, J. Parrow (Eds.), Formal Techniques in Real-Time and Fault-Tolerant Systems. Proceedings, 1996. X, 479 pages. 1996.

Vol. 1136: J. Diaz, M. Serna (Eds.), Algorithms – ESA '96. Proceedings, 1996. XII, 566 pages. 1996.

Vol. 1137: G. Görz, S. Hölldobler (Eds.), KI-96: Advances in Artificial Intelligence. Proceedings, 1996. XI, 387 pages. 1996. (Subseries LNAI).

Vol. 1138: J. Calmet, J.A. Campbell, J. Pfalzgraf (Eds.), Artificial Intelligence and Symbolic Mathematical Computation. Proceedings, 1996. VIII, 381 pages. 1996.

Vol. 1139: M. Hanus, M. Rogriguez-Artalejo (Eds.), Algebraic and Logic Programming. Proceedings, 1996. VIII, 345 pages. 1996.

Vol. 1140: H. Kuchen, S. Doaitse Swierstra (Eds.), Programming Languages: Implementations, Logics, and Programs. Proceedings, 1996. XI, 479 pages. 1996.

Vol. 1141: H.-M. Voigt, W. Ebeling, I. Rechenberg, H.-P. Schwefel (Eds.), Parallel Problem Solving from Nature – PPSN IV. Proceedings, 1996. XVII, 1.050 pages. 1996.

Vol. 1142: R.W. Hartenstein, M. Glesner (Eds.), Field-Programmable Logic. Proceedings, 1996. X, 432 pages. 1996.

Vol. 1143: T.C. Fogarty (Ed.), Evolutionary Computing. Proceedings, 1996. VIII, 305 pages. 1996.

Vol. 1144: J. Ponce, A. Zisserman, M. Hebert (Eds.), Object Representation in Computer Vision. Proceedings, 1996. VIII, 403 pages. 1996.

Vol. 1145: R. Cousot, D.A. Schmidt (Eds.), Static Analysis. Proceedings, 1996. IX, 389 pages. 1996.

Vol. 1146: E. Bertino, H. Kurth, G. Martella, E. Montolivo (Eds.), Computer Security – ESORICS 96. Proceedings, 1996. X, 365 pages. 1996.

Vol. 1147: L. Miclet, C. de la Higuera (Eds.), Grammatical Inference: Learning Syntax from Sentences. Proceedings, 1996. VIII, 327 pages. 1996. (Subseries LNAI).

Vol. 1148: M.C. Lin, D. Manocha (Eds.), Applied Computational Geometry. Proceedings, 1996. VIII, 223 pages. 1996.

Vol. 1149: C. Montangero (Ed.), Software Process Technology. Proceedings, 1996. IX, 291 pages. 1996.

Vol. 1150: A. Hlawiczka, J.G. Silva, L. Simoncini (Eds.), Dependable Computing – EDCC-2. Proceedings, 1996. XVI, 440 pages. 1996.

Vol. 1151: Ö. Babaoğlu, K. Marzullo (Eds.), Distributed Algorithms. Proceedings, 1996. VIII, 381 pages. 1996.

Vol. 1152: T. Furuhashi, Y. Uchikawa (Eds.), Fuzzy Logic, Neural Networks, and Evolutionary Computation. Proceedings, 1995. VIII, 243 pages. 1996. (Subseries LNAI).

Vol. 1153: E. Burke, P. Ross (Eds.), Practice and Theory of Automated Timetabling. Proceedings, 1995. XIII, 381 pages. 1996.

Vol. 1154: D. Pedreschi, C. Zaniolo (Eds.), Logic in Databases. Proceedings, 1996. X, 497 pages. 1996.

Vol. 1155: J. Roberts, U. Mocci, J. Virtamo (Eds.), Broadbank Network Teletraffic. XXII, 584 pages. 1996.

Vol. 1156: A. Bode, J. Dongarra, T. Ludwig, V. Sunderam (Eds.), Parallel Virtual Machine – EuroPVM '96. Proceedings, 1996. XIV, 362 pages. 1996.

Vol. 1157: B. Thalheim (Ed.), Conceptual Modeling – ER '96. Proceedings, 1996. XII, 489 pages. 1996.

Vol. 1158: S. Berardi, M. Coppo (Eds.), Types for Proofs and Programs. Proceedings, 1995. X, 296 pages. 1996.

Vol. 1159: D.L. Borges, C.A.A. Kaestner (Eds.), Advances in Artificial Intelligence. Proceedings, 1996. XI, 243 pages. (Subseries LNAI).

Vol. 1160: S. Arikawa, A.K. Sharma (Eds.), Algorithmic Learning Theory. Proceedings, 1996. XVII, 337 pages. 1996. (Subseries LNAI).

Vol. 1161: O. Spaniol, C. Linnhoff-Popien, B. Meyer (Eds.), Trends in Distributed Systems. Proceedings, 1996. VIII, 289 pages. 1996.

Vol. 1162: D.G. Feitelson, L. Rudolph (Eds.), Job Scheduling Strategies for Parallel Processing. Proceedings, 1996. VIII, 291 pages. 1996.

Vol. 1163: K. Kim, T. Matsumoto (Eds.), Advances in Cryptology – ASIACRYPT '96. Proceedings, 1996. XII, 395 pages. 1996.

Vol. 1164: K. Berquist, A. Berquist (Eds.), Managing Information Highways. XIV, 417 pages. 1996.

Vol. 1165: J.-R. Abrial, E. Börger, H. Langmaack (Eds.), Formal Methods for Industrial Applications. VIII, 511 pages. 1996.

Vol. 1166: M. Srivas, A. Camilleri (Eds.), Formal Methods in Computer-Aided Design. Proceedings, 1996. IX, 470 pages. 1996.

Vol. 1167: I. Sommerville (Ed.), Software Configuration Management. VII, 291 pages. 1996.

Vol. 1168: I. Smith, B. Faltings (Eds.), Advances in Case-Based Reasoning. Proceedings, 1996. IX, 531 pages. 1996. (Subseries LNAI).

Vol. 1169: M. Broy, S. Merz, K. Spies (Eds.), Formal Systems Specification. XXIII, 541 pages. 1996.

Vol. 1170: M. Nagl (Ed.), Building Tightly Integrated Software Development Environments: The IPSEN Approach. IX, 709 pages. 1996.

Vol. 1171: A. Franz, Automatic Ambiguity Resolution in Natural Language Processing. XIX, 155 pages. 1996. (Subseries LNAI).

Vol. 1172: J. Pieprzyk, J. Seberry (Eds.), Information Security and Privacy. Proceedings, 1996. IX, 333 pages. 1996.

Vol. 1173: W. Rucklidge, Efficient Visual Recognition Using the Hausdorff Distance. XIII, 178 pages. 1996.

Vol. 1174: R. Anderson (Ed.), Information Hiding. Proceedings, 1996. VIII, 351 pages. 1996.

Vol. 1175: K.G. Jeffery, J. Král, M. Bartošek (Eds.), SOFSEM'96: Theory and Practice of Informatics. Proceedings, 1996. XII, 491 pages. 1996.

Vol. 1176: S. Miguet, A. Montanvert, S. Ubéda (Eds.), Discrete Geometry for Computer Imagery. Proceedings, 1996. XI, 349 pages. 1996.

Vol. 1177: J.P. Müller, The Design of Intelligent Agents. XV, 227 pages. 1996. (Subseries LNAI).

Vol. 1178: T. Asano, Y. Igarashi, H. Nagamochi, S. Miyano, S. Suri (Eds.), Algorithms and Computation. Proceedings, 1996. X, 448 pages. 1996.

Vol. 1179: J. Jaffar, R.H.C. Yap (Eds.), Concurrency and Parallelism, Programming, Networking, and Security. Proceedings, 1996. XIII, 394 pages. 1996.

Vol. 1180: V. Chandru, V. Vinay (Eds.), Foundations of Software Technology and Theoretical Computer Science. Proceedings, 1996. XI, 387 pages. 1996.

Vol. 1181: D. Bjørner, M. Broy, I.V. Pottosin (Eds.), Perspectives of System Informatics. Proceedings, 1996. XVII, 447 pages. 1996.

Vol. 1182: W. Hasan, Optimization of SQL Queries for Parallel Machines. XVIII, 133 pages. 1996.

Vol. 1183: A. Wierse, G.G. Grinstein, U. Lang (Eds.), Database Issues for Data Visualization. Proceedings, 1995. XIV, 219 pages. 1996.

Vol. 1184: J. Waśniewski, J. Dongarra, K. Madsen, D. Olesen (Eds.), Applied Parallel Computing. Proceedings, 1996. XIII, 722 pages. 1996.

Vol. 1185: G. Ventre, J. Domingo-Pascual, A. Danthine (Eds.), Multimedia Telecommunications and Applications. Proceedings, 1996. XII, 267 pages. 1996.

Vol. 1186: F. Afrati, P. Kolaitis (Eds.), Database Theory - ICDT'97. Proceedings, 1997. XIII, 477 pages. 1997.

Vol. 1187: K. Schlechta, Nonmonotonic Logics. IX, 243 pages. 1997. (Subseries LNAI).

Vol. 1188: T. Martin, A.L. Ralescu (Eds.), Fuzzy Logic in Artificial Intelligence. Proceedings, 1995. VIII, 272 pages. 1997. (Subseries LNAI).

Vol. 1189: M. Lomas (Ed.), Security Protocols. Proceedings, 1996. VIII, 203 pages. 1997.

Vol. 1190: S. North (Ed.), Graph Drawing. Proceedings, 1996. XI, 409 pages. 1997.

Vol. 1191: V. Gaede, A. Brodsky, O. Günther, D. Srivastava, V. Vianu, M. Wallace (Eds.), Constraint Databases and Applications. Proceedings, 1996. X, 345 pages. 1996.

Vol. 1192: M. Dam (Ed.), Analysis and Verification of Multiple-Agent Languages. Proceedings, 1996. VIII, 435 pages. 1997.

Vol. 1193: J.P. Müller, M.J. Wooldridge, N.R. Jennings (Eds.), Intelligent Agents III. XV, 401 pages. 1997. (Subseries LNAI).

Vol. 1194: M. Sipper, Evolution of Parallel Cellular Machines. XIII, 199 pages. 1997.

Vol. 1196: L. Vulkov, J. Waśniewski, P. Yalamov (Eds.), Numerical Analysis and Its Applications. Proceedings, 1996. XIII, 608 pages. 1997.

Vol. 1197: F. d'Amore, P.G. Franciosa, A. Marchetti-Spaccamela (Eds.), Graph-Theoretic Concepts in Computer Science. Proceedings, 1996. XI, 410 pages. 1997.

Vol. 1198: H.S. Nwana, N. Azarmi (Eds.), Software Agents and Soft Computing: Towards Enhancing Machine Intelligence. XIV, 298 pages. 1997. (Subseries LNAI).

Vol. 1199: D.K. Panda, C.B. Stunkel (Eds.), Communication and Architectural Support for Network-Based Parallel Computing. Proceedings, 1997. X, 269 pages. 1997.

Vol. 1200: R. Reischuk, M. Morvan (Eds.), STACS 97. Proceedings, 1997. XIII, 614 pages. 1997.

Vol. 1201: O. Maler (Ed.), Hybrid and Real-Time Systems. Proceedings, 1997. IX, 417 pages. 1997.

Vol. 1202: P. Kandzia, M. Klusch (Eds.), Cooperative Information Agents. Proceedings, 1997. IX, 287 pages. 1997. (Subseries LNAI).

Vol. 1203: G. Bongiovanni, D.P. Bovet, G. Di Battista (Eds.), Algorithms and Complexity. Proceedings, 1997. VIII, 311 pages. 1997.

Vol. 1204: H. Mössenböck (Ed.), Modular Programming Languages. Proceedings, 1997. X, 379 pages. 1997.

Vol. 1206: J. Bigün, G. Chollet, G. Borgefors (Eds.), Audio- and Video-based Biometric Person Authentication. Proceedings, 1997. XII, 450 pages. 1997.

Vol. 1207: J. Gallagher (Ed.), Logic Program Synthesis and Transformation. Proceedings, 1996. VII, 325 pages. 1997.

Vol. 1208: S. Ben-David (Ed.), Computational Learning Theory. Proceedings, 1997. VIII, 331 pages. 1997. (Subseries LNAI).

Vol. 1209: L. Cavedon, A. Rao, W. Wobcke (Eds.), Intelligent Agent Systems. Proceedings, 1996. IX, 188 pages. 1997. (Subseries LNAI).

Vol. 1210: P. de Groote, J.R. Hindley (Eds.), Typed Lambda Calculi and Applications. Proceedings, 1997. VIII, 405 pages. 1997.

Vol. 1212: J. P. Bowen, M.G. Hinchey, D. Till (Eds.), ZUM '97: The Z Formal Specification Notation. Proceedings, 1997. X, 435 pages. 1997.